LEECHES, LICE AND LAMPREYS

Leeches, Lice and Lampreys

A Natural History of Skin and Gill Parasites of Fishes

by

Graham C. Kearn

School of Biological Sciences,
University of East Anglia,
Norwich, United Kingdom

 Springer

A C.I.P. Catalogue record for this book is available from the Library of Congress.

ISBN 1-4020-2925-X (HB)
ISBN 1-4020-2926-8 (e-book)

Published by Springer,
P.O. Box 17, 3300 AA Dordrecht, The Netherlands.

Sold and distributed in North, Central and South America
by Springer,
101 Philip Drive, Norwell, MA 02061, U.S.A.

In all other countries, sold and distributed
by Springer,
P.O. Box 322, 3300 AH Dordrecht, The Netherlands.

Printed on acid-free paper

Cover photograph: The cymothoid isopod parasite *Anilocra pomacentri* attached in its
typical position on the head of the Barrier Reef chromis (*Chromis nitida*). Reproduced with kind
permission of Dr Robert Adlard, Queensland Museum, South Brisbane, Australia.

Every effort has been made to contact the copyright holders of the figures and tables which have
been reproduced from other sources. Anyone who has not been properly credited is requested to
contact the publishers, so that due acknowledgement may be made in subsequent editions.

springeronline.com

All Rights Reserved
© 2004 Springer
No part of this work may be reproduced, stored in a retrieval system, or transmitted
in any form or by any means, electronic, mechanical, photocopying, microfilming, recording
or otherwise, without written permission from the Publisher, with the exception
of any material supplied specifically for the purpose of being entered
and executed on a computer system, for exclusive use by the purchaser of the work.

Printed in the Netherlands.

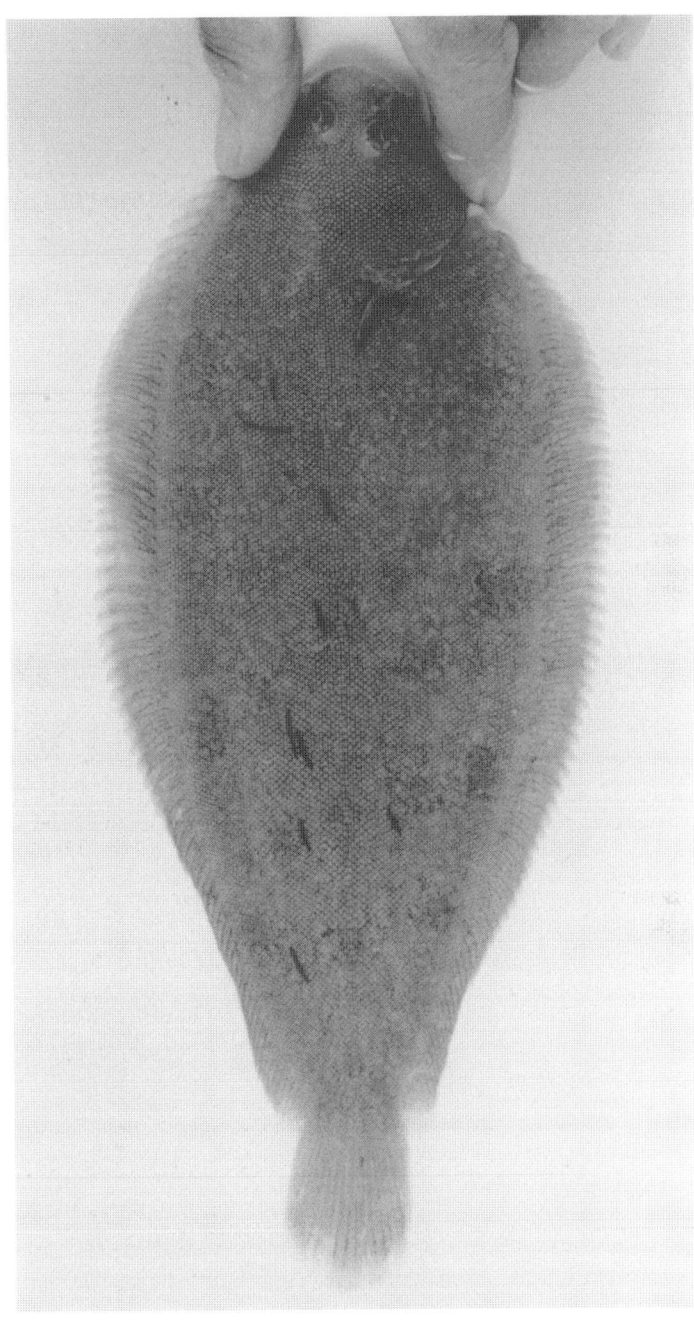

Frontispiece. A common sole (*Solea solea*) from the North Sea infested with the leech *Hemibdella soleae*. Photograph by Sheila Davies.

I dedicate this book to my grandchildren, Megan and Thomas.

"…………whilst this planet has gone cycling on according to the fixed law of gravity, from so simple a beginning endless forms most beautiful and most wonderful have been, and are being evolved."

Charles Darwin (1859). Concluding line from "The Origin of Species by means of Natural Selection".

TABLE OF CONTENTS

Preface	xv
Acknowledgements	xvii
Introduction	1
1 The hosts	2
1.1 Overview	2
1.2 The British Fish Fauna	2
1.3 The external features of fishes	5
1.4 Fish skin	8
1.5 Skin and gill mucus	11
1.6 Fish gills	12
1.6.1 The gills of teleosts	12
1.6.2 The gills of elasmobranchs	16
1.7 Resistance and repair	16
2 'Protozoans'	22
2.1 Introduction	22
2.2 Epizoic ciliates	22
2.3 An ectoparasite – *Ichthyobodo*	27
2.3.1 Free-swimming stage	28
2.3.2 Parasitic stage	29
2.3.3 Pathological effects	30
2.4 A parasite of the epidermis – *Ichthyophthirius*	30
2.4.1 Infective stage	30
2.4.2 Parasitic stages	32
2.4.3 Escape from the host	34
2.4.4 Encysted stage	35
2.4.5 Pathological effects	36
3 Monogenean (flatworm) skin parasites – *Entobdella*	37
3.1 Introduction to monogeneans	37
3.2 The biology of *Entobdella soleae*	39
3.2.1 Attachment and locomotion	40
3.2.2 Food, feeding and egestion	43
3.2.3 Mating	46
3.2.4 Egg assembly	46
3.2.5 Life with little oxygen	47
3.2.6 The egg, hatching and host finding	48
3.2.7 The larva and host invasion	51
3.2.8 Post-invasion migration	52

3.2.9 Host to host transfer and migration of adults and juveniles — 55
3.2.10 Why migrate to the lower surface? — 57
3.3 *Entobdella hippoglossi* — 58
3.4 *Entobdela diadema* — 58
3.5 Host specificity and host switching — 59

4 Other monogenean skin parasites — 61
4.1 Introduction — 61
4.2 *Capsala martinieri* — 61
4.3 Acanthocotylids — 62
4.4 Microbothriids — 66
4.5 Udonellids — 68
4.6 Gyrodactylids — 70
    4.6.1 Attachment — 72
    4.6.2 Feeding — 73
    4.6.3 Consequences of viviparity — 74
    4.6.4 Transmission to new hosts — 75
    4.6.5 Host specificity — 77
    4.6.6 Gyrodactylids in Britain — 78
    4.6.7 The threat of *Gyrodactylus salaris* — 80

5 Monogenean gill parasites – monopisthocotyleans — 82
5.1 Introduction — 82
5.2 Unspecialised gill parasites — 83
5.3 'Dactylogyroideans' — 83
    5.3.1 The British fauna — 83
    5.3.2 Attachment — 85
    5.3.3 Partial endoparasitism – *Amphibdella flavolineata* — 93
    5.3.4 Locomotion — 96
    5.3.5 Copulatory apparatus — 98
    5.3.6 Ecology of ancyrocephalines from mullets — 101

6 Monogenean gill parasites – polyopisthocotyleans — 103
6.1 Introduction — 103
6.2 British polyopisthocotyleans — 103
6.3 Attachment — 105
    6.3.1 Hexabothriid suckers — 107
    6.3.2 Suctorial clamps of *Diclidophora* spp. — 108
    6.3.3 The open clamps of *Cyclocotyla* — 111
    6.3.4 Non-suctorial clamps — 112
    6.3.5 *Anthocotyle merluccii* from hake, *Merluccius merluccius* — 113
6.4 Food and feeding — 114
6.5 Symmetry versus asymmetry — 116
    6.5.1 Reasons for asymmetry in *Axine* — 117
    6.5.2 Are asymmetrical parasites sedentary? — 119
    6.5.3 Maintenance of symmetry — 119
6.6 Eggs, hatching and host finding — 121

    6.6.1 Convergent evolution of dogfish parasites   122
    6.6.2 Rhythmical hatching in *Discocotyle sagittata*   123
    6.6.3 Hatching response to shadows in *Plectanocotyle gurnardi*   123
    6.6.4 Egg bundles and mechanical stimulation in *Diclidophora luscae*   124
    6.6.5 Diapause (?) in *Gastrocotyle trachuri*   125
  6.7 Route to the gills   126
  6.8 Niche restriction and mating   127

7 Leeches   131
  7.1 Introduction   131
  7.2 The British fauna   132
    7.2.1 Marine leeches   132
    7.2.2 Freshwater leeches   135
  7.3 Attachment   137
  7.4 Locomotion   139
  7.5 Host finding   140
  7.6 Feeding   141
  7.7 Mating   145
  7.8 Cocoon assembly   147
  7.9 Some observations on life cycles   149
    7.9.1 Marine leeches   149
    7.9.2 Freshwater leeches   150
  7.10 Leeches as transmitters of diseases   152

8 Siphonostomatoid copepods: (1) Fish lice – caligids   154
  8.1 Introduction to copepods   154
  8.2 General features of caligids   155
  8.3 The adult caligid   159
    8.3.1 The role of the suctorial cephalothorax in attachment   159
    8.3.2 Locomotion   160
    8.3.3. Sensory capabilities   160
    8.3.4 Camouflage   161
    8.3.5 Feeding   161
    8.3.6 Mating   164
    8.3.7 Site preference   166
  8.4 Fate of the eggs   167
  8.5 Nauplius larvae   167
  8.6 The copepodid   168
    8.6.1 Host finding   168
    8.6.2 Settlement and initial attachment to the host   169
  8.7 Chalimus stages   170
  8.8 The problem of moulting   171
  8.9 Host specificity and speciation – *Lepeophtheirus pectoralis*   173
  8.10 'Sea lice' and salmonid fishes   174
    8.10.1 Pathology and host susceptibility   176
    8.10.2 Control   176

9 Siphonostomatoid copepods: (2) pennellids — 178
9.1 *Lernaeocera* — 178
- 9.1.1 Invasion of the first fish host and larval development — 180
- 9.1.2 Mating — 182
- 9.1.3 Establishment and growth of the egg-laying female — 184
- 9.1.4 Ectoparasite or endoparasite? — 185
- 9.1.5 Taxonomic problems and British species — 185

9.2 Eye-maggots – *Lernaeenicus* — 186
- 9.2.1 *Lernaeenicus* in British waters — 187
- 9.2.2 Life cycle — 188
- 9.2.3 The question of pathogenicity — 193

10 Siphonostomatoid copepods: (3) lernaeopodids — 195
10.1 Introduction – changes in attachment — 195
10.2 The salmon gill maggot — 196
- 10.2.1 Occurrence and site of attachment — 196
- 10.2.2 Feeding in the adult female — 197
- 10.2.3 Hatching and invasion of the host — 197
- 10.2.4 Development on the host — 198
- 10.2.5 Mating — 203
- 10.2.6 Relationship with the salmon life cycle — 205

10.3 Other British lernaeopodids — 205

11 Cyclopoid copepods – the anchor worm — 208
11.1 Introduction — 208
11.2 Life cycle — 210
11.3 Pathology and resistance — 213
11.4 The anchor worm in Britain — 213

12 Poecilostomatoid copepods — 214
12.1 Introduction — 214
12.2 Ergasilids — 214
- 12.2.1 British ergasilids — 215
- 12.2.2 Development and attachment of the female — 220
- 12.2.3 Food, feeding and colour — 224

12.3 Bomolochids and taeniacanthids — 226
12.4 The chondracanthids — 228
- 12.4.1 The adult female — 229
- 12.4.2 Where is the male? — 230
- 12.4.3 The life cycle — 232
- 12.4.4 A mesoparasitic chondracanthid – *Lernentoma asellina* — 233

12.5 Philichthyids — 234

13 The common fish louse – *Argulus* — 237
13.1 Introduction — 237
13.2 General morphology — 238
13.3 British branchiurans — 239

TABLE OF CONTENTS

xiii

13.4 Attachment	241
13.5 Host finding	244
13.6 Food and feeding	249
13.7 Reproductive systems and mating	251
13.8 Eggs	257
13.9 Hatching and development	258
13.10 Interactions with other organisms	261
13.11 Relationships	263
14 A mesoparasitic barnacle – *Anelasma*	**265**
15 Isopods	**274**
15.1 Introduction	274
15.2 British isopod fish parasites	275
15.3 Cymothoids	277
15.3.1 General morphology	278
15.3.2 Sex change and mobility	279
15.3.3 Mating	281
15.3.4 Egg incubation and hatching	282
15.3.5 Infecting new hosts	283
15.3.6 Nature of the food	284
15.4 Gnathiids	284
15.4.1 Host invasion	285
15.4.2 Attachment to the fish	285
15.4.3 Mouthparts	286
15.4.4 Feeding and digestion	288
15.4.5 Transmission of microorganisms	290
15.4.6 Predation on gnathiid larvae	291
15.4.7 The free-living reproductive phase in *Paragnathia formica*	292
16 Unionacean molluscs (naiads)	**296**
16.1 Introduction	296
16.2 The adult unionacean	297
16.3 The life cycle	299
16.4 British unionaceans	300
16.5 The swan mussel – *Anodonta cygnea*	301
16.5.1 The glochidium	301
16.5.2 Host finding and establishment on the host	302
16.5.3 Host specificity and infection sites	304
16.5.4 The parasitic stage	307
16.6 The pearl mussel – *Margaritifera margaritifera*	308
16.6.1 The glochidium	310
16.6.2 Infection of the host	311
16.6.3 Survival in the host	312
16.6.4 Development of glochidia	313
16.6.5 Host specificity	313
16.6.6 Overview of the life cycle	313

16.6.7 Threats to the pearl mussel	314
16.7 Some special unionaceans	315
16.8 Acquired immunity to glochidia	316

17 Lampreys — 318
    17.1 Introduction — 318
    17.2 The river lamprey – *Lampetra fluviatilis* — 320
        17.2.1 The ammocoetes — 320
        17.2.2 Metamorphosis (transformation) — 323
        17.2.3 Attachment, feeding and breathing in the adult — 324
        17.2.4 Spawning — 329
    17.3 The sea lamprey – *Petromyzon marinus* — 333
    17.4 Flexibility of lamprey life cycles — 334
        17.4.1 Loch Lomond lampreys — 334
        17.4.2 Sea lampreys in North America — 335
        17.4.3 The brook lamprey – *Lampetra planeri* — 336
    17.5 Other fishes with parasitic tendencies — 337

18 Conclusions — 339
    18.1 Crustacean dominance — 339
    18.2 Life cycles — 339
    18.3 Reproductive biology — 340
    18.4 Attachment and feeding — 341
        18.4.1 Hooks, harpoons and pincers — 341
        18.4.2 Suction — 343
        18.4.3 Glands — 343
        18.4.4 Mesoparasites — 344
        18.4.5 Diet — 346
    18.5 Pathogenesis and host defences — 346
    18.6 Host specificity and speciation — 347
    18.7 Morphology and molecules — 348
    18.8 Habitat selection and niche restriction — 349
    18.9 The future of the British parasite fauna — 350

Appendix 1 Classified list of fishes with common and scientific names — 353

Appendix 2 Classified list of genera of epizoic and parasitic invertebrates — 361

Glossary — 365

References — 375

Index of scientific and common names — 419

Subject Index — 425

PREFACE

This book is about the lives of parasites found on the skin and gills of freshwater and marine fishes. Although many of these parasites are large enough to attract the attention of fishermen, aquarists or naturalists, their biology is well known only to a few professional parasitologists and the book is intended as a reference work for anyone with an interest in finding out more about the natural history of these fascinating, sometimes bizarre and often commercially important animals.

The idea for this book has its origins in the "New Naturalist" series of volumes, published by Collins. These books are scientifically up-to-date accounts of British natural history topics, written in such a way that they appeal to professional biologists as well as to informed naturalists. In 1952, Miriam Rothschild and Theresa Clay published a volume in this series entitled "Fleas, Flukes and Cuckoos". In this book the authors described the natural history of parasites of British birds and included an account of the virtually unknown feather lice or Mallophaga and a study of the special brood parasitism of the cuckoo. This book is no longer in print, but when it was published it was a new venture, representing a considerable departure from the topics covered by previous volumes in the series.

A few years later, in 1959, Sir Alistair Hardy published another volume in the New Naturalist series, namely "The Open Sea, Fish and Fisheries". In this book he clearly recognised the importance of parasites, as the following quotation from the book demonstrates:

> "There must be few adult animals of the size of a small fish (or larger) which do not harbour at least two or three different kinds of parasite. When we consider this, and that every host may carry several of each kind, we begin to realise a fact of nature which at first seems to us most extraordinary: the number of animals we see living freely in the world......is actually far smaller than the number living tucked away in their insides."

Hardy went on to focus for the first time on parasites of fishes, but he had space to deal with them only in a superficial way as part of a chapter devoted to parasitism in general. He summarised the problem as follows:

> "The lives of parasites present as fascinating a field of study to the naturalist as any other; lack of space unfortunately, will only allow the briefest introduction to the subject in a book devoted to the whole of the open sea."

Since that time there has been no comprehensive attempt to describe the biology of fish parasites, despite the fact that the parasite fauna of totally aquatic vertebrates like fishes is substantially different from the parasite fauna of terrestrial vertebrates like birds.

Encouraged by Hardy's words and by my professional interest in fish parasites, the idea of a whole volume devoted to fish parasites began to take shape. This was fuelled by the relatively recent resurgence of interest in parasitism and parasites, with the increasing awareness of the numbers and diversity of parasitic organisms, the influence that they may have on communities and their potential as a force for evolutionary change.

Having decided to write about fish parasites, I was faced with the problem that a book of acceptable size is too small to accommodate the great diversity of parasites, even when this diversity is restricted to parasites of a single major host group such as the birds or fishes. I did not wish to restrict the fish groups even further by limiting the study to either freshwater fishes or to marine fishes - this would inevitably exclude many parasites that are of special interest. I decided therefore to restrict the parasites rather than the hosts by focussing on those parasites that are most likely to be encountered by the aquarist, angler or naturalist. These are the external parasites or ectoparasites, living on the surfaces washed by water. Naturally this includes the gills as well as the skin. Many skin parasites extend their range onto the gills and many gill parasites are closely related to skin parasites. Larger ectoparasites on the skin are visible with the naked eye, and those on the gills may be seen if the gill cover or operculum is raised. Some 'protozoan' ectoparasites are too small to be seen, even with a hand lens, but they may be present in sufficiently large numbers to induce pathological changes that are visible externally. I have excluded all the endoparasites, that is those parasites dwelling inside the host's body, such as tapeworms, roundworms and most 'protozoans', but brief mention has been made of a few that are transmitted by ectoparasites.

The examples of fish parasites referred to in the book are mainly taken from the British fauna, thereby complementing the study of bird parasites by Rothschild and Clay. However, all the major groups of fish ectoparasites have representatives in the British fauna and many of the British examples, or their close relatives, have extensive geographical ranges in the northern hemisphere or beyond. Wherever appropriate I have made reference to interesting parasites not represented in the British fauna. Consequently, the non-British reader should feel comfortable and familiar with the subject matter of the book. For the British reader I have provided for each group of parasites a brief summary of the British fauna.

In writing this book I have tried to produce a well-illustrated text that is up-to-date and sufficiently informative to satisfy the professional biologist, but, at the same time, comprehensible and interesting for the enquiring amateur. Wherever possible I have attempted to explain terms and concepts as they arise in the text. In addition, terms that are used frequently in the text are defined in a glossary. To set the scene, I have included an introductory chapter (Chapter 1) on the general biology of fishes. I have used the common names of fishes where appropriate in the text, but the corresponding scientific names of all the fishes are also indicated and the interrelationships and environmental preferences of the fishes mentioned in the book are given in Appendix 1. The generic names of the parasites and their relationships are presented in Appendix 2.

Graham Kearn
Norwich
July 2004

ACKNOWLEDGEMENTS

I am greatly indebted to many people for help and advice throughout the preparation of this book. I am particularly grateful to the following, who kindly took on the task of reading and reviewing drafts of selected chapters related to their own special interests, as well as helping in other ways: Eugene Burreson, Leslie Chisholm, Roger Lincoln, Peter Maitland, Nigel Merrett, Alan Pike, Ian Whittington, Rodney Wootten and Mark Young. Their advice was invaluable and any errors remaining in the text are entirely my responsibility. I am also grateful to Sheila Davies of the University of East Anglia for her photographic help, to Emma Roberts and other staff at the library of the Centre for Environment, Fisheries and Aquaculture Science, Lowestoft for patiently satisfying endless requests for papers, to Deirdre Sharp at the Library of the University of East Anglia for seeking permission for use of the many illustrations and to Paul Wright also of the University of East Anglia for advice on IT matters. Others who have helped are listed alphabetically as follows: Rob Adlard, David Aldridge, Geoffrey Boxshall, Angela Davies, Bo Fernholm, Bill Gaze, David Gibson, Eileen Harris, Peter Heuch, Tammy Horton, Bozena Koubková, Iveta Matejusová, Andy Shinn and Vaughan Southgate. Comments on my original proposal by anonymous referees also helped to shape the book. My wife Margaret provided useful criticism of the manuscript and invaluable encouragement during the long process of preparation.

INTRODUCTION

Fishes, especially the teleosts or bony fishes, are the most abundant of the vertebrates and most aquatic habitats on the planet have been colonised by them. Their abundance and diversity are matched by their parasites[1]. Practically every tissue and organ in a fish's body provides a substrate or niche for parasites. The skin and gills, those surfaces of fishes that are in contact with water, offer particularly favourable conditions for the establishment and survival of parasitic animals. Many parasites of fishes choose to attach themselves externally to these surfaces (ectoparasites), while others are partly exposed and partly embedded in host tissue (mesoparasites). External parasites of fishes do not have to cope with dehydration, as do the mainly arthropod ectoparasites of terrestrial and aerial vertebrates, and food is much more accessible in the form of living epithelial cells covering the skin and gill surfaces or the rich and superficial supply of blood flowing through the gills. It is these ectoparasites and mesoparasites, many of which are large enough to be visible to the naked eye, that concern us in this book.

Many other parasites live inside the body of the host (endoparasites), where they are hidden from the observer, although they may betray their presence by the effects their activities have on the condition or behaviour of the host, or by the release externally of their offspring (eggs or larvae). Endoparasites reach their internal sites by penetration of the skin or the gill epithelium or by ingestion in food. Some endoparasites may be descended from parasites living on the skin and gills and therefore a thorough knowledge of the biology of external parasites of fishes is important if we are to understand how parasites evolve and progress.

Given the wide distribution and abundance of fishes and the accessibility and vulnerability of their skin and gill surfaces, it is not surprising that interactions between invertebrates and fishes have been frequent in the past and that many of these associations have led to parasitism. The range of invertebrates with representatives that parasitise the external surfaces of fishes is broad and includes 'protozoans', flatworms, annelid worms, many kinds of crustaceans and molluscs (see Appendix 2). Even lampreys, which are agnathan (jawless) vertebrates (see Appendix 1), have become parasitic on some of their gnathostome (jawed) relatives. However, before considering the parasites we need information about the hosts, in particular an appreciation of the range of types of fishes and their habitats, the structure of their skin and gills and the nature of the defences that they are able to mobilise against parasites. These topics will be considered in Chapter 1.

---

[1] In this book the term 'parasite' is restricted to unicellular or multicellular eukaryotic animals that derive benefit from a symbiotic relationship at the expense of their partner (the host).

1

THE HOSTS

1.1 OVERVIEW

Water is a demanding medium for vertebrate life (see Pough *et al.*, 1990; Helfman *et al.*, 1997). Per unit volume, water holds only about one twentieth as much oxygen as air and biological and chemical processes may reduce this further. Warm water contains less oxygen than cool water and as the salt content of water increases the solubility of oxygen decreases. Reduced availability of oxygen is potentially limiting for active animals like fishes. Swimming is also energetically expensive since water is 830 times denser than air and 80 times more viscous. As depth increases so does pressure, while light intensity decreases rapidly with depth and the quality of light changes. Nevertheless, fishes not only cope with these physical changes and constraints but also dominate aquatic habitats on the planet. In the open ocean, fishes are some of the fastest swimming organisms and others have colonised the dark ocean depths, turbulent waters on wave-swept rocky shores, the icy waters of the poles and freshwater lakes and rivers. Although freshwater environments hold a mere 0.01% of all the water on Earth, and lakes and rivers have short histories on a geological time scale, nearly one third of all teleosts (bony fishes; see Appendix 1) live in fresh water.

In 1998, Eschmeyer recognised 23,250 fish species as valid and that about 200 new species are added to this list each year. He estimated that the number of valid fish species could reach 30,000 or 35,000 as poorly sampled geographic areas are studied. The contribution of the different kinds of fishes to this enormous number of species is however strongly asymmetrical. Approximately 96% of fishes are teleosts (see Appendix 1). The elasmobranchs (sharks and rays) are the next most diverse group, but their contribution to the total is a modest 3%. This leaves a few modern survivors of once diverse groups with ancient origins, such as the agnathans (strictly speaking these are not fishes; see p. 318), holocephalans, coelacanths, lungfishes and sturgeons.

Teleosts are therefore the predominant fishes worldwide, excelling in numbers of species, sheer numbers of individuals and in the range of habitats that they occupy. Among the invertebrates, the cephalopod molluscs (squid and cuttlefish) have many fish-like features, but as Marshall (1971) pointed out, their distribution is limited to seawater by their kidney function and to well-oxygenated environments by limitations in the oxygen-carrying capacity of haemocyanin in their blood. This remarkable diversity and success of fishes is matched by their parasites, but in order to understand the challenge to potential fish parasites posed by their hosts we need to know more about this host diversity and about the anatomy and physiology of fishes.

1.2 THE BRITISH FISH FAUNA

According to Wheeler (1978) many of Britain's marine fishes are cool temperate in range and are distributed along the whole length of the Atlantic seaboard. Two important elements contribute to the richness of the fauna. One of these is the Lusitanian element (derived from the Latin name 'Lusitania' for Portugal and much of Spain). Around the British Isles, Lusitanian species are at the northern limit of their range. They may be residents, summer visitors or a combination of the two. The second is the boreal (or Arctic) element, comprising species that normally extend their range south only as far as North Sea coasts. Since the fish fauna south of Britain is richer than that to the north, more Lusitanian representatives contribute to the British fauna than boreal representatives. Our marine fauna is further enriched by the Gulf Stream and the North Atlantic Drift, massive currents of warm water originating in the western Atlantic. These currents bring with them many warm-water fishes such as wreck fish and trigger fishes and make possible the seasonal migrations of fishes of Lusitanian origin, such as tunny, sea bass, mullets and sharks. All of these considerations led Wheeler in 1978 to describe the marine fishes of northern Europe as a "fascinating mixture of common widespread species, migratory species, and rare wanderers". This undoubtedly remains true, although some of the more economically important species have declined as a result of human activities, especially over-fishing.

It has emerged in recent years that the deep-sea fish fauna existing beyond the margins of our western continental shelf is not without significance. This element currently comprises about 178 species of potentially occurring mid-water species and 178 bottom-dwelling fishes (Dr Nigel Merrett, personal communication).

Compared with the marine fauna our freshwater fish fauna is impoverished and to understand why we need first to appreciate that fishes in our freshwaters can be divided into two main categories based on their tolerance to salinity. There are those fishes that are unable to survive in salt water, known as stenohaline species. These are shown in bold type in Appendix 1. This group includes fishes such as pike, *Esox lucius*, and perch, *Perca fluviatilis*, and also the cypriniform fishes such as roach, *Rutilus rutilus*, bream, *Abramis brama* and dace, *Leuciscus leuciscus*. The fishes of the other group are those that can tolerate a range of salinities from fresh water to full seawater; they are known as euryhaline species (underlined in Appendix 1). This group includes migratory (diadromous) fishes like eel, *Anguilla anguilla*, river lamprey, *Lampetra fluviatilis*, sea lamprey, *Petromyzon marinus*, and salmon, *Salmo salar*. The eel breeds in the sea (catadromous), while the lampreys and the salmon breed in fresh water (anadromous). It also includes fish like bass, *Dicentrarchus labrax*, thick-lipped and thin-lipped grey mullets, *Chelon labrosus* and *Liza ramada* respectively, and flounder, *Platichthys flesus*, which are sea fishes capable of penetrating into estuaries and even into fresh water. According to Wheeler (1977), the flounder shows an increasing preference for fresh water as latitude increases, while the three-spined stickleback, *Gasterosteus aculeatus*, shows the reverse trend, with increasing preference for the marine environment in a northerly direction.

The significance of this difference in salinity tolerance is that euryhaline species are able to colonise previously uninhabited rivers via the sea, while stenohaline species obviously cannot do this. The importance of this becomes clear if we consider the effects of the ice ages on the British fauna. It seems likely that the whole of Scotland and all but the most southerly parts of England, Wales and Ireland were purged of life by the great ice sheet that swept south across the country during the last ice age (see Maitland & Campbell, 1992). Re-colonisation of the rivers took place after the northward retreat of the ice that began about 13,000 to 15,000 years ago.

The first fishes to reach the ice-sterilised rivers were euryhaline species arriving from the sea and it is not surprising that migratory euryhaline species like eel, trout (*Salmo trutta*), and salmon, together with sea-going fishes like three-spined stickleback are the most widely distributed of our freshwater fishes. The general consensus is that our stenohaline species came from continental Europe when sea levels were low, exposing a land connection (Doggerland) between the south east of England and the continent, from just north of the river Humber southwards to the Thames. At this time our easterly flowing rivers were either tributaries of the River Rhine or shared a flood plain with this river, permitting an essentially westerly spread of stenohaline fishes from the Rhine system. This migration was terminated when the land bridge submerged, probably about 7500 years ago (Wheeler, 1977). There is some evidence to suggest that a similar temporary link may have occurred between our southern rivers and those of northern France. An alternative suggestion, that some or all of our stenohaline fish species may have colonised British rivers during a previous interglacial period and survived the last glacial period in refugia close to the ice cap, is regarded by Wheeler (1977) as unlikely.

The comparatively short length of time that the land bridge with the continent existed did not allow the full complement of fishes of north-west Europe to reach Britain. Thus, according to Maitland & Campbell (1992), the British Isles has only 55 species of freshwater fishes compared with about 80 in north-west Europe and about 215 in the whole of Europe. Wheeler (1977) pointed out that the known distribution of stenohaline fishes in Britain still provides some support for the idea of entry by way of an eastern and southern land bridge. The fish fauna of eastern England is much richer than that of the western rivers or that of Scottish rivers. Silver bream, *Blicca bjoerkna*, barbel, *Barbus barbus*, burbot, *Lota lota*, bleak, *Alburnus alburnus*, Crucian carp, *Carassius carassius*, spined loach, *Cobitis taenia*, and ruffe, *Gymnocephalus cernua*, were probably all originally confined to river catchment areas in the east of England. The only truly native fishes in the rivers of the west of England, Wales, Scotland and Ireland are euryhaline species such as salmon, trout, eel and sticklebacks, although widespread human introduction of stenohaline species in these areas tends to obscure the picture.

Lampreys, salmonids, smelts (*Osmerus eperlanus*) and sticklebacks are originally marine species that have taken to breeding in fresh water. This environment has the advantage that the vulnerable young stages face fewer predators (see Maitland & Campbell, 1992), but, because fresh water is less productive than the coastal marine environment, young fishes may need to return to the sea to obtain the resources to promote growth and development. The palatability and sporting value of anadromous Atlantic salmon and sea trout have been appreciated for centuries, and this is reflected in the ancient names still used for the distinctive stages in their life cycles. Young salmon (*Salmo salar*) and sea trout (*Salmo trutta*) beginning their lives in fresh water are called 'parr', but as they undergo changes ('smoltification') in preparation for their seaward migration they are referred to as 'smolts'. Fishes returning to fresh water for their first spawning run are 'maidens' and those that survive a spawning run are 'kelts'. 'Mended kelts' are individuals returning to the sea after spawning (see Maitland & Campbell, 1992).

Populations of some of these originally anadromous fishes have been permanently cut off from the sea ('land-locked') by geological processes. Others have found sufficient resources available in fresh water to permit completion of growth and development, enabling them to abandon the sea-going leg of their migration. Since their freshwater environment may still retain a connection with the sea, these populations are not strictly speaking land-locked, although this expression is often used to describe them. In Britain the brown trout (*Salmo trutta*) and the river lamprey (*Lampetra fluviatilis*) (see Chapter 17) have populations confined to fresh water as well as anadromous populations.

Human activity, both intentional and accidental, is responsible for the spread of our native freshwater species to previously uninhabited areas. Humans are also responsible for attempts to introduce exotic species. Many of these attempts failed, but among the successful introductions are zander, *Stizostedion (= Sander) lucioperca*, and bitterling, *Rhodeus sericeus*, from continental Europe and pumpkinseed, *Lepomis gibbosus*, brook charr, *Salvelinus fontinalis*, and rainbow trout, *Oncorhynchus mykiss*, from North America (Maitland & Campbell, 1992).

The relative contributions from the marine and freshwater environments are evident at a glance in Appendix 1. The paucity of freshwater fish families, represented by stenohaline species indicated in bold type, contrasts sharply with the wide variety of marine families, represented by species indicated in normal typeface. It will be obvious to the reader that a major subdivision of fish-like vertebrates, namely the elasmobranchs (sharks and rays), is found only in the sea, while cypriniform bony fishes dominate the freshwater environment.

Most parasites show some degree of host specificity, that is they are restricted to one or to a group of related hosts. At the narrow end of the host specificity spectrum, the parasite may survive on only one or a few closely related host species, while at the broad end the parasite may have catholic tastes and infect a range of families or even orders of fishes. The consequence of host specificity is that in any fish community, different kinds of fishes are likely to have different kinds of parasites and we would expect groups like the uniquely marine elasmobranchs and the uniquely freshwater cyprinids to make distinctive contributions to the parasite fauna. This is true, but, in general, the parasite fauna of our marine fishes is much more diverse than the freshwater parasite fauna, simply because the diversity of our marine fishes is so much greater.

1.3 THE EXTERNAL FEATURES OF FISHES

Most teleosts and many elasmobranch fishes (sharks) have streamlined bodies and are beautifully adapted for swimming (Figure 1.1; Alexander, 1967). Propulsion is usually achieved by lateral movements of the expanded tail fin or caudal fin, with the unpaired dorsal and anal fins providing stability like the feathers of an arrow. In sharks the paired pectoral fins project laterally and horizontally and act as hydrofoils generating lift as the fish swims forward. This lift, augmented by the lift generated by lateral movements of the asymmetrical (heterocercal) tail, counteracts the weight of the fish and enables it to maintain its level in mid-water. Thus when most sharks stop swimming they will sink.

Many teleosts have a gas-filled swimbladder, which makes them neutrally buoyant or weightless, thus freeing the fins from the need to provide lift. During forward swimming the pectoral fins of these fishes are typically held flat against the body thereby minimising drag, and the caudal fin is symmetrical (homocercal).

Although the neutrally buoyant teleost is able to hang motionless in the water, slight adjustments of position are continually required to counteract instability and displacement of the fish by the jets of water leaving the gill cavity. The remarkably flexible fins of teleosts are responsible for making these adjustments by propagating waves along them. The pectoral, caudal and dorsal fins are most often used for this purpose.

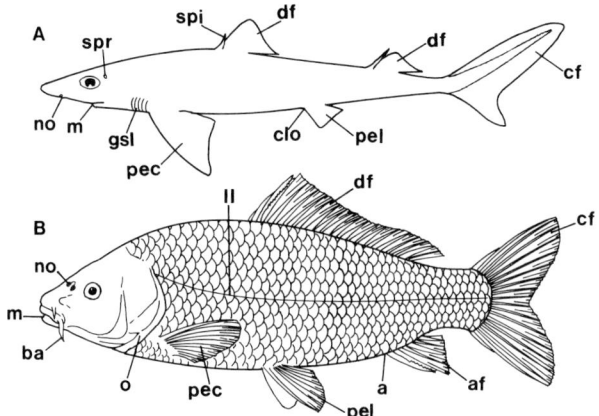

Fig. 1.1. External features of (A) an elasmobranch, the spurdog, *Squalus acanthias*; (B) a teleost, the common carp, *Cyprinus carpio*. a, Anus; af, anal fin; ba, barbel; cf, caudal fin; clo, cloaca; df, dorsal fin; gsl, gill slits; ll, lateral line; m, mouth; no, nostril; o, operculum; pec, pectoral fin; pel, pelvic fin; spi, spine; spr, spiracle. (A) Redrawn from an illustration by Valerie Du Heaume in Wheeler, 1969; (B) redrawn from an illustration by Tenison in Greenwood, 1963.

The combination of neutral buoyancy and the ability to undulate the fins endows the fish with great manoeuvrability. The fish is not only able to hover but also can turn, swim backwards, pitch upwards or downwards and swim vertically up or down with the body held in a horizontal position. In some teleosts, fin undulation is the main mechanism of locomotion. Braking of the forwardly moving fish is another important function for the fins. The pectoral fins have a central role here, but other fins may be involved if pitching or rising is to be avoided as the fish stops.

A characteristic feature of salmonid and coregonid fishes is the presence of a small fleshy adipose fin located dorsally between the more typical dorsal fin and the tail.

Some teleosts have reduced or abandoned the swimbladder. This is understandable in bottom-dwelling fishes like the bullhead, *Cottus gobio*, or the plaice, *Pleuronectes platessa*, where negative buoyancy is needed to stay on the bottom. It is less easy to understand in fast-moving pelagic scombrid fishes like mackerel, *Scomber scombrus*, and tunny, *Thunnus thynnus*, which are denser than seawater and presumably must swim all the time to generate lift in the manner of a shark. Why revert to this energy-expensive way of life? The answer may be related to the threat of predation. Cetaceans such as dolphins and killer whales hunt scombrid fishes. These cetaceans detect prey by echolocation and it has been suggested that fishes with swimbladders would be especially vulnerable since the gas-filled swimbladder is a particularly good reflector of underwater sound (Yalden in Alexander, 1967).

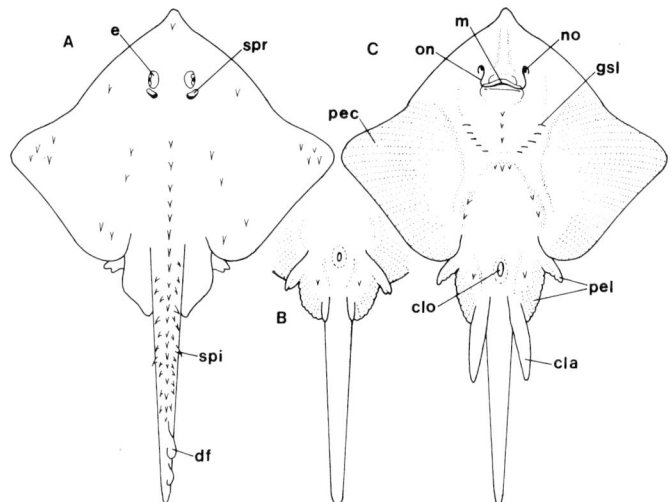

Fig. 1.2. An elasmobranch flatfish, the thornback ray, *Raja clavata*. (A) Dorsal view of female; (B) ventral view of the posterior region of the body of the female; (C) ventral view of male. cla, Clasper (male copulatory organ); e, eye; on, oro-nasal groove. Other lettering as in Fig. 1.1. Based on illustrations by Stebbing in Wheeler, 1978 and by Tenison in Greenwood, 1963.

Elasmobranchs and teleosts have both given rise to bottom-dwelling flatfishes. Many elasmobranchs save the energy expended in continual swimming by finding food on the bottom. Some of these, rays, have become dorso-ventrally flattened (Figure 1.2). The tail is not used for swimming, and dorsal, anal and caudal fins are reduced or absent. Locomotion is achieved by the enormous, undulating, pectoral fins, which propel the fish along the bottom or permit short upward excursions, after which the heavy fish glides back to the bottom. Bottom-dwelling teleosts, having lost their swimbladders, have also become flattened, but curiously this flattening is typically from side to side rather than dorso-ventral. Flatfishes such as the plaice (Figure 1.3A) and the common sole, *Solea solea* (Frontispiece), lie on their left side, while others, such as the turbot, *Scophthalmus maximus*, lie on the right side (Figure 1.3B). In fishes like the plaice the left eye migrates during metamorphosis from the left side of the post-larva to the right side of the bottom-dwelling young. In turbot a similar migration of the right eye takes place. In these teleost flatfishes the dorsal and anal fins run most of the way along the dorsal and ventral edges of the flat body (Figure 1.3). A flatfish like the common sole is able to propel itself gently over the bottom by undulating these fins, but is also capable of driving itself forward and/or upward at greater speed by powerful strokes of the tail.

In teleosts there is a single gill opening on each side of the head covered by a flap or operculum, but in most elasmobranchs there are five separate gill slits on each side, together with a small opening or spiracle just behind each eye (Figures 1.1, 1.2). There are separate gill openings in cyclostomes, but gill ventilation in the parasitic lampreys is specialised and will be dealt with in Chapter 17.

Generally, water is taken in through the mouth into the mouth cavity (buccal cavity) and expelled through the gills, but rays resting on the bottom avoid sucking sand into the gills by taking water in from above via the dorsally positioned spiracles and expelling it through the ventral gill slits (Greenwood, 1963). Teleost flatfishes resting on the bottom or buried in the sediment eject all the water from the gills via the opercular opening on the upper surface, not through the lower opercular opening (Alexander, 1975).

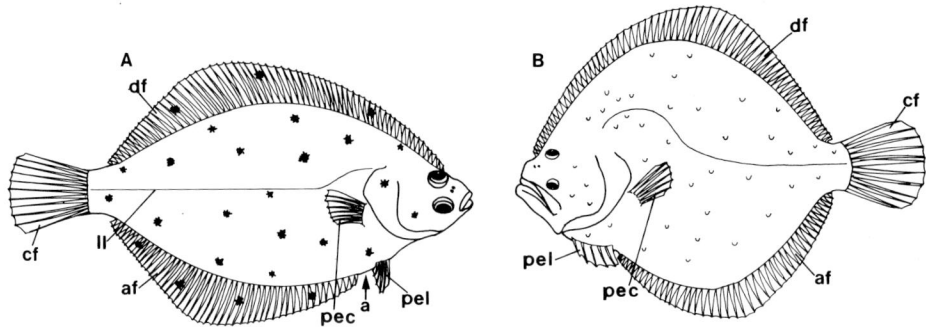

Fig. 1.3. Teleost flatfishes. (A) The plaice, *Pleuronectes platessa*; (B) the turbot, *Scophthalmus maximus*. Labelling as in Fig. 1.1. Based on sketches by Du Heaume in Wheeler, 1969, and Stebbing in Wheeler, 1978.

The nostrils of fishes give access to their olfactory sense organs and are not used for breathing purposes (see Helfman *et al.*, 1997). Typically the single olfactory organ on each side of the head is a sac with a folded lining. The series of ridges so-formed are arranged in parallel or in a radiating pattern. This provides a relatively large surface area to accommodate chemosensory cells. In many teleosts each sac is located anterior to the eye and usually has two openings, an anterior inlet for the olfactory water current and a posterior outlet (Figure 1.1B). Water is propelled through each sac by forward locomotion, by the action of cilia inside the sac or by a pumping action of the sac (see Hara, 2000). In elasmobranchs the relatively large sacs are located on the lower surface of the head, where the mouth is also found (Figures 1.1A, 1.2C). Skin flaps divide each sac, creating an anterior inlet and posterior outlet for water flow. The water current is generated by forward locomotion in active elasmobranchs and is supported by respiratory activity in bottom-dwelling elasmobranchs (Hara, 2000).

Many fishes have a clearly visible line running along each side of the body (Figures 1.1B, 1.3). This lateral line is part of the acoustico-lateralis system and consists of a jelly-filled tube just beneath the skin connected at intervals via lateral line pores to the surface. The tube contains sense organs involved in detecting water displacement and possibly other phenomena associated with pressure changes in water (Pough *et al.*, 1990).

1.4 FISH SKIN

The following account is based mainly on Kearn (1999) and Elliott (2000ab).

Fish skin, like that of other vertebrates consists principally of an outer epidermis and an inner dermis (Figure 1.4). The epidermis consists of layers of epithelial cells (also called keratocytes or Malpighian cells). There may be as few as two layers or tiers in a larva and ten or more in an adult fish. In pelagic fishes the epidermis is often thickest in the dorsal region of the body, but in bottom-dwelling fishes the lower epidermis is often thicker. As in all vertebrates the cytoplasm of the epidermal cells specialises in assembling slender unbranched filaments only eight nanometres in diameter called tonofilaments. Bundles of these filaments course through the cytoplasm of the cell and some are linked to special structures called desmosomes that serve to hold the cells of the epidermis together. Thus the tonofilament network adds strength to the epidermis.

In other vertebrates the basal cell layer of the epidermis is the site of multiplication and production of new epithelial cells, but in fish epidermis, cell division occurs throughout, although most common in the deeper layers. In mammals the outer layer of epithelial cells dies and produces a cornified protective layer, but this rarely occurs in the fish epidermis, although dead cells are regularly sloughed from the epidermal surface. According to Tsai (1996) those of the loach *Misgurnus anguillicaudatus* are sloughed after travelling for about four days from the lower epidermis, while the corresponding cells in humans take 33 – 37 days to make a similar journey.

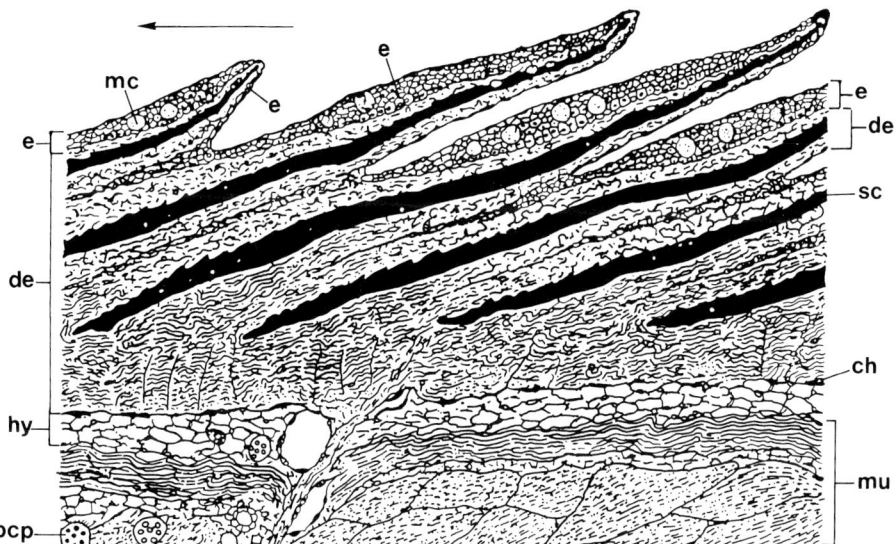

Fig. 1.4. Diagrammatic longitudinal section through the skin of a typical teleost. bcp, Blood capillary; ch, chromatophore; de, dermis; e, epidermis; hy, hypodermis; mc, mucous cell; mu, muscle; sc, scale. Arrow: direction of anterior end of fish. From Elliott, 2000, Copyright © 2000, with permission from Elsevier. Original source: Ward's Natural Science Establishment Inc.

The exposed epithelial cells of fishes, sometimes called pavement cells or squamous cells, differ in some respects from the deeper cells. In these superficial cells a mat of tonofilaments known as a terminal web lies immediately beneath the surface membrane and the exposed cell surface is characterised by the presence of elliptical whorls of micro-ridges, resembling a fingerprint. This exposed surface has an outer coat or cuticle, which is probably secreted by the epithelial cells (see below).

The epidermis also contains goblet (mucous) cells, which make a major contribution to the slime covering the fish (see below). These goblet cells probably differentiate from epithelial cells in the lower layers of the epidermis. Upon reaching maturity and the surface of the skin, the neck of the cell emerges between adjacent epithelial cells. Secretion is released by rupture of the apical cell membrane after which the cell dies.

In addition to epithelial cells and goblet cells two other types of secretory cell are present in the epidermis. These are sacciform cells and club cells (see Elliott, 2000 for details). Generally their functions are poorly understood. Cells containing numerous rod-shaped inclusions (rodlet cells) are also found in the epidermis of many freshwater and marine teleosts, and leucocytes (white blood cells) are also common. The roles of rodlet cells and leucocytes will be discussed further below.

A basement membrane separates the epidermis from the tougher more complex dermis beneath (Figure 1.4). Its toughness is attributed to the presence of a fibrous protein called collagen laid down between the dermal cells. Blood capillaries, lymphatic vessels and nerves occur within the dermis and chromatophores (pigment cells) may also be present.

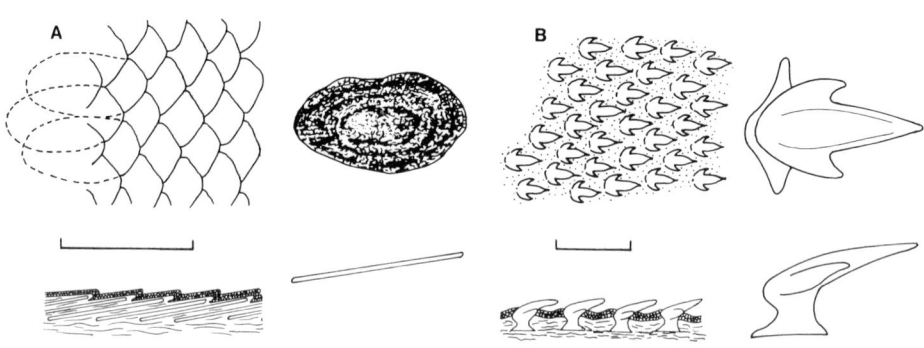

Fig. 1.5. A comparison between (A) the scales of a teleost (cod, *Gadus morhua*) and (B) the denticles (placoid scales) of an elasmobranch (dogfish, *Scyliorhinus canicula*). The skin is shown in surface view above and in diagrammatic section below. One scale of each is shown enlarged. Scale bars relate to surface views : (A) 1 cm; (B) 2mm. Redrawn from Hardy, 1959.

Cyclostomes have no scales but the skin of most other fishes contains scales of one kind or another. Reptiles, birds and some mammals also have scales, but these differ in structure and in origin from those of fishes. The scales of these mainly terrestrial vertebrates, in common with feathers, hair, hooves and claws, are derived from the epidermis and consist largely of the tough protein keratin, while fish scales are entirely, or almost entirely, of dermal origin.

The scales of teleosts are flat plates, typically overlapping and arranged in diagonal rows (Figure 1.1B). They are assembled entirely by the dermis within oblique dermal pockets (Figures 1.4, 1.5A). Each scale consists of collagenous material with superficial calcification and is anchored in the pocket by bundles of collagen fibres attached to the calcified layer. Teleost scales grow continuously as the fish grows and, since growth rates may change with the season, scales of some fishes may contain a record of these changes in the form of concentric growth rings indicating the age of the fish. The scales of teleosts may project above the general skin surface, but a thin layer of epidermis and dermis usually covers projecting edges. Attempts to move the finger forward over the skin of a fish like the common sole are met with resistance because the backwardly projecting edges of the scales are spiny (ctenoid scales). In some teleosts scales are not easy to detect on the surface because they are deeply embedded as in the plaice, or microscopic as in the eel.

Elasmobranchs have denticles (placoid scales), which are tooth-like, isolated structures consisting of dentine on a base of bone (Figure 1.5B). These components are produced by the dermis, but there is an additional contribution from the epidermis in the form of an outer hard enamel layer. Elasmobranch denticles do not grow as the fish grows, new ones being added between existing denticles. The points of denticles project through the epidermis and the exposed regions have no covering of tissue. This gives the skin surface a rough texture to the touch and the crude skins of dogfishes and sharks have been used for polishing wood or metal. However, the denticles of rays are usually scattered and unevenly distributed over the upper surface, leaving smooth denticle-free areas (Figure 1.2). Denticles may be absent as in some stingrays.

A thin single sheet of cells joined by desmosomes called the dermal endothelium provides the inner boundary of the dermis. Beneath this boundary is the hypodermis, which contains deep chromatophores and cells containing fat, together with blood vessels and nerve bundles approaching the skin (Figure 1.4).

1.5 SKIN AND GILL MUCUS

Shephard (1994) has reviewed the production, composition and functions of fish mucus. It is likely to be involved in many of the following: respiration, water and ion balance, excretion, reproduction, disease resistance/protection, and communication, as well as having more specialised functions such as in nest building. The bulk of fish mucus is produced by goblet cells, which are abundant on virtually all fish epidermal surfaces and particularly on the gills (see below). The 'slipperiness' of fish skin is attributable to the high water content of skin mucus and the presence of gel-forming macromolecules, predominantly glycoproteins (mucins) of high molecular weight. Other cells, such as sacciform cells (see above), undoubtedly add their own contributions to fish mucus.

Apparently distinct from the product of goblet cells, is the cuticle. According to Shephard (1994) this is a layer of macromolecular gel, probably secreted by the epithelial (Malpighian) cells and possibly homologous with the glycocalyx found on the surfaces of other cells. Typically the cuticle is about 1 μm thick, but may reach 10 μm. Shephard pointed out that the term 'cuticle' is sometimes used in a general sense for the combined products of epithelial cells, goblet cells and other secretory cells, together with sloughed cells and cell debris. However, Shephard stressed that the products of epithelial (Malpighian) cells and goblet cells are histochemically different. They may remain distinct from one another and this distinction may contribute to the functioning of either or both.

1.6 FISH GILLS

The following account of fish gills and their function is based on Helfman *et al.*, (1997), Kearn (1998) and Olson (2000ab).

The fish gill is a multifunctional organ that deals with the vagaries of the aquatic medium. It is the primary site of respiration and also plays a major role in functions such as osmoregulation (salt and water balance) and acid-base balance.

1.6.1 *The gills of teleosts*

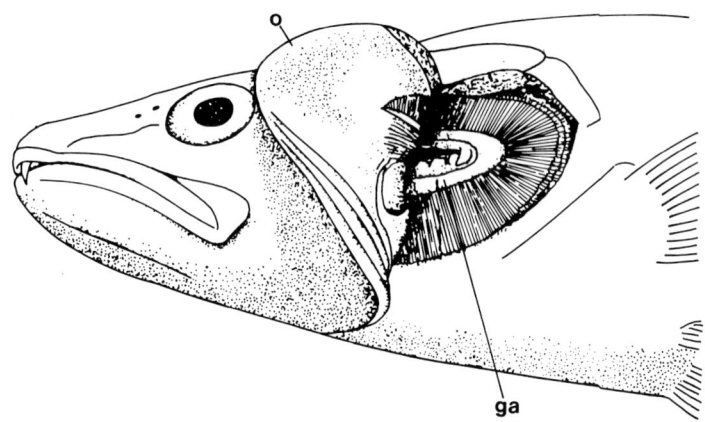

Fig. 1.6. Zander, *Stizostedion* (= *Sander*) *lucioperca* with operculum raised to show gills of the left side. ga, Gill arch; o, operculum. From Rauther, 1937.

Most teleosts have eight gills, four on each side of the head. If the left operculum of a teleost is raised as in Figure 1.6, the first of the four gills on the left side of the head is visible. Each gill has a bony arch projecting from which are two rows of tapering primary gill lamellae (= gill filaments) (Figures 1.6, 1.7). Thus the whole gill or holobranch consists of an anterior and a posterior half-gill or hemibranch, each hemibranch consisting of one of the two rows of primary lamellae. The number of primary lamellae on the gill arch varies from arch to arch, the first and second arches usually carrying the largest number and the anterior hemibranch having a few more than the posterior hemibranch. The length of a primary lamella depends on its position on the gill arch. More primary gill lamellae appear as the fish grows.

These primary lamellae are flattened like a knife blade. A cartilaginous rod extending along the inner edge of each primary lamella (afferent edge; see below) from the gill arch to the tip provides support (Figure 1.8). The primary lamella is the functional unit of the gill and carries two rows of closely spaced, flap-like secondary gill lamellae, which project in opposite directions from the two flat surfaces of the primary lamella (Figures 1.7, 1.8). The gill owes its red colour when fresh to these secondary lamellae, which are thin-walled, contain blood and are the main sites of gaseous exchange. Blood flowing inside the secondary lamellae moves in the opposite

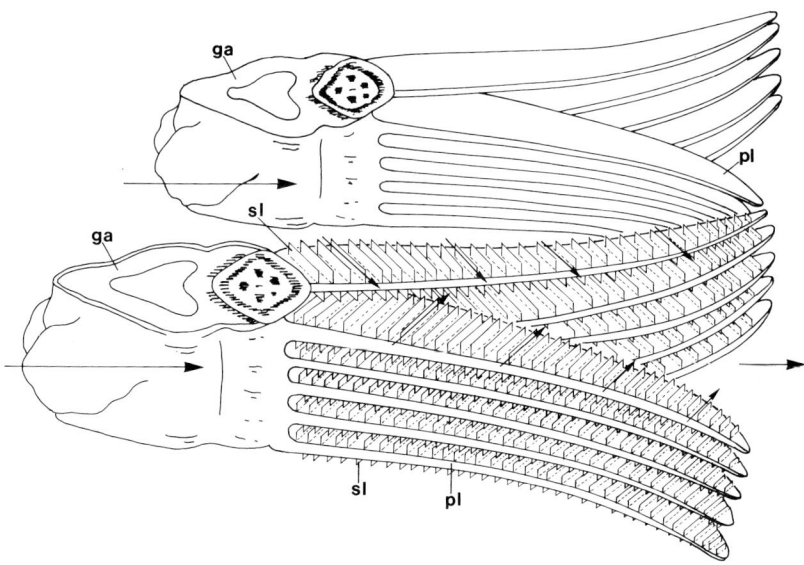

Fig. 1.7. Stereogram showing parts of two adjacent gill arches (ga) of a teleost. pl, Primary lamella; sl, secondary lamella. Arrows show directions of water currents. Modified from an unpublished sketch by Eleanor Skeate.

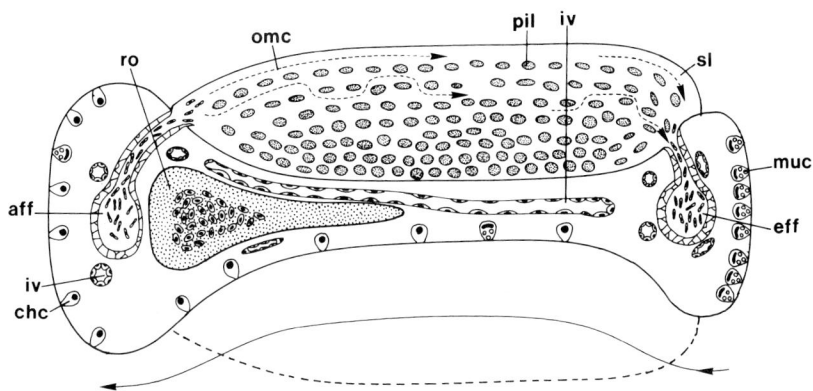

Fig. 1.8. Diagrammatic cross section of a primary gill lamella of a teleost. aff, Afferent (incoming) artery; chc, chloride cell; eff, efferent (outgoing) artery; iv, interlamellar blood vessel; muc, mucous cell; omc, outer marginal channel; pil, pillar cell; ro, cartilaginous rod; sl, secondary lamella. Broken arrow, direction of blood flow; continuous arrow, direction of water flow. A secondary lamella on the opposite side of the primary lamella is out of the plane of section but a broken line indicates its presence. Modified from Olson, 2000b.

direction to the water flowing between the secondary lamellae (Figures 1.8, 1.9), i.e. blood flows from the afferent (inner) border of each primary gill lamella to the efferent (outer) border of the primary lamella. This counter-current arrangement is an important adaptation for maximising gaseous exchange.

The secondary gill lamellae provide an enormous surface area for gaseous exchange. Fish species with sluggish habits have a total gill area of approximately 1.5 – 2 cm^2/g body weight, while for a moderately active fish the corresponding figure is about 2 – 5 cm^2/g and for a very active fish >5cm^2/g.

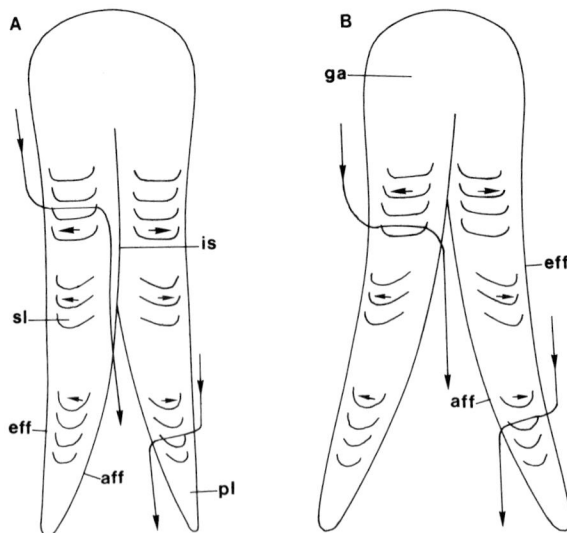

Fig. 1.9. Diagrammatic cross sections through the gill arches from two teleosts. (A) A fish like the brown trout, Salmo trutta, with a well-developed interbranchial septum (is); (B) a fish like the perch, Perca fluviatilis, with virtually no interbranchial septum. aff, Afferent border of primary lamella; eff, efferent border of primary lamella; ga, gill arch; pl, primary lamella; sl, secondary lamella. Long arrows show direction of water flow; short arrows in secondary lamellae show direction of blood flow. See text for further explanation. Modified from Olson, 2000a.

The shapes of secondary lamellae vary between species and even along the length of a single primary lamella (Figure 1.9). Each secondary lamella consists of two layers of epithelium enclosing a thin flat blood space. Post-like pillar cells separate the epithelial layers (Figure 1.8). Since blood pressure in the secondary lamellae is high there is a risk of ballooning and columns of tough collagen fibres linking the two epithelial layers prevent this.

The free outer edges of the secondary lamellae are the best sites for gaseous exchange and there are indications that red blood corpuscles are diverted to these outer margins to take advantage of this, while the blood plasma travels through the inner regions of the lamellae where enzymes in the pillar cells are able to metabolise circulating substances such as hormones.

Physiological studies of teleosts have shown that their gills are ventilated by a double pumping system comprising a buccal (mouth) cavity force pump and an opercular suction pump (Figure 1.10A). These pumps complement each other so that water flow through the gills is more or less continuous. During normal respiration the tips of the primary gill lamellae of adjacent gill arches touch, so that water passing between the gill arches from the buccal cavity to the opercular cavity is forced to pass between the secondary gill lamellae (Figures 1.7, 1.10A). The water current strikes one hemibranch of each gill on its anterior face and the other on its posterior face. The danger of suspended material such as food entering from the buccal cavity and clogging the delicate gills is prevented by spiny gill rakers, which project across and guard the water channels between the gill arches (Figure 1.10A).

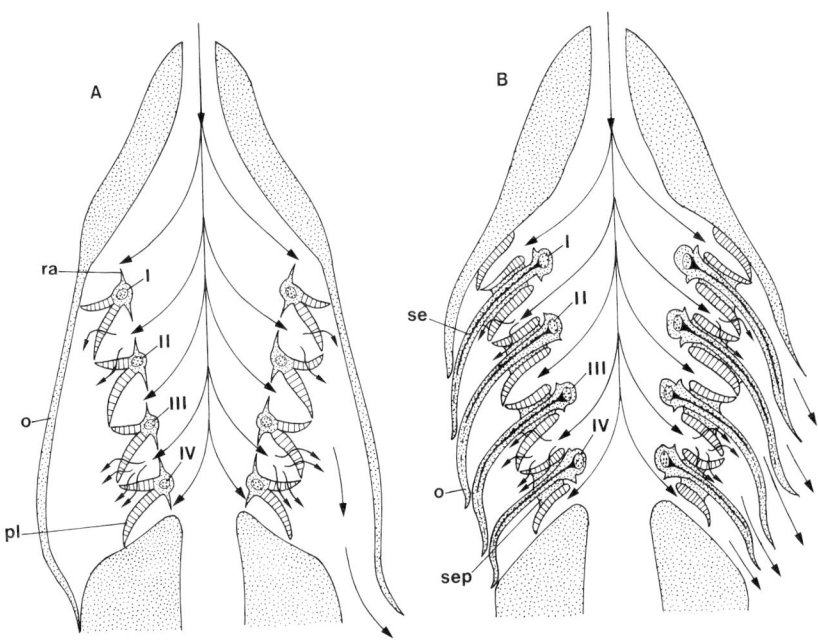

Fig. 1.10. Diagrammatic sections in a horizontal plane through the buccal and gill chambers of (A) a teleost and (B) an elasmobranch showing the arrangement of the gills. The opercula are shown closed on the left and open on the right of each diagram. I – IV, gill arches; o, operculum; pl, primary gill lamella bearing secondary lamellae; ra, gill raker; se, gill septum; sep, septal canal. Large arrows indicate directions of water currents entering and leaving the gill chamber; small arrows indicate the paths of the gill-ventilating currents between the secondary gill lamellae. Modified from Bertin, 1958.

In some teleosts like the trout (*Salmo trutta*) the interbranchial septum extends from the gill arch for one-third to two-thirds of the length of the primary lamella (Figure 1.9A). In these fishes respiratory water that has passed between the proximal secondary lamellae of the gill must flow along the interbranchial septum to reach the opercular cavity. However, in many other teleosts, for example the perch (*Perca fluviatilis*), the septum is greatly reduced, freeing the primary lamellae for most or all of their length and permitting respiratory water to pass unhindered between all the secondary lamellae and directly into the opercular cavity (Figure 1.9B).

The epithelium covering the gills is generally regarded as falling into two categories, one covering the primary gill lamellae and gill arches and the other covering the secondary lamellae. However, both categories of epithelium contain the same three major cell types: pavement cells, chloride cells (ionocytes) and goblet (mucous) cells.

Pavement cells account for about 95% of all cells constituting the gill epithelium. They have no special cytological features that we need to mention, except that like skin epithelial cells most of them have surface microridges. Goblet cells (Figure 1.8) are found scattered over the gill arch, primary gill lamella and basal regions of the secondary lamella. Chloride cells (Figure 1.8) are characteristic of gills and are concerned with salt balance.

1.6.2 *The gills of elasmobranchs*
The similarity between the gills of an elasmobranch and those of a teleost is evident in the comparative diagram in Figure 1.10. An obvious difference is that a single flap or operculum covers the gills on each side of the head of a teleost, while in elasmobranchs each gill slit opens separately on the surface and has its own opercular covering (Figure 1.1). Most elasmobranchs have five openings on each side. Like teleosts, elasmobranchs have primary gill lamellae projecting from a gill arch and carrying rows of blood-filled secondary lamellae (Figure 1.10B). The main difference between teleosts and elasmobranchs concerns the length of the interbranchial septum running between the hemibranchs of each gill. In elasmobranchs this septum is long, projecting beyond the distal ends of the primary lamellae and forming a flap/valve (operculum) for the gill slit immediately posterior to it (Figure 1.10B). Thus the septum forms a complete barrier between hemibranchs and all the respiratory water that has passed between the secondary gill lamellae must pass along a septal canal adjacent to the septum in order to leave the gill (Figure 1.10B).

1.7 RESISTANCE AND REPAIR

Any fish is likely to be assailed by a variety of potential parasites and pathogens. Remarkably, very few of these become established in or on the fish's body. This is because the fish is protected by a complicated defence system, one component of which is a specific humoral immune system similar to our own, operating by way of the blood system.

The basic functions of this immune system in mammals, including ourselves, are well known, in particular the remarkable ability of the body to assemble a great range of specialised protein molecules known as antibodies (immunoglobulins), each of which is specifically adapted for eliminating/neutralising a particular foreign substance or antigen. Parasites may be a source of antigens in the form of structural chemical components, secretory products or excretory material. Clones of white blood cells called lymphocytes assemble each kind of specific antibody. Attachment of antibody to antigen may have many different consequences, one of which may be the destruction of a specific invading organism. The system also has the remarkable capacity of immunological memory. When an animal is exposed to an antigen for the first time, such a 'naïve' animal generates a specific antibody capable of dealing with that antigen. When the same animal receives a second dose of the same antigen, the animal's immune system is already prepared for the challenge and is able to respond more rapidly and effectively.

The specific humoral immune system of fishes is less complex than that of mammals and differs in important ways, but these do not directly concern us here. The interested reader is referred to reviews by Ellis (1982) and by Kaattari & Piganelli

(1996). However there are other components of immune defence in fishes and a basic understanding of these other mechanisms will help the reader to appreciate some of the parasite/host interactions discussed in this book.

The whole body surface and fins of a fish are exposed to water-borne parasites and pathogens, as are the gills and the sensory epithelium of the nostrils. As portals of entry for potentially harmful organisms the skin and the gills differ significantly in importance (see review by Evelyn, 1996). The gills are particularly vulnerable because of the functional need for a large surface area and minimal separation between the fish's blood and the water passing through the gill. In fact only a single layer of fragile epithelial cells separates the blood and water (see above). Several bacteria are known to enter the fish by this route. On the other hand, the skin presents a much more formidable barrier than the gill covering, and only one bacterial fish pathogen, *Vibrio anguillarum*, has been shown to have the capability to penetrate intact skin. Potentially hostile organisms may also gain entry via the gut in the fish's food.

All of these portals of entry have one common feature. Their epithelial covering secretes mucus (see above). Since all fish ectoparasites live in permanent contact with these so-called mucosal surfaces and since mesoparasites (see pp. 1, 185) and endoparasites must pass through a mucosal surface to enter the tissues, we would expect mucus to have an important front-line protective role against such organisms. The finding in fish skin mucus of chemical substances also found in blood serum and known to have defensive capabilities has raised these expectations.

Immunoglobulin has been reported in fish skin mucus. Lysozyme, an enzyme with strong antibacterial properties, may also be present (references in Kearn, 1999) and complement, a complex of several proteins having a variety of protective roles, has also been recorded. In fishes as in mammals, complement may be activated either by antibody/antigen complexes (the so called classical pathway), or by substances of a non-immunological origin (the alternative pathway) (see Yano, 1996). Bakke *et al.* (2002) reported higher levels of complement (C3) in mucus from salmon (*Salmo salar*) resistant to the monogenean *Gyrodactylus salaris* compared with mucus from susceptible salmon (see Chapter 4). All these substances are familiar components of the blood-mediated immune system, but whether they have a functional immunological role in mucus remains to be established. Other substances recorded from fish mucus that may have a defensive role are C-reactive protein, lectin and haemolysin (see Yano, 1996).

Of particular importance is the finding that, after administration of an antigen orally to a fish, antigen-specific antibody appears in skin mucus but little antibody appears in the blood serum (references in Kaattari & Piganelli, 1996). The presence of mucosal antibody without concomitant serum antibody raises the interesting possibility that fishes have a regionally operative immune system and Moore *et al.* (1998) suggested that sequestration of foreign antigens by skin and gill epithelial cells might play a part in the development of this local immunity. Numbers of leucocytes are known to increase greatly when the skin is challenged by damage or infection and Peleteiro & Richards (1988) detected immunoglobulin in lymphocyte-like white cells in the epidermis of the rainbow trout (*Oncorhynchus mykiss*). They also detected immunoglobulin in goblet cells, but were unable to determine how antibody reached these cells.

Initially fishes were thought to have only one type of immunoglobulin, but, more recently, differences in the organisation of immunoglobulins from fish serum and mucus have been identified, and Lobb & Clem (1981) came to the conclusion that the immunoglobulin in skin mucus from the sheepshead seabream, *Archosargus probatocephalus*, is the result of local synthesis and is not derived from serum immunoglobulin. Such a localised integumentary immune system is likely to be of special significance for parasites penetrating mucosal surfaces and for ectoparasites or mesoparasites spending their entire lives in contact with these surfaces.

In the event of the skin being breached by physical injury or by the attachment or feeding activities of large animal parasites like leeches or lice, the fish's epidermis has an important role. It is responsible for the rapid closure of the skin wound, minimising opportunities for entry of micro-organisms and preventing further deterioration in osmotic balance. In experimentally created wounds measuring 5 by 1 mm in the skin of plaice (*Pleuronectes platessa*), Bullock *et al*. (1978a) found that epithelial cells migrating inwards from the edge of the lesion rapidly covered the exposed surface. Fish epithelial cells are among the fastest migrating cells known from *in vitro* studies, moving 10 to 20 times faster than mammalian fibroblasts (Zigmond, 1993, in Elliott, 2000b). Bullock *et al*. (1978a) found that after just 30 minutes at 10°C the epithelial cells at the edge of the wound were flattened. Cell migration in an inward direction was well developed after one hour, with complete closure after nine hours. During cell migration, the epidermis adjacent to the wound became noticeably thinner. Even at 5 °C closure was complete at 12 hours. Thus wound closure in plaice is fast even at low temperatures and is entirely due to migration of pre-existing cells from the periphery of the wound, with no contribution from cell division. This reflects the very slow rate of epidermal cell turnover in plaice (see Bullock *et al.*, 1978b). Wound healing in mammals differs in that coverage is assisted by a burst of epidermal cell division in the wound epithelium, and even at the elevated body temperature of 37°C small wounds take as long as 1 – 5 days to close.

Mechanical injury or invasion by parasites typically elicits a local inflammatory response, which is characterised by the following events: an increase in the numbers of certain leucocytes (monocytes and neutrophils) in the blood circulation, increased blood supply to the affected area, increased permeability of blood capillaries and migration of white blood cells out of the capillaries and into the surrounding tissues (see Secombes, 1996). The activities of macrophages (derived from monocytes) and neutrophils limit the spread of an invader or remove it altogether.

Macrophages are voracious phagocytes, i.e. they ingest and then digest bacteria and debris. Their appetite may be such that the cytoplasm becomes highly vacuolated and 'foamy' in appearance (Secombes, 1996). According to Secombes (1996) fish neutrophils are phagocytic, but Ellis (1982) has suggested that they may also liberate enzymes externally, presumably contributing to the hostility of the environment for pathogens. However, extracellular enzymes may damage host tissue, requiring the activity of the phagocytes to remove host cell fragments.

When phagocytes ingest particles there is an increase in oxygen uptake that is independent of respiration (Secombes, 1996). This is the so-called respiratory burst, which is used in the generation of free radicals known to be toxic for bacteria and protozoan parasites. Some of these free radicals are reactive oxygen species, such as the superoxide anion (O_2^-) and hydrogen peroxide (H_2O_2), and reactive nitrogen species such as nitric oxide (NO). Fish macrophages and neutrophils can generate reactive oxygen species and fish macrophages are known to secrete nitric oxide.

Ellis (1982) noted that extensive host tissue damage may occur in inflammatory responses in fishes and he has suggested that free radicals released into the lesion, particularly by neutrophils, may be responsible for this damage. He also made the interesting suggestion that the black/brown pigment melanin, which is often found in inflammatory lesions and tissues rich in phagocytes, may have the important role of quenching free-radical reactions and hence preventing such self-inflicted damage. He pointed out that macrophages often contain melanosomes (melanin-containing cell inclusions), possibly supplied to the macrophages by melanin-producing cells (melanocytes) and that this melanin may protect the macrophage from internal damage produced by free radicals ingested by the phagocyte. Melanosomes supplied to mammalian epidermal cells by melanocytes are thought to protect against damage by radiation-induced free radicals in the skin. In Ellis' opinion (1982), melanin may have a widespread role in protection of hosts against invasion by parasitic organisms and against potentially self-harmful protective mechanisms of the host's own defence system.

In inflammatory loci vigorous phagocytosis continues for 3 – 4 days, after which the numbers of phagocytic cells decline. The damaged tissues are then repaired. According to Ellis (1982), fish probably have better tissue-regenerating powers than mammals, muscle cells retaining their ability to divide throughout life. If the invader is a parasite too large to remove, encapsulation by host fibre-secreting cells (fibroblasts) may occur.

As noted above, rodlet cells occur in the epidermis and gill epithelium of some fishes. A non-specific defensive function has been attributed to these cells, but this is not the only role attributed to them. Some authors regard them as having physiological functions unrelated to immunity, such as involvement in osmoregulation or sensory reception, and others favour an entirely different view of these cells as intrusive parasitic organisms (references in Elliott, 2000b).

Not all tissue sites in fishes have the capability of destroying or isolating parasites. The brain and the eye have been described as 'immunologically privileged sites', because the presence of a so-called blood/brain barrier effectively excludes components of the immune system from these organs. Consequently a parasite that is capable of finding its way to such a site will be unmolested by the host's immune system. It might be thought that the epidermis would have similar properties since this layer of the skin is beyond the reach of blood capillaries, but, as we have seen above, this is not the case and the epidermis contains numerous components of the immune system.

Several factors are known to influence the general effectiveness of the immune response in fishes. Unlike warm-blooded mammals and birds, temperature has a profound effect on metabolic processes in cold-blooded vertebrates like fishes. One of its effects is on the immune system. Generally speaking the higher the temperature within the physiologically tolerated range of a fish species the faster the onset of the immune response and the greater its magnitude (Ellis, 1982). Low temperatures within the tolerance range delay the onset and reduce or even eliminate the response. The critical temperature for development of the immune response in a fish is related to the natural environmental temperatures favoured by the species. For example, the common carp (*Cyprinus carpio*), which favours relatively warm water, may show no response below 15°C, while in colder water species like the salmonids the response is not inhibited until the temperature falls as low as 4°C (references in Ellis, 1982). Seasonal fluctuations in the expression of the immune response independent of temperature have also been reported (Ellis. 1982).

Suppressive effects on the fish defence system during social interactions have been identified (see Ellis, 1982), but a notorious modulator of the immune response is stress. Wendelaar Bonga (1997) defined stress as a condition in which the dynamic equilibrium of animal organisms (homeostasis) is threatened or disturbed as a result of intrinsic or extrinsic stimuli, commonly defined as stressors. The animal responds by generating a co-ordinated set of behavioural and physiological responses, thought to be compensatory or adaptive, leading to elimination of the threat.

The stress response in teleost fishes shows many similarities to that of terrestrial vertebrates. The differences that do exist are related to the intimate contact of fishes with water. There are two major routes in fish whereby the brain coordinates the stress response: (1) the brain (hypothalamus) – sympathetic nervous system – chromaffin cell axis; (2) the brain (hypothalamus) – pituitary gland – interrenal cell axis (see Wendelaar Bonga, 1997, figure 2). These axes correspond with (1) the brain – sympathetic – adrenal medulla axis and (2) the brain – pituitary – adrenal axis respectively in terrestrial vertebrates. The primary messengers of the two axes in fishes are catecholamines and cortisol respectively. Activation of brain centres by stressors, which may include invasion by parasites, leads to the massive release of these hormones. Their immediate effects are at the tissue level and include increases in cardiac output, improvement in oxygen uptake, mobilisation of energy substrates and disturbance of hydro-mineral balance. Responses then extend to the levels of the organism and population and include inhibition of growth, inhibition of reproduction and, of special interest to the parasitologist, suppression of immune responses, together with reduced capacity to tolerate subsequent or additional stressors.

The inhibitory effects of stress on disease resistance (immunosuppression) by way of cortisol have been demonstrated many times and their adaptive significance is, not surprisingly, questionable. Wendelaar Bonga (1997) suggested that immunosuppression might be advantageous in the short term by preventing over-stimulation of the immune system, but in the long term, chronic immunosuppression seems likely to be disadvantageous by reducing resistance to the establishment of pathogens and parasites. However, this is a rather simplistic picture. Stress modulates the defence system of the fish as it does other physiological systems and while certain aspects of defence may be suppressed by stress others may be enhanced (Ellis, 1981; Wendelaar Bonga, 1997). Catecholamines are known to have both stimulatory and inhibitory effects, the former being less easy to detect but possibly having an adaptive role (Wendelaar Bonga, 1997).

It is clear that our understanding of the stress response in fish and its effects on the immune system is still rudimentary. More detailed exploration of the roles of catecholamines and cortisol in immune modulation is required and the significance and contribution of other chemical messengers known to interact with the immune system, such as cytokines and neuropeptides, need to be assessed (for further details see Wendelaar Bonga, 1997).

Lastly there is one other phenomenon that may have a significant impact on populations of ectoparasites. This is the activity of 'cleaner' organisms. These are fishes or crustaceans that pick off and eat skin and gill parasites from client fishes, usually at special cleaner stations.

2
'PROTOZOANS'

2.1 INTRODUCTION

When I was a student in the 1960s, unicellular organisms were considered to belong to the Phylum Protozoa. Research undertaken since that time, especially in the fields of electron microscopy and molecular biology, has revealed that 'protozoans' (or 'protists') are highly diverse and are organised structurally along a number of distinct lines. These lines are now regarded as separate phyla, although there is no general consensus about how many of these phyla exist or about the interrelationships between them. Although systematists no longer recognise a major group called 'Protozoa', the term 'protozoan' has been preserved and is still widely employed, since it serves a useful purpose as a term indicating any organism at the unicellular level of construction. I will continue to use it in this chapter.

Free-living protozoans are ubiquitous and abundant in the aquatic environment, although they may go unnoticed because of their small size. According to Roberts & Janovy (2000), as many as 45,000 different kinds of protozoans have been identified to date and it is not surprising that contact between these organisms and fishes has produced a range of parasites with many different life styles. Some protozoans occur on the exposed surfaces (skin and gills) of fishes and include epizoic forms (using the host merely as a platform), like the trichodinid and scyphidiid ciliates, as well as forms with an obligatory ectoparasitic life style, such as the flagellate *Ichthyobodo necator*. Many of these organisms are so small that the resolution of the light microscope is not enough to distinguish their structural details and it is necessary to resort to the electron microscope to supplement our knowledge of their anatomy. On the other hand reproductive processes in many endoparasites may produce aggregations of individuals or spore-bearing nodules that are sufficiently large and conspicuous to be visible to the naked eye. Strictly speaking, these eye-catching endoparasites fall outside the remit of this book, but the ciliate *Ichthyophthirius multifiliis* ('white spot') is included because it spends its entire parasitic life inside the epidermis, and its interactions with the epidermis are relevant to our study of the biology of ectoparasites.

2.2 EPIZOIC CILIATES

Trichodinids belong to a group of ciliates (Phylum Ciliophora) known as the peritrichs, which are characterised by differentiation of one pole of the cell for attachment and the opposite pole for collecting food. Trichodinids are disk-shaped or barrel-shaped (Figure 2.1). They spend most of their time with their basal (aboral) surfaces attached to or gliding over suitable substrates, usually the surfaces of living animals, including marine and freshwater fishes. They are propelled by a wreath of locomotory cilia projecting from the edge of the basal disk (Figures 2.2, 2.3A). The disk is supported by a proteinaceous cytoskeleton of considerable complexity but great visual appeal (Figures

2.3B, 2.4). The most conspicuous components are so-called 'denticles', which interlock to form a ring (Figure 2.3B). Each denticle has a blade-like centrifugal projection and a thorn-like centripetal projection. Overlapping the blades and extending outwards are delicate radial pins (Figures 2.2, 2.3B) and fine skeletal rays reinforce the valve-like free border of the disk (Figure 2.3B). Trichodinids can also swim freely, propelled by their cilia.

Fig. 2.1. Typical shapes in side view of trichodinid ciliates from fishes. From Lom & Dykova, 1992, with permission from Elsevier.

The taxonomy of the group used to be confused and the reasons for this have been outlined by van As & Basson (1987). The first described species (type species) of *Trichodina* was *T. pediculus* (see Table 2.1), followed by the establishment of *T. domerguei* and then by other species, all from freshwater fishes.

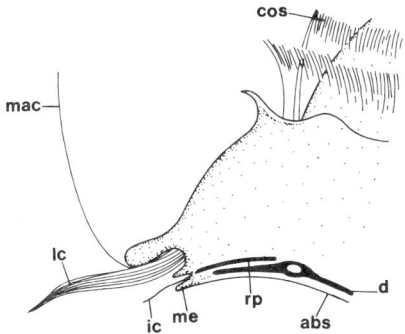

Fig. 2.2. Diagrammatic representation of the body margin of a trichodinid in side view. abs, Aboral surface; cos, cilia of oral surface; d, denticle; ic, inner wreath of cilia; lc, locomotory ciliary wreath; mac, marginal cilium; me, border membrane; rp, radial pin. From Lom & Dykova, 1992, with permission from Elsevier.

However, up to the middle of the last century criteria for distinguishing one trichodinid from another were derived from observations on living animals. The failure of these criteria to provide adequate discrimination led to the synonymisation of all trichodinids from freshwater fishes under the name of the type species, *T. pediculus*. The turning point came in 1958 when Lom developed a silver impregnation technique that created stunning photographs like those in Figure 2.4, clearly revealing the arrangement of the taxonomically important components of the cytoskeleton. Significant differences revealed by this technique re-established *T. domerguei* as distinct from *T. pediculus*, and many new species have been described. According to Lom & Dykova (1992), *Trichodina* contains over 100 species worldwide and they listed seven genera of trichodinids associated with fishes. In spite of this effort, trichodinids of British fishes

Table 2.1. Some Trichodina *spp. recorded from British fishes.*

Species of Trichodina	Host(s)	Micro-habitat	Comments	Authority
T. acuta	Goldfish (*Carassius auratus*)	Skin		Gaze & Wootten, 1998
	Common carp (*Cyprinus carpio*) Rainbow trout (*Oncorhynchus mykiss*) Minnow (*Phoxinus phoxinus*) Brown trout (*Salmo trutta*)			
T. borealis	Plaice (*Pleuronectes platessa*)	Gills	Marine	MacKenzie, 1969
T. branchicola	Tompot blenny (*Parablennius gattorugine*) Shanny (*Lipophrys pholis*)	Gills	Marine	Tripathi, 1948
	Five-bearded rockling (*Ciliata mustela*) Sea scorpion (*Taurulus bubalis*) Three-bearded rockling (*Gaidropsarus vulgaris*) Plaice (*Pleuronectes platessa*) Fifteen-spined stickleback (*Spinachia spinachia*) Tub gurnard (*Trigla lucerna*)			
T. domerguei	Three-spined stickleback (*Gasterosteus aculeatus*)	Skin, rarely gills	Freshwater and marine	Chubb, 1970a Gaze & Wootten, 1998
	Nine-spined stickleback (*Pungitius pungitius*)	Skin		Dartnall, 1973
T. intermedia	Minnow (*Phoxinus phoxinus*)	Gills, skin of fry		Gaze & Wootten, 1998
T. modesta	Bream (*Abramis brama*)	Gills		Gaze & Wootten, 1998
T. nigra	Common carp (*Cyprinus carpio*) Rainbow trout (*Oncorhynchus mykiss*) Brown trout (*Salmo trutta*)	Skin		Gaze & Wootten, 1998
T. pediculus	Three-spined stickleback (*Gasterosteus aculeatus*)	Skin	Possibly accidental infection from *Hydra*	Gaze & Wootten, 1998
T. tenuidens	Three-spined stickleback (*Gasterosteus aculeatus*)	Gills, rarely skin	Freshwater and marine	Chubb, 1970a Gaze & Wootten, 1998
	Nine-spined stickleback (*Pungitius pungitius*)	Gills		Dartnall, 1973

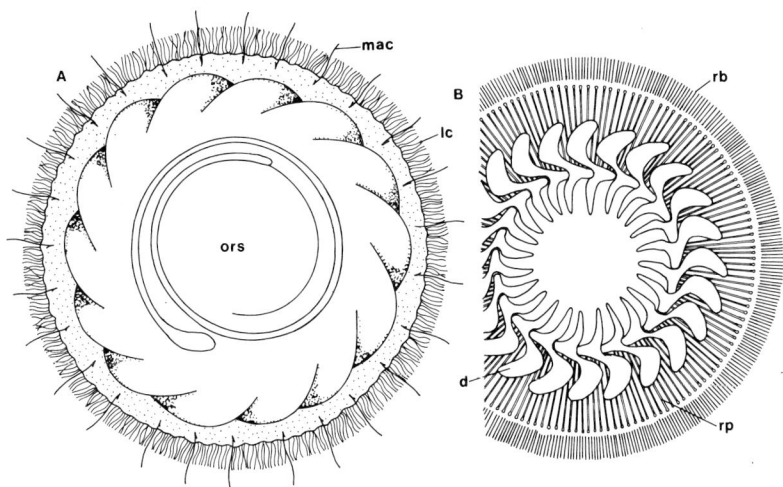

Fig. 2.3. The anatomy of a trichodinid. (A) View of the upper (oral) surface; (B) skeletal parts of the adhesive disk. ors, Oral surface; rb, rods in the border membrane. Other labelling as in Fig. 2.2. From Lom & Dykova, 1992, with permission from Elsevier.

have attracted little attention. Some published records of *Trichodina* spp. on gills and skin of British marine and freshwater fishes are offered in Table 2.1. However, according to Gaze (personal communication), other trichodinids belonging to *Trichodinella*, *Tripartiella* and *Paratrichodina* occur on British freshwater fishes. Yeomans *et al.* (1997) reported *Trichodinella epizootica* on the gills of the three-spined stickleback (*Gasterosteus aculeatus*) in south-east England.

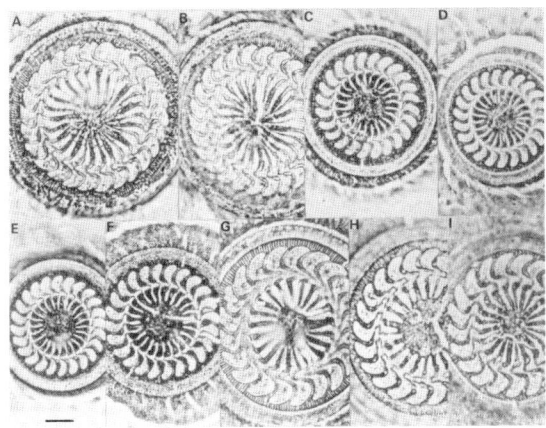

Fig. 2.4. Some British trichodinids impregnated with silver. (A,B) *Trichodina pediculus* from the 3-spined stickleback; (C-F) *T. modesta* from bream; (G) *T. nigra* from brown trout; (H) *T. nigra* from common carp; (I) *T. nigra* from rainbow trout. Scale bar: 10 µm. From Gaze & Wootten, 1998, with permission from the authors and from the Institute of Parasitology, Academy of Sciences of the Czech Republic.

Although many authors regard trichodinids as parasites (e.g. Gaze & Wootten, 1998), what little evidence we have concerning their way of life indicates that they are basically epizoic (Lom & Dykova, 1992). A counter-clockwise spiral array of cilia on the exposed (unattached) oral surface (Figures 2.2, 2.3A) generates a current vortex that collects bacteria and other suitable detrital particles from the surrounding water and transports them to the 'mouth' (cytostome). However, the trichodinid/fish association is not always benign, but the nature of these adverse interactions is poorly understood. Massive proliferation of trichodinids, largely by division of each individual into two daughters may be associated with previously debilitated fishes and may contribute to their death, but some apparently healthy fishes carry large numbers of trichodinids (Wootten, personal communication). Numbers of trichodinids (and other epizoic protozoans such as scyphidiids – see below) are often highest in enriched water, i.e. water containing lots of bacteria for food (Wootten, personal communication; Yeomans *et al.*, 1997).

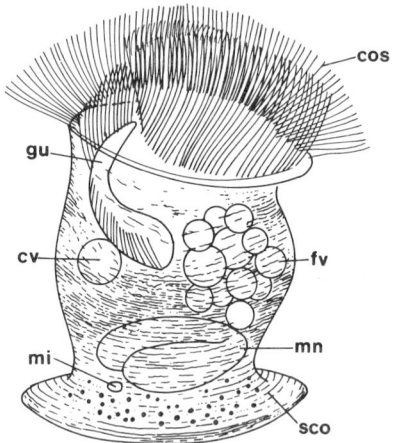

Fig. 2.5. Semi-diagrammatic drawing of *Scyphidia* sp. from a brown trout, *Salmo trutta*. cos, Cilia of oral surface; cv, contractile vacuole; fv, food vacuole; gu, 'gullet'; mi, micronucleus; mn, macronucleus; sco, scopula containing secretory granules. Modified from Pickering *et al.*, 1985.

According to Lom & Dykova (1992), it is the constant attachment and detachment of large numbers of adhesive disks that irritates and eventually damages the host's epidermal cells. Bundles of contractile fibres (myofibres) are attached to the skeletal elements of the disk and contraction of these bundles leads to vaulting of the disk and temporary attachment to the host's surface by suction. It is this vaulting that, according to Lom & Dykova, is potentially damaging, since the disk's sharp rim bites into the host epidermal cell as its surface enclosed by the disk is forcibly lifted. Eventually the epidermal cell is disrupted. Such epithelial damage probably causes death by osmotic imbalance; invasion of the lesion by pathogenic waterborne bacteria and fungi is another potential hazard, but this is rare (Wootten, personal communication). This interaction is not parasitic in the usually accepted sense and whether these organisms have the ability to attack healthy fish epidermal cells has not been established. Indeed, trichodinids seem ill equipped, both in terms of their feeding apparatus and the orientation of their oral surface, to exploit epidermal cells directly as a source of food.

Scyphidiids are peritrichs that lack the aboral cytoskeleton displayed by the trichodinids, having an expanded attachment disk or scopula at the aboral (basal) end (Figure 2.5). The opposite (oral) end of the barrel-shaped organism is heavily ciliated. MacKenzie (1969) described a new species, *Scyphidia* (now *Riboscyphidia*, according to Lom & Dykova, 1992) *adunconucleata*, from the gills of plaice (*Pleuronectes platessa*) caught in Scottish waters, and an unidentified peritrich assigned to the same genus, was found by Wootten & Smith (1980) on the skin of both wild and cultured Atlantic salmon (*Salmo salar*) in eastern Scotland. Pickering *et al*. (1985) also found the latter organism on brown trout (*Salmo trutta*) kept in tanks but supplied with continuously flowing, untreated water from Lake Windermere. Pickering *et al*. found few ciliates on the head and most of those on the body were on the anterior and dorsal regions. Wootten & Smith (1980) regarded the scyphidiid from the salmon as a parasite, but harmful effects were not reported. Another scyphidiid, *Ambiphrya*, is present in Britain on salmonids and on coarse fish (Wootten, personal communication).

Pickering *et al*. (1985) believe that *Scyphidia* sp. cements itself to brown trout. Using the scanning electron microscope, they identified a layer of homogeneous material, which they regarded as cement, around the attached scopula. Granules within the scopula may be secretory bodies containing the adhesive. The scopula of a scyphidiid from the surface of a mollusc also appears to have secretory ability (see Lom & Corliss, 1968).

Occasionally Pickering *et al*. (1985) found two specimens of *Scyphidia* sp. sharing a common platform of cement on the surface of brown trout. They interpreted these as daughter cells produced by division of one parent individual. They also observed numerous unattached free-swimming stages (telotrochs), which are presumably responsible for infecting new hosts, but these stages were not studied.

A high degree of host specificity is a feature of some, but not all *Trichodina* species (see Table 2.1 and van As & Basson, 1987) and *Scyphidia* sp. consistently failed to infect brook charr (*Salvelinus fontinalis*) kept in the same tanks as readily infected brown trout (Pickering *et al*., 1985). Pottinger *et al*. (1984) claimed that infection with *Trichodina* or with *Scyphidia* led to a reduction of mucus-secreting (goblet) cells in brown trout epidermis. These observations indicate that the relationships between some epizoic animals and their hosts may be more intimate than anticipated hitherto, or that the life styles of some of these organisms may have shifted towards commensalism or parasitism in ways that may not have been recognised. Further work on the biology of these organisms promises to be interesting.

2.3 AN ECTOPARASITE – *ICHTHYOBODO*

Whether mobile peritrich ciliates like *Trichodina* are parasites in the strict sense is debatable, but there is no doubt about the parasitic nature of the relationship between the bodonid flagellate *Ichthyobodo necator* (often incorrectly called *Costia necatrix*) and fishes. The freely swimming stages of flagellate protozoans (Phylum Mastigophora) are propelled by organelles called flagella, which are basically similar in structure to cilia but move differently. A flagellum undulates while a cilium performs whiplash-like strokes. Individual protozoans typically carry only one or a few flagella, while ciliates carry many cilia capable of beating in a co-ordinated manner.

The parasitic stage of *I. necator* attaches itself to and exploits the superficial epithelial cells of the skin or gills and is found on practically all freshwater fishes (Lom & Dykova, 1992). It is dangerously pathogenic, especially to young fish and to fish with lowered resistance. It appears that the presence of *Ichthyobodo* on marine fishes went unnoticed until relatively recently. Needham & Wootten (1978) were the first to record the parasite in the sea off Scotland on salmon smolts, but since these fishes had come from fresh water it was assumed that the parasites had also originated in fresh water. However, later records of the parasite on marine fish such as the dab, *Limanda limanda*, many kilometres off shore in the North Sea, indicated that there are well-established marine populations of *Ichthyobodo* (see Diamant, 1987). That these marine forms may constitute a separate race or species is suggested by apparent differences in the attachment apparatus of marine and freshwater forms (see Roubal & Bullock, 1987) and by the failure of flagellates from a freshwater source to transfer to marine fishes (see Lom & Dykova, 1992). Molecular techniques have since indicated that '*Ichthyobodo necator*' comprises several sibling species, including a species on salmon parr in freshwater, a species preferring the gills of salmon in both fresh water and the sea and a marine species on cod (*Gadus morhua*) (Todal et al., 2004).

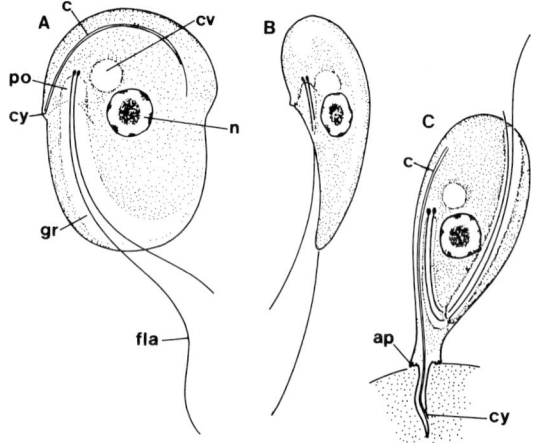

Fig. 2.6. Ichthyobodo necator, (AB) free-swimming stage viewed from two different directions; (C) parasitic stage attached to a host epithelial cell. ap, Attachment plate; c, cytostomal canal; cv, contractile vacuole; cy, cytostome; fla, flagellum; gr, groove; n, nucleus; po, pocket. From Joyon & Lom, 1969, with permission.

2.3.1 *Free-swimming stage*

The free, non-feeding form of '*I. necator*' is a roughly saucer-shaped cell (Figure 2.6AB), measuring in preparations about 8 – 12 µm long and 6 – 10 µm in width (Joyon & Lom, 1969). The concave surface has a curved peripheral groove leading in a clockwise direction to a shallow pocket containing at its base two basal bodies (kinetosomes), each of which gives rise to a single flagellum. Four flagella may be present if the cell is dividing (Robertson, 1985). The two flagella, one longer than the other, extend from the pocket along the groove and drive the organism through the water, the shorter of the two flagella being the most active. When moving rapidly the organism rotates about its longitudinal axis. The cytostome ('mouth') opens at the edge of the pocket and communicates with a narrow cytostomal canal lined by microtubules that extends into the cytoplasm and follows a curved clockwise path near the cell surface (Joyon & Lom, 1969; Lom & Dykova, 1992). There is a single central nucleus and nearby, in freshwater forms, a contractile vacuole.

2.3.2 Parasitic stage

Unlike the peritrich ciliates, it is the cytostomal region of *Ichthyobodo* that is attached to a fish epidermal cell. Robertson (1985) referred to records of as many as 15 parasites attached to one epidermal cell, but one to three is more usual. More rarely the parasite attaches itself to a mucous cell in the epidermis or to a chloride cell in the gills (Roubal & Bullock, 1987).

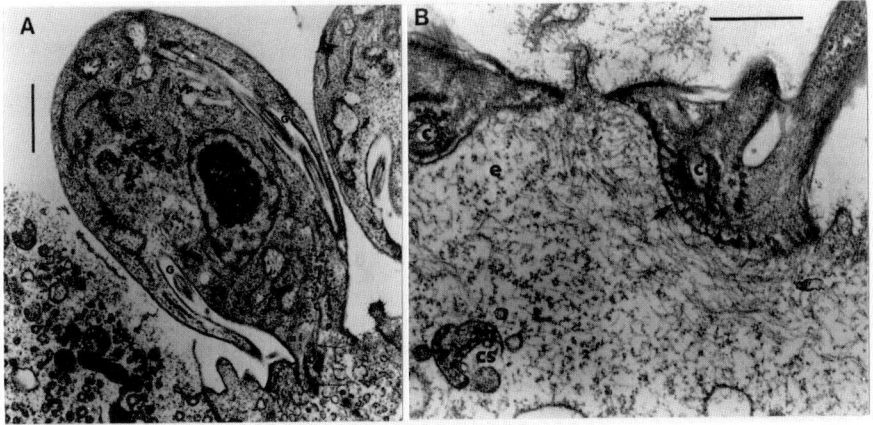

Fig. 2.7. Electron micrographs of sections through *Ichthyobodo necator* attached to host epithelial cells. (A) Whole organism; (B) enlarged view of attachment plate (indicated by the arrow) and cytostomal root (cs) embedded in epidermal cell (e). c, Cytostomal canal; G, groove containing flagellum. Scale bars: 1 μm. (A) From Joyon & Lom, 1969, with permission; (B) from Lom & Dykova, 1992, with permission from Elsevier.

In order to achieve this attachment attitude the free-swimming stage must undergo considerable torsion. It transforms from a saucer-shape to a pear-shape, with the cytostome opening at the tip of the 'stalk' of the pear (Figure 2.6C). The region immediately surrounding the cytostome area becomes an attachment plate that provides a stable attachment zone by fusion between the cell membranes of parasite and host (Figure 2.7; Diamant, 1987). The organism then assembles a cytostomal tube, which penetrates deep into the host epidermal cell (Figures 2.6C, 2.7) – as much as 5 μm according to Diamant (1987). The cytostomal canal within this tube provides communication between the embedded cytostome and the freely exposed cell body. In the attached form the pocket containing the roots and proximal regions of the flagella is particularly long (Figure 2.6C). While the organism is attached the flagella are immobile (Joyon & Lom, 1969).

According to Lom & Dykova (1992) transformation from the swimming to the attached form takes only a few seconds. The reverse transformation from the fixed to the swimming form also occurs and is equally rapid (15 seconds according to Joyon & Lom, 1969).

It has been suggested that host cell contents, perhaps pre-digested, are withdrawn into the parasitic cell body via the cytostomal tube (Robertson, 1985). Vacuolation and degenerative changes have been recorded inside the host cell (Diamant, 1987, Robertson, 1985, Roubal *et al.*, 1987) and there is evidence of preferential ingestion of DNA (Robertson *et al.*, 1981). However, according to Roubal *et al.* (1987), the parasite obtains this DNA from the cytoplasm after leakage from the nucleus since they claim that the nucleus is not directly attacked.

The parasite multiplies asexually by longitudinal division while attached to a host cell (Robertson, 1985). There have been reports that the parasite can survive on detached and decaying cells at the bottom of tanks or ponds containing fish. This has not been confirmed and, according to Robertson (1985), is unlikely. A more likely possibility according to Robertson, but also not yet confirmed, is that the parasite can survive adverse conditions by detaching and enclosing itself in a resistant cyst wall.

2.3.3 Pathological effects
Hyperplasia of the affected host epidermal cells is extensive, with almost complete disappearance of mucous cells in the same area (Robertson *et al.*, 1981). A feature of costiasis (a popular name for the disease) is a grey/white slimy appearance attributable to this hyperplasia. Destruction of the integrity of the epidermal cells leads to uptake of water (spongiosis) and the whole plaque of hyperplastic epidermal cells may be shed. It is most likely that death is the result of osmoregulatory breakdown, dilution of the blood and circulatory failure (Lom & Dykova, 1992).

2.4 A PARASITE OF THE EPIDERMIS – *ICHTHYOPHTHIRIUS*

The ciliate *Ichthyophthirius multifiliis* (Phylum Ciliophora) is a parasite living inside the epidermis covering the skin, gills and buccal cavity of fishes. The parasite reveals itself in the form of superficial white spots reaching about 1 mm in diameter (exceptionally 1.5 mm). *I. multifiliis* is the bane of aquarists and fish farmers, being notoriously contagious and pathogenic and causing 'white spot disease' or 'ich' in freshwater fishes. Matthews (1994) refers to deaths attributed to this parasite of 18 million killifish, including *Orestias agassii* (= *agassizii*), a commercially important species, in Lake Titicaca, Peru. The Chinese were apparently aware of the disease in cyprinid fishes at least as early as the Sung Dynasty (AD 960 – 1127) and ichthyophthiriasis was well known in Europe during the Middle Ages and was probably introduced into Britain with cultured carp (*Cyprinus carpio*) (references in Matthews, 1994 and Ewing & Kocan, 1992).

2.4.1 Infective stage
The stage of the parasite infective to the fish is a ciliated, free-swimming organism called a theront (Figure 2.8), measuring 25 – 70 by 15 – 22 µm (Lom & Dykova, 1992).

A cytostome communicates with a pre-buccal cavity in the leading third of the cell. Typical cellular organelles are present such as surface cilia, numerous mitochondria providing energy for the cell's activities, a large macronucleus, a small micronucleus and a contractile vacuole, the last-named being responsible for removing excess water entering the cell by osmosis. Special features of the theront include the following: an apical elevation or 'perforatorium' described by Ewing & Kocan (1992) as a fused ridge of cilia; a trailing caudal cilium at least three times longer than the cilia covering the rest of the cell; many so-called mucocysts opening on the surface; a dense body or organelle of Lieberkühn associated with the pre-buccal cavity.

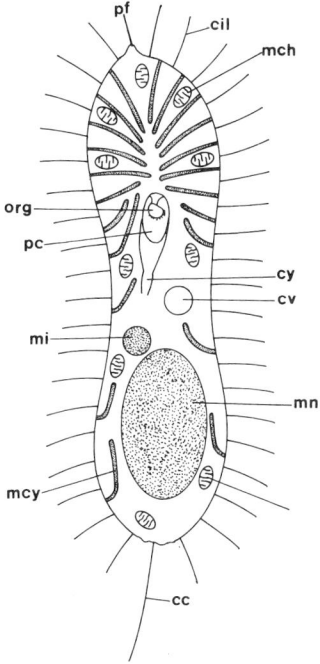

Fig. 2.8. Diagram showing the main anatomical features of the theront of *Ichthyophthirius multifiliis*. cc, Caudal cilium; cil, cilium; cv, Contractile vacuole; cy, cytostome; mch, mitochondrion; mcy, mucocyst; mi, micronucleus; mn, macronucleus; org, organelle of Lieberkühn; pc, prebuccal cavity; pf, perforatorium. Scale bar: 10 μm. Modified from Matthews, 1994.

According to Matthews (1994), the theront swims continuously, spinning anticlockwise on its longitudinal axis, the trailing cilium perhaps serving to maintain directional stability. Slow cruising alternates with sudden bursts of speed, during which the theront wildly gyrates and may alter course. It is during such a burst of speed that contact with the fish host usually occurs. The theront swims towards the light, and consequently swims towards the surface (Lom & Čerkasovová, 1974), where it may come under the influence of chemical attractants from the skin of potential fish hosts. Lom & Čerkasovová (1974) and Buchmann & Nielsen (1999) found that substances with high molecular weights (proteins) in fish blood serum were attractive to theronts.

Buchmann & Nielsen showed that they swim towards the source in a diffusion gradient. It is known that serum proteins find their way into epidermal skin mucus (see, for example, Hines & Spira, 1974a), so it is likely that mucus released by the epidermis is the source of attraction for theronts. In fact, Buchmann *et al.* (1999) went so far as to suggest that theronts home in on mucous cells and actually invade the skin via the mucous cell openings, which lie between the epidermal cells. They quote as evidence for this that *I. multifiliis* invades localities where mucous cell densities are highest, such as the dorsal, pectoral and pelvic fins.

Others do not support the idea that theronts enter the epithelium via mucous cells or between epidermal cells. Ewing *et al.* (1985) and Matthews (1994) described disruption of one or more epithelial cells as the theront penetrates, but it is not clear how this disruption/penetration is brought about. The term 'perforatorium' used for the specialised leading end of the invasive theront gives the impression that its function is known, but Ewing *et al.* (1985) were unable to determine its role in the penetration process. Mucocysts are known to discharge just before and during theront entry and this, and the arrangement of these mucocysts at the leading end of the theronts, were regarded by Matthews & Matthews (1984) as strong evidence for their role in digesting a pathway into the host. Some support for this comes from the observation of Ewing & Kocan (1992) that the mucocyst secretion induces a change in the plasma membranes of host epithelial cells as soon as the secretion makes contact with them. Enzymes that might have a digestive role in penetration have been identified in the vicinity of invading parasites, in water containing theronts and in vesicles inside theronts (see Matthews, 1994), but there is no unequivocal evidence that any of these are involved in penetration. Ewing *et al.* (1985) placed a different emphasis on the role of the mucocyst secretion in penetration, regarding it as an adhesive, serving to attach the newly arrived theront to the host's surface. Geisslinger (1987) claimed that some of the cilia of the theront had adhesive properties.

The organelle of Lieberkühn has also been implicated in penetration by the theront, but there seems to be some disagreement in the literature about the timing of involvement of this structure. According to Ewing & Kocan (1992) it disappears minutes after invasion of the skin, while Matthews (1994) records its discharge 4 – 6 hours after parasite invasion. This needs to be resolved before any assessment of its possible involvement in penetration can be made.

Theronts live for up to 30 hours at 23 – 24 °C. Twelve hours after emerging from the cyst (see section 2.4.4 below) their infectivity is about 34%, but after 20 hours has decreased to slightly more than 1%.

2.4.2 Parasitic stages
Ewing & Kocan (1986, 1992) described the events of penetration and establishment of the parasite in the gills of the channel catfish, *Ictalurus punctatus*. Within five minutes the theront has completed its penetration and initial migration within the gill epithelium, often stopping close to the cartilage supporting the primary gill lamella. The parasite does not breach the basement membrane and enter the dermis. Debris released from the disrupted host cells is 'mopped up' during the next 35 minutes or so by ingestion via the cytostome, leading to the formation in the cytoplasm of the young feeding stage or trophont of many debris-containing vesicles (food vacuoles) in which the host material is digested. Intact epithelial cells cover the parasite externally within 45 minutes of

entry. According to Ewing & Kocan (1986), migration is resumed during the next day or so, the parasite moving from the region of entry to another location in the primary lamella near a major blood vessel. More than 80% of the parasites were found adjacent to the afferent gill vessel (see also below).

As early as 40 minutes after exposure to theronts the epithelial cells surrounding the trophont are relatively normal in appearance, suggesting that the trauma of invasion, at least for the host, is short-lived. However, Ewing *et al.* (1986) found that, for the parasite, the first 10 minutes after exposure of the channel catfish at 21°C are critical for the survival of the parasite. Within this period the parasite population that gained entry to the host declined by 50%, while no further losses occurred during the next 35 minutes. According to Ewing *et al.* (1986), considerable reserves are likely to be expended entering the host and moving through the gill epithelium and this is reflected in a reduction of the theront's organelles during this period. The only parasites to survive are those that are able to renew these reserves and replace their organelles by ingesting host material from disrupted epithelial cells.

Within the epidermis the trophont grows steadily, increasing its volume about 3000 times and its diameter to 1 mm or more (Figure 2.9). Trophonts in the surface tissue of the skin appear larger than they are because they are flattened in the epidermis or crowded with others in a burrow. Lipid (fat) bodies accumulate in the trophont and are likely to be important sources of energy for later non-feeding stages (Ewing & Kocan, 1992). Mucocysts also increase in number significantly during the first three days in the fish at 21°C and these are destined to produce a cyst wall after departure of the trophont from the host.

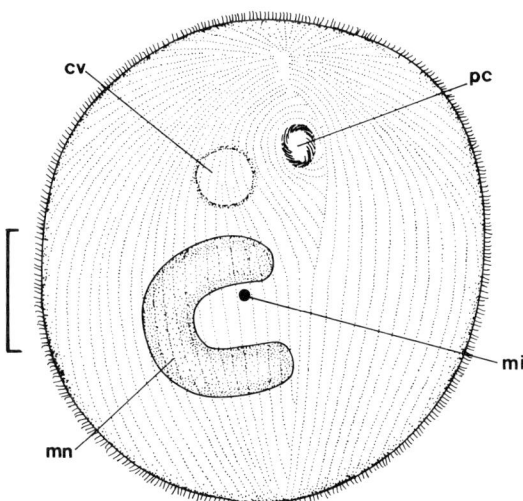

Fig. 2.9. The mature trophont of *Ichthyophthirius multifiliis*. Lettering as in Fig. 2.8. Scale bar: 250 μm. From Lom & Dykova, 1992, with permission from Elsevier.

A pre-buccal tube leads to the 'mouth'. This tube is lined by conspicuously beating cilia, which are responsible for collecting food particles and transporting them to the mouth. McCartney *et al.* (1985) described the dislodgement of host tissue by the parasite's cilia and Lom & Dykova (1992) talk of 'mechanical action' of the trophont releasing host cell debris for ingestion. However, the viewpoint of Ewing & Kocan (1987) is more realistic; they stated that precisely how the trophont destroys and feeds on fish epithelial cells during the course of development is unknown. It seems somewhat unlikely that mechanical (ciliary) action alone would be sufficient to destroy a healthy epithelial cell and it is possible that substances released by the parasite may promote disintegration of host cells.

As long ago as 1893, Stiles (in Ewing *et al.*, 1988) suggested that 'galleries' (longitudinal clusters of trophonts) might result from reproduction within the epidermis of the fish host. This idea then fell out of favour, clustering being explained either as an aggregation of independently migrating parasites or as the result of multiple invasions through the same opening in the epithelium. Almost a century later, the idea that trophonts might reproduce was resurrected and has received support from experimental infections of channel catfish by Ewing *et al.* (1988). They found that the population of trophonts in fins increased five-fold from day 3 post-exposure to day 5, in spite of precautions taken to prevent reinfection. Moreover, at day 3, 100% of parasite loci in the fins and gill arches contained solitary trophonts, while on days 4 and 7, 10% and 69% respectively of fin loci contained clusters of parasites. A similar pattern was detected in the gills. These observations were interpreted as evidence that trophonts undergo multiplication by cell division in the fish's epithelium.

2.4.3 Escape from the host
When normal development is completed, after a period of about seven days (at 21 °C), the cytoplasm of the large trophont contains fat, glycogen and protein reserves and leaves the still-living host spontaneously. Ewing & Kocan (1987) showed that the parasite becomes increasingly separated from the overlying host epithelial cells as a prelude to exit, and they recognised a loose association between this process and the discharge of the many contractile vacuoles present in the cytoplasm of the parasite at this stage. Ewing & Kocan (1992) suggested that the contents of the vacuoles might have digestive properties, aiding release of the parasite from the enclosing host tissue. However, like so many features of the biology of this parasite, precisely how the trophont escapes from the host is not understood.

Thus, when fully mature, trophonts leave 'healthy' hosts, but, in addition, the mature trophont has the interesting ability to abandon a dead or dying host (Ewing & Kocan, 1987). This is not simply a case of passive release of trophonts by disintegration of the epithelium of the dead fish, but is an active and often rapid process on the part of the parasite, triggered presumably by some cue provided inadvertently by the moribund host. Ewing & Kocan (1987) found that, five days after infection of channel catfish with *I. multifiliis* at 21°C, a dozen free trophonts could be collected a mere 20 minutes after killing a host by pithing. However, for trophonts of four days growth and three days growth, 60 minutes and 90 minutes respectively were required to achieve a comparable result. Thus older trophonts respond more quickly to the unknown stimulus from the host, indicating an active role on the part of the parasite.

Ewing & Kocan (1992) raised the possibility of a link between the ability of mature trophonts in the gills to abandon a dead or dying host and the preferred location of trophonts near the afferent gill blood vessel (see above). They argued that, since this vessel carries systemic blood returning from the fish's body, its oxygen tension would be expected to fall as a result of the imminent death of the fish and parasites close to the vessel might be able to detect such a change and respond accordingly.

2.4.4 *Encysted stage*

After escape, the trophont settles on a convenient substrate such as weed or the bottom, and then secretes a cyst wall (Figure 2.10), presumably by discharging its mucocysts (Ewing *et al.*, 1983). According to MacLennan (1937), this cyst wall is made of protein and consists of two layers. The outer ectocyst is secreted first and is a clear, extremely sticky, membranous layer, adhering readily to debris or to glassware or needles. This stickiness is sufficient to prevent dislodgement from the substrate, even in fairly strong currents of water. A thin layer of debris and bacteria accumulates on the exposed surface of the ectocyst after a few hours. MacLennan described the endocyst as consisting of fine anastomosing fibrils, arranged singly or in bundles.

The encysted stage or tomont slowly rotates within the cyst wall and absorbs its buccal apparatus. It then begins to divide repeatedly, the divisions at first being precisely synchronous, and produces, depending on size and temperature, 250 to more than 2000 small, ciliated, so-called tomites. After the last division the tomites develop into theronts, which break out through the cyst wall.

Ewing *et al.* (1986) found that the length of time spent by the trophont within the tissue of the channel catfish is critical for survival after leaving the host and for successful reproduction of the encysted tomont. Parasites spending less than three days in host tissue were less likely to survive after departure from the host and more likely to produce abnormal theronts after reproduction. Thus, a trophont is less likely to survive or reproduce if the host dies one or two days after invasion by the parasite.

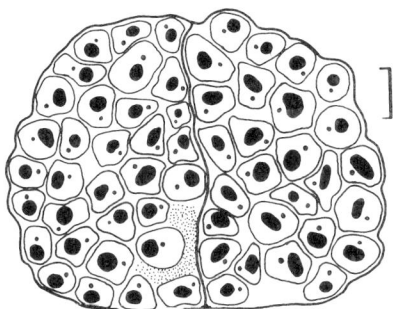

Fig. 2.10. Section through a cyst of *Ichthyophthirius* after the completion of all divisions. Cilia omitted. Scale bar: 10 μm. From MacLennan, 1937. Copyright © (1937, John Wiley). Reprinted by permission of Wiley-Liss, Inc., a subsidiary of John Wiley & Sons, Inc.

Ewing *et al.* (1986) found that theront production varied significantly with parasite size and with temperature and length of residence by the trophont in the host. Theront production per tomont increased with each extra day spent by the trophont in the host and was significantly greater at 24°C than at 21°C. After five days in the host at 24 °C each tomont produced on average 562 theronts compared with 240 at 21°C. More theronts were produced by larger tomonts and the production of theronts by tomonts of equal diameter was greater at 24°C than at 21°C.

The total length of the life cycle at 23 – 24 °C is 2.5 – 7.5 days, but longer at lower temperatures.

2.4.5 *Pathological effects*

Accounts of the pathological effects of *I. multifiliis* are given by Hines & Spira (1974b) and by Lom & Dykova (1992). The destruction of gill and skin epithelia by the parasite has two major consequences. First, respiration may be impaired to the extent that infected fishes may die if the ambient oxygen concentration falls only slightly. Secondly, the fish may lose its ability to osmoregulate. The latter is probably the main cause of mortality, which commonly occurs when the parasite breaks out of the skin (Wootten, personal communication). However, fishes are not entirely defenceless against the depredations of *I. multifiliis*. If the fish receives a sub-lethal dose of theronts and is protected from further infection for the next three weeks, a strong resistance develops against further infection (Hines & Spira, 1974a). If the immune fish is then maintained in an infective environment, the resistance persists for at least eight months, the fish being refractory to infection when exposed to numbers of parasites that would kill previously unexposed ('naïve') fishes.

Hines & Spira (1974a) found that blood serum from immunised fishes immobilised free-swimming stages of *I. multifiliis in vitro*, by stopping ciliary movements. Further work showed that this *in vitro* immobilisation was attributable to parasite-generated antibodies present both in blood serum and in skin mucus, which bind to antigenic proteins in the membranes covering the cilia (see Clark & Dickerson, 1997). This led to the supposition that the immune mechanism involved the immobilisation of theronts approaching the immune host too closely by antibodies present in skin mucus. Paradoxically, this was not supported by *in vivo* experiments by Cross & Matthews (1992), in which theronts were not immobilised and were able to invade the caudal skin of an immune common carp (*Cyprinus carpio*). Surprisingly however, within two hours, 79% of these successful penetrants had disappeared. Since no parasite material was detected at penetration sites, it was concluded that the parasites had been forced to abandon the host prematurely.

It seems therefore, that we have here a novel immune mechanism for dealing with a skin parasite – rather than killing the parasite, antibody binding to it may influence its behaviour, inducing it to leave the host. Clark *et al.* (1995) attributed the failure of an immune host to immobilise theronts *in vivo* to the low level of antibody in its skin mucus. They postulated that the same antibody might have different effects at different concentrations: immobilisation at high concentrations and forced exit from the host at low concentrations.

3

MONOGENEAN (FLATWORM) SKIN PARASITES – *ENTOBDELLA*

3.1 INTRODUCTION TO MONOGENEANS

Monogeneans are flatworms (platyhelminths) belonging to the Phylum Platyhelminthes, a major sub-division of the Animal Kingdom. In contrast with protozoans, flatworms are metazoans, i.e. they have multicellular bodies. On an evolutionary scale of anatomical advancement their rather compact bodies, lacking a skeleton or blood system, are regarded as relatively primitive, although their three-layered (triploblastic) construction (ectoderm on the outside, endoderm lining the gut and mesoderm between) places them ahead of sponges and the cnidarians (sea anemones, jellyfish and relatives).

There are many different kinds of free-living platyhelminths and although this is not a monophyletic assemblage its diverse members are commonly referred to as 'turbellarians'. Freshwater habitats may yield black or white planarians (triclad 'turbellarians') measuring a centimetre or so in length and gliding gracefully and effortlessly over stones or waterweeds, propelled by a ventral 'sole' of beating cilia. Similar, but often larger, polyclads may be found beneath stones on a rocky shore. There is also an abundant microfauna, invisible without a microscope, living between sand grains on sandy shores, and tiny platyhelminths capable of swimming freely using their cilia make up a large proportion of this 'interstitial fauna'.

A few 'turbellarians' have adopted a parasitic life style. For example, the triclad *Micropharynx parasitica* is attached by a posterior adhesive zone to the skin of *Raja* spp. and probably feeds on the ray's epidermis (Ball & Khan, 1976). In addition to the largely non-parasitic 'turbellarians' there are three major groups of platyhelminths that are entirely parasitic. Two of these, the tapeworms (Cestoda) and the flukes (Digenea) are endoparasites, but the third group, the Monogenea, comprises a range of highly specialised skin and gill parasites of fishes. The biology of these remarkable ectoparasites will be considered in this and the following three chapters.

Every major group of fish-like vertebrates is parasitised by monogeneans. One of the two sub-divisions of the Monogenea, the Monopisthocotylea, includes a range of skin parasites, but there are many other monopisthocotyleans living on the gills. Members of the other sub-division, the Polyopisthocotylea, are with few exceptions gill parasites. There are major differences between the members of these two sub-divisions and separate origins have been suggested for the two groups, i.e. the monophyly of the Monogenea has been questioned (see Chapter 6).

Monogeneans have been recorded on 'primitive' jawless agnathans, on holocephalans, on the coelacanth, *Latimeria chalumnae*, and on sturgeons. They are widespread on sharks and rays (elasmobranchs), but achieve their greatest diversity on the teleosts, where the 'explosive' radiation of the hosts has been matched by extensive co-evolutionary radiation of their monogenean parasites. Remarkably, some monogeneans (the polystomatids) appear to have kept pace with the evolution and emergence of amphibians from fish-like ancestors, and survive today inside the bladders of frogs. However, this is not the end of the adaptability of monogeneans since some frog parasites have managed to switch permanently to unrelated hosts sharing their freshwater environment, establishing themselves in the bladders or mouths of freshwater turtles. Another, *Oculotrema hippopotami*, is perhaps the most surprising, having found a niche beneath the eyelids of the hippopotamus.

It is clear from their wide distribution on fish-like vertebrates that monogeneans originated a very long time ago and have mostly retained their close association with fishes. Llewellyn (1963 and Llewellyn, J., 1965) offered evidence to suggest that the Polyopisthocotylea (see Chapter 6) has been in existence since before the Ordovician period, pushing the origins of the Monogenea back into the remoteness of the early Palaeozoic period. With such a unique and unimaginably long relationship between monogeneans and fishes and the superb adaptations to parasitism that have evolved during this protracted and intimate association, there can be no better choice for the introductory chapters of this book.

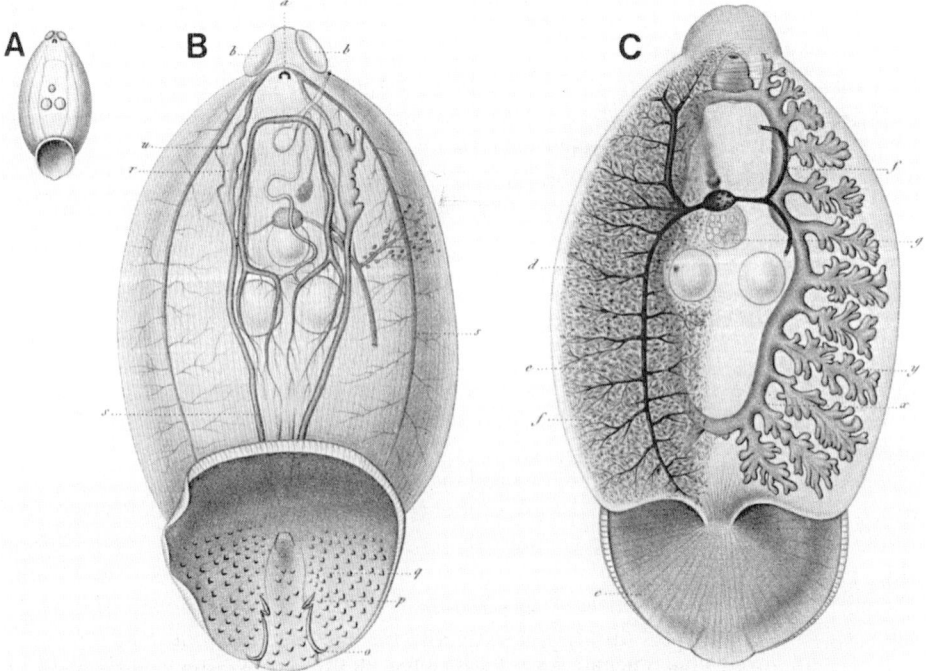

Fig. 3.1. Early drawings by van Beneden (1858) of *Epibdella* (= *Entobdella*) *hippoglossi* from the skin of *Pleuronectes* (= *Hippoglossus*) *hippoglossus* (the halibut). (A) Approximately natural size; (B) enlarged in ventral view: (C) enlarged in dorsal view.

One of the largest monogeneans is the monopisthocotylean *Entobdella hippoglossi* on the skin of the halibut, *Hippoglossus hippoglossus*. Given the size of this parasite (see Figure 3.1A), it is not surprising that it attracted attention. Müller described it as early as 1776, although he regarded it as a leech and named it *Hirudo hippoglossi*. Van Beneden's illustrations of *Epibdella* (= *Entobdella*) *hippoglossi*, published in 1858, are reproduced in Figure 3.1. Van Beneden & Hesse did not describe its much smaller relative, *Phyllonella* (= *Entobdella*) *soleae* from the skin of the common or Dover sole (*Solea vulgaris* = *Solea solea*) until 1864. *Entobdella hippoglossi* and *Entobdella soleae* occur in Britain and there is a third British species, *Entobdella diadema*, found on the stingray, *Dasyatis pastinaca*. *E. diadema* is intermediate in size and was first described as *Epibdella diadema* in 1902 by Monticelli from *Trygon* (= *Pteroplatytrygon*) *violacea* in the Mediterranean Sea.

These three British representatives of *Entobdella* illustrate well the phenomenon of host specificity (see Chapter 1) to which repeated reference will be made in this book. In British waters *E. soleae* is restricted to the common sole, *Solea solea*, apart from a few records from the sand sole *Pegusa* (= *Solea*) *lascaris*. *E. hippoglossi* is found only on halibut and *E. diadema* only on the stingray *Dasyatis pastinaca*. The distribution of all three parasites extends beyond British waters, where they have been recorded on a few other related hosts. For example, *E. soleae* occurs on Senegalese sole, *Solea senegalensis*, as well as on *S. solea* on the Portuguese continental coast (Carvalho-Varela & Cunha-Ferreira, 1987).

In spite of the early discovery of these three species of skin parasites, it was not until the 1960s that we began to appreciate just how remarkable they are in their adaptations to life on fish skin. *Entobdella soleae* has contributed most to this understanding because of the ease with which the host and its parasite can be maintained in laboratory aquaria. I propose to begin by describing the biology of *E. soleae* and its British relatives. With this as a background, it will be easier to appreciate the adaptations of other monopisthocotylean skin parasites, which will be introduced in Chapter 4. Chapters 5 and 6 will be concerned with monopisthocotylean and polyopisthocotylean gill parasites respectively.

3.2 THE BIOLOGY OF *ENTOBDELLA SOLEAE*

The following account of the biology of *Entobdella soleae* is condensed from Kearn, 1998, where relevant references will be found. The common sole (*Solea solea*) lives on sandy or muddy bottoms (Wheeler, 1969) and *E. soleae* is a permanent resident on the skin. Infestation levels are low. In a small sample of soles from the North Sea about 50% were infested with an average of 2 (1 – 5) parasites (Kearn *et al.*, 1993) and, in a 'naked eye' search of soles caught near Plymouth, 3 (1 – 9) adult parasites per infested fish were recorded (Kearn, 1971a). Most adult parasites are confined to the sole's lower surface. With a little practice the larger adults are easy to spot, since their five or six mm long flat bodies appear faintly yellow against the unpigmented white background of the sole's lower surface. On the other hand, when the living parasite is detached and viewed with a microscope by transmitted light, the parasite is semi-transparent, permitting the observer a clear view of the internal anatomy and of events taking place inside the body.

The life cycle of *E. soleae*, like those of other monogeneans, is uncomplicated, involving no other organisms as intermediate hosts (Figure 3.2). The eggs are tetrahedral, an unusual shape for an egg, and are a rich brown colour. They would be conspicuous against the white lower skin of the sole, but are never found attached there.

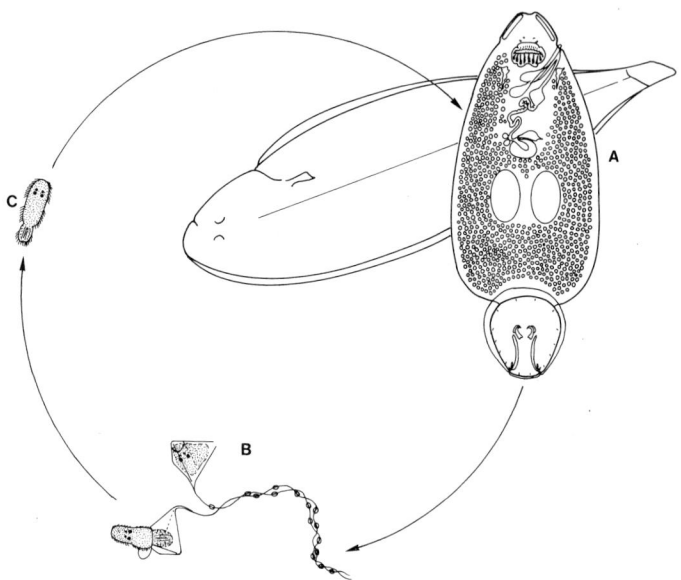

Fig. 3.2. The life cycle of the monogenean *Entobdella soleae*. The adult parasite (A) inhabits the lower surface of the common sole (*Solea solea*). Tetrahedral eggs attached to sand ballast on the sea-bed (B) liberate free-swimming ciliated larvae (oncomiracidia) (C), which typically invade the upper surface of the sole and migrate to the host's lower surface. Modified from Kearn, 1986a.

Although each egg has a long stalk bearing adhesive droplets like a row of beads, this glue will not stick to sole skin and freshly laid eggs of the parasite leave the host. However, the glue readily adheres to sand grains, which act as ballast, preventing the eggs being swept away by tidal and other currents from areas inhabited by soles. After about a month spent developing on the sea bottom, a larva propelled by cilia escapes from the tetrahedral egg capsule. This larva must find a sole in order to survive and grow to reproductive age.

In aquaria the parasite has one other way of establishing itself on new hosts. Juvenile and adult parasites regularly transfer themselves from one sole to another and this may be an important means of colonising new soles in the sea.

3.2.1 *Attachment and locomotion*

E. soleae attaches itself to the relatively flat skin of its host by a muscular, saucer-shaped, posterior haptor (Figures 3.3, 3.4). The haptor is attached principally by suction, but the holdfast is also anchored at its posterior margin by hooks, two of which, called anterior hamuli, are large enough to penetrate into the dermis. The haptor is packed with muscles (Figure 3.4), some running vertically from the dorsal to the ventral surface and others following a circular path around the saucer. These intrinsic muscles are capable of altering the shape of the saucer and generating suction on their own. However, the two anterior hamuli, which run in an antero-posterior direction in the haptor, and two curved, centrally located, so-called accessory sclerites, which are orientated obliquely in a dorso-ventral direction (Figures 3.4, 3.5), are part of an independent suction-generating apparatus of considerable sophistication.

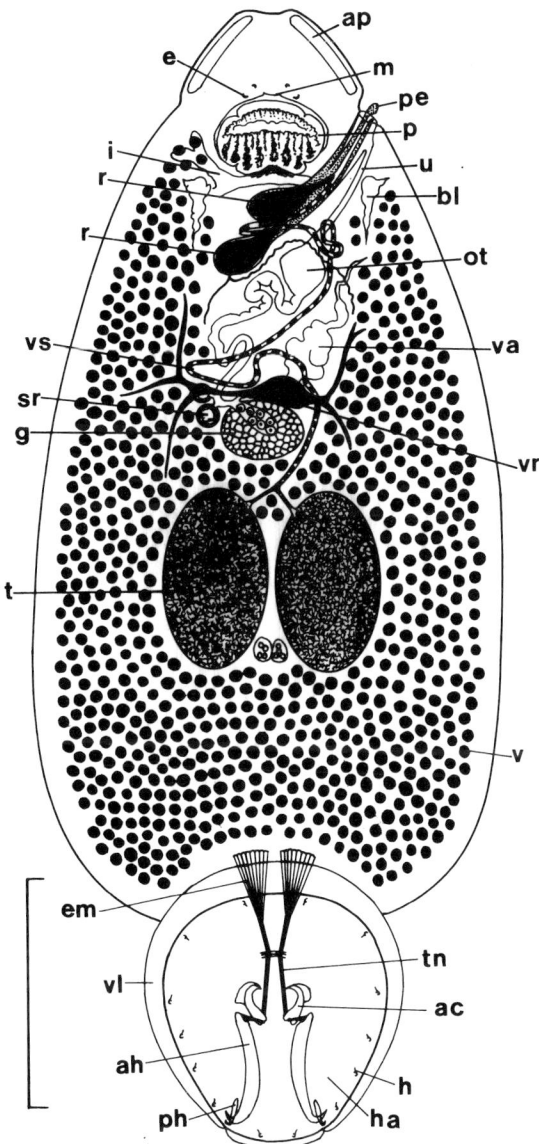

Fig. 3.3. The anatomy of an adult specimen of *Entobdella soleae* in ventral view. ac, Accessory sclerite; ah, anterior hamulus; ap, adhesive pad; bl, bladder; e, eye; em, extrinsic hook-operating muscle; g, germarium; h, hooklet; ha, haptor; i, intestine; m, mouth; ot, ootype; p, pharynx; pe, penis; ph, posterior hamulus; r, reservoirs for spermatophore jelly; sr, seminal receptacle; t, testis; tn, tendon; u, uterus; v, vitellarium; va, vagina; vl, marginal valve; vr, vitelline reservoir; vs, vas deferens. Scale bar: 1 mm. From Kearn, 1998, Copyright © 1998, with kind permission of Kluwer Academic Publishers.

Suction is achieved by lifting the anterior ends of the long shaft-like anterior hamuli (Figure 3.5). These shafts are embedded in the roof of the saucer and as they are raised they also lift the roof and produce suction. The shafts require powerful muscles to lift them, but the problem is where to house these muscles and how to apply this lift in a haptor in which space for large muscles is severely restricted, especially in a dorso-ventral direction. The problem is solved by a piece of neat biological engineering. The large elongated muscles are housed in the body and their force is exerted via long tendons that run into the haptor through fluid-filled canals. The ends of the tendons are attached to the shafts, but by passing through grooves in the central accessory sclerites they are able to change their direction and pull in a roughly dorsal direction on the ends

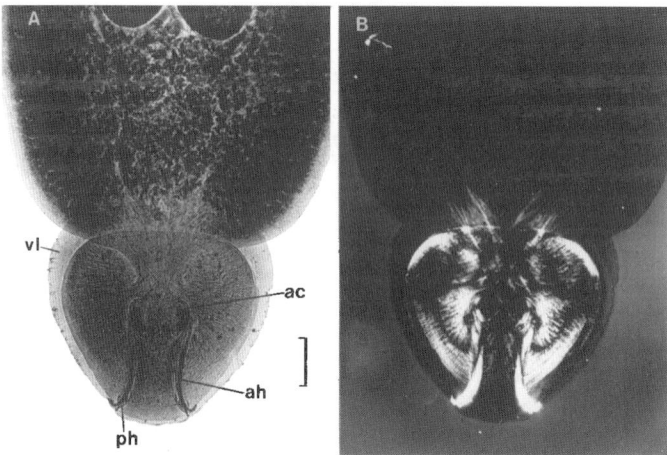

Fig. 3.4. The posterior region of the body of the same specimen of *Entobdella soleae* seen in (A) bright field illumination and (B) polarised light. Note strong birefringence in (B) indicating the presence of muscle. The anterior hamuli are also birefringent. Labelling as in Fig. 3.3. Scale bar: 250 μm. From Kearn, 1999, with kind permission of Cambridge University Press.

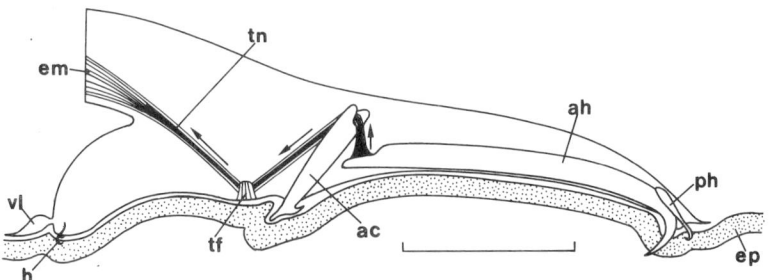

Fig. 3.5. A diagrammatic parasagittal section through the haptor of *Entobdella soleae* attached to host skin. The arrows show the direction of movement of the tendon (tn) when the extrinsic muscle (em) contracts. ep, Host epidermis; tf, transverse fibres. Other lettering as in Fig. 3.3. Scale bar: 250 μm. Modified from Kearn, 1971a.

of the shafts. This is not the only contribution made by the accessory sclerites. The efficient functioning of the apparatus requires the accessory sclerites to be prevented from sinking into the skin of the host when force is applied via the tendons. This is achieved partly by the curved 'foot' of the accessory sclerite, which protrudes from the ventral surface of the haptor and by virtue of its shape resists penetration into the skin, and partly by the host's bony scale, which provides an impenetrable plate beneath the haptor.

When suction is generated, efficiency will also be reduced by influx of seawater beneath the edge of the saucer and by any inward movement of the periphery. The first problem appears to be solved by the presence of a marginal valve to keep out seawater. Any tendency for the haptor to move inward is resisted by 14 tiny hooklets, which are arranged at intervals around the haptor edge and are capable of pinning the haptor to the host's skin.

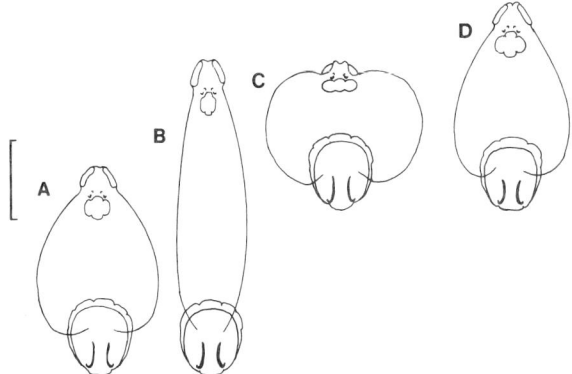

Fig. 3.6. (A-D) Successive stages of a locomotory 'step' in *Entobdella soleae*. See text for explanation. Scale bar: 1 mm.

E. soleae is not permanently attached by the haptor. On the contrary, the animal is very mobile and frequently changes its location on the host in the manner of a leech or looper caterpillar (Figure 3.6 A-D). Locomotion involves extending the body, attaching the head temporarily by two sticky pads (Figure 3.7), releasing the haptor and reattaching it close to the head. The adhesive released by the sticky pads is remarkable since it sticks with great tenacity to wet, slimy fish skin. It resembles commercial epoxy resins like 'Araldite' in that stickiness appears to develop after interaction between two secreted components. Another important feature of this system is that the attached pads can be released instantaneously when the haptor has been firmly reattached.

3.2.2 Food, feeding and egestion

Blood-feeding parasites reveal the nature of their diet either by the red colour of the gut if feeding has been recent or by the presence of a dark brown pigment, haematin, which is the indigestible remnant of earlier blood meals (see Chapter 6). Gut contents of this nature are not found in *E. soleae*. The parasite feeds on the skin's superficial epidermis, which is not supplied with blood vessels.

E. soleae has no jaws or teeth with which to harvest epidermal cells and relies for this purpose on secretion from its large glandular pharynx (Figure 3.7). The parasite feeds by everting this organ through the widely open mouth and applying it to the skin surface (Figure 3.8A). It is assumed that the pharyngeal secretion is applied to the skin via papillae, which project from the lining of the pharynx (Figure 3.7).

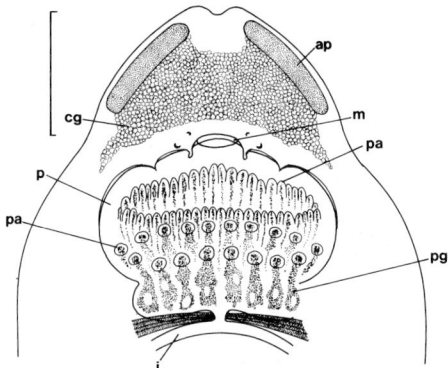

Fig. 3.7. Diagram of the anterior region of *Entobdella soleae* showing the pharynx in retracted position; ap, Anterior adhesive pad; cg, cement glands; i, intestine; m, mouth; p, pharynx; pa, pharyngeal papilla; pg, pharyngeal gland cell. Scale bar: 250 µm. From Kearn, 1994, Copyright © 1994, with permission from Elsevier.

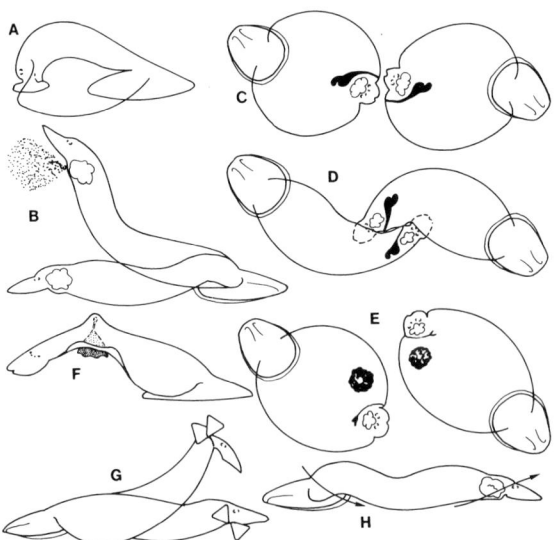

Fig. 3.8. The behaviour of *Entobdella soleae*. Sketches showing (A) feeding, (B) egestion, (C-F) consecutive stages of mating and mutual insemination, (G) behaviour of parasite with eggs tethered to the body, (H) breathing movements (arrows show direction of water flow). In (E) and (F) a spermatophore is attached to each individual ventrally. In (F) the parasite is shown undergoing body contractions that serve to withdraw sperm from the spermatophore. (B) From Kearn *et al.*, 1996, with permission from Taylor & Francis (see http://www.tandf.co.uk); (F) from Kearn, 1970, with permission of Cambridge University Press; (A, C-E, G, H) from Kearn, 1998, Copyright © 1998, with kind permission of Kluwer Academic Publishers.

Whether the pharyngeal secretion initiates digestion of the epidermal cells or simply causes their dissociation is not known, but the epidermal material is are rapidly sucked up by muscular contraction of the pharynx and pumped into the parasite's branched intestine. Feeding lasts for about five minutes; the pharynx is then detached and tucked back into the head. The circular feeding wounds are clearly visible in the host's skin (Figures 3.9A, 3.14A), especially after preservation in formalin. The epidermis enclosed by the everted pharynx is completely eroded, but the dermis beneath appears undamaged (Figure 3.9B).

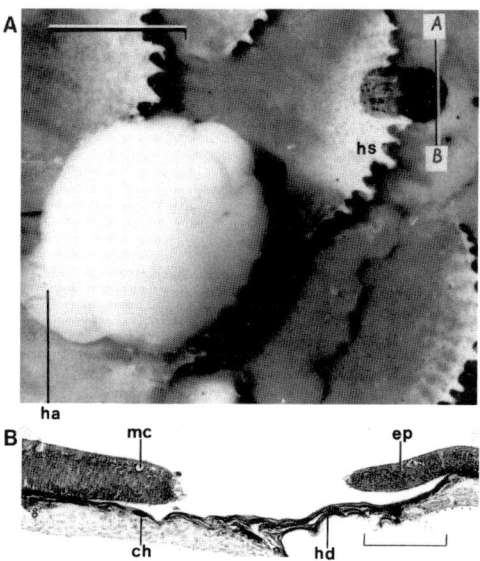

Fig. 3.9. Feeding in *Entobdella soleae*. (A) Parasite preserved in Bouin's fluid immediately after feeding. Note dark feeding wound at top right. AB, plane of histological sections taken through the feeding wound. One of these sections is shown in (B). ch, Chromatophore; ep, host epidermis; ha, haptor; hd, host dermis; hs, host scale; mc, mucous cell. Scale bars: (A) 1 mm; (B) 150 µm. From Kearn, 1963a, with permission of Cambridge University Press.

Fish epidermis is a highly suitable food for a permanent skin parasite because of its important role in wound healing (see Chapter 1). Wounds of the size made by *E. soleae* are likely to be closed rapidly by inward migration of undamaged epidermal cells from the edges of the wound, minimising danger of skin infection by micro-organisms and osmoregulatory problems. Mitosis (cell division) occurs at all levels in teleost epidermis (p. 9) and sooner or later the cells removed by the parasite will be replaced.

Thus, the relatively small populations of *E. soleae* on individual wild soles (infrapopulations) (see above) pose no threat to their hosts and, indeed, there is a wide safety margin in terms of infrapopulation size and the potential for serious damage to the host. In aquaria, soles can support much larger populations of parasites with no obvious ill effects. However, there is a danger in the restricted space of the aquarium that high survival rates of larvae and elevated invasion success will lead to an exponential rise in numbers of adults on captive hosts. The situation is rapidly reached in which the many hundreds of parasites strip off the epidermis faster than the host can replace it, osmoregulation is compromised, infection by micro-organisms takes place and host death ensues. In a fish-farming situation this would be a serious problem and it would be necessary to treat soles regularly (e.g. with fresh water, which rapidly kills *E. soleae*, see below) to limit the population size of the parasites.

After digesting and absorbing the host's epidermal tissue, the parasite must dispose of the indigestible material. Platyhelminths have no anus, so waste material must be ejected via the mouth. In order to do this *E. soleae* swallows seawater. The pharynx in its retracted mode repeatedly pumps seawater into the intestine and then, when fully inflated, the seawater and the debris suspended in it are forced out through the mouth (Figure 3.8B). Thus, the parasite uses seawater to flush out its intestine prior to the next meal.

3.2.3 Mating
Entobdella soleae, like the majority of parasitic platyhelminths, is hermaphrodite, i.e. each individual has both male and female reproductive organs (Figure 3.3). The male organs become functional before the female organs, a phenomenon known as protandry. Thus, *E. soleae* is able to donate and receive sperm before it is capable of producing eggs and presumably is able to keep this sperm alive until the reproductive system becomes fully functional. Experiments with isolated parasites indicate that self-insemination does not take place in *E. soleae* (see also below).

E. soleae has two testes and a long coiled vas deferens, which provides accommodation for large quantities of sperm (Figure 3.3). The vas deferens leads to a fleshy penis, the tip of which is capable of protrusion. There is a vagina, but its aperture leads to a tube that is so narrow that without enormous expansion could not possibly accommodate the substantial male organ. There is a much more spacious uterus with a large aperture shared with the penis canal. This uterus is destined to become the terminal region of the functional female reproductive tract and serves for brief storage and as an exit route for the eggs (see below). Although large enough to accommodate the penis, there is no record of insemination via this route in *E. soleae* and observations on mating in action refocus on the vagina as the port of entry for sperm.

After a brief 'courtship' in which some tactile interplay takes place between two individuals, their anterior regions interlock, so that the penis tip of each parasite is adjacent to the vaginal opening of the other (Figure 3.8CD). What follows is difficult to observe, but it seems highly unlikely that the penis is inserted into the vagina. The two individuals remain interlocked for no more than a few seconds and, after separation, each parasite carries a jelly-like spermatophore attached externally to its ventral surface in the region of the vaginal opening (Figure 3.8E). Almost immediately after separation each parasite undergoes vigorous contractions of the body, which serve to suck the dark central mass of sperm from the spermatophore into the vagina (Figure 3.8F). Proximally the vagina is more spacious (Figure 3.3) and sperm withdrawn from the spermatophore is stored here, the rest of the spermatophore being eventually discarded. Since sperm exchange can take place between protandrous individuals in which only the male system is functional, the vagina, which is normally regarded as part of the female reproductive system, must become operational at the same time as the male system. However, parasites with a fully mature female reproductive system can also donate and accept spermatophores.

3.2.4 Egg assembly
The ovary in monogeneans consists of two separate organs, both of which communicate with the oviduct. There is a single compact germarium producing egg cells and a diffuse follicular vitellarium producing vitelline cells (Figure 3.3). Vitelline cells can be

regarded as modified egg cells that make no genetic contribution to the offspring, but perform two important roles. They contain droplets of material destined to produce the eggshell and they provide nutrients for the growth and development of the embryo. Mature vitelline cells from the vitelline follicles are carried via ducts to a storage reservoir, located anterior to the germarium.

Mature egg cells enter a separate chamber in the germarium and it is here that fertilisation takes place. When and how spermatozoa reach this fertilisation chamber is not known. The proximal end of the vagina, where immobile spermatozoa are stored, opens into the vitelline reservoir, which in turn communicates with the oviduct and the germarium. However, there are usually two, occasionally one or three, spherical seminal receptacles opening off the oviduct where active spermatozoa are stored, and it may be this cache that supplies the fertilisation chamber.

Egg assembly begins with the release of a single, already fertilised egg cell from the germarium. As this cell progresses along the oviduct a stream of vitelline cells is released behind it from the vitelline reservoir. Egg cell and vitelline cells enter an egg mould or ootype where the vitelline droplets are released and fuse around the cell mass to produce the shell. Four pads in the ootype press the soft shell of the egg capsule into a tetrahedral shape and excess shell material trailing behind the egg capsule becomes the egg stalk. Adhesive droplets are added at intervals along the stalk and as time passes the colourless shell material becomes first yellow, then brown as its principal protein component is converted to a stiffer, chemically and physically resistant material called sclerotin.

Eventually the egg leaves the ootype and enters the uterus. Eggs may be laid singly or they may be retained in the uterus and laid in groups of two or three. Typically eggs are laid in two stages. First, the egg capsule or capsules are squeezed out of the uterus, but remain attached to the parasite temporarily by their egg stalks, which are still lodged within the uterus. Parasites with eggs partly extruded in this way frequently perform movements in which the anterior region of the body is moved abruptly upwards and backwards, with the egg capsules uppermost (Figure 3.8G). When such movements are performed by parasites attached to the underside of the host, the egg capsules would undoubtedly snag in the underlying sediment. Further movement of the body would then drag the egg stalks out of the uterus and promote attachment of the adhesive droplets on the egg stalks to sand grains.

3.2.5 Life with little oxygen
The habitat of the common sole creates problems for the parasite. Sand and mud at the sea bottom contain little oxygen. Organic detritus sinks to the bottom where it undergoes bacterial decomposition, which consumes oxygen. The problem is exacerbated by the sole's habit of burying itself in the sediment for long periods of time and by a preference on the part of adult parasites for the lower surface of the sole. Thus, the adult parasites are likely to experience low ambient oxygen levels when the host is resting on the bottom or buried, and higher levels when the host is active and feeding. It is known that resting soles eject all their respiratory water through the upper opercular opening (Yazdani & Alexander, 1967), so that parasites on the lower surface will not experience even a refreshing host respiratory current. The situation may be particularly critical for the parasite's eggs, since they spend the whole of their roughly four week

period of embryonic development, during which demand for oxygen is likely to be high, in this anoxic environment.

Adult parasites in stagnant seawater on the lower surface of a resting sole would rapidly use up any available ambient oxygen and the parasite solves the problem by creating its own water current. Transverse undulations pass successively down the body, driving away oxygen-depleted water in an anterior direction and drawing fresh, at least partially oxygenated, seawater over its body from a posterior direction (Figure 3.8H). The parasite is able to adjust its rate of undulation and the amplitude of the waves to match its requirements, as we can demonstrate in the laboratory. When a parasite is subjected to low ambient oxygen (as when the host is buried), its breathing movements are relatively fast and the body undulations of relatively large amplitude. When the oxygen level in the water is raised (as when the host is swimming) the breathing movements slow down and the amplitude of the waves is reduced. There is evidence too that the parasite can respond to low ambient oxygen levels by changing body shape. Extending itself in the plane of its flat body increases the surface area available for the inward diffusion of oxygen. The corresponding reduction in body thickness reduces diffusion distances to internal sites.

Breathing movements are not an option for the embryos of *E. soleae* developing in eggs attached to sand grains at the bottom. More oxygen may be available at the surface of the sediment than deeper down, but it is more than likely that eggs of *E. soleae* will sink into the sediment because of the sand ballast attached to their egg stalks and the shifting of the sediment by water movements. However, a striking feature of *E. soleae* eggs is their tetrahedral shape (Figure 3.2). This may favourably influence the uptake of oxygen by the developing embryo because the tetrahedron has a greater surface area/volume ratio and shorter surface to centre diffusion distances than eggs of the same volume and more conventional ovoid shapes.

3.2.6 The egg, hatching and host finding
Four weeks is a long time to spend at the sea bottom and there is a danger that during this incubation period small predators living in or on the sediment will eat the eggs of *E. soleae*. The free-living polychaete worms and amphipod crustaceans that serve as food for their host (Wheeler, 1969) are two such groups of potential predators. However, sclerotin, the main component of the eggshell of *E. soleae*, is a remarkably tough protein, which resists attack even by powerful protein-splitting enzymes such as mammalian trypsin (Llewellyn J., 1965; Kearn, 1975). So, even if the eggs were eaten by a predator, provided they were swallowed whole without cracking of the shell, they would be likely to survive passage through the gut and after evacuation would be able to complete their development and hatch. Macdonald (in Kearn, 1975) demonstrated that this is indeed so – intact eggs of *E. soleae* which had passed through the guts of various crustaceans completed their development and hatched.

The larva or oncomiracidium (Figure 3.10) escapes from the tetrahedral egg through an opening created by detachment of one of the four corners of the tetrahedron (Figure 3.2). This cap or operculum is held in place by cement, which is thought to be dissolved by a secretion produced by the larva. After its escape from the egg, cilia propel the larva through the water. Hatching may occur beneath the surface of the sediment, the larva temporarily joining the interstitial fauna (see above), but the larva's survival depends partly on its ability to find its way upwards out of the sediment.

Freshly hatched larvae are photopositive, using their four conspicuous eyes to orientate themselves, and if sufficient natural light penetrates into the sediment they are likely to swim upwards and reach open water. However, the larva also swims upwards in total darkness, indicating that they respond to gravity, but the nature and location of the gravity receptors in the body are unknown.

Fig. 3.10. Electronic flash photomicrograph of a living oncomiracidium of *Entobdella soleae*. Note three bands of ciliated epidermal cells (c), four conspicuous eyes and folded haptor. Only the hooklets are fully developed. Scale bar: 50µm.

While the larva remains inside the eggshell it is quiescent and uses little energy. Hatching will use up some of its meagre resources, as will escaping from the sand if this is necessary. The larva must find a sole within 24 hours or run out of energy and die. When the larva reaches the surface of the sediment it may be lucky enough to make contact with a resting sole, but the chances of this happening are small. Clearly, the survival of the parasite will be greatly enhanced if the parasite has some way of determining whether a sole is nearby before committing itself to hatching. One of the ways in which a sole reveals its presence is by releasing chemical substances from its body into the water. Skin mucus is probably an important source of such chemicals. Remarkably, fully developed eggs of *E. soleae* hatch within a few minutes after treatment with sole mucus. Presumably, a chemical component of the sole's mucus passes through the eggshell, activates the larva and initiates hatching behaviour, but the identity of this chemical stimulant remains unknown.

If fully developed eggs of *E. soleae* are placed on the upper surface of a sole and kept under observation, they hatch within a few minutes and many of them attach themselves immediately to the host's skin without swimming. The ciliated epidermal cells on the surface of the larva then become detached and disperse, with their cilia still beating. We learn two things from this exercise: first that the larvae are infective as soon as they hatch and, secondly, that the larvae are capable of infecting their host without swimming.

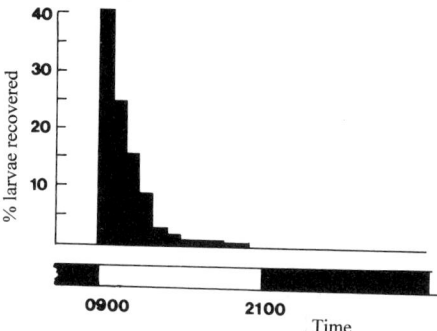

Fig. 3.11. The daily hatching pattern of *Entobdella soleae* cultured in seawater free of host contamination at 14°C. The histogram summarises observations made over a period of 15 days, with fluorescent lighting turned on and off abruptly (LD 12:12). White panel, period of illumination; black panel, period of darkness. Larvae collected in each of the 24 daily one hour periods have been added together and given as hatching percentages. From Kearn, 1986a, Copyright © 1986, with permission from Elsevier.

But what happens if no sole settles near the fully developed eggs? By chance a sole may not settle on or near the eggs within the life span of a resting, fully developed unhatched larva. Rather than relying entirely on a sole settling on or near the eggs, some eggs hatch spontaneously, giving the larva the opportunity to seek a host actively. However, this hatching is timed to maximise the chances of meeting a resting host. Soles have a daily activity pattern, resting during daylight hours by partly burying themselves in the sediment and moving and feeding at night (Kruuk, 1963). Consequently, the chances of a larva locating and attaching itself to a sole may be greater during the day when the host is a stationary target than during the night when the host is swimming about. So, spontaneous hatching takes place soon after dawn (Figure 3.11), that is, at the time when soles are inactive. Most hatching is completed during the first four hours of daylight and spontaneous hatching during the rest of the daylight period and during the night is rare. By hatching early in the day the larvae have the whole of the period of daylight to locate a resting host.

The behaviour of these spontaneously emerging larvae also favours host location. As already stated, freshly hatched larvae are photopositive and are likely to swim upwards in the sea. Such movement away from the bottom seems, at first sight, highly disadvantageous, but there is little advantage in remaining on the bottom if there is no sole nearby and, after swimming upwards for a time, the larva abruptly becomes photonegative and returns to the bottom. The descending larva is likely to be shifted horizontally by currents and will undoubtedly return to a different place on the bottom. It may be fortunate enough to make contact with a resting sole in this new locality, but if not, its alternating upward and downward (photopositive and photonegative) swimming behaviour will continue until it is successful in finding a resting host or exhausts its supply of energy.

So, *E. soleae* maximises opportunities for locating new hosts by a combination of two hatching strategies that take advantage of the behavioural features of the host. If there are no hosts nearby the parasite rations out hatching, larvae emerging in small numbers on successive days at a time on each day when hosts are most likely to be inactive and most vulnerable. The remaining unhatched larvae are capable of detecting chemical cues from a host settling close by and respond by hatching within a few minutes in large numbers. Since soles may be moderately active during the hours of daylight on dull days (Wheeler, 1969), soles may make contact with eggs at any time during the day or the night. The ability of the fully developed eggs to respond to host mucus at any time ensures that the parasite can take advantage of these opportunities whenever they arise.

In this context, it is interesting that soles seem capable of switching mucus secretion on and off. Buried soles are reluctant to leave the sediment and when forced to do so often have a layer of sand grains attached firmly to the skin (Kearn, 2002). These grains do not become free until the reluctant fish has been active for a few seconds. This suggests that mucus secretion is switched off when the sole is buried, thereby concealing its presence from predators hunting with their olfactory sense, and switched on when the sole resumes its activity. Soles may liberate more skin mucus at the time when they become stationary or bury themselves than at other times. It is conceivable that when a sole leaves the sand, it may not maximise mucus flow until it has moved away from its resting site.

3.2.7 The larva and host invasion

The larvae of *E. soleae* (see Figure 3.10) are only 1/4 mm (250 µm) in length (just visible to the naked eye). The disk-shaped posterior attachment organ or haptor has a ring of well-developed hooklets, but, because the edges of the haptor of the free-swimming larva are turned inwards, these hooklets are not immediately operational. In fact, sticky areas on the head of the larva effect initial attachment and the hooklets are not deployed until the haptor has been unfolded and planted on the skin. The larva's chemical sense, which may already have been employed in promoting hatching, has another important role at the initial attachment stage. It is likely to determine whether or not the larva will commit itself to the potential host with which it has made contact by jettisoning its only means of swimming – its ciliated epidermal cells.

Detached scales from the common sole carry small patches of skin, to which larvae of *E. soleae* readily attach themselves and shed their ciliated cells (Figure 3.12). The larvae respond in the same way to Agar jelly that has been in contact with the skin of the common sole - presumably the jelly is impregnated with the chemical substances from sole skin that are recognised by the larva and promote attachment. Given a choice of scales or skin from a variety of other flatfishes such as dab (*Limanda limanda*), lemon sole (*Microstomus kitt*), solenette (*Buglossidium luteum*) and rays (*Raja* spp.), the larvae of *E. soleae* show a strong preference for scales of the common sole. These experiments suggest that the larvae identify their host by specific and as yet unknown chemical substances produced by the skin. They are unlikely to establish themselves on other flatfishes such as plaice and rays that fail to produce these specific signals, and larvae making contact with such non-host fishes most probably retain their ciliated cells and are able to continue swimming.

There is no evidence at present to indicate that free-swimming larvae are attracted to the common sole (Kearn, 2002). They may not recognise their host until they make contact with it (contact chemoperception), and the chemical signals to which they respond may be an integral part of the host's surface epithelial cells and not components of skin mucus. This strategy would be expected if it is confirmed that buried/resting soles switch off their mucus flow (see above), since contact chemoperception would permit host recognition at times when skin mucus was not being produced.

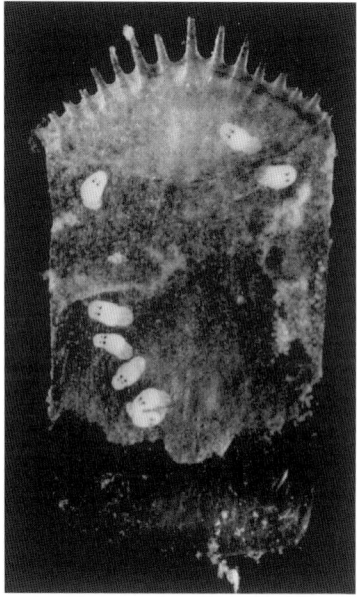

Fig. 3.12. Eight larvae of *Entobdella soleae* freshly attached to a detached dorsal scale from the common sole. Note that host epidermis has migrated almost to the posterior edge of the scale lamina. Scale preserved in Bouin's fluid. From Kearn, 1971a, with permission from University of Toronto Press.

Larvae swimming up through the sediment may make contact with the lower surface of a resting sole, but those larvae that break free from the sediment into open water are more likely to establish themselves on the upper surfaces of the host. Soles kept in aquaria and soles freshly collected at sea have more newly attached larvae on their upper surfaces than on the lower surfaces. This difference might reflect greater attractiveness of the upper surface and a preference expressed by the larvae, but simple tests in which larvae are offered equal numbers of detached, skin-bearing scales from the upper and lower surfaces of a common sole, show that the larvae have no such preference. They settle equally readily on upper and lower skin. It is likely, therefore, that this apparent preference for the upper surface is a reflection of the fact that this surface is generally more accessible than the lower surface, which, for much of the time, is in contact with or below the surface of the sediment.

3.2.8 Post-invasion migration

This brings us to another interesting feature of the parasite's biology. Although freshly attached parasites are more common on the upper surface, most adult parasites are encountered on the lower surface. Differential mortality offers one possible explanation for this phenomenon. Parasites establishing themselves on the upper surface may have a greatly reduced chance of completing their development and reaching maturity compared with parasites invading the host from below. Migration offers an alternative explanation. Larvae alighting on the upper skin of the sole may be capable of finding their way to the lower surface where they reach sexual maturity.

It has been demonstrated experimentally that migration from the upper to the lower surface does indeed take place. This was achieved by killing any parasites from previous infections by immersion of the fish in fresh water for 10 – 20 minutes. A small circular patch of upper skin of the uninfected sole was then exposed to larvae, using a Perspex cylinder to delimit the infection area. The host was killed after 40 days, and, after rendering the parasites opaque and clearly visible by immersion in formalin, the positions of all the parasites on the fish were determined. Many of the parasites were already on the lower surface (Figure 3.13A^1), demonstrating their ability to make the journey from the upper to the lower surface.

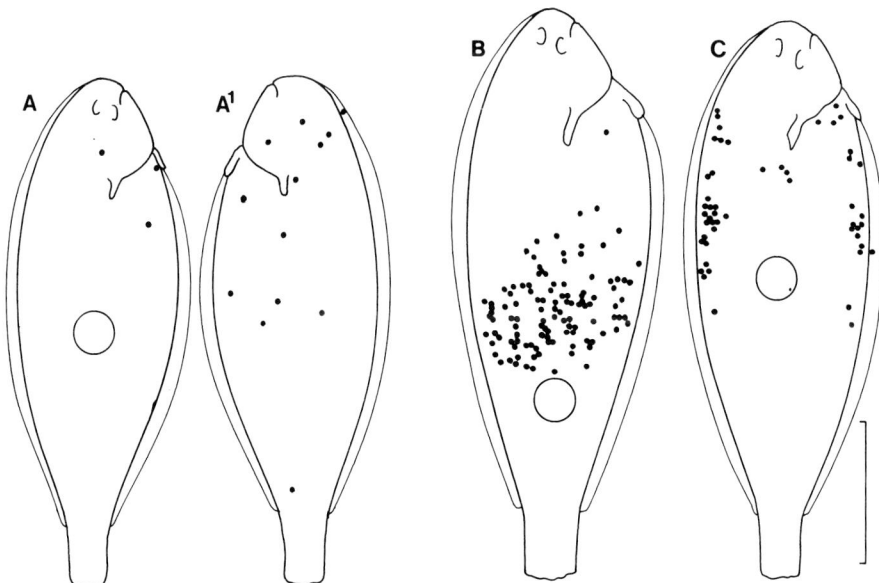

Fig. 3.13. The distribution of *Entobdella soleae* (indicated by dots) on three common soles (*Solea solea*), (AA^1) 40 days and (B, C) nine days after experimental exposure to oncomiracidia. Each ring shows the area of the sole exposed to larvae. Upper host surface (A) and lower host surface (A^1) shown for the single 40-day-old infection; dorsal host surfaces only shown for the two nine-day-old infections. Scale bar: 5 cm. From Kearn, 1984, Copyright © 1984, with permission from Elsevier.

This is a remarkable migration for such a small, relatively simple flatworm. It is not the movement itself or the distance covered that is particularly special, since the parasite is able to move rapidly and repeatedly like a leech using the sticky pads on the head and the suctorial haptor (Figure 3.6). It is the parasite's ability to find its way that is of outstanding interest. A parasite moving at random on the sole's upper surface has little chance of reaching the lower surface (Kearn *et al.*, 1993) and yet many parasites achieve this in as little as 40 days, indicating that parasites are using directional clues to find their way to the lower surface. But, before we give some thought to the nature of these clues, there is one other important feature of the migration that is relevant. If we repeat the infection experiment just described, but terminate it after just nine days, none of the parasites have reached the lower surface but most of those on the upper surface have moved forwards on the host (Figure 3.13BC).

This forward migration is not eliminated if we keep the sole in total darkness for the 9 days following experimental infection, showing that light and 'vision' are not used for direction finding by the parasite, but there are other more likely clues. Water currents usually pass in an antero-posterior direction over the sole's body. Soles can shuffle backwards using the fins that run down the edges of their bodies, but rarely do so - most of the time they move forwards. In addition, the gill ventilating current leaving the right gill chamber via the opercular opening on the upper surface washes backwards over the sole. The ctenoid scales also provide a clue since their spiny projecting edges all point towards the tail of the sole (Figure 3.14).

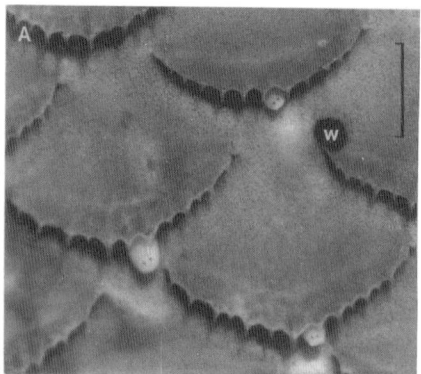 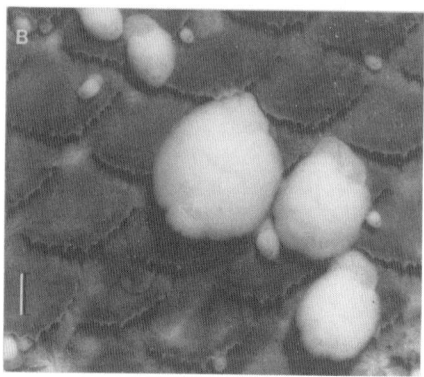

Fig. 3.14. The typical orientation of (A) juvenile parasites and (B) adults, on a heavily infested common sole from an aquarium. The head of the sole is in the direction of the top of the figure and the tail of the sole in the direction of the bottom. w, Feeding wound of large juvenile or adult. Scale bars: 0.5 mm.

Both of these directional signposts, water currents and scale orientation, could explain another behavioural feature of parasites of all ages and sizes. Most parasites orientate themselves with the haptor anteriorly placed with respect to the fish and with the body projecting in the direction of the fish's tail (Figure 3.14). Many of these parasites tuck the haptor beneath the projecting spiny edge of a scale. Parasites can be experimentally detached and induced by manipulation to attach themselves the wrong way round, i.e. with the body pointing towards the sole's head. These disorientated parasites soon shift their positions and, usually within a few minutes, have readopted the typical orientation with the body pointing towards the tail. An additional observation demonstrates that water currents are not essential for this rapid reorientation. If the experiment is repeated with a freshly killed sole, reorientation occurs readily in the absence of any currents from host locomotion or respiration. This indicates that the parasites have an 'awareness' of scale orientation, which probably operates through their sense of touch. That this sense is well developed is evident from the parasite's recoiling response when the edges of the body are touched, and the electron microscope reveals that these body edges are abundantly supplied with cilia-based sensilla that may be touch receptors. Searching movements, in which the parasite swings the head from side to side while attached by the haptor, could provide the parasite with sensory information about the proximity and orientation of scales.

It is difficult to establish experimentally that the scales provide the only directional cues for forward migration on the common sole. This is because the migration is protracted over a period of days and it is not possible to eliminate water currents over such a long period of time. However, one other observation adds a little extra weight to the idea of direction finding by reference to the scale pattern. Migrating parasites on experimentally infected soles often accumulate along the margins of the upper surface (Figure 3.13C). The scales are arranged in diagonal rows (Figure 3.14), providing natural routes in an antero-lateral direction across the surface of the sole. Migrating parasites may move forwards in a zigzag fashion using these routes, rather than migrating directly forwards towards the head of the host. If this interpretation is correct then contact with the scales seems likely to be more important than water currents for orientation and, in any case, the elevated projecting edges of the scales probably shelter small parasites from water currents flowing backwards on the sole.

The lower surface of the common sole is the destination for migrating parasites and yet by migrating forwards the parasites take the longest possible route to reach that surface. A possible explanation for this is that transfer to the lower surface is difficult or impossible at locations along the body margins fringed by the anal and dorsal fins. These fins are thin, flexible and motile and such a transfer would demand considerable agility and flexibility on the part of the parasite. However, these marginal fins terminate on or near the head (Figures 3.13, 3.15), so that access to the fish's lower surface is unrestricted in the anterior region. Indications from experimental infections support this explanation (Kearn, 1984), but whether the fins present an insurmountable barrier or are merely difficult to negotiate is hard to ascertain.

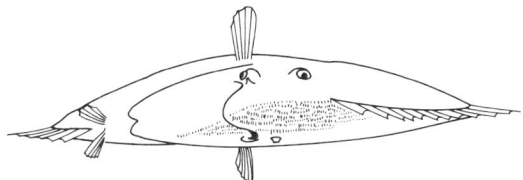

Fig. 3.15. En face view of a resting common sole (*Solea solea*). Illustration by Peter Stebbing, from Wheeler, 1978.

3.2.9 *Host to host transfer and migration of adults and juveniles*

It was mentioned earlier that juvenile and adult parasites readily change hosts in aquaria and it is for this reason that adult parasites are sometimes found on the upper surfaces of captive soles. There is no evidence that adult parasites can backtrack and return to the upper surface of their host. A sole with adult parasites on its lower surface may settle on top of another sole, providing the opportunity for the adult parasites to transfer to the upper surface of the sole beneath. Adult parasites placed experimentally on the upper surface of an uninfected recipient host move in an anterior direction on the host's upper surface over a period of days and then transfer to the lower surface (Figure 3.16). Thus, adult parasites retain the ability, so well developed in newly attached larvae, to find their way from the upper to the lower surface of the host.

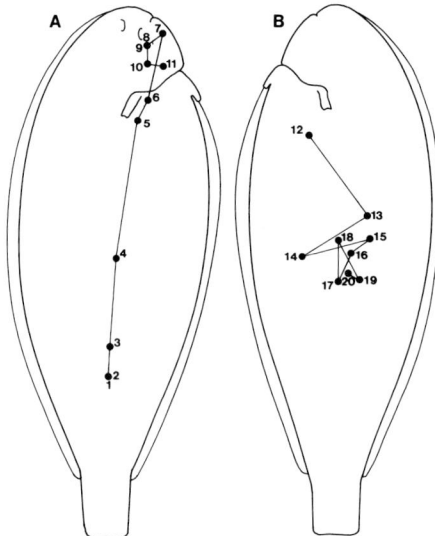

Fig. 3.16. The successive daily positions over 20 days of an adult *Entobdella soleae* (A) on the upper and (B) on the lower surface of the same common sole (*Solea solea*). The parasite was implanted at position 1 on the upper surface. Modified from Kearn, 1988.

It is not known whether adult or juvenile parasites transfer from host to host in the wild, but they are likely to have a brief opportunity to do so during host mating, which has been observed on several occasions in aquaria (Baynes *et al.*, 1994). Pairing takes place at night on the bottom of the aquarium, the male shuffling beneath the female. The two fishes then stay close together and swim together with closely synchronised body movements towards the surface, where they remain in bodily contact for up to 70 seconds, during which period synchronised spawning and sperm release are thought to occur.

Although captive soles appear to spawn at the surface in shallow tanks, this does not necessarily imply that spawning in the wild takes place at the sea surface. This would require a greatly extended vertical migration of the pair. However, Baynes *et al.* (1994) recalled interesting reports by de Veen (1967) of soles drifting with the tidal flow at the surface of the North Sea. This phenomenon was recorded at night and during the spring period when soles breed. De Veen believed that these soles, which were regarded as separate individuals, were taking advantage of faster-flowing surface currents to transport them to the spawning grounds, but as Baynes *et al.* pointed out, intimately paired soles are likely to be regarded as single fishes from the side of a ship and de Veen's observers may have witnessed surface spawning. If this turns out to be correct, then the period of opportunity for transfer of parasites from fish to fish during host mating may be longer than indicated by aquarium observations and the direction of transfer during pairing episodes will be from females to males.

3.2.10 *Why migrate to the lower surface?*
Why do parasites of all ages have this remarkable instinctive ability to find their way to the lower surface? Selection pressures to maintain this behavioural feature must be strong, and yet adult parasites readily feed on the upper skin and oxygen availability would probably be higher. However, the lower surface offers some advantages. First, parasites on the lower surface would be better situated to attach their eggs to sand grains. The characteristic behaviour performed by parasites carrying partly expelled eggs tethered by egg stalks still lodged in the uterus (see Figure 3.8G), would effectively attach the eggs by their adhesive droplets to sand grains beneath the host. Eggs released by parasites on the upper surface are likely to be carried away by water currents from sandy/muddy areas inhabited by soles before the opportunity arises for sand grains to become attached as ballast to the sticky egg stalks.

A second possible advantage for parasites on the host's lower surface is that in this situation they are less likely to be taken by predators. In the case of the sole/*Entobdella* association, we have no evidence that predation is a threat to the parasite, but there is good evidence that other skin parasites, including monogeneans, are frequently taken by so-called 'cleaner organisms' (fishes, crustaceans) in coral reef communities (Kearn, 1998, p. 101; Grutter, 2002). 'Cleaning' is less frequently reported in temperate seas, but this may reflect the less pleasant environment for prolonged fish watching compared with coral reefs. The pigmentation of some parasitic crustaceans and particularly their ability to change colour may conceal them from potential predators and wrasse have been used to remove caligid copepods (sea lice) from farmed salmon (see Chapter 8). Van Beneden (1856) noted pigmentation in *Epibdella* (=*Benedenia*) *sciaenae*, a large monogenean related to *Entobdella soleae* collected from the skin of the meagre, *Sciaena aquila* (= *Argyrosomus regius*) caught off the coast of Belgium. Moreover, he remarked on the failure of the human observer to recognise these pigmented parasites *in situ*.

Restriction of parasite activities to the lower surface of the host may have other, perhaps more significant, advantages related to mating. Infection levels of soles are low - on average two or three parasites per fish. The chances of these parasites meeting for mating seem small, and self-insemination seems a likely option in these circumstances. Curiously, this is not the case. In a series of experiments in each of which a single immature parasite was induced to attach itself to an uninfected sole and allowed to reach maturity, no evidence for self-insemination was found. Eggs, if laid at all by these parasites, were not viable. Thus, cross insemination seems to be essential for the production of viable eggs, and individuals readily exchange spermatophores when they come into contact, even when detached from the host.

By restricting their distribution to the host's lower surface, parasites greatly reduce the area of the fish's skin that needs to be searched to find a mate. Nevertheless, this restriction is unlikely to be enough. Kearn *et al.* (1993) considered the random movement of two adult parasites on the lower surface of a relatively small (20 cm long) common sole and calculated that they are likely to make contact during their 120-day sexually active lifetime. However, it is advantageous for a parasite to mate as soon as it is mature to maximise reproductive output, and yet random movement cannot ensure early insemination of newly mature individuals. The answer to this conundrum may be that movement of parasites is not random. Parasites may be able to find each other more directly and quickly by responding to chemical messages (pheromones) released into the water by potential partners.

There is no evidence yet that pheromonal attraction does take place, but if it does what better place for this system to operate than on the lower surface of a sole. The fish spends much of its life resting on or buried in the sediment, thereby trapping a thin layer of seawater between its lower surface and the bottom. We know that when soles are resting they shunt all their respiratory water through the upper opercular opening (Yazdani & Alexander, 1967), so the thin layer of water beneath the sole will be undisturbed by currents. This will provide an ideal environment for the horizontal outward diffusion of pheromones from the thin flat bodies of the parasites. Body undulation (breathing movements, see above) and the ability of the parasite to rotate the body about the attached haptor, might also play a part in the dispersal and detection of pheromones. There are many unicellular glands and sensilla along the parasite's body margins that might provide a source of these chemicals and a system for their detection. The upper surface of the sole is a highly unsuitable place for such a pheromonal system to operate because of the exposure to sea currents.

3.3 ENTOBDELLA HIPPOGLOSSI

Morphological differences between *E. soleae* from the common sole, *Solea solea* (Soleidae) and *E. hippoglossi* from the halibut, *Hippoglossus hippoglossus* (Pleuronectidae) are small and include overall size (*E. hippoglossi* is bigger), the shapes of their accessory sclerites (cf. Figures 3.1 and 3.3), and the absence of adhesive droplets on the egg stalks of *E. hippoglossi*. However, there are some surprising differences in hatching. In *E. hippoglossi* hatching takes place during darkness not during daylight (Kearn, 1974). More specifically, the larvae emerge during the first two or three hours following dusk. Moreover, the eggs of *E. hippoglossi* do not hatch if treated with mucus from the halibut. It is most likely that this behavioural divergence between these two close relatives reflects fundamental behavioural differences between their two flatfish hosts. The halibut is the largest flatfish and, in fact, is one of the largest fishes in British waters, reaching a length of 2.5 m (Wheeler, 1969, 1978). It is an active predator with a wide-ranging diet including crustaceans, cephalopod molluscs and fishes. It is found on a wide range of bottoms including sand, gravel and rocky ground, but mid-water species found in its gut indicate that it also makes extensive vertical forays. The diet and way of life of the halibut are consistent with a diurnal pattern of activity and feeding, and the nocturnal hatching pattern of *E. hippoglossi* also points to a nocturnal resting phase on the part of the host. The reason why halibut mucus fails to stimulate hatching is more obscure, but the wide range of bottoms visited by the host and the more frequent excursions into open water may reduce the frequency and duration of the host's contacts with the parasite's eggs.

3.4 ENTOBDELLA DIADEMA

Entobdella diadema is somewhat intermediate in size between *E. soleae* and *E. hippoglossi*. This parasite is of special interest for three reasons: first, its host is an elasmobranch flatfish, the stingray *Dasyatis pastinaca*, not a teleost; secondly, its eggs are induced to hatch by shadows; thirdly, its spermatophores are known and are different from those of *E. soleae*.

Fully developed eggs can be stimulated to hatch simply by passing a hand between them and the source of illumination. Hatching is extremely rapid, larvae emerging in seconds rather than minutes as in *E. soleae* (see Kearn, 1982). The stingray host has a maximum length of over a metre and is a bottom dweller with a preference for sandy or muddy bottoms and shallow water (Wheeler, 1978). In well-lit waters it casts a strong shadow and any stingray cruising slowly enough over the bottom or resting briefly is likely to promote rapid hatching and become infected with larvae.

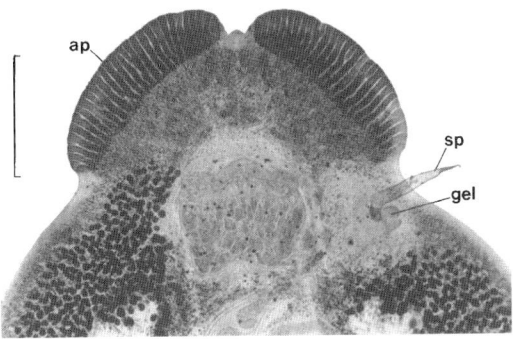

Fig. 3.17. The anterior region of *Entobdella diadema* in ventral view. ap, Adhesive pad; gel, genital lips; sp, spermatophore. Scale bar: 1 mm.

The spermatophores of *E. diadema* are elongate and spindle shaped, measuring about 0.6 mm in length and 0.15 mm in width (Figure 3.17; Llewellyn & Euzet, 1964). Unlike the jelly-like spermatophores of *E. soleae*, those of *E. diadema* appear to have an outer casing, which splits under pressure liberating masses of active sperm. The common genital opening shared by the male reproductive canal and the uterus is provided with a pair of muscular lips (Figure 3.17). The vaginal opening is adjacent to this genital opening and Llewellyn & Euzet suggested that the lips might be used to manoeuvre the spermatophore into the vagina of a mating partner. In the specimen illustrated in Figure 3.17, the proximal end of the spermatophore is lodged in the vaginal opening.

3.5 HOST SPECIFICITY AND HOST SWITCHING

As our knowledge of the biology of *E. soleae* and *E. hippoglossi* has increased, the significance of the phenomenon of host specificity has become clearer. Morphologically, there seems to be no reason why parasites like *E. soleae* should not survive on any flatfish. We might expect the haptor or the pharynx to function as well on halibut skin as on sole skin. Certainly the haptor of *E. soleae* functions well on glass and on other smooth inert surfaces, and parasites will attempt to feed, unsuccessfully of course, on glass. However, the results of some simple transfer experiments do not support these expectations (Kearn, 1967a). A detached specimen of *E. soleae* has no difficulty in attaching itself to the skin of another common sole and remaining attached.

Similarly, the parasite can usually attach itself to non-host fishes such as the solenette (*Buglossidium luteum*: Soleidae) and plaice (*Pleuronectes platessa*: Pleuronectidae), but remains in place on these alien hosts for little more than 24 hours. The parasite performs better on the skin of an elasmobranch flatfish (*Raja* sp.), remaining attached for 2 – 8 days. The implication is that *E. soleae* has some kind of attachment problem on solenette and plaice, but not on rays or on glass. Since the duration of attachment to ray skin and to glass is similar, this indicates that *E. soleae* may be unable to feed on ray skin. Thus, there seem to be subtle differences in the skin of apparently similar fishes. The haptor and the pharynx of the parasite may be able to function only on hosts whose skin provides the appropriate physical and/or chemical signals.

The much greater differences between the behaviour patterns of potential flatfish hosts place demands on their skin parasites and reinforce host specificity. In terms of habitat and behaviour, the common sole and the halibut have little in common and survival of their skin parasites demands specific behavioural adaptations of their eggs and larvae to ensure infection of new hosts.

Morphologically, *E. soleae* and *E. hippoglossi* are much closer to each other than to *E. diadema*. However, the differences between *E. soleae* and *E. hippoglossi* on the one hand and *E. diadema* on the other, are relatively small compared with the enormous differences between their teleost (common sole and halibut respectively) and elasmobranch (stingray) hosts, which are unrelated and morphologically distinct. It is probable that teleost flatfishes have evolved relatively recently compared with flat elasmobranchs. We would expect much bigger differences between *E. soleae/E. hippoglossi* and *E. diadema*, if all three parasites had evolved from a common ancestor parasitising the ancient forerunner of teleost and elasmobranch fishes (phylogenetic speciation). It seems much more likely that this is another example of host switching or ecological transfer, like the colonisation of the hippopotamus by polystomatid monogeneans (see above). *Entobdella* has probably colonised flat teleosts relatively recently, transferring to these 'newcomers' from flat elasmobranchs (see Llewellyn, 1982).

Host specificity and host-switching will be considered further in Chapter 4.

4
OTHER MONOGENEAN SKIN PARASITES

4.1 INTRODUCTION

Species of *Entobdella* are not the only monopisthocotylean monogeneans living on the skin of fishes. Many are large enough to be seen with the naked eye. *Capsala martinieri* (Figure 4.1A), a relative of *E. soleae*, and in the same family (Capsalidae), is one of the largest monogeneans, exceeding *E. hippoglossi* in size. Members of the Acanthocotylidae, Microbothriidae and Udonellidae, although smaller than *Capsala* are not difficult to see. At the other end of the spectrum are the tiny members of the Gyrodactylidae. These are especially diverse and abundant on freshwater fishes, but also occur on marine hosts. The gyrodactylids appear to have 'broken the mould' as far as monogeneans are concerned, having a reproductive pattern entirely different from that of the rest of the monogeneans, and, as a consequence of this, some unique biological features.

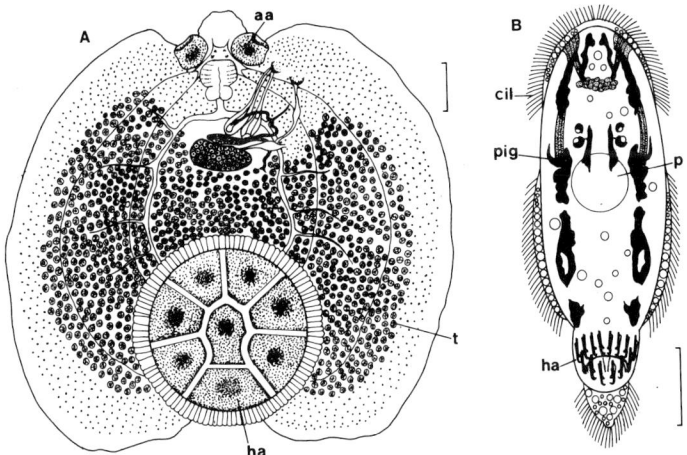

Fig. 4.1. (A) Adult (ventral view) and (B) larva of the monogenean *Capsala martinieri*. cil, Cilia; ha, haptor; p, pharynx; pig, body pigment; t, testis. Scale bars: (a) 2 mm; (b) 100 μm. Modified from (A) Sproston, 1946, (B) Kearn, 1963b.

4.2 *CAPSALA MARTINIERI*

The oceanic sunfish, *Mola mola*, occasionally visits the southern shores of Britain. In spite of the fact that one of its skin parasites, *C. martinieri*, reaches 2 cm or more in diameter (Figure 4.1A), it is very difficult to distinguish the parasite *in situ*. The close application of its thin, flat body to the host's skin helps to make it inconspicuous, but the main reason for its invisibility is the presence throughout the body of a network of brown/black pigment, which matches the background of host skin. This provides effective camouflage, at least to the human eye. Reports of the sunfish being 'cleaned' (= deparasitised) by fishes and by seabirds are on record (references in Kearn, 1976), indicating that camouflage might be an advantage for *C. martinieri*.

The circular haptor of *C. martinieri* is also large, reaching 8 mm in diameter in adults. No hooks have been found in the adult (hooklets are present in the larva; see Figure 4.1B), so the adult haptor must generate suction using its own muscles, not operating as a single sucker like *Entobdella* but organised into separate suctorial units or loculi. There are eight of these, one central and seven peripheral (Figure 4.1A).

C. martinieri lays tetrahedral eggs, liberating ciliated oncomiracidia already containing some of the camouflage pigment widespread in the adult's body (Figure 4.1B). Nothing is known about the fate of the eggs or about how these tiny larvae find their way to a fish that is thought to spend most of its life near the surface or in mid-water (Wheeler, 1969).

4.3 ACANTHOCOTYLIDS

Like *E. soleae*, acanthocotylid monogeneans are epidermis feeders (see Kearn, 1963a). They are occasionally encountered in naked-eye searches of the ventral surfaces of five of the commonly caught rays in the Plymouth area, England, namely *Raja clavata*, *R. montagui*, *R. brachyura*, *R. naevus* and *R. microocellata* (see Kearn, 1967b; Llewellyn et al., 1984). A comparative study by Kearn (1967b) of samples of parasites from the first four of these rays, revealed no significant anatomical differences between them, and they were all regarded as *Acanthocotyle lobianchi* Monticelli, 1888 (Figure 4.2A), a parasite described by Monticelli from the ventral surface of the thornback ray, *R. clavata* caught at Naples. Llewellyn et al. (1984) also regarded parasites collected from *R. microocellata* as *A. lobianchi*. Monticelli (1899) described a second species, which he called *A. oligoterus* from *R. clavata* at Naples, but Kearn (1967b) regarded this as a synonym of *A. lobianchi*.

The rather broad host specificity of *A. lobianchi*, embracing several species of *Raja*, contrasts sharply with the seemingly narrow host specificity of two other less commonly encountered species of *Acanthocotyle* found at Plymouth. Like *A. lobianchi*, *A. elegans* and *A. greeni* (see Figure 4.2B) occur on the thornback ray, but have not been found on any of the other common rays known to harbour *A. lobianchi* at Plymouth (Llewellyn et al., 1984). The reasons for these differences in host preference are unknown.

Ciliary propulsion plays an important part in host location by the oncomiracidium of *Entobdella soleae*. However, swimming is not an essential prerequisite for invasion of the host. Larvae emerging from eggs of *E. soleae* placed directly on host skin (*Solea solea*) and hatching in response to stimulation by host skin mucus are capable of attaching themselves to host skin without swimming (Chapter 3). It is this second option that has been developed in *Acanthocotyle*, to the extent that their larvae are unable to swim, having lost their ciliated epidermal cells (Kearn, 1967b). Hatching is not induced in *A. lobianchi* by shadows cast by the host, as it is in the stingray parasite *Entobdella diadema*. The banana-shaped eggs of *A. lobianchi* (see Figure 4.3A), like those of *E. soleae*, respond to chemical cues in host skin mucus (Macdonald, 1974), but unlike *E. soleae*, fail to hatch in the absence of host mucus.

Although unhatched larvae can survive for more than 80 days, they will ultimately die if they receive no chemical signal from the host. Also, since the larva of *Acanthocotyle* is unable to swim at any time, the host must be virtually in contact with the eggs for successful infection to take place.

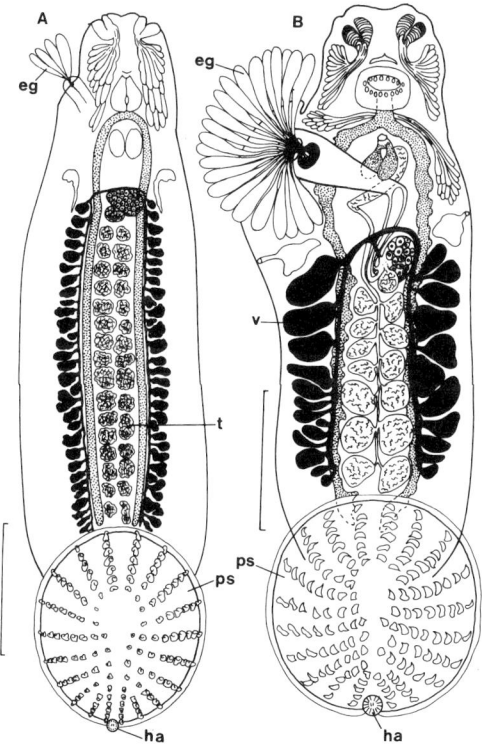

Fig. 4.2. (A) *Acanthocotyle lobianchi* and (B) *A. greeni*, both in ventral view. eg, Egg; ha, haptor; ps, pseudohaptor; t, testis; v, vitellarium. Scale bars: (A) 1 mm; (B) 0.5 mm. (A) Modified from Dawes, 1947; (B) from Macdonald & Llewellyn, 1980, with permission of Cambridge University Press.

The chances of successful invasion are improved by speeding up hatching. After appropriate stimulation the larva of *E. soleae* takes several minutes to escape from the egg (Kearn, 1975), while the larva of *Acanthocotyle* (and that of *E. diadema*, see Chapter 3) emerges in less than a second after stimulation (Figure 4.3B). It is assumed that in *Acanthocotyle* the cement holding the egg lid or operculum in place is weakened in advance, so that on receipt of the hatching stimulus the larva merely has to extend its body to push off the operculum and thrust the head out of the egg like a jack-in-the-box. Examination of the fully developed unhatched egg reveals that the excessively long larva has to be folded to fit inside (Figure 4.3A), so that part of the extension on hatching may simply involve straightening of the crumpled larva. Contraction of circular body muscles probably make further extension possible, increasing the resting length by two or three times (Figure 4.3B) and enabling the sticky adhesive areas on the head to reach host skin, as long as it is not too far away. Since the egg is attached to sand grains by a short stalk provisioned with a droplet of glue (Figure 4.3A), the larva should have no difficulty in withdrawing its haptor from the heavily weighted egg capsule.

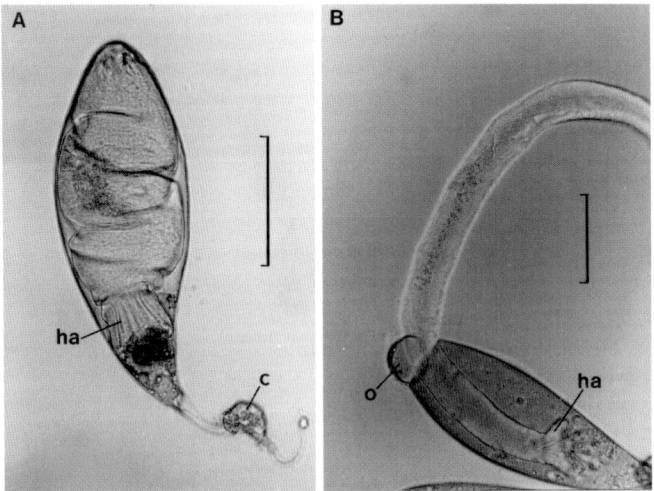

Fig. 4.3. (A) Egg of *Acanthocotyle lobianchi* containing a fully developed larva about 40 days old. Note body folds. (B) Egg with emerging larva, photographed with electronic flash immediately after stimulation of the unhatched egg with host mucus. c, Cement on egg stalk; ha, haptor; o, operculum. Scale bars: 50 μm. (B) From Kearn, 1986a, Copyright © 1986, with permission from Elsevier.

Most free-swimming larvae of *E. soleae* settle on the upper surface of their flatfish host (Chapter 3). The larvae of *A. lobianchi* are unable to swim, and most will establish themselves on the ray's lower surface, as a direct consequence of rays resting on top of eggs on the sea bottom. However, like soles (*Solea solea*), rays may partly bury themselves when they come to rest on the bottom, using their 'wings' to toss sand onto their upper surface. In this way some eggs may be thrown onto the upper surface, where the larvae may hatch and attach themselves. This interpretation of events is supported by the distribution of parasites on wild caught rays, determined by scraping their upper and lower surfaces and searching the scrapings. The lower surfaces yield large numbers of parasites ranging in size from newly attached larvae to adults up to 5 mm in length, while relatively few parasites are found in scrapings from the upper surfaces (Kearn, 1967b).

A. elegans and *A. greeni* have other features of interest. Adult specimens of *A. elegans* live on the upper surface of the thornback ray. In naked-eye searches at Plymouth, seven out of 11 fishes (64%) were infected, with one fish harbouring nearly 50 parasites and the rest up to 10 parasites each (Kearn, 1967b). No adults were found on the lower surfaces. This distribution raises the possibility that the larva of *A. elegans* reaches the upper surface by swimming, but the larvae, like those of *A. lobianchi*, lack ciliated cells. The most likely explanation is that most larvae of *A. elegans* attach themselves to the ray's lower surface and then migrate to the upper surface, but it is hard to understand why *A. elegans* makes this journey. Thornback rays may have a 'wingspan' of over 60 cm, so these small parasites, which are less than 3.5 mm long when mature, may have to cover considerable distances to reach the upper surface.

Some kind of competitive interaction between *A. elegans* and *A. lobianchi* might be the driving force behind the segregation of these two parasites, but, if this is so, interspecific competition has not led to a similar displacement of *A. greeni*, which is found on the lower surface of the thornback ray. Macdonald & Llewellyn (1980) examined two thornback rays with wingspans of 45 cm and 67 cm and found three and six adult specimens of *A. greeni* respectively on their lower surfaces.

An intriguing feature of *A. greeni* is that up to 80 eggs are retained by the adult parasite until at least some of these eggs are ready to hatch (Figure 4.2B). The opening of the uterus is located on a fleshy 'arm'. The ovoid egg capsules are exposed outside the body, but the bases of their short stalks are fused together and rooted just inside the uterus. The whole egg bundle is sheltered dorsally by a shield-like extension of the body margin. In *A. lobianchi* and in *A. elegans* no more than eight eggs are retained in this way before the bundle is shed (Figure 4.2A). Macdonald & Llewellyn suggested that the unciliated larvae of *A. greeni* developing in the older eggs of the bundle are stimulated to hatch by host mucus, the larvae attaching themselves to the same host as that of their parent (autoinfection). After release of the egg bundle, larvae hatching from younger eggs will have the opportunity to infect new hosts.

Most of the risks to survival inherent in the dissemination of eggs and the invasion of new hosts are avoided by autoinfection, and this may have been an important factor in the reduction of egg output in *A. greeni*. Macdonald & Llewellyn (1980) calculated that this parasite probably makes no more than three eggs per day, compared with two eggs per hour in *E. soleae* (Chapter 3).

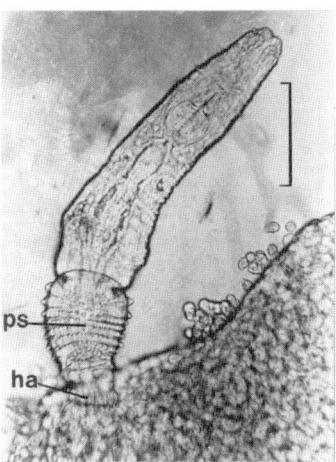

Fig. 4.4. A living juvenile specimen of *Acanthocotyle lobianchi* attached to ray epidermis by the haptor (ha). The pseudohaptor (ps) is developing from the posterior region of the body, but is not yet functional. Scale bar: 50 μm.

It has been established that the chemical cue to which the larvae of *A. lobianchi* respond is urea (Kearn & Macdonald, 1976). This is a waste product of protein metabolism that is present in the skin mucus of most fishes, but usually in very small amounts. The mucus of elasmobranchs contains relatively high concentrations of urea because these animals retain this substance as part of a strategy to raise the osmotic pressure of the blood and avoid the osmotic loss of water to the sea. There is not enough urea in the mucus of the common sole (*Solea solea*) to hatch eggs of *A. lobianchi* and the chemical hatching factor for *E. soleae* remains unknown (Kearn, 1986ab).

The skin of the thornback ray has extensive flat scaleless areas and, like *E. soleae*, *A. lobianchi* relies on suction for attachment. This is generated by contraction of intrinsic muscles in the attachment disk and there is no equivalent of the extrinsic muscle/sclerite system of *Entobdella*. However, the ventral surface of the disk of *A. lobianchi* carries radial rows of rather squat sclerites (Figure 4.2A), each with a projecting twisted blade and these will undoubtedly provide additional purchase as the disk is spread out onto the smooth skin and then contracted. The surprising feature of this attachment disk is that it is not the true haptor. It is a false haptor or pseudohaptor derived from the posterior region of the body of the developing parasite (Figure 4.4; Kearn, 1967b). The true haptor is retained on the posterior margin of the fully developed pseudohaptor as a tiny disk bearing 16 hooklets (Figure 4.2), similar to those on the margin of the haptor of *E. soleae*.

4.4 MICROBOTHRIIDS

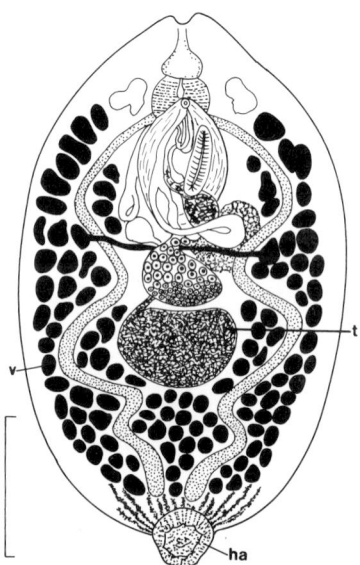

Fig. 4.5. The microbothriid monogenean *Leptocotyle minor*. ha, Haptor; t, testis; v, vitellarium. Scale bar: 0.5 mm. Modified from Sproston, 1946.

Hooks and other sclerites have been abandoned by the microbothriid monogeneans, which are parasites of sharks. A common member of this group in British waters is *Leptocotyle minor*, found on the skin of the dogfish, *Scyliorhinus canicula*. Two others have been recorded occasionally at Plymouth: *Pseudocotyle squatinae* on monkfish,

Squatina squatina, and *Microbothrium apiculatum* on spurdog, *Squalus acanthias* (see Llewellyn *et al*., 1984).

Adult *Leptocotyle minor* are 2 or 3 mm long with a small hookless haptor (Figure 4.5). The haptor is small because it has to fit on one of the dogfish's denticles (Figures 1.5, 4.6). As described in Chapter 1, the plate-like distal portion of the dogfish denticle protrudes through the host's epidermis and has no covering of skin. Hooks would be unable to penetrate the naked, hard, exposed surface of the denticle and *Leptocotyle* has opted for cement, not suction, to attach itself to this surface (Kearn, 1965). Gland cells lying in a transverse band just anterior to the haptor secrete the cement (Figure 4.6).

The use of cement does not prevent the parasite from moving its haptor to a new denticle by a rather slow-motion leech-like 'step'. Like *Entobdella* and *Acanthocotyle*, *L. minor* secretes an anterior adhesive. This is used to fix the head region temporarily to a distant denticle, while the haptor is slowly stripped away like adhesive tape, moved forward and reattached behind the head. *L. minor* can cement its haptor to glass and leaves a patch of cement behind at the attachment site when it is removed. *L. minor* is an epidermis feeder, gaining access to this food source between the denticles (Kearn, 1965).

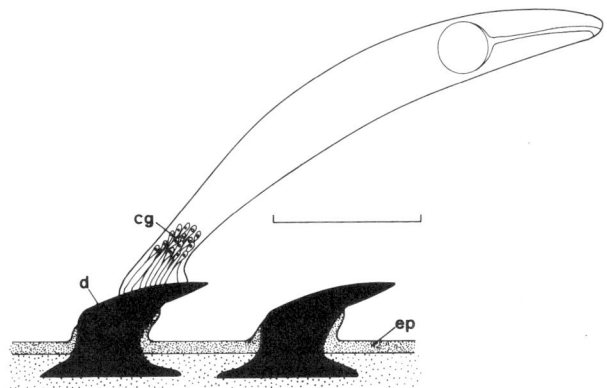

Fig. 4.6. Diagram illustrating the mode of attachment of *Leptocotyle minor* to a denticle (d) in the skin of the dogfish, *Scyliorhinus canicula*. cg, Cement glands; ep, host epidermis. Scale bar: 0.5 mm. From Kearn, 1994, Copyright © 1994, with permission from Elsevier.

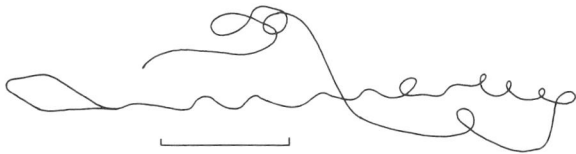

Fig. 4.7. The egg of *Leptocotyle minor*. Scale bar: 200 µm. From Whittington, 1987a, with permission of Cambridge University Press.

The eggs of *L. minor* are rather conventional in shape (Figure 4.7) but at one end the eggshell is prolonged to form a very slender stalk of great length (almost 3 mm). The egg stalk has no adhesive properties and, in fact, serves just the opposite function –

it keeps the egg in suspension in open water (Whittington, 1987a). The long slender stalks promote lifting of the eggs off the bottom by water turbulence and retard sinking of suspended eggs, in the same way that the silk threads produced by small spiders permit them to remain airborne. The eggs contain a fully developed larva after 16 – 18 days at 13 – 14 °C (Whittington, 1987a). When exposed to alternating 12-hour periods of light and darkness at 13 – 14 °C, in seawater free of contamination by fishes, the eggs fail to hatch, like those of *Acanthocotyle*, and survive for up to 88 days. Unlike *Acanthocotyle* however, the fully developed larva of *L. minor* is ciliated. Whittington (1987a) found that dogfish body washings prepared by immersing a dogfish in approximately 14 litres of fresh seawater for 1 hour, stimulated hatching of fully developed eggs of *L. minor* in 10 – 15 seconds. As in *Acanthocotyle*, further studies identified urea as the most likely stimulatory component of dogfish body washings, but *L. minor* is much more sensitive to this substance than *A. lobianchi*. The effective threshold concentration for *L. minor* lies between 0.0005 and 0.001 μ moles urea/ ml seawater compared with between 0.002 and 0.017 μ moles urea / ml for *A. lobianchi*.

To appreciate the significance of this interesting combination of features we must consider the biology of the host. The dogfish, *Scyliorhinus canicula*, is a bottom dweller, foraging on sand, gravel and mud for bottom living invertebrates and occasionally fishes (Wheeler, 1978). The eggs of *L. minor* are susceptible to water turbulence such as convection currents, and unattached eggs on the seabed will be sent into suspension by the activities of the dogfish. Fully developed eggs drifting close to the bodies of dogfishes will come under the influence of host skin secretions, but, unlike the relationship between *Acanthocotyle* and rays, host skin secretions will be diluted before they reach the eggs. Prolonged intimate contact between egg and host is not likely to take place, unless the egg stalk becomes entangled on the denticles. *Leptocotyle* copes with these constraints by responding to a much lower concentration of urea than *Acanthocotyle* and by retaining its ciliated cells and its ability to swim, features which will enhance its chances of making contact with a host.

4.5 UDONELLIDS

The absence of hooks on the posterior attachment organs of platyhelminth fish parasites creates a problem for systematists. Without this unequivocal evidence of monogenean ancestry, there is always the possibility that the hookless organism has evolved from one of the many free-living groups of flatworms ('turbellarians') and that the resemblance to a monogenean is due to convergence. Kearn & Gowing (1990) regarded six tiny spicules around the margin of the posterior adhesive organ of the larva of *L. minor* as vestiges of hooklets, supporting a monogenean origin for the microbothriids. No such clues have been found in the posterior attachment organs of udonellids. Consequently, their affinities have been long debated, with arguments advanced in favour of 'turbellarian', monogenean and separate status (see Rohde, 1994). Molecular data now place them among the monogeneans (Littlewood *et al.*, 1998; Olson & Littlewood, 2002).

Hooks are of no use in the microbothriid haptor because these parasites attach themselves to the hard denticles of their shark hosts. Choice of a similar hard substrate has presumably led to loss of hooks in the udonellids, but they attach themselves not directly to their fish hosts but to the surfaces of copepods (Figure 4.8A), which are themselves fish ectoparasites (Chapter 8). As in the microbothriids, the haptors of udonellids are supplied with glands (Figure 4.8B) and, according to Ivanov (1952) are attached to the hard exoskeletons of the copepods by cement, not by suction.

Only one species, *Udonella caligorum* (see Figure 4.8), has been reported in British waters, on a variety of copepod and fish hosts (Dawes, 1946). There are records as far apart as Plymouth on *Caligus labracis* and *C. centrodonti* parasitising ballan wrasse, *Labrus bergylta* (see Marine Biological Association, 1957), and the west coast of Scotland on *Caligus elongatus* collected from the buccal cavities of cod, *Gadus morhua* (see Kabata, 1973).

It is unlikely that there is any kind of parasitic relationship between *Udonella* and its copepod partner. *Udonella* most probably feeds directly and independently on the epidermis of the fish host. Kabata (1973) noted that many udonellids on the copepod carapace occupy marginal positions, from which the fish host's skin could easily be reached. Dark pigment found by Olivier *et al.* (2000) in the gut of *U. myliobati* on the copepod *Lepeophtheirus myliobati* was thought to be melanin from ingested epithelial cells of the shark *Carcharias taurus*. There is no evidence that *Udonella* ingests blood,

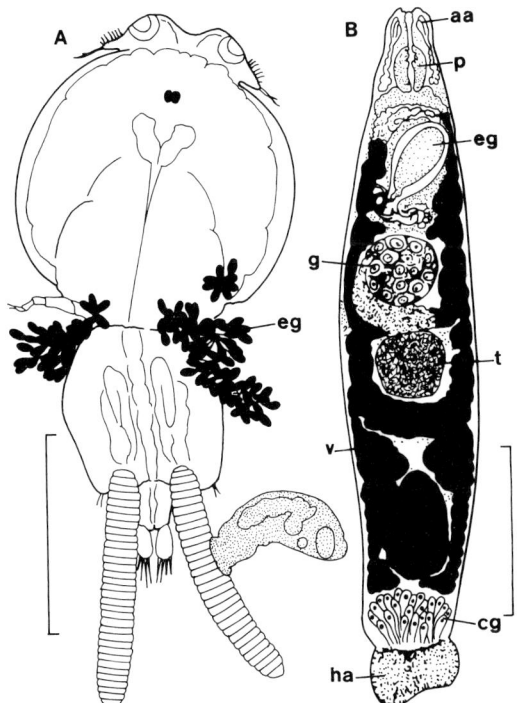

Fig. 4.8. Udonella caligorum. (A) A caligid copepod carrying a single *Udonella caligorum* on one of its two egg sacs and bunches of *Udonella* eggs on its body. (B) Adult. aa, Anterior adhesive sac; cg, cement glands; eg, egg; g, germarium; ha, haptor; p, pharynx; t, testis; v, vitellarium. Scale bars: (A) 1 mm; (B) 0.5 mm. From Kearn, 1998, Copyright © 1998, with kind permission of Kluwer Academic Publishers.

in spite of the fact that some are carried into the gill cavity by their copepods. The homogeneous and granular gut contents are consistent with a diet of fish epidermis, as is the pharynx, which is protrusible like that of *Entobdella* and contains gland cells opening into the lumen.

Caligus with attached *Udonella* have been caught in the plankton, swimming free of their fish hosts (e.g. Baylis & Jones, 1933). Thus, *Udonella* exploits the mobility of copepod fish parasites, relying on these crustaceans for dispersal and for infection of new fish hosts. The phenomenon whereby organisms are transported between hosts by another organism is known as phoresy. The copepods also provide a platform for the eggs (Figure 4.8A), which in *U. caligorum* are usually cemented to the copepod's genital segments. The larvae of *Udonella* do not need to swim to reach new hosts and are unciliated.

There is little information on how udonellids spread to new copepods. They are able to move like other monogeneans, using eversible adhesive sacs on the head for temporary attachment to the substrate while relocating the haptor (Nichols, 1975). Tufft (in Nichols, 1975) observed transfer of flatworms between copepods on starry flounder, *Platichthys stellatus*, and also reported that *Udonella* is capable of attaching itself temporarily to the fish's surface, later shifting onto a new copepod. Adult udonellids appear to prefer female copepods (Nichols, 1975; Timofeeva, 1977), which retain their egg sacs until their nauplius larvae hatch (see Chapter 8). Nichols (1975) observed synchronous hatching of udonellid and copepod eggs, but no *Udonella* larvae established themselves on the freshly hatched nauplii.

4.6 GYRODACTYLIDS

Gyrodactylids are among the smallest monogeneans, most of them ranging in length from 0.4 to 0.8 mm (Figure 4.9). In terms of the number of species, the most prominent genus in the Gyrodactylidae is *Gyrodactylus*: Bakke *et al.* (2002) recognised 402 valid species and stressed the remarkably low morphological diversity of this assemblage. Many live on the skin, but others spend all or part of their lives in the buccal, pharyngeal and gill cavities. Nevertheless, whatever their habitat they are skin feeders and have no obvious specialisations for life in these cavities. Consequently, those living on the gills and adjacent internal sites will be considered alongside the skin parasites in this chapter rather than in the next. The most notorious member of the genus currently is *G. salaris*, which in spite of its small size, is highly pathogenic to some strains of salmon (*Salmo salar*).

Gyrodactylids have the widest host range of any monogenean family, being found on 19 orders of bony fishes inhabiting marine and freshwater habitats (Bakke *et al.*, 2002). Gyrodactylids have not been found on agnathans or on acipenserids (sturgeons). Gyrodactylids from teleosts occasionally use chondrichthyans as transport hosts, but, apart from this, chondrichthyans are rarely parasitised. A few gyrodactylids are found on amphibians, and *Isancistrum* occurs on squid of the genus *Alloteuthis* (see Llewellyn, 1984). *Isancistrum* is the only monogenean parasitising an invertebrate, although phoretic relationships with invertebrates do occur (see udonellids above and p. 112). The ancestors of *Isancistrum* were presumably fish parasites that succeeded in switching hosts (see below and Chapter 3). Host specificity in gyrodactylids ranges from species restricted to a single host to *Gyrodactylus alviga*, which infects 16 host species belonging to nine orders (see Bakke *et al.*, 2002).

Gyrodactylus spp. are well known to zoology students because of their prominence in most zoological textbooks, but as an example of a typical monogenean there could be no worse choice, because the highly specialised reproductive biology of these animals sets them apart from their relatives. They have a unique lifestyle and many features that are unusual for monogeneans. Most members of the Gyrodactylidae are viviparous (*Gyrodactylus*, *Isancistrum* and 14 other genera contain exclusively viviparous species; see Boeger *et al.*, 2003), i.e. they do not lay eggs but give birth to living individuals, which are usually fully developed. In contrast, a few members of the

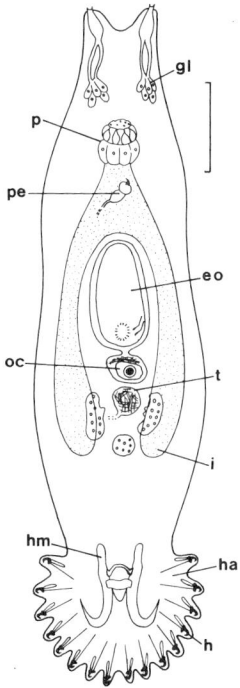

Fig. 4.9. Gyrodactylus pungitii. eo, Embryo in embryo sac; gl, gland cells; h, hooklet; ha, haptor; hm, hamulus; i, intestine; oc, oocyte; p, pharynx; pe, penis; t, testis. Scale bar: 100 μm. Modified from Malmberg, 1970.

family lay eggs (oviparous), as do most other monogeneans. The obvious consequence of the absence of free-swimming larvae in the viviparous gyrodactylids is that the invasion of new hosts is the task of fully developed parasites, typically by direct transfer during host-to-host contacts (Figure 4.10 and see below). This single reproductive change from oviparity to viviparity has also had a significant influence on interaction with the host's immune system, on pathogenesis, on parasite speciation and possibly even on parasite size.

The parent *Gyrodactylus* retains the embryo inside a sac usually identified as the uterus but which is more likely to be equivalent to the egg mould or ootype of oviparous monogeneans (Figure 4.9). Here the embryo grows into a daughter. As it develops the embryo becomes folded, but its posterior end is readily identified by the presence of developing hooks. Sixteen identical hooklets appear around the edge of the haptor disk and two hooks of a different shape (hamuli) appear a little later in the centre of the disk. By the time the embryo is fully-grown, the central hamuli are significantly larger than the peripheral hooklets. Von Nordmann observed embryonic hooks inside the parent's body as long ago as 1832, but he regarded them as part of the gut. Their true nature was recognised by von Siebold in 1849 (references in Harris, 1993). Immediately prior to birth, the daughter is as large as the parent (Figure 4.10).

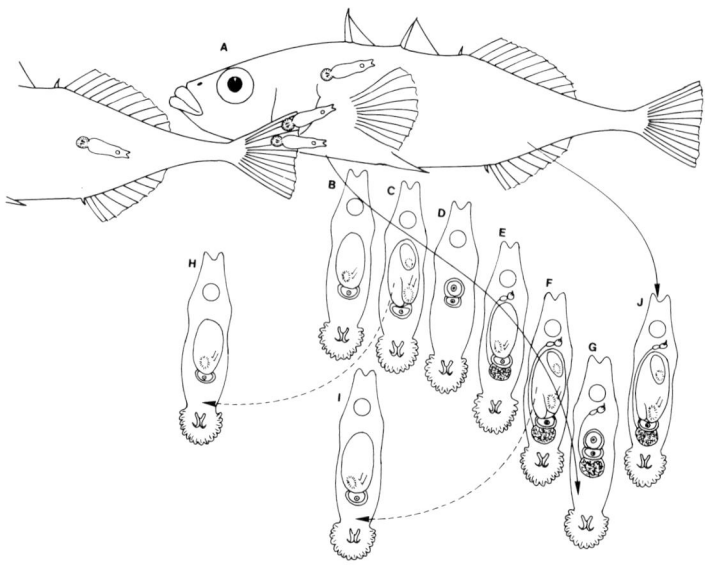

Fig. 4.10. The life cycle of *Gyrodactylus*. (A) Parasites transfer from fish to fish when hosts make contact. (B) – (G) Successive stages of a new-born individual: (C) gives birth to first daughter (H); (D) oocyte enters empty embryo sac; (E) male system develops, providing opportunity for mating with other individuals (J); (F) gives birth to second daughter (I). From Kearn, 1994, Copyright © 1994, with permission from Elsevier.

4.6.1 *Attachment*

Gyrodactylids rely entirely on their hooks for attachment (Figure 4.9), apparently without supplementation by suction or cement. The gyrodactylid haptor does not attach well to clean glass and any feeble binding that does occur may be due to deposits of anterior adhesive secretion left behind during locomotion (Lester, 1972).

It has been suggested that the hooklets of monogeneans may have evolved in response to the special challenges of attachment to fish epidermal cells (Kearn, 1999). Each hooklet has its own set of tiny muscles, giving it the independence to perform vigorous gaffing movements, which ensure that the curved blade penetrates a host epidermal cell. The uniform shapes and limited size range of hooklets throughout the Monogenea may have been imposed by the unique and conservative cytoarchitecture of superficial epithelial cells of fish skin. The blades of the hooklets are just the right size to be rooted firmly in the tangled 'terminal web' of tough filaments lying just beneath the cell surface (see Chapter 1), limiting mechanical damage to a small puncture. Although the purchase achieved by one hooklet seems feeble in relation to the overall task of attachment, the gyrodactylid spreads the load by employing 16 hooklets in a circle.

In gyrodactylids the hooklets have the added support of the large hamuli, which are potentially capable of penetrating through the epidermis into the tough fibrous dermis. However, there is some doubt as to whether the hamuli are functional in some gyrodactylids. Harris (1982) claimed that skin parasites like *Gyrodactylus gasterostei* do not use their hamuli to pierce the skin, the hamuli being covered by parasite tegument, whereas gill parasites such as *G. rarus* use their hamuli, creating deep wounds. Shinn *et al.* (2003) also came to the conclusion that attachment in gyrodactylids is mainly the responsibility of the hooklets, the hamuli only rarely penetrating the host's skin. However Shinn *et al.* claimed that the unattached hamuli have a supportive role to play in the deployment of the hooklets. They proposed that rotation of the hamuli serves to apply tension to the hooklets, permitting the hooklets to relax their muscles without disengaging their hooklet points.

It is hard to believe that the relatively large and sharp-pointed hamuli do not at any time enter host skin. Shinn *et al.* (2003) alluded to the possibility that sudden forces applied to the parasite may lead to hamulus penetration. Stresses and strains tending to dislodge a skin parasite are likely to vary enormously, depending on whether the host is resting or active. When the host is resting these stresses and strains will be weak and in this situation deployment of the hooklets may be more than adequate to provide sufficient security. However, a sudden burst of speed by the host will generate strong forces and the extra security provided by the prompt penetration of the large hamulus hooks into the host's skin may be essential to prevent dislodgement.

The linking of the two hamuli by two bars, one dorsal and the other ventral (Figure 4.9), is an unusual arrangement. This may confer sufficient rigidity to permit the two linked hamuli to function as a 'drag anchor', the points of the hamuli being driven into host tissue by forces exerted on the parasite or by momentum as the whole haptor is moved in an anterior direction in contact with the host. The hook points of the anterior hamuli of *E. soleae* may be forced into host skin in the same way (Chapter 3).

4.6.2 Feeding

Gyrodactylus feeds in a similar manner to *E. soleae* (references in Sterud *et al.*, 1998). Enzymes from the pharynx are assumed to disaggregate host epidermal cells and begin digestion, the resulting fine homogeneous suspension being sucked up and pumped into the intestinal tubes or caeca by the pharynx (Cable *et al.*, 2002a). The surface of the lining of the intestine is active, engulfing droplets of intestinal contents by a process called pinocytosis (see also p. 115). Digestion continues inside the intestinal lining and, after absorption of released nutrients, the residual waste is expelled into the intestinal lumen.

The two unbranched intestinal caeca run alongside the embryo sac (Figure 4.9), so that digested food material has the shortest distance to travel to supply the needs of the developing embryo. Harris (1982, in Sterud *et al.*, 1998) recorded feeding every 15 – 30 minutes in *G. gasterostei* from the three-spined stickleback, *Gasterosteus aculeatus*; one feeding wound was found to be 20 – 30 µm in diameter, extending down to the basement membrane. Mo (1994, in Sterud, *et al.*, 1998) quoted an observation by Malmberg to the effect that feeding gyrodactylids may penetrate the basement membrane and attack the dermis.

4.6.3 *Consequences of viviparity*

The retention of embryos greatly reduces reproductive output in gyrodactylids – a new embryo cannot be accommodated in the embryo sac until the parent has given birth to the previous occupant. Speeding up embryonic development is helpful. *G. turnbulli*, a parasite of the guppy, *Poecilia reticulata*, a popular aquarium fish, gives birth two or three times during a period of seven days (Scott, 1982, in Harris, 1993), but this is painfully slow compared with an egg output of two per hour in the oviparous *E. soleae*.

Gyrodactylids have responded to the pressures for increased reproductive output in unique ways. Before the daughter is born, a second embryo begins to develop inside the daughter (Figure 4.10), and, in some species, a third embryo appears within the second. Thus one individual may encapsulate two or even three further generations and when the first of these generations is born, the second generation within it already has a head start. Gyrodactylids also divert resources from their own reproductive system into the development of their embryos, delaying the appearance of the male reproductive system until after the first birth and the commencement of development of the second daughter (Figure 4.10). This phenomenon in which the female reproductive system develops first is known as protogyny and differs from protandry found in the oviparous monogeneans like *Entobdella*, in which the male system is the first to develop. A remarkable gyrodactylid, *Gyrodactylus gemini*, found by Ferraz *et al.* (1994) on the skin of an imported Amazonian fish, *Semaprochilodus taeniurus*, doubles its reproductive output in a unique way by accommodating two embryos side by side in the embryo sac.

The telescoping of generations leads to an exponential increase by autoinfection in the parasite population on a single susceptible host fish (infrapopulation), since initially parasites are being born faster than they are leaving the host. This rapid increase in numbers of parasites on susceptible hosts may pose a severe threat to host survival and there appear to be two ways in which this threat is moderated. First, the size of feeding wounds and hence the damage inflicted on the host is proportional to parasite body size. Gyrodactylids may have reduced the impact of high infection intensity by promoting a reduction in size of individual parasites (Kearn, 1998). This may explain why most gyrodactylids, although small (see above), have a pair of large hooks or hamuli in addition to 16 small hooklets (most other monogeneans of similar size subsist with just hooklets). Reduction in overall body size may have taken place in gyrodactylids without reduction or loss of hamuli (the squid parasite *Isancistrum*, is an exception, having no hamuli). The observations indicating that the hamuli of some gyrodactylids may not be functional (see above) are consistent with this idea. Secondly, in many gyrodactylid/host interactions, parasite population increase is eventually limited by a host response, such that most or all of the parasite infrapopulation is eliminated, although not necessarily killed and lost from the suprapopulation. It has been pointed out by Bakke *et al.* (1992) that gyrodactylids forced to leave the host by the host's immune response may reattach themselves to the same or to a different host individual.

How the host controls the parasite population is not well understood, but there are indications that epidermal mucous cells have an important part to play (see Chapter 1 and Sterud *et al.*, 1998). The continuous secretion of this mucus and the toxic effects of its chemical components seem to be important in defence against infective agents, including gyrodactylids (Sterud *et al.*, 1998; Harris *et al.*, 1998). There is a correlation between resistance to *G. derjavini* in salmonid fishes and high mucous cell densities (Buchmann & Uldal, 1997), and the work of Buchmann & Bresciani (1998) indicates that this parasite species, in the later stages of infection, avoids areas with a high density of mucous cells. Elevated levels of complement (C3) have been reported in blood and mucus of salmon (*Salmo salar*) resistant to *G. salaris* (see Bakke *et al.*, 2002).

After eliminating a primary infestation many fishes are resistant to the establishment of a secondary infestation for a few weeks (see Kearn, 1998). We tend to associate such a pattern of acquired immunity with blood-mediated production of specific antibodies in response to primary invasion by a parasite species (see Chapter 1). This mechanism ensures that hosts are prepared in advance to counter any further invasion by the same parasite species. However, attempts to immunise fishes against gyrodactylids by injecting antigens from gyrodactylids into them, have not been successful, the treated hosts failing to become resistant (see, for example, Lindenstrøm & Buchmann, 2000). The mechanism involved in the acquisition of resistance against gyrodactylids remains to be elucidated.

In this context it is interesting to note that the primary response of fishes to gyrodactylids, i.e. the response of so-called 'naïve' (previously uninfected) fishes, appears to be non-specific. Richards & Chubb (1996) found that guppies (*Poecilia reticulata*) that have supported a primary infestation with *G. turnbulli* are partially resistant to *G. bullatarudis* and *vice versa*. On a broader front, Larsen *et al.* (2002) found that brown trout fry (*Salmo trutta*) infected with live nematode larvae (*Anisakis* sp.) in their body cavities, had significantly lower levels of infestation with *G. derjavini*. Thus, activation of the fish's immune system by a totally unrelated parasite appears to generate a non-specific response capable of eliminating or controlling gyrodactylids.

4.6.4 *Transmission to new hosts*
Entobdella soleae can infect new hosts (*Solea solea*) by two routes: 1) by way of freely disseminated eggs and the ciliated, free-swimming larvae that emerge from them and 2) by direct transfer of juvenile and adult parasites from host to host when hosts make physical contact. Gyrodactylids have abandoned the first option and rely heavily but not entirely on the second option (Figure 4.10). The numbers of parasites on individual hosts increase rapidly, partly as a result of the telescoping of generations and partly because they do not face the hazards to which free-swimming larvae are exposed. The dispersion of this burgeoning population of parasites over the fish's body will ensure that the parasites are able to take advantage of any host contacts, wherever and whenever they occur, for transmission to new hosts.

Gyrodactylids have anterior adhesive glands with ducts opening into two eversible sacs, one on each of the two protruding head lobes (Figure 4.9). These lobes are thrust forward when the parasite performs 'searching' movements and provide firm but temporary attachment to a new host. Together with the haptor they also provide alternate attachment of the body extremities during leech-like locomotion.

Each anterior head lobe of a gyrodactylid terminates in a projecting 'spike organ' (Figure 4.11), which is assumed to have a sensory function. This spike is dorsal to the adhesive sac, about 4 µm long and is retractable. It consists of a bundle of 10 modified cilia, each separately innervated but exposed at the tip. Lyons (1969) believed that the spikes detect chemicals. They are situated at the anteriormost tip of the parasite and are protruded at times when the parasite is 'searching', i.e. swinging the extended body from side to side with the haptor attached. They seem unlikely to be touch-sensitive organs because they are retracted when the head lobes make contact with the substrate during locomotion. Each spike is surrounded by at least five permanently extended single sensilla that seem more likely to be touch receptors. These features all indicate that these prominent 'spike' organs may be chemosensory structures used in host discrimination.

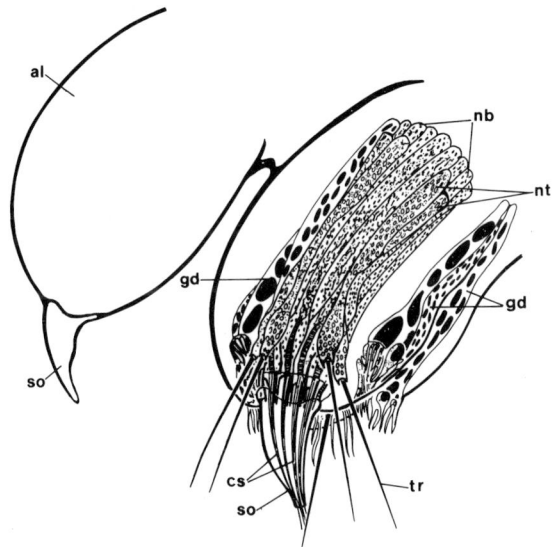

Fig. 4.11. Diagrammatic reconstruction of the anterior head lobes (al) of *Gyrodactylus*. Details of the 'spike' organ (so), associated touch receptors (tr) and ducts of the adhesive glands (gd) are shown for the right lobe only. cs, Sensory cilia of 'spike' organ; nb, nerve bundle to 'spike' organ; nt, nerves to touch receptors. From Lyons, 1969, with permission of Cambridge University Press.

The importance of contacts between hosts for transmission of gyrodactylids is emphasised by the reductions in transmission that occur when a barrier such as a net is erected, which prevents contact between infected and uninfected hosts but not passage of parasites (references in Bakke *et al.*, 1992). However, a small proportion of parasites do succeed in transferring in these situations, presumably by crawling through the mesh and waiting on the bottom until picked up by the recipient hosts. Harris (1982 in Bakke *et al.*, 1992) estimated that 10% of transmission occurs by this indirect route in *G. gasterostei*. Transmission via the substrate may be especially important for bottom-dwelling hosts, while host-to-host contacts seem likely to be more significant for pelagic hosts. However, gyrodactylids establishing themselves on fishes caged in mid-water presumably drift to these fishes after detachment from freely ranging infected hosts (see Bakke *et al.*, 1992). *Gyrodactylus rysavyi* from the Nile catfish, *Clarias gariepinus*, is capable of directional swimming by flexing the body (El-Naggar *et al.*, 2004).

Cable *et al.* (2002b) described a novel transmission mechanism in *G. turnbulli* parasitising guppies (*Poecilia reticulata*). Recently dead fishes typically float at the surface and their parasites abandon the corpses and hang motionless by the haptor from the surface film. This behaviour increases the likelihood that the detached parasites will meet a new, living host since guppies are surface feeders.

Successful re-establishment of detached gyrodactylids on new hosts depends on their ability to survive starvation. Cable *et al.* (2002a) found that starved *G. gasterostei* begin feeding within one minute after reattachment to a three-spined stickleback (*Gasterosteus aculeatus*), and the ultrastructural evidence indicated that the normal digestive processes in the intestinal lining resumed. However, periods of starvation in excess of 24 hours induced pathological changes in the gut lining and dramatic changes in the embryo, indicating that, in these circumstances, resources are diverted from supporting embryo development to survival and maintenance of the parent. At 10 °C, detached *Gyrodactylus gasterostei* may survive for up to 89 hours, but there is no guarantee that parasites are capable of re-establishment on new hosts after prolonged starvation.

Physical contact between fishes of the same species takes place during mating, during agonistic displays and while shoaling. Opportunities for transfer of parasites from one species of host to a different species (interspecific transfer) are less common, occurring in mixed-species shoals and in interactions between predators and prey. Scavenging on dead fishes may also encourage interspecific transmission and the record of the salmon parasite *G. salaris* on the flounder, *Platichthys flesus*, may be an example of this (Mo, 1987, in Bakke *et al.*, 1992). However, the frequency of interspecific transfers may have been underestimated. Some authors, such as Malmberg (1970, in Bakke *et al.*, 1992), regarded gyrodactylids as narrowly host specific, and this appeared to be confirmed by Bakke *et al.* (1992) who found that 74% of what they regarded as 319 valid species of *Gyrodactylus* occurred on a single host species. However, when Bakke *et al.* (1992) revised their estimate by restricting their calculation to 76 gyrodactylid species that had been more intensively studied in the field and in the laboratory, only 30% were recorded from a single host species. Some gyrodactylids are capable of reproducing on several host genera (e.g. *G. salaris* on *Salmo, Oncorhynchus, Salvelinus, Thymallus*), while on others they may be unable to reproduce (e.g. *G. salaris* on eel, *Anguilla anguilla*). Nevertheless, hosts on which reproduction cannot occur may still have a role in dispersal of parasites (*G. salaris* survives on the eel for 8 days and is probably unable to feed on this host).

There are indications that behavioural flexibility on the part of gyrodactylids may be important in the transmission process. In *G. salaris*, parasites that had not yet given birth for the first time were less likely to transfer to a new host than worms that had already given birth at least once (Harris, 1993; Harris *et al.*, 1994). There is evidence too that when a host dies, the behaviour of its gyrodactylids changes (Bakke *et al.*, 1992). Transmission from dead hosts appears to be more efficient than from living ones, at least in *G. salaris* and *G. bullatarudis* (references in Bakke *et al.*, 1992), and detached specimens and specimens attached to dead hosts increase their 'searching' activity and reduce their host-preference selectivity (Bakke *et al.*, 1992). Gyrodactylid species that inhabit the gill chamber or other internal sites must leave these sites and migrate onto the skin to take advantage of host-to-host contacts for transmission.

4.6.5 Host specificity

Gyrodactylid parasites of salmonids, and particularly *G. salaris*, have been studied intensively in recent years and this work has greatly contributed to our understanding of host specificity. These findings have been summarised in the excellent review of Bakke *et al.* (2002).

Gyrodactylids are limited by host responses that reduce or eliminate their infrapopulations. These responses lie on a continuum ranging from susceptibility, in which the parasite population increases unchecked until the host dies, through ability to respond, in which initial growth of the parasite population is stopped, to innate resistance, in which the parasite population never grows and sooner or later disappears. The mechanism underlying this continuum of response appears to be the same, at least in salmonids, but there is great variation in the kind of response mounted by individual fishes belonging either to the same strain or to different host species. This variation seems to be genetically controlled, probably by a single polygenic mechanism, innate resistance being a more vigorously expressed version of the same genotype as that of a fish responding to infection. Individual variation of host response to *G. salaris* occurs in salmon strains (*Salmo salar*) of Norwegian and Baltic origins (see below), but in the Norwegian strains, the 'centre of gravity' on the response continuum is shifted towards the susceptible end while in the Baltic strains it is nearer the resistant end. This host response mechanism, which is still poorly understood, will determine whether or not the fish will support an infection, i.e. it will determine the host specificity of the parasite.

The suitability of the epidermis of a potential host fish as a food source may contribute to host specificity in gyrodactylids. Using electron microscopy, Cable & Harris (unpublished, in Bakke *et al.*, 2002) detected differences in the digestive cycles of *G. gasterostei* on its natural host, the three-spined stickleback, *Gasterosteus aculeatus*, and on the nine-spined stickleback, *Pungitius pungitius*. Although experimental starvation has significant effects on the survival and reproduction of gyrodactylids, parasites on innately resistant hosts do not seem to experience direct starvation, since *Gyrodactylus salaris* kept for six days on innately resistant brown trout reproduced at normal rates when returned to salmon (Bakke *et al.*, 2002). Thus food quality rather than overt starvation is likely to determine the success of parasites on different hosts.

4.6.6 *Gyrodactylids in Britain*

The gyrodactylid fauna of Britain has received less attention than it deserves. Only 20 species of British gyrodactylids from 10 freshwater and two marine fish species are listed in Table 4.1, all except one attributed to *Gyrodactylus*. This is not likely to be a reflection of the paucity of our gyrodactylid fauna, but a general lack of enthusiasm on the part of most British parasite taxonomists for studying these small parasites. The reason for this is the morphological conservatism of the group, species being distinguished by relatively minor morphometric differences and by subtle differences in shapes of the sclerites. Morphological variation and lack of specificity may add further complications by blurring the limits of species. It has also been reported that the hard parts of the haptor of the same species may differ markedly in size depending on the environmental temperature (Ergens, 1976; Ergens & Gelnar, 1985).

The paucity of our knowledge of the British gyrodactylid fauna and the rewards that diligent research on these animals offers, are emphasised by two studies. Harris (1985) collected gyrodactylids from five species of freshwater fishes from just two sites in West Sussex (Table 4.1). He identified 11 species of *Gyrodactylus*, one of which (*G. rogatensis*) was a previously undescribed species; four were new records for Britain and one (*G. pungitii*) was recorded from its type host for the first time in Britain. The second study, by Huyse & Volckaert (2002), is concerned with gobies, *Pomatoschistus*, collected in the English Channel and on the continental shores of the North Sea. *Gyrodactylus rugiensis* was originally described from *P. minutus* and *P. microps*, but

Table 4.1. Some British fishes and their gyrodactylid monogeneans. 'G' = Gyrodactylus; 'Gy' = Gyrodactyloides. Gill parasites are also included since they feed on epithelium like their skin-parasitic relatives and have no obvious adaptations for life on the gills (see also Chapter 5).

Host	Parasite	Site on host	Locality	Authority
Eel (*Anguilla anguilla*)	*G. anguillae*	Gills	Devon	Kennedy & Di Cave, 1998
Atlantic salmon (*Salmo salar*)	*G. derjavini*	Skin	Scotland	Shinn *et al.*, 1995, 1998
	G. caledoniensis	Skin	Scotland	Shinn *et al.*, 1995, 2001
	Gy. bychowskii	Gills	Scotland	Bruno *et al.*, 2001
Brown trout (*Salmo trutta*)	*G. truttae*	Skin	Scotland	Shinn *et al.*, 1995, 1998
Grayling (*Thymallus thymallus*)	*G. thymalli*	Fins	Various, England	Denham & Long, 1999
Pike (*Esox lucius*)	*G. lucii*	Fins	Hampshire, Wiltshire	Denham & Long, 1999
Minnow (*Phoxinus phoxinus*)	*G. aphyae*	Skin, fins	Wales	Chubb, 1964
			Sussex	Harris, 1985
	G. laevis	Gills	Wales	Chubb, 1964
			Sussex	Harris, 1985
	G. limneus	Skin, fins	Wales	Chubb, 1964
			Sussex	Harris, 1985
	G. macronychus	Skin, fins	Wales	Chubb, 1964
			Sussex	Harris, 1985
	G. minimus	Gills	Sussex, Bedfordshire	Harris, 1985
Stone loach (*Noemacheilus barbatulus*)	*G. pavlovskyi*	Skin, fins	Sussex	Harris, 1985
	G. sedelnikowi	Skin, fins	Sussex	Harris, 1985
Five-bearded rockling (*Ciliata mustela*)	*G. medius**	Skin, gills	Swansea	Srivastava & James, 1967
Three-spined stickleback (*Gasterosteus aculeatus*)	*G. arcuatus*	Gill arches, pharynx lining, skin	Various, England, Wales, Scotland	Harris, 1985
	G. gasterostei	Skin, fins	Wales	Powell, 1966 (in Harris, 1985)
			Sussex	Harris, 1985
Nine-spined stickleback (*Pungitius pungitius*)	*G. pungitii*	Skin, fins	Sussex	Harris, 1985
Bullhead (*Cottus gobio*)	*G. rogatensis*	Skin, fins	Sussex	Harris, 1985
Plaice (*Pleuronectes platessa*)	*G. unicopula*	Gills	Scotland	MacKenzie, 1970
Leopard-spotted goby (*Thorogobius ephippiatus*)	*G. quadratidigitatus*	Skin, fins, gills	Dorset	Longshaw *et al.*, 2003

*May be an undescribed species, according to Bakke *et al.*, 2002.

Huyse & Volckaert recorded two new host species, namely *P. pictus* and *P. lozanoi*. However, their molecular and morphometric analyses revealed that the parasites on these four fish species are not genetically uniform and conceal two cryptic species, namely *G. rugiensis* on *P. microps* and a new species, *G. rugiensoides*, on the other three hosts. There is also evidence of a genetic distinction between parasites identified as *G. micropsi* on *P. microps* and *G. micropsi*–like parasites on *P. minutus* and *P. lozanoi*. *G. longidactylus* is an additional species, which seems to be strictly specific to *P. lozanoi*. Huyse *et al.* (2003) reported the presence of other gyrodactylid species on gobies, but these have not yet been described and named. Since *P. minutus*, *P. microps* and *P. pictus* are present in British coastal waters and *P. lozanoi*, itself a cryptic species, is likely to be present, it is conceivable that their gyrodactylid parasites are also present.

According to Huyse *et al.* (2003), *G. anguillae* on the eel (*Anguilla anguilla*), may be derived relatively recently by host switching from gyrodactylids on European gobies (*Pomatoschistus*). Kennedy & Di Cave (1998) believe that this British parasite has been ousted from the gills of the eel by competition with introduced gill-parasitic monogeneans belonging to *Pseudodactylogyrus*.

Switching of parasites between unrelated hosts, a phenomenon that has already been described in the polystomatids and in *Entobdella* (Chapter 3), appears to have been especially common in gyrodactylids (see Bakke *et al.*, 2002; Ziętara & Lumme, 2002; Huyse *et al.*, 2003). The most spectacular example is *Isancistrum*, the ancestors of which switched from fishes to cephalopod molluscs. There are other less spectacular but equally interesting examples. Co-evolution (phylogenetic speciation) is indicated by the primary distribution of species belonging to the so-called *G. wageneri* group of gyrodactylids on cyprinid teleosts, but host switching has also taken place from cyprinids to gasterosteids, percids and cottids sharing the same habitat (see Harris, 1985, 1993). Some of the recipient hosts in these switching events are predators feeding on cyprinids, and predator-prey contacts may provide important opportunities for host switching.

Harris (1985) emphasised the similarity between the gyrodactylids of stenohaline fishes, such as the minnow (*Phoxinus phoxinus*) and the stone loach (*Noemacheilus barbatulus*) in Britain, and those from western Europe. This corroborates the hypothesis that these fishes colonised eastern England via the Doggerland land bridge, immediately after the last glaciation (see Chapter 1). Harris also interpreted the presence of *G. gasterostei* and *G. pungitii* on the sticklebacks *Gasterosteus aculeatus* and *Pungitius pungitius* respectively, as evidence that, at least in southern England, these sticklebacks colonised Britain via the same land bridge. Wootton (1976) suggested that these euryhaline fishes colonised Britain from the sea, but if these two gyrodactylids turn out to be stenohaline, like many of their relatives, it is difficult to understand how these parasites could have survived their hosts' migrations prior to colonising Britain.

4.6.7 *The threat of* Gyrodactylus salaris
Unfortunately some hosts fail to limit infrapopulation growth of gyrodactylids and are eventually overwhelmed and die. *Gyrodactylus salaris* has achieved great notoriety because of the damage it has done in Norway to salmon parr, both in fish farms and in the wild.

G. salaris was first detected in Norway on the Atlantic salmon, *Salmo salar*, in 1975 (Johnsen, 1978). The epidemic in the Norwegian salmon strains probably originated with several introductions of *G. salaris* on the resistant Baltic salmon in the 1970s (Bakke *et al.*, 2002). In 1990, 34 infected rivers in Norway and about 35 infected hatcheries were reported (Anonymous, 1990). Papers published in the 1980s (see Bakke *et al.*, 1992) estimated that Norwegian catches of adult salmon returning to spawn in the rivers had declined by up to 520 tonnes per year (20% of the total catch). *G. salaris*, together with 'acid rain', took the blame for this decline (Johnsen & Jensen, 1986, 1988). Individual salmon parr, which rarely exceed 15 cm in length, may support in excess of 10,000 parasites (Sterud *et al.*, 1998). In heavily infected fishes, Sterud *et al.* recorded a marked decline in host epidermal thickness and in mucous cell density, which they attributed to the heavy grazing pressure of the parasites. These changes in epidermal structure are likely to be linked with the pathogenicity of *G. salaris*. Epidermal damage disturbs the osmotic integrity of the fish and promotes invasion by opportunistic secondary pathogens such as the fungus *Saprolegnia*, leading ultimately to death.

The parasite is now present on the Swedish west coast, on the Arctic White Sea coast of Russia, in Finland, Denmark, Germany and Spain (Bakke *et al.*, 2002). Bakke & MacKenzie (1993) showed that Scottish pre-smolt salmon are as susceptible as Norwegian salmon to *G. salaris* and suffer similar levels of mortality, but, in 1989-1991 Shinn (in Bakke & MacKenzie, 1993) examined more than 2000 salmonid fish from over 250 localities around Britain and found no *G. salaris*. There can be little doubt that the introduction of this parasite into Britain would have disastrous consequences for the salmon farming industry and probably also for wild populations. Therefore, the frequent importation of fish and fish products into Britain, generated by salmonid aquaculture in mainland Europe, poses a real threat and Bakke & MacKenzie recommended that all possible measures should be taken to avoid introduction of this parasite species.

5

MONOGENEAN GILL PARASITES – MONOPISTHOCOTYLEANS

5.1 INTRODUCTION

The buccal and gill chambers of fishes have been colonised several times by monogeneans, suggesting that these sites offer substantial improvements in life style for skin parasites. However, it is by no means clear what these advantages are. Ready access to host blood does not seem to have been a significant factor, since, with the exception of the polyopisthocotylean monogeneans, most monogenean gill parasites are epidermis feeders. Predation pressure from 'cleaner' fishes and other organisms that subsist on a diet of fish ectoparasites, may be less than on the body surface, especially in the gill chamber, and there will be more opportunity to meet conspecifics for mating in the more compact living space. In general, no special challenge faces monogenean colonisers from the outer skin surface. The buccal and branchial cavities offer extensive, relatively flat areas of epidermis for attachment and feeding, and ambient water currents are not necessarily stronger than they would be on the skin surface of a rapidly moving fish, although gill-ventilating currents are likely to be more or less continuous. This situation is reflected in the lack of obvious specialisations in many monopisthocotylean monogeneans (gyrodactylids; capsalids) living in the buccal and gill chambers. In contrast, the haptors of the 'dactylogyroideans' (see Appendix 2) and the polyopisthocotyleans have undergone extensive morphological adaptations for life on the gills.

Like the capsalids, 'dactylogyroideans' have two pairs of large hooks or hamuli, but in typical 'dactylogyroideans' only two of the four hamuli point in a ventral or ventro-lateral direction, the other two pointing dorsally or dorso-laterally (Figure 5.2). Some 'dactylogyroideans' have undergone extensive speciation in parallel with the enormous expansion and diversification of teleost fishes. Particularly spectacular is the 'explosive' expansion on cyprinid fishes of the dactylogyrines (*Dactylogyrus*, *Neodactylogyrus*) in which the ventrally orientated hamuli are reduced. In fact Gibson, Timofeeva & Gerasev (1996), following Gusev (1985) in rejecting *Neodactylogyrus* in favour of *Dactylogyrus*, listed over 900 nominal species of *Dactylogyrus* worldwide, compared with a mere 400 or so in *Gyrodactylus* (see Chapter 4). This makes *Dactylogyrus* the largest helminth genus.

The polyopisthocotyleans have evolved unique organs functioning as clamps for attachment to the gills and, in addition, have undergone changes related to their exploitation of blood as a source of food. All of these changes set the polyopisthocotyleans apart from the monopisthocotyleans and earn them a separate place in this book (Chapter 6).

5.2 UNSPECIALISED GILL PARASITES

Gyrodactylids are small, agile, epidermis-feeders, for which the buccal and gill chambers may simply be extensions of the host's body surface. Even those gyrodactylids that spend most of their lives in the gill chamber (see Table 4.1) continue to feed on epidermis and have no obvious haptoral modifications. Behavioural adaptations seem more likely, since the parasites may have to leave the gill chamber at some stage in order to take advantage of host-to-host contacts for transmission. However, knowledge of such behavioural events is limited and there is little to add on gill parasitism to the account of gyrodactylids as skin parasites given in Chapter 4.

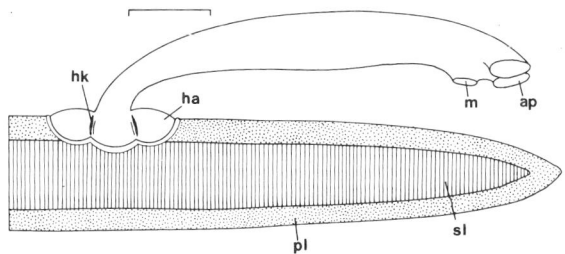

Fig. 5.1. Adhesive attitude of an adult specimen of the capsalid monogenean *Trochopus pini.* ap, Anterior adhesive pad; ha, haptor; hk, hooks (hamuli); m, mouth; pl, edge of primary lamella; sl, secondary lamellae. Scale bar: 1 mm. From Kearn, 1971b, with permission of Cambridge University Press.

Some capsalids have established themselves in the gill chamber, but, like the gyrodactylids, have undergone little modification for life on the gills. In British waters *Trochopus pini* is found on the tub gurnard (*Trigla hirundo* = *T. lucerna*). It is readily recognisable as a relative of *Entobdella soleae*, having a disk-shaped haptor with similar hook armature and an epidermis-feeding habit (Kearn, 1971b). However, the versatility and physical flexibility of its haptor enable it to occupy a specialised niche. Like *Capsala* the haptor is subdivided into separate muscular loculi, which operate as independent suctorial units and by folding the haptor transversely it is able to mould itself to the curved border of a primary gill lamella, without encroaching on the respiratory surfaces (Figure 5.1). *Trochopus* retains the ability to attach itself to flat surfaces, unfolding the haptor to make this possible (Kearn, 1971b).

5.3 'DACTYLOGYROIDEANS'

5.3.1 *The British fauna*
Five families of 'dactylogyroideans' have representatives on British fishes (Table 5.1), but the meagre (*Argyrosomus regius*), host of the calceostomatid *Calceostoma* sp., is rare and the parasite is unlikely to be encountered in British waters.

Interest in dactylogyrines in Britain was late in developing compared with eastern Europe, where the popular practice of farming carp was threatened by the activities of these parasites. In fact in 1947 Dawes reported the complete absence of species of *Dactylogyrus* or *Neodactylogyrus* in Britain, but it emerged later that this was far from the truth. The lack of records was merely a reflection of lack of parasitological effort and with the upsurge of interest in freshwater fish parasites in the 1960s and 1970s dactylogyrines were found to be widespread (Table 5.2).

Table 5.1. 'Dactylogyroideans' on British freshwater and marine fishes.

Parasite family or sub-family	Parasite species	Host	Authority
Tetraonchidae	Tetraonchus monenteron	Pike (Esox lucius)	Chubb, 1970b
	Tetraonchus borealis	Grayling (Thymallus thymallus)	Davies, 1967 (in Kennedy, 1974)
Diplectanidae	Diplectanum aequans	Bass (Dicentrarchus labrax)	Dawes, 1947
	Pseudodiplectanum kearni	Thickback sole (Microchirus variegatus)	Oliver, 1980
Dactylogyridae			
(a) Dactylogyrinae	Dactylogyrus spp.	See Table 5.2	
	Neodactylogyrus spp.	See Table 5.2	
(b) Ancyrocephalinae	Ergenstrema labrosi	Thick-lipped grey mullet (Chelon labrosus)	Anderson, 1981ab
	Ligophorus angustus		
	Ergenstrema mugilis	Thin-lipped grey mullet (Liza ramada)	Anderson, 1981a
	Ancyrocephalus paradoxus	Perch (Perca fluviatilis)	Dawes, 1947
Calceostomatidae	Calceostoma sp.	Meagre (Argyrosomus regius)*	Llewellyn et al., 1984
Amphibdellidae	Amphibdelloides maccallumi	Electric ray (Torpedo nobiliana)	Llewellyn, 1960
	Amphibdella flavolineata		

* Rare in British waters.

Table 5.2 is unlikely to include all the British dactylogyrines. Several British cyprinids are not included in Table 5.2 and since these fishes support other dactylogyrines in continental Europe, it is possible, indeed probable, that some of these parasites await discovery in Britain. *Dactylogyrus vastator* as its name implies is particularly damaging to common carp fry (*Cyprinus carpio*). It has been reported in Ireland by Kane (1966) but has not so far been found in Britain, although *D. anchoratus* and *D. extensus* are present here on carp (Table 5.2).

It is worth noting that ruffe, *Gymnocephalus cernuus*, host to *D. amphibothrium* and *D. hemiamphibothrium*, is a percid not a cyprinid fish. This relationship may be the consequence of a single host-switching event from a cyprinid fish to the ruffe, followed by speciation of the parasite on its percid host.

Prior to 1981 there were no records of 'dactylogyroidean' parasites on mullets caught in British waters. Then Anderson (1981b) examined adult thick-lipped grey mullets, *Chelon labrosus*, at Plymouth and found the gills to be infested exclusively with a 'dactylogyroidean' (ancyrocephaline) monogenean, *Ligophorus angustus*. At a later date she also examined some juvenile *C. labrosus* and although ancyrocephaline monogeneans were present, quite unexpectedly they turned out to be exclusively a previously undescribed parasite, which Anderson (1981a) named *Ergenstrema labrosi*. Anderson explored the reasons for this dramatic change of parasites and her findings will be considered below. Anderson (1981a) also recorded *E. mugilis* for the first time in Britain on thin-lipped grey mullets, *Liza ramada*, from the estuary of the River Yealm in south Devon. Llewellyn *et al.* (1984) omitted *E. mugilis* from their checklist of monogenean parasites of Plymouth hosts.

Table 5.2. Dactylogyrine monogeneans recorded from the gills of British freshwater fishes

Parasite	Host(s)	Authority
Dactylogyrus (single haptoral bar)		
D. amphibothrium	Ruffe (*Gymnocephalus cernuus*)	Kearn, 1968
D. anchoratus	Common carp (*Cyprinus carpio*)	Shillcock, 1972 (in Kennedy, 1974)
D. auriculatus	Bream (*Abramis brama*)	Kearn, 1968
D. cordus	Dace (*Leuciscus leuciscus*)	Davies, 1967 (in Kennedy, 1974)
D. extensus	Common carp (*Cyprinus carpio*)	Said, personal communication
D. gobii	Gudgeon (*Gobio gobio*)	Shillcock, 1972 (in Kennedy, 1974)
D. hemiamphibothrium	Ruffe (*Gymnocephalus cernuus*)	El-Naggar & Kearn, 1980
D. similis	Roach (*Rutilus rutilus*)	Rizvi, 1964 (in Chubb, 1965)
D. sphyrna	Roach (*Rutilus rutilus*)	Rizvi, 1964 (in Chubb, 1965)
D. vistulae	Chub (*Leuciscus cephalus*)	Davies, 1967 (in Kennedy, 1974)
	Roach (*Rutilus rutilus*)	
	Rudd (*Scardinius erythrophthalmus*)	Kennedy, 1975
Neodactylogyrus* (two haptoral bars)		
N. crucifer	Roach (*Rutilus rutilus*)	Rizvi, 1964 (in Chubb, 1965)
N. folkmanovae	Chub (*Leuciscus cephalus*)	Davies, 1967 (in Kennedy, 1974)
N. nanus	Roach (*Rutilus rutilus*)	Davies, 1967 (in Kennedy, 1974)
N. phoxini	Minnow (*Phoxinus phoxinus*)	Shillcock, 1972 (in Kennedy, 1974)
N. prostae	Chub (*Leuciscus cephalus*)	Davies, 1967 (in Kennedy, 1974)
N. suecicus	Roach (*Rutilus rutilus*)	Mishra & Chubb, 1969
N. tuba	Dace (*Leuciscus leuciscus*)	Davies, 1967 (in Kennedy, 1974)
N. wunderi	Bream (*Abramis brama*)	Mishra & Chubb, 1969

* *Neodactylogyrus* has been rejected in favour of *Dactylogyrus* by Gusev (1985) (see Gibson *et al.*, 1996).

Elasmobranch fishes have almost entirely escaped the attentions of dactylogyroideans. The exceptions are the amphibdellids on the gills of electric rays, species of *Torpedo* (Table 5.1). The ancestors of these parasites most probably colonised their elasmobranch hosts by host switching from a teleost.

5.3.2 Attachment
The dorsally pointing orientation of two of the four hamuli of typical 'dactylogyroideans' (Figure 5.2) is an adaptation for attachment between two adjacent secondary gill lamellae. Moreover, in many 'dactylogyroideans' two pairs of hooklets have also been relocated on the dorsal surface of the haptor (Figure 5.2). The ventrally pointing hamuli and hooklets attach the parasite to one secondary gill lamella, while the dorsally pointing hamuli and hooklets attach the parasite to the adjacent secondary lamella (Figure 5.3).

This development has created an opportunity for the parasites to exploit a large number of attachment sites within the gill cavity, but, because these sites are small, there is a restriction on the overall size of 'dactylogyroideans'. Consequently, although a fish may be infected with very large numbers (Llewellyn, 1960, recorded more than 2000 specimens of *Amphibdelloides maccallumi* on the gills of a single electric ray, *Torpedo nobiliana*), they are individually small, and a good lens or low power microscope is needed to see them.

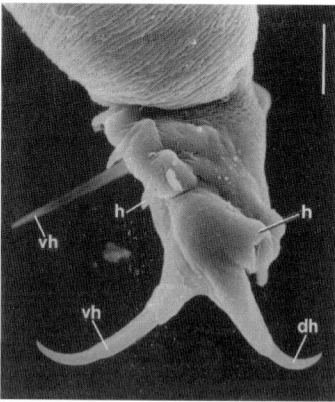

Fig. 5.2. Scanning electron microscope view of the left side of the haptor of the 'dactylogyroidean' monogenean *Tetraonchus monenteron* showing the orientation of the hamuli and hooklet-bearing papillae. dh, Dorsal hamulus; h, hooklet-bearing papilla; vh, ventral hamulus. The dorsal hamulus on the parasite's right side is hidden by the haptor. Scale bar: 10 µm. Photograph by R. Evans-Gowing.

'Dactylogyroidean' hamuli are superbly shaped for the task they have to perform, with broad roots for the attachment of muscles and tendons, merging into a beautifully curved and tapering hook capable of protrusion from the haptor (Figures 5.2, 5.4). Until quite late in the last century the shapes and sizes of these hooks were the

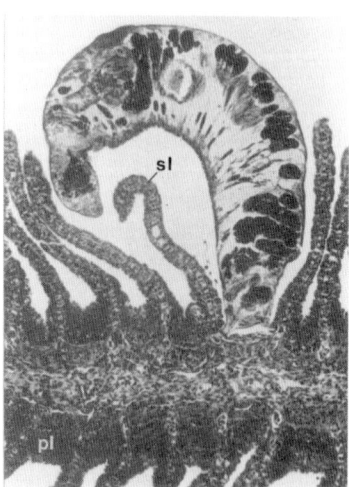

Fig. 5.3. A section through *Amphibdelloides maccallumi* attached between two secondary gill lamellae (sl) of the electric ray, *Torpedo nobiliana*. pl, Primary lamella. From Llewellyn, 1960, with permission of Cambridge University Press.

main focus for parasitologists and little attention was given to their mode of operation. This attitude stemmed from the popularity of systematics, which demanded strongly flattened specimens for the accurate determination of the shapes and sizes of their sclerites. The shortest route for the systematist was to simultaneously flatten and preserve the parasites. Living unflattened specimens were rarely studied. The consequence of flattening is that the sclerites are forced into unnatural positions, rendering any attempts to relate structure to function difficult.

A significant development occurred in 1960 when Llewellyn published an account of the anatomy and function of the haptor of *Amphibdelloides maccallumi*. His observations were made on living and relaxed specimens, as well as on flattened specimens mounted on slides and he also cut sections through wax-embedded individuals.

(a) Amphibdelloides maccallumi *on the electric ray* Torpedo nobiliana. In *A. maccallumi*, the deeply embedded proximal roots of the four large hamuli are flattened plates (Figure 5.4A). On each side, the flat roots of one ventral and one dorsal hamulus are virtually in contact, with their hooked regions pointing respectively ventrally and

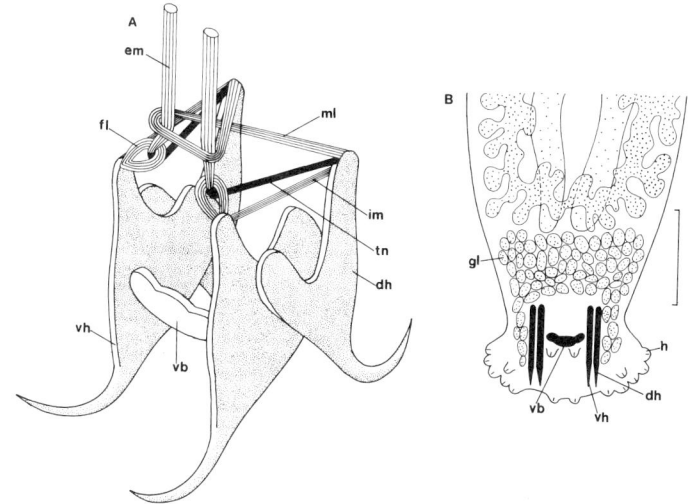

Fig. 5.4. Amphibdelloides maccallumi. (A) Stereogram showing the principal attachment apparatus; (B) the haptor in ventral view. em, Extrinsic muscle; dh, dorsal hamulus; fl, fibrous loop; gl, gland cells; h, hooklet-bearing papilla; im, intrinsic adductor muscle; ml, muscle loop; tn, tendon; vb, ventral bar; vh, ventral hamulus. Scale bar for (B): 200μm. (A) Modified from Llewellyn, 1960; (B) from Llewellyn, 1960, with permission of Cambridge University Press.

dorsally, the dorsal hamulus flanking (lying outside) the ventral hamulus (Figure 5.4). Between the two ventral hamuli is a short transverse bar, which is connected to the tegument by fibres. Fibres also connect the ventral hamuli with the bar and the ventral hamulus on each side is connected in turn to the associated dorsal hamulus. According to Llewellyn (1960), this complicated arrangement of fibres probably fulfils tethering and bracing functions and allows the ventral and dorsal hamuli of each lateral pair to counter-rotate about their regions of mutual attachment in the planes of their flat superimposed plate-like surfaces. The consequence of counter-rotation is that the points of the ventral and dorsal hamuli are thrust outwards from the haptor and impale the secondary gill lamellae in contact with the haptor's ventral and dorsal surfaces (Figure 5.3).

This counter-rotation is brought about in a remarkable way by a pair of extrinsic adductor muscles lying anterior to the haptor and running in a longitudinal direction with respect to the parasite's body (Figure 5.4A). At its posterior end each of these two muscles grades into a tendon, which passes through a fibrous loop attached to the anterior extremity of the outer root of the ventral hamulus. This loop is reminiscent of a pulley or fairlead, since as each tendon passes through the loop, it changes direction by almost 90°. It then runs in a dorsal direction across the haptor and attaches to the anterior extremity of the outer root of the dorsal hamulus. As a consequence of contraction of each extrinsic adductor muscle, the anterior extremities of the outer roots of the ventral and dorsal hamuli will be drawn towards each other, counter-rotating the hamuli and driving the hook points outwards and, according to Llewellyn, often right through the secondary lamellae. Such a sophisticated means of operating the hooks is unexpected in flatworms, which are usually regarded by zoologists as rather low on the evolutionary ladder, but, as we will see shortly, another 'dactylogyroidean' has taken this sophistication a stage further.

The attachment apparatus of *A. maccallumi* has one or two refinements. On each side there is an intrinsic muscle running directly between the anterior extremities of the outer roots of the ventral and dorsal hamuli, i.e. parallel with the dorso-ventrally running distal length of tendon (Figure 5.4A). Contraction of these intrinsic muscles will augment the action of the extrinsic adductor muscles and provide additional thrust to drive the hooks into or through the secondary lamellae. After initial hook penetration, sustained contraction of the extrinsic muscles may no longer be necessary, and the contraction of the intrinsic muscles may be sufficient to hold the hamuli in position, except perhaps when the host is strongly ventilating its gills. A curious muscle loop, attached at each end to the anterior extremity of the outer root of the dorsal hamulus and encircling the two extrinsic muscles near the origins of the tendons, may have a similar function. Contraction of the loop would grip the two extrinsic muscles simulating a 'catch' mechanism, permitting the extrinsic muscles to relax while maintaining the hooks in the protracted position. As Llewellyn (1960) has suggested, the relaxation of the extrinsic adductor muscles would permit exploratory movements of the body proper to be made more freely.

In addition to the four large hamuli, *A. maccallumi* has 16 tiny hooklets (8 pairs), each of which is mounted on a papilla. On each side of the haptor there are five marginal papillae and one submarginal papilla (Figure 5.4B). Two papillae are located centrally on the ventral surface of the haptor and two papillae are located on the posterior margin of the haptor, between the two lateral pairs of hamuli. Llewellyn (1960) described these papillae as muscular and labile. In *Chauhanellus australis* from the Australian blue catfish (*Arius graeffei*), Kearn et al. (2002) observed that the hooklets are particularly active during locomotion, just prior to protrusion of the hamuli from the newly seated haptor. Thus, the hooklets may have an important function in establishing initial attachment of the haptor in a new site. By pinning the secondary lamellae to the haptor surface they will facilitate penetration of the lamellae by the thrusting hamulus points.

(b) Tetraonchus monenteron *on pike*, Esox lucius. *Tetraonchus monenteron* is a common parasite found only on the pike. As its generic name indicates, it has four large hooks (hamuli) and in this and other respects it resembles *A. maccallumi*. Like *A. maccallumi* the hamuli are arranged in lateral pairs, each comprising one ventral and one dorsal hamulus, with the dorsal hamulus lying outside (flanking) the ventral hamulus (Figures 5.2, 5.5, 5.6). The lateral pairs of hamuli are associated with a transverse bar, with the plane of each pair of hamuli usually orientated obliquely as shown in Figure 5.5A. There are 16 hooklets, with a similar distribution to those of *A. maccallumi*, except that two of the six laterally situated hooklets on each side are dorsally orientated (Figures 5.2, 5.5B).

A pair of extrinsic adductor muscles operates the hamuli as in *A. maccallumi*, but there is one important difference. There is a loop, like that of *A. maccallumi* attached to the outer root of the ventral hamulus, but there is a bigger loop attached to the anterior extremity of the dorsal hamulus (Figure 5.6). The tendon from each muscle follows a much more circuitous route in *T. monenteron*. First the tendon passes beneath a band of transverse fibres linking the anterior extremities of the outer roots of the ventral hamuli. Next the tendon threads its way through the dorsal loop and then through the ventral loop, from which it returns to the anterior extremity of the dorsal hamulus and attaches itself close to the dorsal loop.

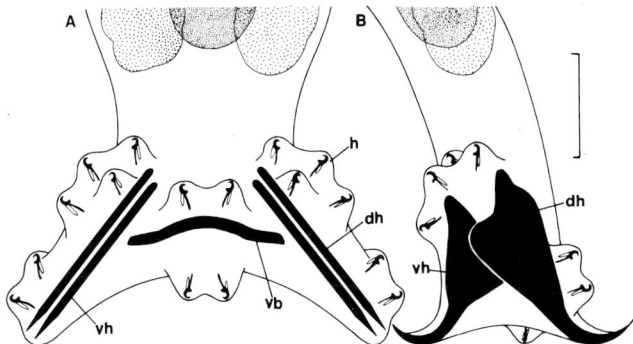

Fig. 5.5. The haptor of *Tetraonchus monenteron* in (A) ventral view and (B) lateral view. dh, Dorsal hamulus; h, hooklet-bearing papilla; vb, ventral bar; vh, ventral hamulus. Scale bar: 50 μm. Modified from Kearn, 1966.

In specimens of *T. monenteron* freshly removed from pike, counter-rotation of the hamuli of each lateral pair, resulting in outward thrusting of the hook points, was frequently observed. This indicates that the laterally situated hamuli are bound to each other as well as to the bar and tegument, in a similar way to *A. maccallumi*. Contraction of each extrinsic muscle will produce this counter-rotation by drawing together the roots of the ventral and dorsal hamuli, thereby forcing the hook points into the secondary gill lamellae. Kearn (1966) observed that in sections through parasites attached to the gill (i.e. with the hamuli fully adducted) the fibrous loops are in contact with each other.

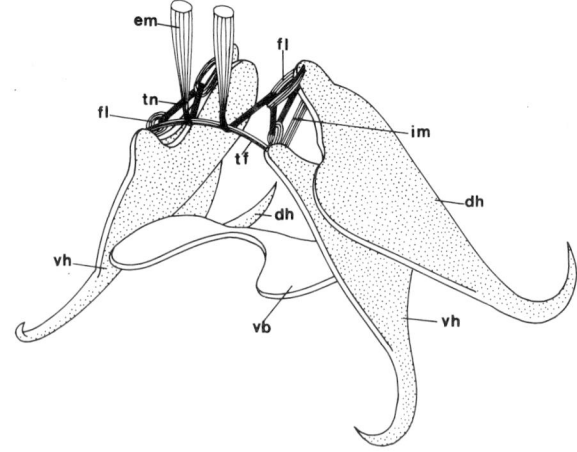

Fig. 5.6. The principal attachment apparatus of *Tetraonchus monenteron*. dh, dorsal hamulus; em, extrinsic adductor muscle; fl, fibrous loop; im, intrinsic adductor muscle; tf, transverse fibres; tn, tendon; vb, ventral bar; vh, ventral hamulus. From Kearn, 1966, with permission of Cambridge University Press.

The additional complexity of the attachment apparatus of *T. monenteron* probably confers a significant mechanical advantage compared with that of *Amphibdelloides maccallumi*. A rough comparison of the mechanical features of the haptors of the two parasites can be made by analogy with systems of pulleys. Kearn (1966) reduced each system to its basic features, namely a single pair of counter-rotating hooks together with their loops, operating muscle and tendon (Figure 5.7). *A. maccallumi* has a single 'pulley' (one loop) and *T. monenteron* three 'pulleys' (two

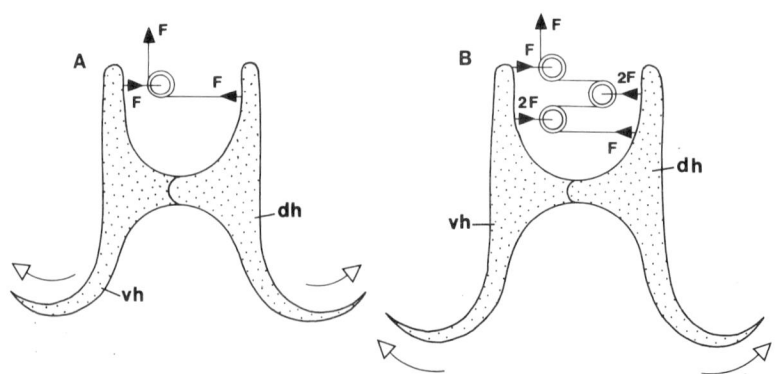

Fig. 5.7. Models representing the attachment apparatus of (A) *Amphibdelloides maccallumi* and (B) *Tetraonchus monenteron*. A force F is exerted on the free end of each tendon. See text for explanation. dh, Dorsal hamulus; vh, ventral hamulus. From Kearn, 1966, with permission of Cambridge University Press.

loops and the transverse fibres beneath which the tendon passes before reaching the loops). Kearn also made the assumption that the tendons lie in the plane of movement of the hamuli, that there are no frictional effects and that the axes of rotation of the hooks are in the same relative positions in the two parasites. In this model, the force acting on each hamulus of *A. maccallumi* will be equal to the force applied by the muscle (Figure 5.7A), but in *T. monenteron* the corresponding force exerted on each hamulus will be three times the force exerted by the muscle (Figure 5.7B). In simple terms, for the same effort the hooks of *T. monenteron* will be inserted with greater force, or, alternatively, less effort will be needed by *T. monenteron* to insert the hooks with the same force.

What possible advantage this could have for *T. monenteron* is obscure. It seems unlikely that the secondary gill lamellae of pike are tougher to penetrate than those of electric rays. However, the system would require a relatively 'weak' tonic contraction of the extrinsic muscles to maintain the hooks in the protracted position.

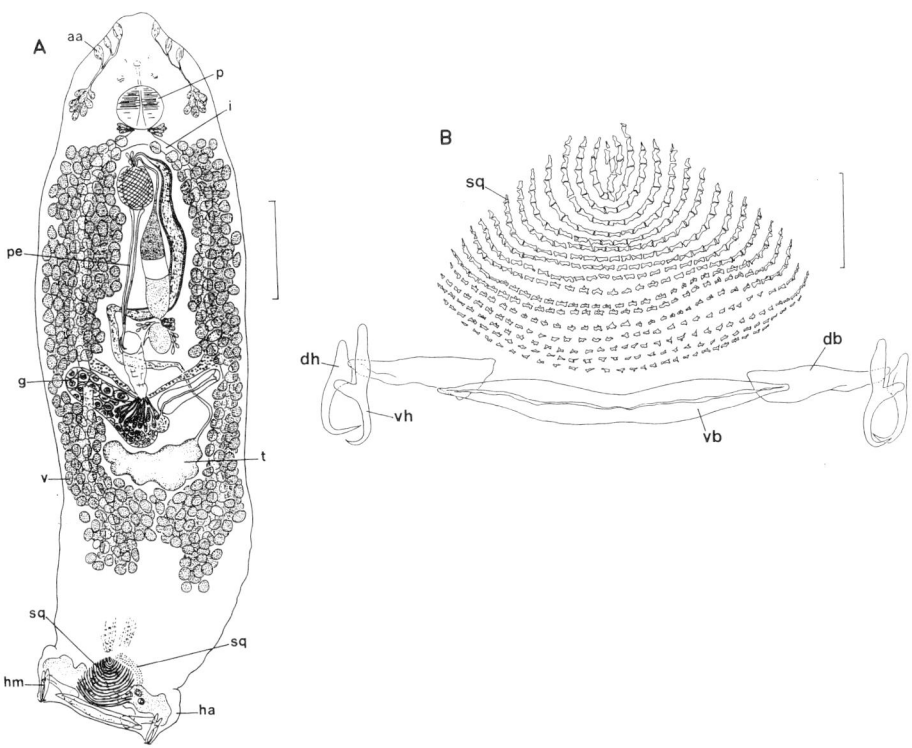

Fig. 5.8. Diplectanum aequans (A) in ventral view (dorsal squamodisc seen by transparency through haptor); (B) haptor sclerites (only one squamodisc is shown). aa, Anterior adhesive sac; db, dorsal bar; dh, dorsal hamulus; g, germarium; ha, haptor; hm, hamuli; i, intestine; p, pharynx; pe, penis; sq, squamodisc; t, testis; v, vitellarium; vb, ventral bar; vh, ventral hamulus. Scale bars: (A) 150 μm; (B) 50 μm. Reproduced with permission from Oliver, G., 1968, Vie et Milieu, Série A, Biologie Marine, volume 19, pp. 95 – 138.

Most 'dactylogyroideans' have glands associated with the haptor (as in *A. maccallumi*, see Figure 5.4B). Kearn & Gowing (1989) found two different glandular systems in *T. monenteron*, one having several openings on the haptor surface and the other releasing its secretion around the points of the hamuli. Presumably these secretions have some role in attachment, perhaps as adhesives, or, in the case of the hamulus secretion, to lubricate the hamulus points or to soften host tissue in preparation for penetration. The haptor is also equipped with a set of short protruding hairs. As the electron microscope reveals, these are modified cilia and are assumed to be sensory, perhaps providing the parasite with information on the position of the haptor during locomotion (see below) and controlling hooklet activity and hamulus protrusion.

(c) Diplectanum aequans *on bass*, Dicentrarchus labrax. The haptor has undergone interesting further developments in the diplectanids. Llewellyn *et al*. (1984) reported two marine representatives in Britain: *Diplectanum aequans* (Figure 5.8) on bass, *Dicentrarchus labrax*, and *Pseudodiplectanum kearni* on the thickback sole, *Microchirus variegatus*. According to Paling (1966) bass may harbour between 30 and 250 parasites per fish.

The most significant feature of these parasites is the presence of disk-shaped friction pads or squamodiscs, one on the ventral side in the region of the haptor peduncle and another identical one on the dorsal side (Figure 5.8). When the parasite is in position with the haptor in the slot between two adjacent secondary lamellae, the pads are in contact with the surfaces of the lamellae and rows of small teeth on the pads resist any tendency for the haptor to withdraw from its slot. The prominence of the squamodiscs is reflected in some displacement and reduction of the hamulus apparatus. The ventral and dorsal hamuli are retained but are relatively small and have been displaced somewhat in a lateral direction by the large centrally located squamodiscs (Figure 5.8B). The two lateral pairs of hamuli maintain their connection with each other via three elongated bars, one ventral and two dorsal. There is no system of extrinsic muscles, tendons and loops operating the hamuli. According to Paling (1966), relatively inconspicuous muscles running between the hamuli and the bars serve to counter-rotate the hamuli and impale gill tissue. In *P. kearni* the reduced ventral hamuli have lost their strongly hooked distal regions.

(d) *Dactylogyrines*. The dactylogyrines have reduced their ventral hamuli to vestigial spicules, which most probably serve no function (Figure 5.9). The dorsal hamuli and their supporting bar remain at full size and fully functional. An increase in the importance of the hooklets, including the development of extensions to their handles, appears to have compensated for the loss of anchorage provided by the demise of the ventral hamuli. In addition, in species of *Neodactylogyrus*, some of the hooklets on the ventral surface are associated with a small ventral bar, which may provide firm attachment for the enhanced musculature operating these hooklets (Kearn, 1968). Whether this ventral bar is a new acquisition, which has appeared since the ventral hooklets gained greater prominence, or whether it is the persistent ventral bar originally associated with the functional ventral hamuli, is an intriguing question.

Most 'dactylogyroideans', including many dactylogyrines, have no obvious host reaction at their site of attachment on the gills. It is likely that these parasites are free to move to a new site, in the manner described below for *T. monenteron*, using their anterior adhesive glands for temporary attachment while disengaging the hamuli and relocating the haptor. However, *Dactylogyrus extensus* on the gills of common carp (*Cyprinus carpio*) evokes a host response in the form of an overgrowth of host tissue at

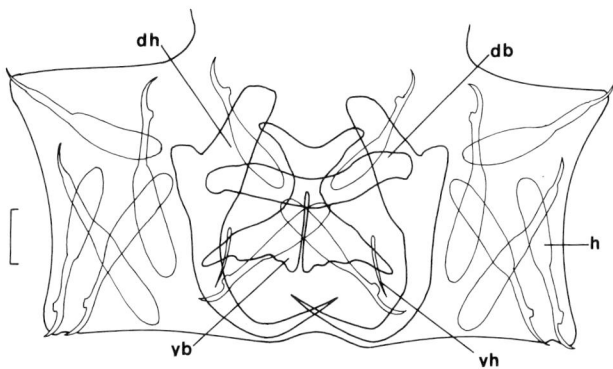

Fig. 5.9. The haptoral sclerites of *Neodactylogyrus crucifer* from roach, *Rutilus rutilus*. db, Dorsal bar; dh, dorsal hamulus; h, hooklet; vb, ventral bar; vh, ventral hamulus (vestigial). Scale bar: 10 μm. From Kearn, 1994, Copyright © 1994, with permission from Elsevier.

the site of attachment. The haptors of young parasites are freely located between the secondary gill lamellae, but as these young parasites grow the gill epithelium undergoes excessive proliferation (hyperplasia). This is eventually so extensive that the secondary lamellae are no longer distinguishable and the haptor of the parasite is entirely enclosed by the tumour-like growth. Curiously, in *D. anchoratus* found on the gills of the same carp host, this does not occur.

Mr Ashraf Said (personal communication) found that some *D. extensus* with encapsulated haptors were dead, suggesting that they remain encapsulated until the end of their lives, but whether the death of the parasite is precipitated prematurely by the host's reaction is not known. It is possible that the vigorous host response reflects a parasite/host relationship that has been established relatively recently, and that, over an extended period of time, adjustment will take place, with weakening of the host's response This may have already happened in *D. anchoratus* from the same host. Alternatively, *D. anchoratus* may escape encapsulation by regularly changing its attachment site.

One other possibility is worth consideration. We may be wrong in assuming that *D. extensus* is a passive focus for the host's anti-parasite response. It may be that *D. extensus* actively provokes the host, the ensuing encapsulation of the haptor ensuring permanent, secure attachment with virtually no expenditure of energy, apart from that required to stimulate the host's defences.

5.3.3 Partial endoparasitism – Ampibdella flavolineata

In addition to *Amphibdelloides maccallumi* on the electric ray *Torpedo nobiliana*, up to six specimens of a larger parasite, *Amphibdella flavolineata*, were found on the same host by Llewellyn (1960). *A. flavolineata* is not attached between secondary gill lamellae. Its body projects from the relatively smooth strip of gill mucosa lying between the interbranchial septum and the zone occupied by the secondary gill lamellae (Figure 5.10). Its bulky haptor is totally buried and expanded in the subcutaneous tissues of the host. The haptor is connected to the projecting body by a narrow peduncle or stalk, which is constricted by the superficial tissues of the host. The haptoral papillae bearing the hooklets are inflated (Figure 5.11) and because they project in various directions provide an effective rooting system. The two median pairs of papillae are especially large, the two posterior median papillae projecting in a posterior direction and the two anterior median papillae extending ventrally. The six anterolateral papillae are carried on an expanded flange that is obliquely orientated.

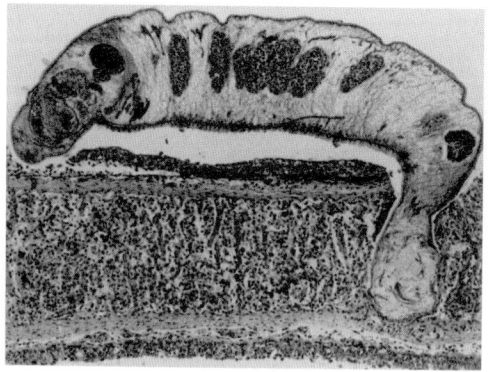

Fig. 5.10. A section through *Amphibdella flavolineata* 'rooted' subcutaneously in the gill tissue of the electric ray, *Torpedo nobiliana*. Cf. Fig. 5.3. From Llewellyn, 1960, with permission of Cambridge University Press.

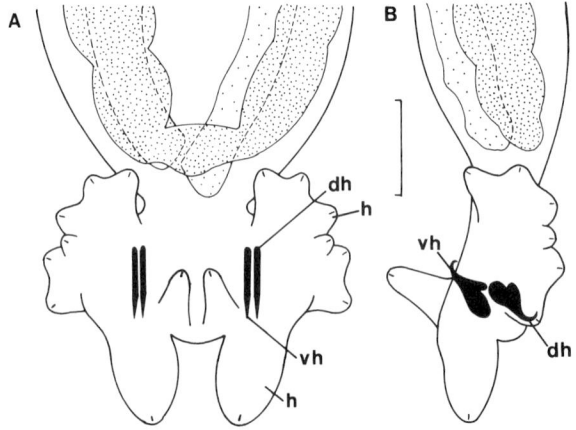

Fig. 5.11. The haptor of *Amphibdella flavolineata* in (A) ventral view and (B) lateral view. dh, Dorsal hamulus; h, hooklet-bearing papilla; vh, ventral hamulus. Cf., Fig. 5.4B. Scale bar: 200μm. From Llewellyn, 1960, with permission of Cambridge University Press.

Firm rooting of the parasite in the gill has permitted reduction of the hamulus apparatus. Ventral and dorsal hamuli are present in *A. flavolineata*, but they are smaller than those of *A. maccallumi*. There are no supporting bars or associated fibres in *A. flavolineata*, no loops attached to the ventral hamuli and no muscles connecting the hamuli of the left side of the parasite with those of the right. The muscles that are present run longitudinally, one relatively weak muscle bundle from each hamulus running posteriorly to the tegument between the posterior median papillae and somewhat larger bundles on each side running anteriorly to beyond the peduncle. Haptor glands are also absent.

There is no hyperplasia associated with the embedded haptor of *A. flavolineata* and, in fact, there is little evidence of any host reaction. Thus the haptor has not become embedded as it has in *Dactylogyrus extensus* by overgrowth of host tissue (see above). A possible explanation is that the haptor has made its own way into the host's tissue, but, as Llewellyn (1960) has pointed out, there are no haptor glands and no other equipment that could be used to tunnel into the gill. There is however, another explanation, which is totally unexpected and surprising. This is that the parasite may have reached its present location from the opposite direction, i.e. from inside the body of the host.

That this is likely to be the correct explanation stems from the finding of juvenile amphibdellids inside the heart of electric rays. Llewellyn (1960) found three apparently unattached juvenile amphibdellids among the loosely woven muscle strands in the ventricle of the heart of the electric ray. Llewellyn compared these with specimens of *A. flavolineata* embedded in the gill mucosa and found them to be similar. The heart parasites differed in their smaller size, in the absence of enlarged hooklet-bearing lobes, in the absence of vitellaria and in the rudimentary state of the germarium. The heart parasites and the gill forms had the following in common: the size and shape of the male copulatory sclerites and features of the male reproductive system and vagina, the shape of the hamuli and the absence of a transverse bar and haptor gland. There can be little doubt that the parasites in the heart were young *A. flavolineata* in which the female system was not yet functional. However, Llewellyn found that in each heart parasite the seminal vesicle was full of sperm, showing that the testis was mature. Protandry, i.e. maturation of the male reproductive system before the female system, is a feature of most monogeneans (but not gyrodactylids; see Chapter 4) and *A. flavolineata* seems to conform to this pattern. Moreover, Llewellyn found sperms in the seminal receptacle of his heart parasites, indicating that copulation had probably already taken place, as it does in protandrous individuals of *Entobdella soleae* (see Chapter 3). Two copulating juveniles of the related parasite *Amphibdella torpedinis* were found in sections through the heart of its Mediterranean host the marbled electric ray, *Torpedo marmorata* (preparation by A. Raibaut of the Station Biologique, Sète, France; reported by Llewellyn, 1960).

Thus the life cycle of *Amphibdella* is, as far as we know, unique among monogeneans, with an endoparasitic phase in the blood system of the host followed by a mesoparasitic phase on the gills. The fact that protandry is a feature of most monogeneans may have been an important factor favouring the development of such a life cycle. The host's nutrient-rich blood will promote parasite development, and in the confined spaces of the blood system opportunities to meet a partner for cross-insemination will be greater. However, eggs laid by the blood-dwelling parasite have no means of escape from the host and are likely to perish. While in the early male phase the parasite can exploit the positive advantages of life in the blood without the danger of wasting eggs. By the time the female reproductive phase is reached and egg laying begins the parasite has established communication with the outside world via the gills and is able to release its eggs into the gill-ventilating current from the emergent anterior end of the body. Ruszkowski (1931) claimed to have found eggs and egg-laying adult specimens of *Amphibdella* in the blood of *Torpedo*. However, Llewellyn (1960) found no eggs or egg-laying adults of *A. flavolineata* and *A. torpedinis* in the blood systems of their respective hosts and suggested that Ruszkowski had chanced upon an incident of precocious egg laying. How these amphibdellids reach the blood system is unknown, but, as Llewellyn (1960) has suggested, it seems most likely that the oncomiracidium gains access by penetrating the thin-walled gill epithelium.

Most monopisthocotyleans appear to be epidermis feeders, including *Amphibdelloides maccallumi* (see Kearn, 1963a), but the blood-dwelling juveniles of *Amphibdella* are likely to be blood feeders. There is no record of the nature of the gut contents of these juveniles, but fully mature, gill-parasitic *Amphibdella torpedinis* were observed to contain red gut contents by Llewellyn (in Kearn, 1963a). Only one out of four preserved specimens of *A. torpedinis*, in which Llewellyn had observed red fluid while alive, was positive when tested spectroscopically by Kearn (1963a) for evidence of host haematin, but it is conceivable that the adults take no more blood after emerging on the gills. It is also possible that the blood-dwelling juveniles are able to absorb nutrients through the general body tegument.

5.3.4 *Locomotion*
Many 'dactylogyroideans' are able to move from one locality to another on the gills. Their attachment sites are the slots between secondary gill lamellae and relocation demands precise orientation of the haptor and the ability to manoeuvre it into the new

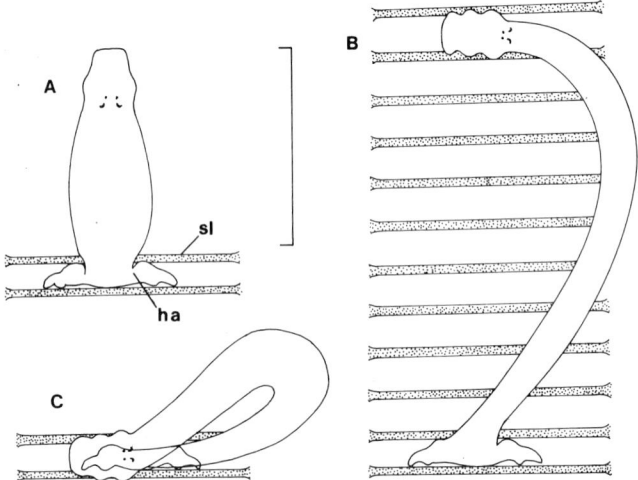

Fig. 5.12. Locomotion in *Tetraonchus monenteron*. (A) Resting position; (BC) two stages in changing location on the same side of a primary lamella. Only the free (distal) edges of the secondary gill lamellae (sl) are shown. ha, Haptor. Scale bar: 0.5 mm. From Kearn, 1987, with permission from the American Society of Parasitologists.

slot. Locomotion of *T. monenteron* takes place frequently on excised primary gill lamellae of the pike (*Esox lucius*), particularly during the first 90 minutes after removing the gills (Kearn, 1987). First, the parasite elongates the body and, if relocation is to take place on the same side of the primary lamella, the body curves so that the longitudinal axis of the head region lies parallel with the secondary gill lamellae at the new site (Figure 5.12AB). The parasite then everts the six adhesive sacs on the head region and, using the adhesive secretion produced by these sacs, attaches each lateral border of the head to the distal, free borders of two adjacent secondary gill lamellae. Thus, the head of the parasite forms a bridge between the two adjacent secondary lamellae and holds the slot between the lamellae open in readiness to receive the haptor. The haptor is then detached, brought forward rapidly and inserted into the slot between the secondary lamellae, so that the transverse axis of the haptor and the longitudinal axis of the head are temporarily parallel (Figure 5.12C). With the haptor snugly located in the new site, and perhaps temporarily anchored by the hooklets, the hamuli are able to

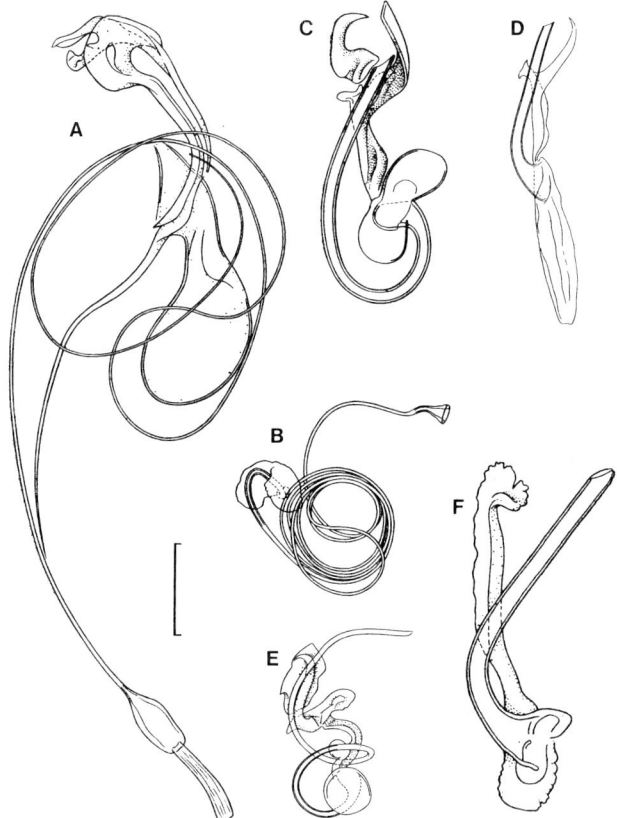

Fig. 5.13. Copulatory sclerites of some British species of *Dactylogyrus* and *Neodactylogyrus*, drawn to approximately the same scale (Scale bar: 20 μm). (A) Male copulatory sclerites and (B) vagina of *D. auriculatus*. (C-F) Male copulatory sclerites of *N. crucifer, D. amphibothrium, D. sphyrna, D. extensus*. From Gusev, 1985, with permission from Nauka Publishers.

counter-rotate as described above and impale the bases of the two adjacent secondary lamellae. The head region is then released and the body returns to its relaxed position (Figure 5.12A). Parasites also move frequently in the same way from one side of a primary lamella to a site on the opposite side of the same primary lamella.

It is interesting that *T. monenteron* moves in a similar way on a glass substrate. The head region is first attached to the glass by the adhesive sacs and the haptor is then brought forward and inserted beneath the median region of the head, i.e. beneath the bridge formed by the attached head. After the head is released, the haptor remains attached only briefly, probably held in position temporarily by anterior adhesive secretion remaining on the glass. This performance on glass suggests that it is the position of the attached head rather than the secondary gill lamellae that determines the orientation and site of re-attachment of the haptor during locomotion.

5.3.5 *Copulatory apparatus*
Parasites on fish gills are subjected to strong and continuous water currents (see Chapter 1). The ever-present danger of dislodgement by these currents has exerted strong selection pressure on monogeneans and has led to the evolution of surprisingly sophisticated and effective organs of attachment. Mating is an activity that is also likely to be disrupted by gill-ventilating currents and the male copulatory organs of 'dactylogyroideans' are more complicated than those of many skin parasites like *Entobdella* or *Leptocotyle*. There is usually a sclerotised (hardened) penis tube, accompanied by a matching sclerotised vagina with an internal diameter just large enough to accommodate the tube (Figure 5.13AB). The penis tube is often flanked by one or more accessory sclerotised structures, sometimes of great complexity (see Figure 5.13) and glands may also be present.

The diversity of the male copulatory organs of 'dactylogyroideans' is illustrated by a small sample from British dactylogyrines in Figure 5.13. It is generally supposed that evolutionary pressures for reproductive isolation by way of interspecies incompatibility of genitalia have generated this diversity, but surprisingly little is known about how these male copulatory organs function. The male copulatory apparatus has been studied in only two 'dactylogyroideans', namely in *Amphibdelloides maccallumi* by Llewellyn (1960) and in the ancyrocephalines *Ergenstrema labrosi* and *Ligophorus angustus* by Llewellyn & Anderson (1984).

(a) Amphibdelloides maccallumi. The relative positions of the male copulatory apparatus and the vagina of *A. maccallumi* are shown in Figure 5.14A. During mating, separation of the participating individuals is prevented by pincers (Figure 5.14B), which grasp the dorso-lateral surface of the co-copulant's body, where the vaginal opening is situated. The pincers are composed of two sclerites, the one having a strong single-pointed, distal hook and the other a broader termination bearing five pointed protuberances.

The sclerites are housed in the common genital atrium, which also receives the uterus. Prior to copulation it is likely that the pincers are pushed out through the opening of this genital atrium by contraction of protractor muscles (Figure 5.14B), two of which run to the base of each sclerite. There are muscles that would serve to open the jaws of the pincers and a muscle running from one sclerite to another, which would be in a position to close the pincer jaws and grip the body of the co-copulant. In *A. maccallumi* there is a disk-shaped fibrous pad close to the vaginal opening (Figure 5.14A). Llewellyn (1960) suggested that this pad provides a toughened target for the pincers; by gripping the tough pad, damage to the soft underlying organs is minimised. The penis tube is long, narrow and sclerotised. It passes distally through the pincer sclerite with the single-pointed terminal hook. The penis tube is flexible and is enclosed in a sleeve of muscle running parallel with the tube and connecting the sclerotised disk at the proximal (basal) end of the penis tube to the penis-bearing pincer sclerite. Contraction of the muscle sleeve would result in the protrusion of the penis tube from the pincer sclerite and its insertion into the sclerotised vaginal tube. Withdrawal of the penis tube would be brought about by contraction of retractor muscles attached to its basal disk.

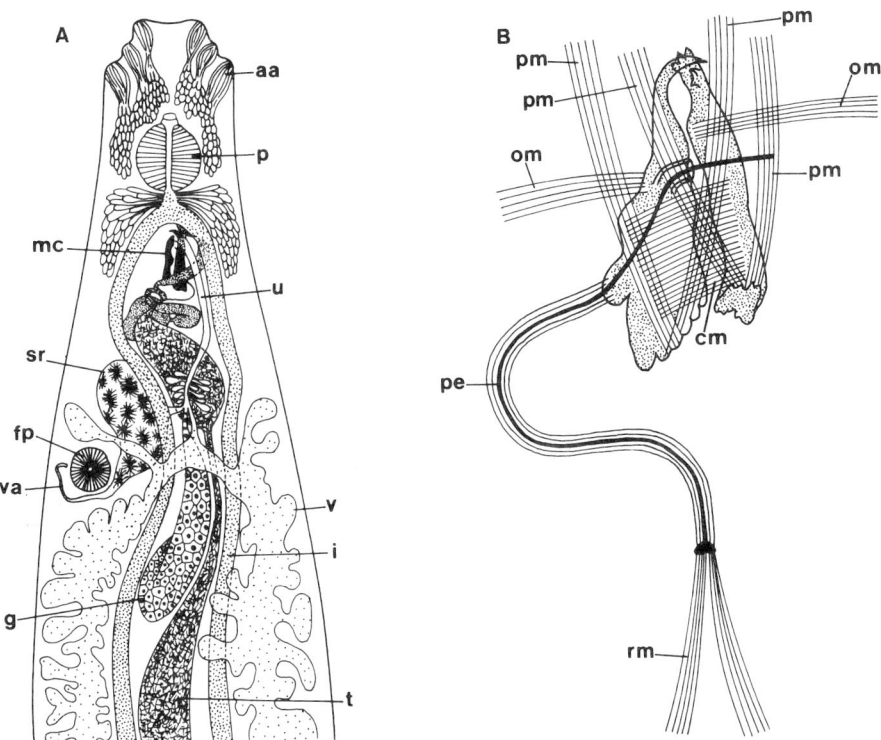

Fig. 5.14. (A) Anterior region of the body of *Amphibdelloides maccallumi* showing relative positions of the male copulatory apparatus (mc) and vagina (va). (B) The male copulatory apparatus of *A. maccallumi*. The pincers are displaced slightly to permit illustration of muscles. The penis tube is shown solid for ease of interpretation. aa, Anterior adhesive sac; cm, closing muscle; fp, fibrous pad; g, germarium; i, intestine; om, opening muscle; p, pharynx; pe, penis tube; pm, protractor muscle; rm, retractor muscle; sr, seminal receptacle; t, testis; u, uterus; v, vitellarium. From Llewellyn, 1960, with permission of Cambridge University Press.

As Llewellyn & Anderson (1984) pointed out in a later study (see below), this arrangement is strongly reminiscent of an engineer's Bowden cable, for example the flexible cable-release system of a camera. The main difference is that in the monogenean version, the power is supplied from within by contraction of the muscle sleeve surrounding the penis tube, while in the cable release it is supplied from the outside by finger pressure.

The narrow sclerotised vaginal tube opens into a spacious seminal receptacle where sperm is stored (Figure 5.14A). Llewellyn (1960) observed that sperms in the seminal receptacle are aggregated around spherical or elongated masses of an acidophilic substance, which may serve to nourish the sperms and prolong their storage life.

(b) *Ancyrocephalines from mullets.* Llewellyn & Anderson (1984) studied the copulatory apparatus of the ancyrocephalines *Ergenstrema labrosi* and *Ligophorus angustus* from the thick-lipped grey mullet, *Chelon labrosus*. The male copulatory apparatus of *E. labrosi* comprises a slender penis tube about 200 µm long, with a uniform external diameter of approximately 1 µm for most of its length (Figure 5.15). At its proximal end the penis tube expands and receives the sperm duct from a storage reservoir or seminal vesicle. At the distal end the penis tube passes through a beak-like accessory sclerite about 25 µm long, which opens through the ventral tegument and is tethered to that tegument by short fibres. As in *A. maccallumi*, protrusion of the penis tube resembles the operation of a Bowden cable. There is a sleeve of muscle enclosing the tube and running along its length from its expanded proximal base to the proximal end of the beak-like accessory sclerite. Contraction of this muscle sleeve would cause the penis tube to slide out through the accessory sclerite and protrude from the animal. In fact, Llewellyn & Anderson (1984) observed this event in a living parasite under pressure between a coverslip and a slide, as much as 40 µm (20% of the total length of the penis tube) emerging from the tip of the beak.

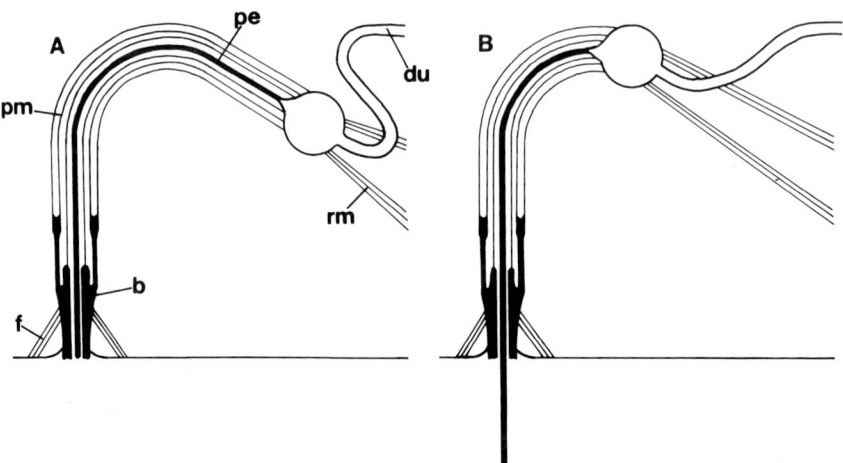

Fig. 5.15. The protrusion of the penis tube of *Ergenstrema labrosi* in diagrammatic longitudinal section. (A) Retracted; (B) protruded. b, 'Beak' of accessory sclerite; f, fibres; pm, protractor muscle; rm, retractor muscle; pe, penis tube (shown solid for ease of interpretation); du, duct from seminal vesicle. From Llewellyn & Anderson, 1984, with permission of Cambridge University Press.

Unlike *A. maccallumi* there are no pincers in *E. labrosi* or in *L. angustus*. Two mating individuals are, as far as is known (mating has not been described), connected together merely by the penis tube lodged in the vagina.

5.3.6 *Ecology of ancyrocephalines from mullets*
As mentioned above, Anderson (1981b) found a striking difference between the ancyrocephaline monogeneans parasitising young and old thick-lipped grey mullets, *Chelon labrosus*, in the Plymouth area, England. In her preliminary samples, she found that young mullets were exclusively infested with *Ergenstrema labrosi*, while old hosts carried a different ancyrocephaline, *Ligophorus angustus*. She then embarked on a more intensive sampling programme. She collected young fish, 3 – 20 cm in total length and up to four years old, from St John's Lake, an inlet of the Tamar Estuary. Adult fishes measuring between 20 and 40 cm in length and aged between four and nine years were caught in the open sea and in the Tamar Estuary. Fishes up to two years old harboured only *E. labrosi* and fishes five years old and older were parasitised only by *L. angustus*. Fishes between two and four years old were infested with both species of parasite.

Anderson was aware that this changing pattern of infection with host age could be a reflection of the changing immunological status of the fish. It is possible that young fishes have an innate resistance to infection with *L. angustus* but not with *E. labrosi*, and that as the hosts age they reject *E. labrosi* and become susceptible to *L. angustus*. However, Anderson (1981b) demonstrated that this is not the case, since in the laboratory she was able to infect young mullets with *L. angustus* and old mullets with *E. labrosi*.

The parasite changeover as the hosts grow older seems most likely to be related to changes in salinity as mullets move from tidal pools out into the estuary and the open sea. The salinity at St John's Lake, the habitat of young mullets, fluctuates on a daily basis from 10 to 30 $^{0}/_{00}$, while in the estuary and open sea the salinity is constant and higher. According to Anderson (1981b), the movement from one habitat to another occurs when the fishes are 2 – 4 years old, so there is a strong correlation between this migration and the changeover of parasites. Anderson found that *E. labrosi* survived best in 30 – 60% seawater, while *L. angustus* survived best in 50 – 100% seawater. In *E. labrosi* the rate of egg-laying was adversely affected by salinities greater than 50% sea water and ceased in 100% sea water, while *L. angustus* laid most eggs in 100% sea water. However, these observations were made on parasites detached from the host, whereas the old mullet that acquired *E. labrosi* in an aquarium (see above) was kept in 100% sea water and became infected by larvae from eggs laid by parasites on young mullets. Thus, at least in the laboratory, *E. labrosus* can maintain itself in 100% seawater, albeit at a low level. Perhaps in the open sea, this level of propagation is not sufficient to maintain a viable population of *E. labrosi* on old mullets.

Anderson (1981a) made another interesting discovery with regard to *Ergenstrema labrosi*. She had expected to find them on the gill rakers since the only other species of the genus, *E. mugilis*, is restricted to this habitat and does not occur on the gills of its host the thin-lipped grey mullet, *Liza ramada*. *E. labrosi* was indeed found on the gill rakers of its host, but, surprisingly was also found on the gill lamellae in 50% of the fishes examined. Parasites in these two distinct microhabitats were identical. In 18% of the fishes examined, the gill rakers only were infected and in 27% the gill lamellae only were infected. Such a dual habitat distribution is rare in monogeneans. Roubal & Quartararo (1992) found the monogenean *Anoplodiscus cirrusspiralis* on the fins and in the nasal capsules of the snapper, *Pagrus* (= *Chrysophrys*) *auratus* in Australia.

A comparison of the distribution of *E. labrosi* and *E. mugilis* on the gill rakers of their respective hosts revealed that most of the former occur in the middle region of the gill arch while most of the latter occur in the dorsal and ventral thirds of the arch. Anderson (1981a) concluded that not only does *E. labrosi* occur in a previously unrecorded microhabitat (the gill lamellae) for parasites of the bispecific genus *Ergenstrema*, but also differs from *E. mugilis* in its distribution among the gill rakers, the common habitat of the two parasites.

6

MONOGENEAN GILL PARASITES – POLYOPISTHOCOTYLEANS

6.1 INTRODUCTION

The polyopisthocotylean monogeneans, most of which are fish gill parasites, are so different from their monopisthocotylean relatives that separate origins have been suggested for the two groups, i.e. the monogeneans have been regarded as paraphyletic rather than monophyletic (Mollaret *et al.*, 1997; Justine, 1998). There can be little doubt from consideration of the differences between them that they have been separate for a very long time. However, the larvae of polyopisthocotyleans share with those of monopisthocotyleans some basic features, such as 16 hooklets, each with a shield- or gutter-shaped sclerite (domus), two pairs of eyes and three bands of propulsive ciliated cells, indicating divergence from a common ancestor, albeit a very long time ago.

    This monogenean divergence must have taken place before the modern groups of fishes (holocephalans, elasmobranchs, chondrosteans, teleosts; see Appendix 1) had emerged from the ancestral fish stock, because each modern group of fishes is parasitised by a unique group of polyopisthocotylean monogeneans. Thus, as the ancient fish stock radiated and differentiated into the fishes we recognise today, each newly emerging group of hosts inherited a sub-set of polyopisthocotylean monogeneans, which, in isolation, developed its own unique features. This provides an excellent example of co-evolution between hosts and parasites, the consequence of which is that related hosts have related parasites (so-called 'Fahrenholz's Rule').

    This is exciting enough, but it appears that co-evolution did not end with the fishes. It is generally accepted that amphibians have evolved from fish ancestors, although there is some disagreement as to which group of ancient fishes provided the raw material for this important evolutionary development. Polyopisthocotylean fish parasites apparently had sufficient evolutionary flexibility to survive the fundamental upheaval that accompanied the host changes from a fully aquatic to an amphibious life-style. These adaptable parasites are represented today by polystomatid polyopisthocotyleans, many of which inhabit the bladders of anuran amphibians (frogs), for example, *Polystoma integerrimum* in the bladder of the common frog, *Rana temporaria*.

6.2 BRITISH POLYOPISTHOCOTYLEANS

Polyopisthocotylean monogeneans are well represented on British fishes and many of the more common species and their hosts are listed in Table 6.1. Worldwide, most polyopisthocotyleans are marine, but a few families have freshwater representatives. In Britain, discocotylid and diplozoid polyopisthocotyleans parasitise freshwater fishes (Table 6.1).

Table 6.1. A selection of British polyopisthocotylean monogeneans and their hosts. All the parasites are marine except for the Discocotylidae and the Diplozoidae. Host scientific names illustrate the range of host specificity (see text). Records from Dawes, 1947, Rasheed, 1983, 1984, Llewellyn et al., 1984, Pascoe, 1987, Whittington & Kearn, 1990, Gannicott & Tinsley, 1997.

Parasite	Host
Hexabothriidae	
Hexabothrium appendiculatum	Dogfish (*Scyliorhinus canicula*)
	Nursehound (*S. stellaris*)
Rajonchocotyle emarginata	Blonde ray (*Raja brachyura*)
	Thornback ray (*R. clavata*)
	Small-eyed ray (*R. microocellata*)
	Spotted ray (*R. montagui*)
	Cuckoo ray (*R. naevus*)
Mazocraeideans	
Mazocraeidae	
Kuhnia scombri	Mackerel (*Scomber scombrus*)
Kuhnia sprostoni	Mackerel
Grubea cochlear	Mackerel
Diclidophoridae	
Diclidophora merlangi	Whiting (*Merlangius merlangus*)
Diclidophora minor	Blue whiting (*Micromesistius poutassou*)
Diclidophora luscae	Bib (*Trisopterus luscus*)
Diclidophora esmarkii	Norway pout (*Trisopterus esmarkii*)
	Poor cod (*Trisopterus minutus*) (?)
Diclidophora denticulata	Saithe (*Pollachius virens*)
Diclidophora pollachii	Pollack (*Pollachius pollachius*)
Diclidophora palmata	Ling (*Molva molva*)
Diclidophora phycidis	Forkbeard (*Phycis blennoides*)
Cyclocotyla chrysophryi	Red sea-bream (*Pagellus bogaraveo*)
Paracyclocotyla cherbonnieri	Baird's smooth-head (*Alepocephalus bairdii*)
Microcotylidae	
Axine belones	Garfish (*Belone belone*)
Atrispinum labracis	Bass (*Dicentrarchus labrax*)
Microcotyle donavini	Ballan wrasse (*Labrus bergylta*)
Gastrocotylidae	
Gastrocotyle trachuri	Scad (*Trachurus trachurus*)
Pseudaxine trachuri	Scad
Plectanocotylidae	
Plectanocotyle gurnardi	Red gurnard (*Aspitrigla cuculus*)
	Grey gurnard (*Eutrigla gurnardus*)
	Tub gurnard (*Trigla lucerna*)
	Streaked gurnard (*Trigloporus lastoviza*)
Anthocotylidae	
Anthocotyle merluccii	Hake (*Merluccius merluccius*)
Discocotylidae	
Discocotyle sagittata	Brown trout (*Salmo trutta*)
	Rainbow trout (*Oncorhynchus mykiss*)
Diplozoidae	
Diplozoon paradoxum	Bream (*Abramis brama*)
Paradiplozoon homoion	Roach (*Rutilus rutilus*)

It can be seen that a high degree of host specificity is maintained. Should a parasitologist specialising in monogeneans be given a polyopisthocotylean monogenean separated from its fish host, the identity of the host would in many cases be revealed immediately (see Table 6.1). When more than one host species is parasitised by a polyopisthocotylean, the hosts usually belong to the same genus (e.g. *Hexabothrium appendiculatum, Rajonchocotyle emarginata, Diclidophora esmarkii*, see Table 6.1) or more rarely the same family (*Plectanocotyle gurnardi*).

Table 6.1 also includes *Grubea cochlear* from mackerel, *Scomber scombrus*. In spite of considerable parasitological attention paid to its common host over the years, this parasite was unknown in British waters until Whittington & Kearn in 1990 collected a single specimen at Plymouth. In 1993 Lyndon & Vidal-Martinez (1994) found that nine out of 60 mackerel (15%) caught in Lyme Bay off Exmouth in Devon were infected with adult *G. cochlear* at intensities of 1 – 3 parasites per fish. This may indicate that the parasite is extending its range northwards into British waters.

There are other examples of fauna changes. The marine scad, *Trachurus trachurus*, appears to have moved northwards into the Plymouth region in the first half of the last century (see Llewellyn, 1962). It is not known whether these fishes brought with them an abundant monogenean population (*Gastrocotyle trachuri* and *Pseudaxine trachuri*) or whether the newly arrived fishes were colonised by local parasite stock. In 1979 a species of *Diclidophora* appeared in the Plymouth area on the gills of poor cod, *Trisopterus minutus*, a fish that had been free of monogeneans in this area for at least 25 years (Llewellyn *et al.*, 1980). Llewellyn *et al.* attributed this to an influx of Norway pout, *Trisopterus esmarkii*, into the Plymouth area from western and northern regions and the transfer of *Diclidophora esmarkii* from this fish to the poor cod. Later work by Tirard *et al.* (1992) revealed another possibility. They proposed that poor cod from a southern population had moved into the Plymouth area and brought with them a previously undescribed gill parasite specific to poor cod. This parasite has not been described and it is not yet known which of these explanations is correct.

These examples illustrate not only that the parasite fauna is changing but also that these changes can be rapid.

Natural levels of infestation with polyopisthocotyleans usually present no threat to individual fishes or to populations, except perhaps when fishes are competing for limited resources (see p. 114), but in fish farms or on introduced alien hosts these levels may be exceeded, with potentially serious consequences. Heavy infestations with the polyopisthocotylean *Discocotyle sagittata* pose such a threat in trout hatcheries to the introduced rainbow trout, *Oncorhynchus mykiss* (see Gannicott, 1997 in Rubio-Godoy & Tinsley, 2002).

6.3 ATTACHMENT

The reader may already be dismayed by the length and complexity of the names of the two subdivisions of the Monogenea. However, if we break down the words 'monopisthocotylean' and 'polyopisthocotylean' into their separate units and delve into ancient Greek to determine their meaning, then these long names become useful, because they emphasise an important difference between the two groups. 'Mono-' is derived from 'μόνος', single, and 'poly-' from 'πολύς', much, in plural πολλοί, -αί, -ά, many. 'Opistho-' is derived from 'ὄπισθεν' meaning behind and 'cotyle' from 'κοτύλη' a cup. As the names indicate, the monopisthocotyleans have a single sucker-cup (haptor) at the rear of the body while the polyopisthocotyleans have many sucker-cups at the rear. Although the oncomiracidia of both groups begin their short lives with a single haptor cup armed with up to 16 tiny hooklets, subsequent development follows different pathways. As development proceeds in the monopisthocotyleans, the haptor grows, retaining its cup shape in some and losing it in others, but always maintaining its integrity as a single adhesive organ. In polyopisthocotyleans the haptor becomes functionally subdivided usually into six or eight separate suckers or clamps, each of which is assembled during post-larval development at the site of one of the peripherally situated hooklets.

There are reasons to believe that in early polyopisthocotyleans these sub-units were muscular suckers, similar to those that are used today by polystomatid polyopisthocotyleans to attach themselves to the lining of the urinary bladders of their amphibian hosts. Suction in these sub-units was probably generated by contraction of short radial muscles running between the inner surface of the sucker (the cup lining) and the outer surface of the cup wall. By spreading attachment over a wider area using multiple suckers, a relatively large polyopisthocotylean parasite could be supported in the gill cavity of a fish, since this 'spread-eagle' arrangement would confer stability and resist detachment by powerful and more or less continuous gill-ventilating currents. Many monopisthocotylean gill parasites are so small that a lens or microscope is needed to see them, but polyopisthocotyleans are generally much larger and are usually detectable with the naked eye.

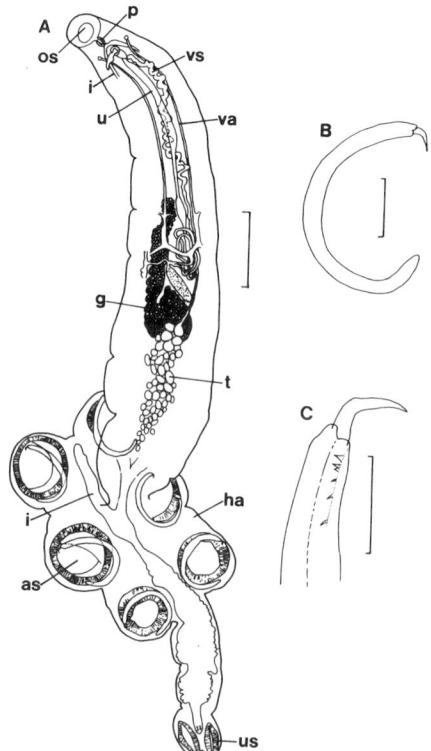

Fig. 6.1. (A) A typical hexabothriid monogenean, *Squalonchocotyle catenulata*. (B) Hook from one of the armed suckers. (C) Enlarged distal region of sucker hook. as, Armed sucker; g, germarium; ha, haptor; i, intestine (part only shown); os, oral sucker; p, pharynx; t, testis; u, uterus; us, unarmed sucker; va, vagina; vs, vas deferens. Scale bars: (A) 1 mm; (B) 300 μm; (C) 100 μm. From Guberlet, 1933.

6.3.1 Hexabothriid suckers

Hexabothriid monogeneans are highly specialised polyopisthocotylean parasites of elasmobranch fishes (Table 6.1). They attach themselves inside the gill chamber by means of eight suckers, two of which are small and isolated on a haptoral appendage (Figure 6.1A). Each of the large suckers is armed with a curved skeletal bar or sclerite (Figure 6.1BC), but the small suckers are unarmed. Euzet & Maillard (1976) have described how the sclerite of the large sucker enhances the suctorial efficiency of the organ. This sclerite is an integral part of the sucker wall, following a somewhat eccentric course across its surface (Figure 6.2AB). The muscular wall of the sucker is interrupted along the route of the sclerite, which is enclosed in a sheath (Figure 6.2C). Distally the sclerite has a prominent hook (Figure 6.1BC) and part way along the sclerite there is a gap in the sheath. A prominent extrinsic protractor muscle enters the sheath at this point (Figure 6.2A, pm), turns in the direction of the proximal (blunt) end of the sclerite and eventually attaches to this proximal end. When the muscle contracts it will effectively push out the hook and drive it into host tissue. Then, as the muscle contracts further, the proximal end of the sclerite will be lifted, raising the roof of the sucker and reducing the pressure inside it. This will draw a plug of soft host tissue into the sucker cavity (Figure 6.2C). Running around the opening of the sucker is a sphincter muscle (Figure 6.2C, sm). Contraction of this muscle will close the sucker opening and grip the base of the host tissue plug, providing added security of attachment and possibly permitting relaxation of the extrinsic protractor muscle attached to the sclerite.

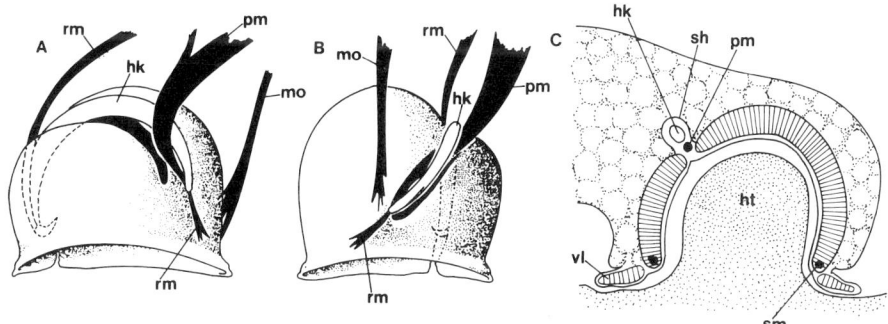

Fig. 6.2. Diagrammatic views of armed suckers and associated muscles of a hexabothriid monogenean. (A) Whole sucker from the left side of the haptor in postero-lateral view; (B) whole sucker from the right side in posterior view. Sleeves enclosing hooks not shown. (C) Diagrammatic section through an armed sucker attached to the host. hk, Hook; ht, plug of host tissue; mo, muscle for sucker orientation; pm, protractor muscle; rm, retractor muscle; sh, sheath; sm, sphincter muscle; vl, marginal valve. From Euzet & Maillard, 1976, with permission from Professor L. Euzet.

As suction is generated, there will be a tendency for seawater to flow into the sucker cavity between the circular rim of the sucker opening and the host's surface. This is prevented and the efficiency of the suctorial mechanism maintained by a marginal valve running around this rim (Figure 6.2C, vl). Other muscles associated with the sucker are less prominent, reflecting their less powerful roles. There are retractor muscles involved in withdrawing the hooked sclerite (Figure 6.2AB, rm) and muscles involved in orientating the sucker (Figure 6.2AB, mo).

6.3.2 Suctorial clamps of Diclidophora *spp.*
Hexabothriid suckers are well suited for attachment to soft and smooth non-respiratory surfaces in the gill chamber, but application of these relatively large suckers to the respiratory surface may be damaging for the host, at best interfering with the host's gaseous exchange and at worst rupturing thin-walled secondary gill lamellae. Monopisthocotylean monogeneans inhabiting the gills minimise damage to the host's respiratory surfaces by remaining small enough to sit between two adjacent secondary lamellae, impaling these lamellae with counter-rotating large hooks (hamuli). This permits the host to support relatively large numbers of these small parasites with minimal interference with host gaseous exchange.

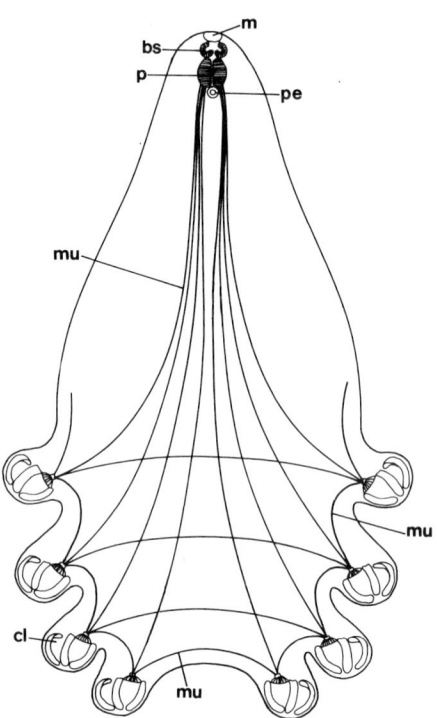

Fig. 6.3. Clamps and clamp musculature of *Diclidophora phycidis.* bs, Buccal sucker; cl, clamp; m, mouth; mu, muscle; p, pharynx; pe, penis. Modified from Llewellyn & Tully, 1969.

The mazocraeidean polyopisthocotylean monogeneans, which parasitise teleost fishes, have evolved a way of attaching themselves to the gills that is entirely different and is capable of supporting parasites much larger than monopisthocotyleans. They have modified the basically suctorial haptoral sub-units of their ancestors to create clamps, each of which is capable of grasping one to three host secondary gill lamellae. The rigidity needed to create such a clamping mechanism has been achieved by acquiring a skeleton of extra sclerites, which outline and support the two separate jaws of each clamp. Typically these gill parasites are equipped with four pairs of clamps, which are sufficient to support a relatively large monogenean (many are several millimetres in length), but, as we will see shortly, there are tendencies both to reduce the clamp numbers and to increase them.

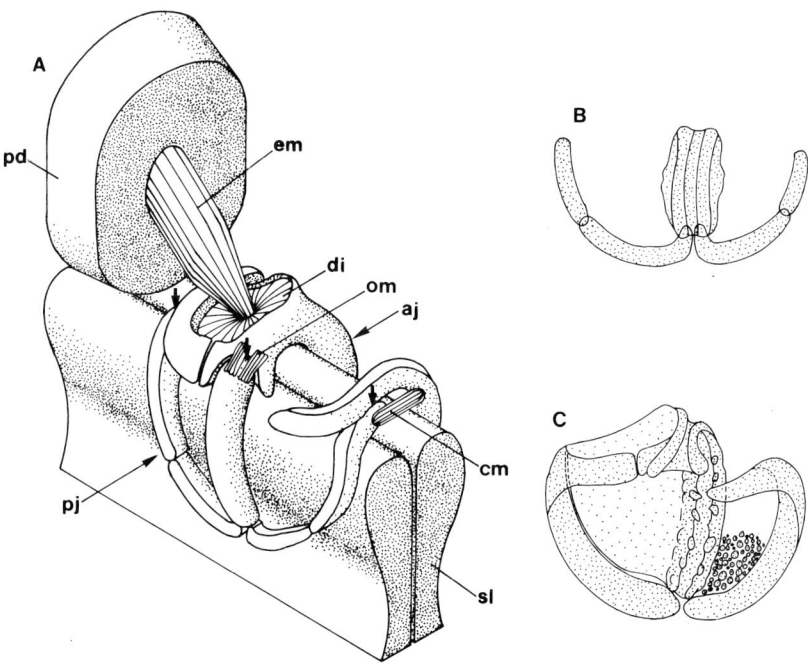

Fig. 6.4. The clamp of *Diclidophora*. (A) Stereogram of a clamp grasping two secondary gill lamellae (sl) of the host. aj, Anterior ('fixed') jaw; cm, closing muscle of clamp; di, diaphragm; em, extrinsic muscle; pd, peduncle; pj, posterior ('moveable') jaw. Arrows, articulations between anterior and posterior jaws. (BC) The supporting sclerites of the posterior ('moveable') jaw and the anterior ('fixed') jaw respectively. From Llewellyn, 1958, with permission of Cambridge University Press.

Species of *Diclidophora* have four pairs of clamps (Figure 6.3). Llewellyn (1958) made the interesting discovery that these organs are basically suckers in which the suction pressure generated is converted into a clamping action. A three-dimensional representation of a single clamp grasping two secondary gill lamellae of the host is shown in Figure 6.4A. The fleshy covering of the clamp jaws has been removed, revealing only the clamp skeleton and some of the muscles associated with it. There are two separate jaws, an anterior jaw, which will be regarded as fixed (much of this is out of view in Figure 6.4A behind the gill lamellae), and a posterior jaw, which will be regarded as moveable. The moveable jaw (Figure 6.4B) consists of five sclerites. Four of these support the semi-circular peripheral edge of the jaw, while the fifth provides a median brace. There are three points of articulation with the fixed jaw (Figure 6.4A, arrows). The periphery of the fixed jaw is supported by just two curved sclerites (Figure 6.4C) and there is also a median sclerite. However, an extra and highly significant feature of the fixed jaw is a ring at the hinge line, created by contributions from the inner peripheral sclerite and the median sclerite. Stretched across the ring is a diaphragm (Figure 6.4A). A prominent extrinsic muscle (Figure 6.4A, em) is inserted on the central region of this diaphragm.

Llewellyn (1958) studied living specimens of *D. merlangi* forcibly detached from the gills of whiting, *Merlangius merlangus*. In *D. merlangi* each clamp is mounted on a stalk or peduncle and Llewellyn repeatedly observed the extension of the peduncle with the jaws of the clamp opening, followed by the rapid snapping together of the jaws and the withdrawal of the peduncle. According to Llewellyn, short muscles running outside the hinge between the median sclerite of the posterior jaw and the ring, open the clamp (Figure 6.4A, om). The open clamp then embraces two or three secondary lamellae and is closed by three short muscles running between the moveable and the fixed jaws just below the hinge points (one is shown in Figure 6.4A, cm). These closing muscles permit the jaws to grasp the lamellae lightly, being too weak to make a significant contribution to attachment, but they do ensure that the marginal valve running along the edges of the jaws is in contact with the gill lamellae.

The stage is now set for the *tour de force*. The powerful extrinsic muscle contracts and lifts the diaphragm. The marginal valves seal the edges of the jaws and prevent influx of seawater. Seawater inside the clamp acts as a very efficient hydraulic transmitter and suction is generated. Since the two jaws are freely hinged this suction is converted into a clamping action, which is proportional to the pull of the extrinsic muscle.

The extrinsic muscle attached to each clamp diaphragm subdivides. In *D. phycidis* branches of this extrinsic muscle run to the diaphragms of adjacent clamps and other branches run transversely across the body to link the two opposite clamps of a pair (Figure 6.3). Another muscle branch from each clamp follows an anterior course in the body and, together with the anterior branches from the other clamps, joins the pharynx. Llewellyn (1956a) with reference to *Plectanocotyle gurnardi* (see below) stressed the efficiency of the inter-clamp arrangement of muscles, since each attached clamp will provide a firm origin for the muscles inserted on other clamps. He also doubted that the longitudinal clamp muscles could make a significant contribution to clamp function, since they could do so efficiently only when the anterior end of the parasite is attached, otherwise the effects of their contraction would be to shorten the body rather than to close the clamps. Llewellyn (1956a) suggested that the longitudinal muscles might serve to manoeuvre the anterior end of the body.

It has been shown by Maule *et al.* (1989) that water turbulence elicits rapid and strong contractions of these longitudinal muscles in *D. merlangi*. It has been suggested that the 'cough' reflex, in which the host temporarily closes the operculum and rapidly and briefly reverses water flow through the gills as a means of removing debris from the gill cavity, creates the kind of turbulence that would elicit this response. According to Maule *et al.* (1989) and Halton *et al.* (1998), the rapid shortening of the parasite's body brought about by contraction of these muscles would minimise the likelihood of dislodgement from the gills during this violent event. They do not explain how body shortening would provide added security, although the concomitant reduction of surface area might be advantageous. It is also possible that the violent and sudden contraction of the parasite's body would make some contribution to increased clamp suction pressure and more powerful clamping to the gill, in spite of the fact that the anterior end of the animal is unattached.

6.3.3 *The open clamps of* Cyclocotyla

Cyclocotyla (= *Choricotyle*) *chrysophryi* (Figure 6.5) from the red sea-bream, *Pagellus bogaraveo*, has a very different adhesive attitude from its relatives. Llewellyn (1956b) observed a single living specimen and found it attached to the outer surface of the gill with its individual pedunculate attachment organs spread out and spanning several primary lamellae. Although the attachment organs of this parasite contain supporting sclerites, the organs function as open suckers not clamps, enabling them to attach to flat non-respiratory surfaces like the borders of the primary lamellae or the gill arch. Llewellyn (1956b) observed that when forcibly detached the parasite readily reattached.

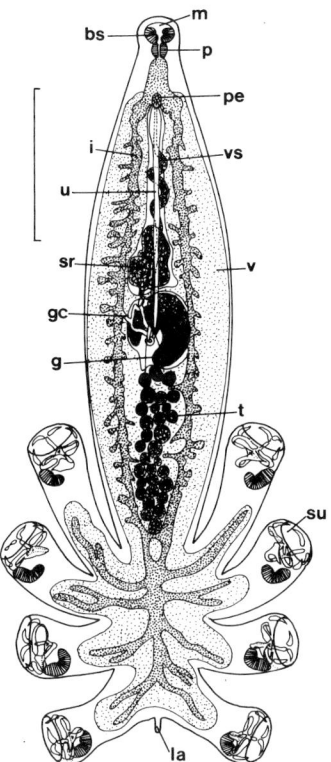

Fig. 6.5. Cyclocotyla (= Choricotyle) chrysophryi. bs, Buccal sucker; g, germarium; gc, genito-intestinal canal; i, intestine; la, lappet bearing hooks; m, mouth; p, pharynx; pe, penis; sr, seminal receptacle; su, haptoral sucker; t, testis; u, uterus; v, vitellarium; vs, vas deferens. Scale bar: 1 mm. From Llewellyn, 1941, with permission of Cambridge University Press.

Pascoe (1987) found *Paracyclocotyla cherbonnieri* attached to the gill arch, gill rakers, the edges of the primary lamellae and the inside of the operculum of Baird's smooth-head, *Alepocephalus bairdii*, caught in deep water in the area of the Rockall Trough. Moreover, detached specimens readily attached to a glass surface by means of their haptoral suckers. There are records of related species living in the mouth cavities of their hosts and they have also been found on the carapaces of cymothoid isopod parasites (Sproston, 1946), which may provide transport to new fish hosts (phoresy; see Chapter 4).

All these observations bear witness to the surprising mobility of these monogeneans, in contrast with the suspected sedentary behaviour of those polyopisthocotyleans that clamp themselves to secondary gill lamellae (see below). Like *Diclidophora*, the ancestors of *Cyclocotyla* may have been sedentary with functional suctorial clamps, which have reverted secondarily to open suckers permitting mobility (see Llewellyn, 1970). The shapes and the disposition of the sclerites in the attachment organs of *C. chrysophryi* are consistent with this notion. Alternatively, open, sclerite-supported suckers like those of *Cyclocotyla* may be the forerunners of the sucker clamps of parasites like *Diclidophora*.

6.3.4 Non-suctorial clamps
The clamps of other mazocraeidean monogeneans operate on a different principle. *Plectanocotyle gurnardi* attaches itself to the secondary gill lamellae on one side only of

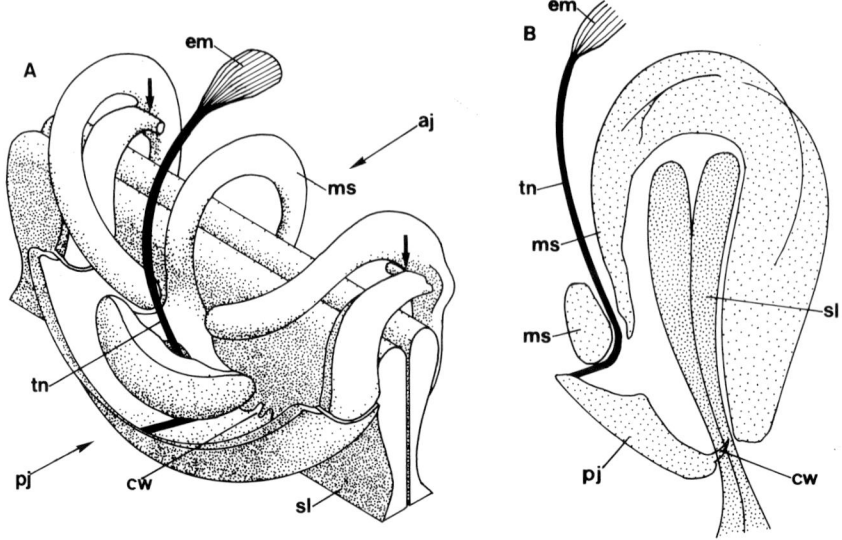

Fig. 6.6. The clamp of *Plectanocotyle gurnardi*. (A) Stereogram of whole clamp grasping two secondary gill lamellae; (B) diagrammatic section through the centre of an attached clamp in the plane of the median S-shaped sclerite. aj, Anterior ('fixed') jaw; cw, claw; em, extrinsic muscle; ms, S-shaped median sclerite; pj, posterior ('moveable') jaw of clamp; sl, secondary gill lamella; tn, tendon. Arrows, articulations between anterior and posterior jaws. (A) From Llewellyn, 1956a, with permission of Cambridge University Press; (B) based on photographs in Llewellyn, 1956a.

a primary gill lamella of its gurnard host by means of three pairs of postero-ventrally directed clamps (Figure 6.7A). These clamps have no structure equivalent to the ring and diaphragm of *Diclidophora*. Each clamp of *Plectanocotyle* consists of two jaws that are hinged together, with the hinge axis lying approximately transversely in the body (Figure 6.6A). Thus, as in *Diclidophora*, one jaw, which can be conveniently described as the 'fixed' jaw, is anterior and the other, which we will refer to as the 'moveable' jaw, is posterior. The free border of the moveable jaw is supported by three separate but articulating sclerites (Figure 6.6A). The free border of the fixed jaw is similarly supported, but only by two sclerites, which meet in the centre of the jaw. At the hinge line each of these sclerites continues, curving through the hinge line and travelling in an oblique direction into the proximal region of the posterior jaw. These extensions do not meet each other but are separated by an S-shaped median sclerite (Figure 6.6A, ms). The curved upper part of the 'S' extends about halfway into the anterior jaw and the curved lower part extends three quarters of the way into the posterior jaw. This curved lower part is elaborated to form a crucial component of the clamping mechanism. It channels and directs the course of a long slender tendon originating from the extrinsic clamp-operating muscle. The tendon passes through a perforation in this region of the S-shaped sclerite and then, guided in a groove on the inner surface of the same sclerite, changes direction prior to attachment to the edge of the moveable jaw.

Figure 6.6B represents diagrammatically a section through the centre of the clamp in the plane of the S-shaped median sclerite. It can be seen that contraction of the extrinsic muscle will close the clamp by pulling the two jaws together, thereby gripping the enclosed secondary gill lamellae. Small anteriorly directed claw-like extensions on the inner edge of the moveable jaw are likely to counter any tendency for the jaws to slip (Figure 6.6AB).

The clamps of *Kuhnia scombri* from mackerel (*Scomber scombrus*), *Discocotyle sagittata* from brown trout (*Salmo trutta*), *Paradiplozoon homoion* (identified as *Diplozoon paradoxum*) from roach (*Rutilus rutilus*) and *Gastrocotyle trachuri* from scad (*Trachurus trachurus*) operate on the same principle as the clamps of *Plectanocotyle* (see Llewellyn, 1957, Llewellyn and Owen, 1960, Owen, 1963, Llewellyn, 1964, respectively). *K. scombri*, *D. sagittata* and *P. homoion* each have eight clamps, but *G. trachuri* has 25 to 35 clamps arranged in a single row along one side only of the body (the reason for this asymmetry will be considered below). Bovet (1967) pointed out that the wall of the clamp in diplozoids (Table 6.1; Bovet worked on parasites from bream, *Abramis brama*, and roach) contains radial fibres running from the inner to the outer surface. He regarded these radial fibres as muscles and that their contraction was important in the initial phase of clamp operation, serving to suck the secondary gill lamellae of the host into the clamp cavity prior to closure of the clamp jaws. If this interpretation is correct, then suction may still have a minor role in the operation of these clamps and these radial fibres may be a persistent feature inherited from ancestors with unarmed haptoral suckers, like the modern polystomatids.

6.3.5 Anthocotyle merluccii *from hake*, Merluccius merluccius
Anthocotyle merluccii from the gills of hake is strikingly different from any of the other monogeneans we have considered so far (Figure 6.7E). It has four pairs of clamps, but the clamps of the anteriormost pair are huge compared with the others. The explanation for this is that these two giant clamps grip the border of a primary gill lamella, while the six small clamps grip the secondary gill lamellae in the typical polyopisthocotylean manner (Llewellyn, 1956b).

According to Llewellyn (1970), the clamp of *Anthocotyle* operates on a principle that we have not yet encountered. The clamp has no major suctorial component like *Diclidophora* and no extrinsic muscle/tendon/fair-lead system like *Plectanocotyle*. It is closed directly by intrinsic muscles running between the proximal regions of the peripheral sclerites of the anterior and posterior jaws.

6.4 FOOD AND FEEDING

The observer is struck immediately by the presence in most polyopisthocotyleans of scattered dark-brown pigment, which is rarely seen in skin- or gill-parasitic monopisthocotyleans. Closer examination reveals that this pigment is associated with the branched intestine of the parasite, but there are no pigment cells (chromatophores) in the host's gills that might provide a source of this gut pigment. A search of the gills of fresh fishes sometimes reveals an important clue to the origin of this pigment. Occasionally a living polyopisthocotylean will be found with red gut contents, indicating that the parasite has recently ingested host blood.

Llewellyn (1954) obtained evidence in support of the blood-feeding habit. Using the light microscope he found identical, dark brown to black, roughly spherical pigment granules with diameters of $0.5 - 0.8$ µm, both in the gut lumen and in the cells lining the intestine. After consideration of the physical and chemical properties of the pigment and the results of a variety of histochemical tests and spectroscopic examination, Llewellyn came to the conclusion that this extracellular and intracellular pigment is haematin, an indigestible residue derived from the digestion of haemoglobin from the host's blood.

All polyopisthocotylean parasites of fishes are blood-feeders, with the possible exception of *Concinnocotyla australis* from the Australian lungfish, *Neoceratodus forsteri* (see Kearn, 1998). It is likely that polyopisthocotyleans acquired their taste for blood and their gill-dwelling habit very early in their evolutionary history. This diet is reflected in anatomical features at the anterior end of the parasite. The mouth is terminal or sub-terminal (ventral) and in some polyopisthocotyleans (e.g. hexabothriids) is associated with circular and other muscle fibres arranged to form an oral sucker (Figure 6.1A). In others the buccal cavity contains a pair of compact, muscular, buccal suckers (Figures 6.3, 6.5). The pharynx of polyopisthocotyleans is relatively small and muscular and lacks the large gland cells typical of monopisthocotylean skin feeders like *Entobdella soleae*.

The only account of feeding that seems to be based on actual observations is that of Bovet (1967) on diplozoids. He described the opening of the mouth and its application to the gill lamella, the rotation of the buccal suckers to face anteriorly and their use to fix the parasite to the branchial surface. He claimed that the pharynx is protruded and used to rupture the epithelium of the gill, and that blood is then ingested by peristaltic contractions of the pharyngeal muscles. Llewellyn (1954) noted the presence of bright red gut contents in some specimens of *Axine belones* and *Diclidophora luscae* examined within $1 - 2$ hours of the capture of their hosts, but was unable to find any intact host red blood cells (erythrocytes) within the gut. However, in a specimen of *Discocotyle sagittata* taken from the gills of a brown trout (*Salmo trutta*) and fixed within a few minutes of the fish's death, he found cells that were undoubtedly partly digested erythrocytes. Thus, disruption of erythrocytes (haemolysis) appears to be underway within minutes of ingesting host blood and in *D. merlangi* there is evidence that secretion from glands located anterior to the pharynx (pre-pharyngeal glands) is responsible for this haemolytic activity (Halton *et al.*, 1974).

Llewellyn (1954) rightly suspected that the cells lining the parasite's gut engulf the haemoglobin and other material released from the disrupted erythrocytes. Droplets of gut contents are taken into the cells lining the gut by the process of pinocytosis, already described with reference to gyrodactylid skin feeders (Chapter 4). Halton (1974) showed that the number of pinocytotic vesicles at the surface of the cell increases significantly when host haemoglobin is present in the adjacent gut lumen. Digestion then takes place inside the gut cells. Llewellyn (1954) suggested that the globin fraction of the haemoglobin molecule is the chief nutriment of the parasite, since deposits of ferric iron (from the haematin molecule) were found only within the gut lumen and its lining cells, not in the body at large. The advent of the electron microscope provided general support for the observations made with the light microscope by Llewellyn.

Sproston (1945) suggested that haematin, the waste product of intracellular digestion of haemoglobin, accumulates in *Kuhnia scombri*, so that old parasites have abundant deposits, but the observations of Llewellyn (1954) were more consistent with the periodic elimination of haematin via the mouth (there is no anus). He found fully-grown specimens of *Diclidophora merlangi* without pigment and he observed the elimination of haematin via the mouth in living parasites kept in seawater. According to Llewellyn (1954), the egesta consisted of (a) free haematin from the lumen, presumably ejected by the gut lining cells at the end of the digestive process, and (b) whole epithelial cells laden with haematin granules and sloughed off into the lumen. Later, Halton (1976) confirmed with the electron microscope that haematin cells are sloughed off and egested, but he found that this is a rare event. The unloading of residual haematin by the gut cells presumably proceeds concurrently with sequestration of food from the gut lumen and intracellular digestion. The small size of the pinocytotic vesicles of the haematin cell relative to the granules of free haematin in the lumen is likely to preclude uptake of extruded haematin from the lumen.

Bovet (1967) observed that egestion (regurgitation of waste) in diplozoids takes place not via the pharyngeal lumen but via a unique canal, not found in monopisthocotyleans, linking the intestine with the mouth cavity (bucco-intestinal canal). Wiskin (1970) observed this canal in the hexabothriid *Rajonchocotyle emarginata* and repeatedly observed the regurgitation and evacuation of haematin via this route both in juvenile and in adult parasites. Blood feeding seems to impose a one-way system of flow through the pharyngeal lumen, denying the parasite use of this route for egestion, but the reasons for this arrangement are obscure.

In addition to the pharyngeal by-pass mentioned above, there is a second unique tubular connection in polyopisthocotyleans, namely the genito-intestinal canal, providing a curious link between the intestine and the reproductive system (Figure 6.5). Llewellyn (1972) suspected a relationship between this connection and the blood-feeding habit. He pointed out that other blood feeders harbour symbiotic micro-organisms that release otherwise unavailable nutrients from indigestible components of the diet (see, for example, leeches, Chapter 7). He suggested that the genito-intestinal canal provides a means whereby symbiotic micro-organisms from the gut might gain access to the reproductive system for incorporation in the eggs, ensuring that the next generation of parasites possessed the appropriate and essential gut symbionts. However,

Morris & Halton (1975), using transmission electron microscopy, failed to find microorganisms in the intestine of *D. merlangi*.

Polyopisthocotyleans are large and it needs to be asked whether their blood-feeding activities harm their fish hosts. There is a general feeling amongst parasitologists that polyopisthocotyleans are not harmful in natural situations (see, for example, Frankland, 1955). However, a statistical analysis by Anderson (1974) of a population of *Diplozoon paradoxum* living on bream (*Abramis brama*) in a gravel pit in Dagenham in Essex, revealed evidence that under certain circumstances they may be pathogenic. During the winter, food organisms may be scarce and, although not all bream cease feeding completely during the winter months, a severe reduction in feeding activity occurs. Anderson (1974) noted that during these winter months fish with above average burdens of parasites had a lower than expected body weight, whereas at other times this relationship was not apparent. It seems likely therefore that when fish are competing for limited resources, they are unable to support large numbers of blood-feeding parasites without suffering some loss of condition. Gannicott (1997, in Rubio-Godoy & Tinsley, 2002) blamed anaemia induced by the feeding activities of *Discocotyle sagittata* for high mortality rates in rainbow trout (*Oncorhynchus mykiss*) reared in fish farms in the Isle of Man.

6.5 SYMMETRY VERSUS ASYMMETRY

In spite of the fact that the various polyopisthocotyleans occupy broadly similar habitats in the gill chambers of their respective hosts, there is considerable variation in the shapes of the bodies and haptors of these parasites. Some maintain symmetrical bodies and haptors while others display various degrees of asymmetry.

Plectanocotyle gurnardi, *Kuhnia scombri*, *Discocotyle sagittata* and *Diclidophora* spp. are examples of symmetrical gill parasites (Figure 6.7A-D respectively). They have three or four pairs of clamps, those on one side being mirror images of those on the other side, and, with the exception of *Diclidophora* spp., one or more pairs of small hooks on the posterior margin of the haptor.

In *Atrispinum labracis* and *Microcotyle donavini* (Figure 6.7F) clamps greatly exceed four pairs. Clamps appear to be added continually as the parasite grows, perhaps throughout life. Winch (1983) found a specimen of *A. labracis* measuring 4 mm in length (preserved) with 63 pairs of clamps and Sproston (1946) found one individual of *M. donavini* with 55 pairs. The basic symmetry of the haptor is maintained, with hooks retained from larval and post-larval stages situated between the two lateral rows of clamps and indicating the posterior end of the haptor. However, the longitudinal axis of the body is inclined with respect to the haptor axis, thereby imparting an asymmetrical aspect to the whole animal (Figure 6.9F). According to Winch (1983) this asymmetry is permanent and irreversible in *A. labracis*.

A further development has occurred in *Axine belones* from the garfish, *Belone belone*. Like *Atrispinum* and *Microcotyle*, *Axine* has many clamps and the haptor and body axes are inclined to each other at an angle of about 30° (Figure 6.7G). This asymmetry is a permanent feature produced by differential growth. Examination of the haptor reveals that all of the clamps (50 – 70) are arranged in a single row along one margin of the body. Persistent hooks are located about mid-way along the row of clamps, suggesting that this single clamp row represents clamps from both sides of the body. One way in which this arrangement could have come about is by suppression of growth of one side of the body. However, if this were the case, as Llewellyn (1956b) has pointed out, we would expect clamps in front of the hooks to be mirror images of those behind the hooks. Llewellyn observed that these two sets of clamps (anterior and posterior to the hooks) are not mirror images of each other and all face in the same direction. How this is brought about remains unknown.

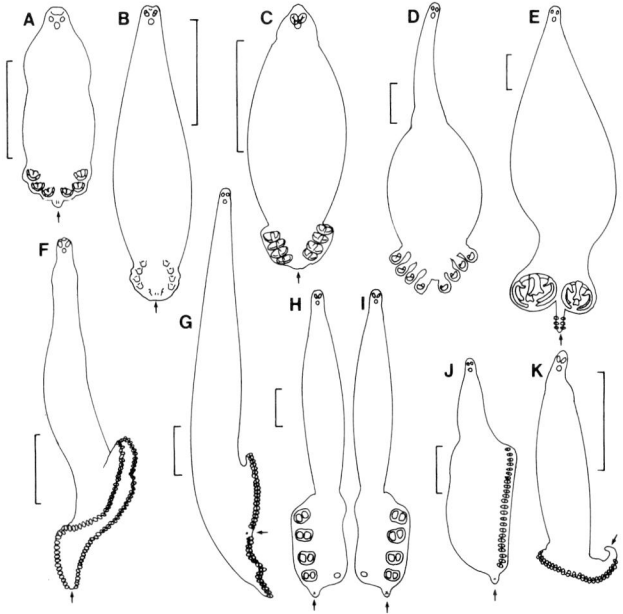

Fig. 6.7. Clamp disposition and body shapes of some mazocraeidean polyopisthocotyleans. (A) *Plectanocotyle gurnardi*; (B) *Kuhnia scombri*; (C) *Discocotyle sagittata*; (D) *Diclidophora merlangi*; (E) *Anthocotyle merluccii*; (F) *Microcotyle donavini*; (G) *Axine belones*; (H,I) *Grubea cochlear*, showing 'left-footed' and 'right-footed' individuals; (J) *Gastrocotyle trachuri*; (K) *Pseudaxine trachuri*. Arrows: positions of hooks. Scale bars: 1 mm. (A, B, D, E, G, H, I, J). From Kearn, 1994, Copyright © 1994, with permission from Elsevier.

In *Gastrocotyle trachuri* and *Pseudaxine trachuri*, asymmetry has taken another step. These two parasites, which share the same host, the scad (*Trachurus trachurus*), resemble *Axine* in having a single row of 20 – 30 clamps on one side of the body (Figure 6.7JK). However, they differ from *Axine* in having their hooks situated at the posterior end of the clamp row, suggesting that *Gastrocotyle* and *Pseudaxine* have suppressed clamp development on one side of the body. Llewellyn (1959) provided confirmation of this. He showed that the free-swimming larvae and the early pre-clamp stages of *G. trachuri* on the gills are symmetrical. In older post-larvae, clamps develop in a posterior-anterior sequence, but on one side of the body only.

6.5.1 Reasons for asymmetry in Axine

Using the lateral vaginal opening as a reference point (it is always on the parasite's left side), Llewellyn (1956b) noticed that in a sample of 100 specimens of *A. belones* the clamps were on the parasite's right side in 87 specimens and on the left side in the remaining 13 cases. Thus, we need to understand not just why *Axine* is asymmetrical but also why some individuals have clamps only on the right side ('right-footed') and others only on the left side ('left-footed').

Llewellyn (1956b) recognised two important features of the adhesive attitude of *A. belones*. These are that the parasite is always attached with the haptor closest to the gill arch of the host, the garfish (*Belone belone*), and that the clamps are attached to the gills at a position upstream relative to the gill-ventilating current, with the anterior mouth-bearing end downstream. Such an adhesive attitude goes some way towards meeting two major ecological demands, namely the need to resist dislodgement from the host and to extract blood from the highly vascular secondary lamellae in the face of a strong and virtually continuous gill-ventilating current. It has been shown in Chapter 1, that the gill-ventilating current flows between the gill arches and, striking the external surfaces of the gill (the exposed surfaces of the inner and outer hemibranchs), passes between the secondary lamellae into the space between the hemibranchs. Here it turns through $90°$ and, flowing roughly parallel with the primary lamellae, enters the opercular cavity and leaves the fish via the opercular opening.

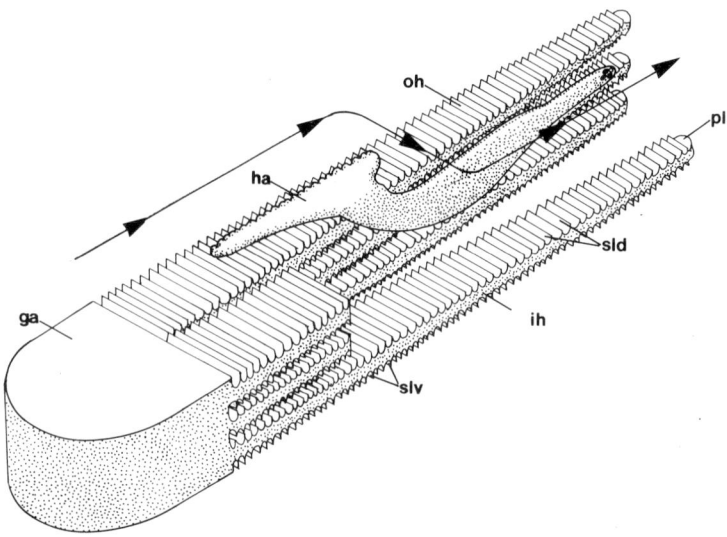

Fig. 6.8. Diagrammatic stereogram of *Axine belones* on the gills of garfish, *Belone belone*. ga, Gill arch; ha haptor; ih, inner hemibranch; oh, outer hemibranch; pl, primary lamella; sld, secondary lamellae on dorsal side of primary lamella; slv, secondary lamellae on ventral side of primary lamella. Arrows, direction of gill-ventilating current. From Llewellyn, 1956b, with permission of Cambridge University Press.

A diagrammatic representation of a portion of a gill with *Axine* attached to it is shown in Figure 6.8. Indicated on it is the path taken by the host gill-ventilating current (see also Figure 1.7). The clamps of the parasite are attached to the outer ends of the secondary gill lamellae on the dorsal face of the primary lamella, i.e. they are attached where the gill stream begins its journey between the secondary gill lamellae and flows towards the inter-hemibranch space. The body of the parasite being inclined as we have seen at about 30° to the line of clamps, projects from between the primary lamellae of the outer hemibranch into this space. Where it crosses the inner border of the primary lamella, the body is bent at right angles to the plane of the haptor. Thus, the body of a polyopisthocotylean like *Axine*, although large (up to 8.9 mm in length, according to Sproston, 1946) is hidden from view between the hemibranchs of the gill and it is necessary to part the primary gill lamellae to see it.

The adhesive attitude of *Axine* means that the longitudinal axis of most of the body, i.e. the part lying between the two hemibranchs of the gill, lies parallel with the path of the gill-ventilating current. Thus, the parasite is streamlined, minimising interference with the host's gill flow and reducing turbulence, which would place an increased load on the parasite's haptor. Even if this does not present a serious threat to the parasite's security, it would be likely to demand a greater expenditure of energy from the parasite to maintain its grip on the gill.

The parasite illustrated in Figure 6.8 is a 'left-footed' individual. If consideration is given to the adhesive attitude of a second parasite attached on the opposite side (the ventral side) of the primary lamella to which the parasite in Figure 6.8 is attached, it will be obvious that this second parasite would need to be 'right-footed' to be compatible with the currents flowing around it. Similarly a parasite attached to the dorsal surface of the inner hemibranch of the gill illustrated in Figure 6.8 would likewise need to be 'right-footed'. Thus, the direction of asymmetry is dependent upon the particular site of attachment of the parasite and the pattern of water flow over this site (see also Figure 6.7HI). Since there are presumably equal numbers of sites suitable for 'right-' and 'left-footed' individuals, it is curious that most individuals (87%) of *Axine belones* are 'right-footed'.

6.5.2 *Are asymmetrical parasites sedentary?*

The question arises as to how parasites like *Axine* co-ordinate the movement of so many clamps. One possibility is that many, possibly all, polyopisthocotyleans with clamps in excess of four pairs never change their position on the gills, merely using each new clamp as it becomes functional to grasp extra free secondary lamellae. The asymmetry of polyopisthocotyleans like *Axine* is a permanent feature produced by differential growth and once established in their microhabitat they are most unlikely to be able to move to a new site. Even if a parasite like *Axine* could shift to the adjacent surface of the neighbouring primary lamella on the hemibranch or to the opposite surface of its own primary lamella, its asymmetrical shape would be incompatible with the new location and totally opposed to the direction of the gill-ventilating current. The best that *Axine* could achieve would be to shuffle its clamps along the surface of the primary lamella. This has not been observed in polyopisthocotyleans with many clamps like *Axine*, but Bovet (1967) described such a shuffling movement in the 'diporpa' larva of diplozoids (see also below).

Obtaining food is no problem for a sedentary polyopisthocotylean because of the plentiful local supply of blood. Finding a mate is not so easily solved and we will return to this later.

6.5.3 *Maintenance of symmetry*

Having gained an appreciation of why so many gill parasites display asymmetry, the question arises as to how some polyopisthocotyleans maintain their symmetry. We must examine the symmetrical parasites one by one because no single answer is applicable to all of them.

The four clamps on each side of the haptor of *Kuhnia scombri* (Figure 6.7B) can be accommodated side by side on the same side (either dorsal or ventral) of a primary gill lamella. Llewellyn (1957) examined the locations of about two dozen specimens on the gills of mackerel (*Scomber scombrus*) and found all of them attached with the body axis parallel to the primary lamella near its proximal end. In this location the parasite is in the lee of the gill arch and is sheltered by an interbranchial septum lying between the inner and outer hemibranchs of the gill. Consequently, *Kuhnia* is unlikely to be subjected to a strong unilateral water current in this locality, permitting the parasite to retain its bilateral symmetry (Llewellyn 1957, 1966). This contrasts with the highly asymmetrical body of *Grubea cochlear* (Figure 6.7HI), which inhabits the same host. *Grubea* has four clamps on one side of the haptor only and a single small clamp on the opposite side, but the parasite is much larger than *Kuhnia* and presumably is exposed to uninterrupted and much stronger unilateral currents beyond the shelter of the interbranchial septum.

The brown trout (*Salmo trutta*) also has an interbranchial septum, but, according to Llewellyn & Owen (1960), *Discocotyle sagittata* (Figure 6.7C) is attached in such a position that its body projects distal to this septum and hence would be exposed to a strong unilateral current. Llewellyn & Owen found that the secondary gill lamellae of trout up to 30 cm long are too small to accommodate two clamps of *Discocotyle* side by side. The parasite solves this problem by attaching the four clamps on one side of its body anterior to those on the other side. Sometimes the leading clamps are those of the left side, sometimes those of the right side. The temporary asymmetry achieved is always such that the body of the parasite projects towards the interbranchial space, i.e. its body is inclined in the direction taken in that particular locality by water currents passing through the gill. Unlike *Axine* and other parasites with a permanently asymmetrical body, the asymmetry of *Discocotyle* is reversible and the clamps of relaxed detached specimens invariably resume their symmetrical positions.

Diplozoids are related to *Discocotyle* and they display a similar facultative (impermanent) asymmetry, with the clamp row on one side of the body attached in front of the other on the surface of the primary lamella. However, as Bovet (1967) pointed out, this attitude cannot be explained as in *Discocotyle* by lack of space on the surface of the primary lamella, since there is ample width to accommodate the two rows of clamps side by side (Figure 6.9). Nevertheless, because of this attitude the body of the parasite projects towards the interbranchial space. This reduces interference with the gill-ventilating current as in *Discocotyle*, but, more importantly, it permits the parasite to connect by permanent body fusion with a partner (see below for further consideration of this remarkable phenomenon). In 71% of fused pairs of *Diplozoon paradoxum* on bream the two partners are attached to opposite faces of the same primary lamella.

Plectanocotyle gurnardi (Figure 6.7A) is also small, but Llewellyn (1966) attributed its symmetry to the fact that, unlike most of its relatives, it is not sessile. The anterior end of this parasite has adhesive properties, readily sticking to a needle, and it is presumably capable of leech-like locomotion. As it moves from site to site on the gills it will be exposed to currents from various directions, like its gill-parasitic monopisthocotylean relatives, so that it is not obliged to become asymmetrical.

Difficulties in co-ordinating the actions of several clamps during locomotion may have led to the reduction of clamps from four to three pairs in *Plectanocotyle*.

None of the explanations above can account for the symmetry of *Diclidophora* spp. (Figures 6.3, 6.7D). With their wide bodies and clamps mounted on stalks or peduncles, they are far too large to be accommodated on one side of a primary lamella. In fact, these features give them sufficient reach to be able to attach to both sides of the

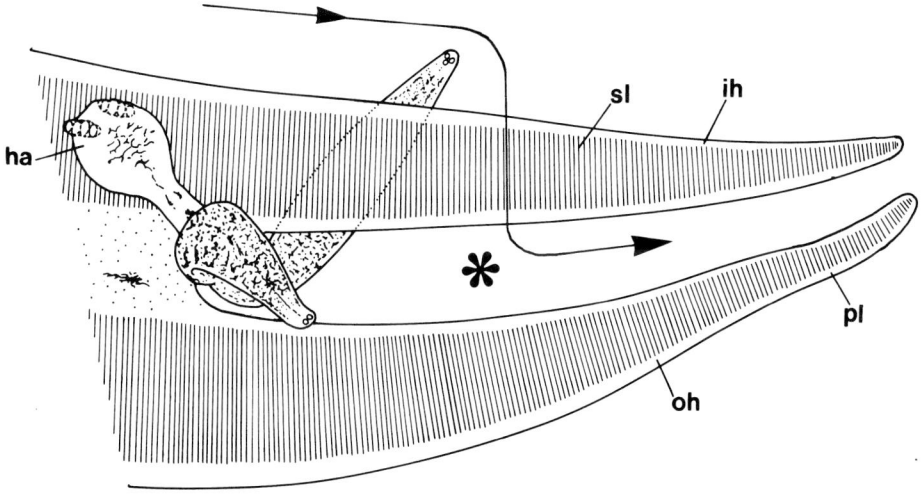

Fig. 6.9. A pair of fused *Diplozoon* adults attached to the gill. ha, Haptor (haptor of other individual out of view on other side of primary lamella); ih, inner hemibranch; oh, outer hemibranch; pl, primary gill lamella; sl, secondary gill lamellae. Star, interbranchial space. Arrows, direction of gill-ventilating current. From Bovet, 1967.

same primary lamella as in *D. luscae* or to span up to five primary lamellae as in *D. merlangi*. Most species lie along the inner (afferent) border of the primary lamella with the haptor near the gill arch (Llewellyn & Tully, 1969). The body of *D. luscae* is folded ventrally along the mid-line, so that the clamps of the left side attach to secondary lamellae on one side of the primary lamella and the clamps of the right side attach to the secondary lamellae on the opposite side of the same primary lamella. The special feature of this adhesive attitude is that wherever the parasite attaches itself on the gills, the gill-ventilating current will irrigate both sides of the body equally and body symmetry can be maintained.

6.6 EGGS, HATCHING AND HOST FINDING

The eggs of most polyopisthocotyleans are fusiform (spindle-shaped) or spheroidal (Figure 6.10). Tetrahedral eggs, like those of *Entobdella* (Chapter 3), are unknown in polyopisthocotyleans. Some eggs of polyopisthocotyleans have no stalks, while others have stalks at one or at both ends (Figure 6.10). There is a wide range of hatching options in polyopisthocotyleans and the examples that follow have been chosen to illustrate this range.

122   CHAPTER 6

6.6.1 Convergent evolution in dogfish parasites
In Chapter 4 we considered the biology of the egg and oncomiracidium of the monopisthocotylean monogenean *Leptocotyle minor* and the remarkable adaptations for infection of the dogfish, *Scyliorhinus canicula*. Whittington (1987ab) also studied in detail the egg and oncomiracidium of a polyopisthocotylean, *Hexabothrium appendiculatum*, from the same host. *L. minor* and *H. appendiculatum* are distantly related and yet their eggs, egg hatching and oncomiracidial behaviour patterns are strikingly similar. This is a spectacular example of convergent evolution, in which the eggs and larvae of two unrelated parasites, faced with the same problems of infecting the same host, have experienced the same selection pressures and have solved these problems in similar ways. Once established on the host, their microhabitats are different, with *L. minor* on the skin and *H. appendiculatum* on the gills, and their biology then diverges.

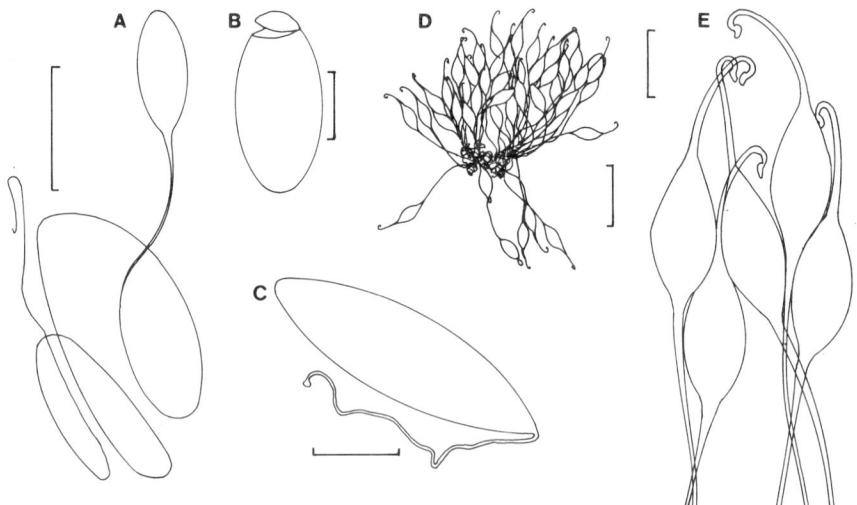

Fig. 6.10. The eggs of some polyopisthocotyleans. (A) *Hexabothrium appendiculatum*; (B) *Discocotyle sagittata*; (C) *Plectanocotyle gurnardi*; (D) *Diclidophora luscae*, whole egg bundle; (E) a few eggs of *D. luscae* enlarged. Scale bars: A, 200 µm; B,E, 100 µm; C, 50 µm; D, 500 µm. From: (A) Whittington, 1987a, (B) Owen, 1970, (C) Whittington & Kearn, 1989, with permission of Cambridge University Press. (DE) Drawn from photographs in Kearn, 1986a.

The fusiform egg of *H. appendiculatum* has a single, amazingly long and slender stalk (mean length almost 3 mm; egg length about 200 µm) (Figure 6.10A). The egg of *L. minor* is strikingly similar (Figure 4.7). As in *L. minor*, water turbulence generated by the activities of dogfish swimming near the bottom is likely to interact with the long stalk, lifting it and the egg off the bottom. Frictional resistance to sinking is also provided by the long stalk, keeping the egg in suspension and increasing its chances of making contact with the host. The egg sinks slowly in still seawater at a rate of 1.26 cm/minute, removal of the stalk increasing the sedimentation rate by 38%.

The effectiveness of this strategy is greatly increased, again like *L. minor*, because fully developed eggs of *H. appendiculatum* hatch rapidly (within 25 – 50 seconds) when exposed to seawater from a tank containing a dogfish or to washings prepared by squirting water over a dogfish's skin. Like *L. minor* the chemical hatching factor in these washings is urea, a substance known to be present in significant amounts in elasmobranch secretions (see p. 66). The low threshold of this response, again shared with *L. minor*, is consistent with the likelihood that drifting eggs of *H. appendiculatum* will experience brief exposure to relatively dilute secretions from swimming dogfish. Eggs kept in clean seawater, i.e. in the absence of chemical or any other signals from the host, remain unhatched and may still be alive after 88 days at about 13°C.

As in *L. minor*, the fleeting and less intimate contacts between the eggs of *H. appendiculatum* and the dogfish demand that the oncomiracidium is capable of swimming and ciliated cells are retained. The similarity with *L. minor* extends also to oncomiracidial behaviour, which is characterised by a brief free-swimming life (typically no more than 30 minutes), a tendency to remain in the vicinity of the eggs and a readiness to attach to any available substrate (Whittington, 1987b). All of these features are likely to promote the rapid and successful establishment of larvae on the host that provided the chemical hatching stimulus.

6.6.2 Rhythmical hatching in Discocotyle sagittata

The eggs of the freshwater parasite *Discocotyle sagittata* are ovoid with no stalks (Figure 6.10B). The native host is the brown trout, *Salmo trutta*, but the introduced rainbow trout, *Oncorhynchus mykiss*, is also readily infected. Gannicott & Tinsley (1997) collected eggs from parasites on the latter host and incubated them in dechlorinated water uncontaminated by trout. They found that hatching is rhythmical with a 24-hour cycle, most larvae emerging during the first two hours of darkness. This hatching pattern may confer advantages for host finding since the trout host is relatively inactive at night and therefore more accessible to slow-moving infective oncomiracidia. However, Gannicott & Tinsley found that hatching could be induced during the hours of daylight by treating the eggs with trout skin mucus or with host gill tissue. This suggests that trout occasionally rest during daylight and that chemical cues from the host, hitherto unidentified, are able to override the normally strict nocturnal hatching rhythm.

6.6.3 Hatching response to shadows in Plectanocotyle gurnardi

The fully developed eggs of *Plectanocotyle gurnardi* are fusiform with a short stalk at the end opposite the operculum (abopercular end) (Figure 6.10C). When eggs aged 13 days or more are shadowed, many hatch within 1 – 2 seconds by rapidly extending the body and pushing off the weakly attached operculum. A remarkably similar hatching response to shadows occurs in the monopisthocotylean *Entobdella diadema* (Chapter 3).

The gurnard hosts (see Table 6.1) of *P. gurnardi* are round-bodied and therefore will not cast such extensive shadows as the stingray hosts of *E. diadema*. However, gurnards forage for food most actively during daylight, moving slowly over the bottom and exploring the sediment with finger-like, independent pectoral fin rays (see Whittington & Kearn, 1989). These fishes will undoubtedly cast shadows and stimulate eggs in the sediment as they work. Freshly hatched larvae swim vertically upwards and are likely to make contact with the host. If they fail to do so they become photonegative, return to the bottom and attach themselves by anterior glandular secretions to the substrate. If undisturbed they will remain attached, but will immediately resume swimming if a shadow is cast over them. A newly arrived potential host is likely to produce such a shadow and the prompt response of the larva will offer another opportunity for host contact. The sit-and-wait strategy presumably conserves energy and may explain why the oncomiracidia of *P. gurnardi* retain their ciliated cells for longer periods of time than most other oncomiracidia (see Whittington & Kearn, 1989).

Mechanical disturbance also stimulates attached oncomiracidia of *P. gurnardi* to renewed swimming, but curiously, these larvae tend to swim downwards irrespective of light direction, a response that seems to relate more to predator evasion than to host location.

6.6.4 *Egg bundles and mechanical stimulation in* Diclidophora luscae

An unusual feature of the eggs of *Diclidophora luscae* from the gills of the bib, *Trisopterus luscus*, is that each egg is not laid as it is assembled. The fusiform eggs, each about 250 µm long, are retained inside the uterus until on average 58 eggs (46 – 68) have accumulated (Whittington & Kearn, 1988). These eggs are then laid in a single bundle, which is held together by the entangled abopercular stalks (average length 700 µm) (Figure 6.10D). The eggs also have a hooked opercular stalk, on average 164 µm long (Figure 6.10E). The egg bundles are roughly cone-shaped, with the 'apex' of the cone formed by the entangled abopercular stalks and the hooks of the opercular stalks projecting outwards from the 'base' of the cone (Figure 6.10D). Whittington & Kearn (1988) found that these bundles sink rapidly at a speed of 17.52 cm/minute, with the 'apex' of the cone invariably pointing downwards. Single eggs removed from the bundles with as much of the abopercular stalks as possible sank much more slowly at 1.98 cm/minute.

Whittington & Kearn (1988) observed that egg bundles exposed to alternating experimentally imposed 12-hour periods of light and darkness, with minimum disturbance and no shadowing, began to hatch after 30 – 34 days at 13 – 14°C, and most oncomiracidia emerged during the two hours before and the two hours after 'dusk'. However, when the fully developed egg bundles were subjected to vigorous disturbance during the light or the dark periods, either by agitation with a needle or by jets of seawater, they began to liberate large numbers of ciliated oncomiracidia within 10 seconds. Eggs continued to hatch for up to five minutes after disturbance ended. Shadowing and host skin secretions appeared to have no stimulatory effect on hatching (see Whittington & Kearn, 1988 and Macdonald, 1975, respectively).

The egg bundles of *D. luscae* do not become attached to the gills or to any other part of the body of the bib and are likely to sink rapidly after laying (Whittington & Kearn, 1988). When they reach the sea bottom, the free, hook-like opercular stalks may act as grappling hooks, tethering the bundles to outgrowths or encrustations of animal or plant origin (Kearn, 1986a). The bib spends most of its time on or near the bottom, especially around rocks or wrecks (see Macdonald, 1975) and as the egg bundles are disturbed by the activities of the foraging fish, hatching will take place, followed by host infection. Chances of contacting the host are further improved by the behaviour of the freshly hatched larva, which swims vigorously upwards then sinks passively. Upward swimming and passive sinking are then repeated, thereby keeping the larva in suspension in the fish's living space, while at the same time minimising the expenditure of energy.

Another possible advantage of laying eggs in bundles combined with mass hatching triggered by host-generated mechanical disturbance, is that a single host fish is likely to be infected by many oncomiracidia from a single egg bundle and hence from a single parent parasite. As suggested by Llewellyn (1981), this could lead to mating between siblings, the survival value of which may be the conservation of highly specialised and useful characters.

Rapid sinking of the egg bundles and anchorage by the 'grappling hook' stalks might also prevent egg bundles being swept away from areas suitable as habitats for bib. In fact, Kearn (1986a) has suggested that it may be advantageous for the parasite to release its egg bundles only when the host is visiting a favoured resting or feeding site, where the tethered egg bundles would be likely to meet bib at other visiting times. If these visits have a daily regularity, then egg laying may have a corresponding daily regularity. Unfortunately, it is not yet known whether *D. luscae* has a rhythmic daily laying pattern, but Macdonald & Jones (1978) identified a laying rhythm in a diplozoid[1] from southern barbel, *Barbus meridionalis*, in France. They found that more eggs were laid by this parasite during the night than during the day and that this would lead to deposition of larger numbers of eggs in the nightly resting-places of the diurnally active hosts. The timing of egg hatching, which takes place at dusk, enhances the effectiveness of this egg-laying pattern.

6.6.5 *Diapause (?) in* Gastrocotyle trachuri

During the course of several years of study at the Plymouth Laboratory of the Marine Biological Association of the United Kingdom, Llewellyn (1962) became aware of an interesting phenomenon. During the summer months of July, August and September he found only adult specimens of *Gastrocotyle trachuri* and *Pseudaxine trachuri* on the gills of their hosts, young scad (*Trachurus trachurus*). On the other hand, young fish caught in May harboured larvae, juveniles and adults. He attributed this pattern to some kind of temporary reproductive inactivity, leading to a cessation of infection. How reproductive inactivity is achieved is unknown, but Llewellyn suggested either that eggs may enter a state of diapause or that egg assembly itself may be suspended.

According to Llewellyn, this cessation of reproductive activity 'anticipates' a change in host behaviour. Young scad leave the bottom, where infection occurs, in July and become plankton feeders. Thus the parasites avoid production of offspring at a time when hosts are absent, but they resume reproductive activity in August, thereby 'anticipating' the return of potential hosts, young scad, to the sea bottom in October. If, as seems likely, the behavioural cycle of the host is orchestrated by hormonal changes, then these same changes may regulate the parasites' reproductive cycle – as blood feeders the monogeneans would have access to circulating host hormones.

---

[1] Regarded by Macdonald & Jones (1978) as a subspecies, namely *Diplozoon* (now *Paradiplozoon*) *homoion gracile*, as distinct from the subspecies *P. homoion homoion* on roach. Whether this subspecific distinction is valid has not yet been resolved (see Matejusová *et al.*, 2001).

6.7 ROUTE TO THE GILLS

Some monopisthocotylean gill parasites are known to attach themselves initially to the outer body skin of their hosts, later migrating to the gills (see, for example, Kearn, 1968). This invasion pattern requires sufficient behavioural flexibility to enable these skin invaders to find their way to the gills. It is also essential for the young skin parasite to replace the resources used in host finding and to fuel itself for the energetically costly process of migration. Since these parasites feed on superficial epidermis, food is accessible to the smallest skin invaders. Using a fluorescent dye to stain living oncomiracidia of the polyopisthocotylean *Heterobothrium okamotoi*, Chigasaki et al. (2000) detected newly attached larvae on the body skin as well as on the gills of the host, the tiger puffer (*Takifugu rubripes*). However, it is not known whether these skin invaders are able to reach the gills and there is now a considerable body of evidence to suggest that many polyopisthocotylean oncomiracidia gain direct access to the gills, most probably by being passively drawn in and carried to the gills by the host's gill ventilating current. It is perhaps not surprising that the evidence outlined below points in this direction, because unless polyopisthocotyleans are able to switch from epidermis feeding to blood feeding during their early life, initial feeding on blood would be impossible, dermal blood vessels being inaccessible to tiny newly-attached parasites.

Frankland (1955) described the attachment of larvae of *Diclidophora denticulata* to the gills of saithe, *Pollachius virens*, and observed the shedding of their ciliated cells. Bovet (1967) observed that the ciliated oncomiracidium of *Diplozoon paradoxum* suddenly stops swimming when it enters the inhalant respiratory current of the host and is carried passively into the buccal cavity and to the gills. He induced larvae to attach to gills by irrigating with jets of water and noted that they lost their ciliated cells, but he never observed an oncomiracidium crawling on the surface of the host.

Paling (1969) found that the oncomiracidia of *Discocotyle sagittata* failed to attach themselves to scales of brown trout (*Salmo trutta*). On the other hand, oncomiracidia introduced into the mouth of the host via a pipette established themselves on the gills and underwent development. Paling was convinced that passive infection of the gills via the mouth takes place. Gannicott & Tinsley (1998) have reinforced this recently. Within two hours of hatching, they found newly invaded larvae of *D. sagittata* without their ciliated cells on the gills of rainbow trout (*Oncorhynchus mykiss*). An important additional observation was that these larvae already contained red gut contents, indicating that larvae are capable of feeding on blood soon after hatching. No larvae were found on the operculum or buccal lining, supporting the notion that larvae in suspension are drawn passively through the buccal cavity to the gills and Rubio-Godoy & Tinsley (2002) found that the proportions of early stages of *D. sagittata* on the gill arches of rainbow trout corresponded with the relative amounts of water flowing over the arches.

Wiskin (1970) found small specimens of *Rajonchocotyle emarginata* little larger than oncomiracidia in the gill mucus of heavily infected thornback rays (*Raja clavata*) from an aquarium. She found pigment regarded as haematin in the smallest parasites, indicating early blood feeding, and found no small parasites in mucus from the skin and buccal cavity of a heavily infected ray. Whittington & Kearn (1989) also found no specimens of *Plectanocotyle gurnardi* in scrapings from the skin surfaces of gurnards that seemed likely to have been exposed to frequent invasion by oncomiracidia. On the other hand, they did find immature parasites on the gill arch and base of the gill, indicating that newly invaded larvae carried into the mouth by the gill-ventilating current may alight here. Llewellyn (reported by Whittington & Kearn, 1989) examined gills and scrapings from the body surfaces of many gurnards caught at the time of year when invading larvae might be expected, and found small parasites (some without clamps) on the gills, but none in the body scrapings. The turbulence of water passing through the gills may promote attachment, since Whittington & Kearn (1989) observed that many oncomiracidia of *P. gurnardi*, subjected to turbulence by being drawn into a pipette, attached themselves by the anterior ends to the rim of the pipette aperture or to its inner surface.

6.8 NICHE RESTRICTION AND MATING

Adult polyopisthocotyleans are not randomly distributed over the available surfaces of the four pairs of teleost gills. *Diclidophora merlangi* on the whiting (*Merlangius merlangus*), for example, shows a strong preference for the first (anteriormost) gill, which, according to Llewellyn (1956b), accommodates 80.3% of adult parasites, with 12.1% on gill 2, 1.5% on gill 3 and 6.1% on gill 4. Macdonald & Caley (1975) obtained similar figures for the distribution of 417 adult *D. merlangi* on the gills of 310 whiting caught off the northeast coast of Scotland. They also observed that the numbers of parasites on the left and right sides of the fish were almost exactly the same, but the outer hemibranchs (all gills included) harboured almost all the parasites (94.3%) compared with only 5.7% on the inner hemibranchs. The microhabitat of *Kuhnia sprostonae* on the mackerel (*Scomber scombrus*) is even more restricted since it is found only on the pseudobranch, a gill vestige on the inside of the operculum.

The mechanism behind this phenomenon of niche restriction and its advantages for the parasite have been much debated, but the answers to these questions seem most likely to be found in the sedentary way of life of most adult polyopisthocotyleans. Cross insemination is desirable for monogeneans and if relatively few sedentary adults are present on each fish and are dispersed over the gills many individuals may be denied the opportunity to mate with conspecifics. Parasites must rely on the extensibility and manoeuvrability of the free anterior region of the body, where the penis opening is situated, to reach a co-copulant. Any mechanism that favours aggregation of adult parasites is likely to have a selective advantage.

It seems unlikely that the invading free-swimming oncomiracidia have sufficient manoeuvrability and are discerning enough to swim directly to the restricted zone where the adult parasites aggregate. Llewellyn (1956b) suggested that infective larvae are involuntarily swept over the gills and that the number of passively drifting larvae received by each gill will be proportional to the volume of water passing over it. In this way passive infection could produce an aggregated distribution in parts of the gill cavity receiving larger volumes of water. Statistical analyses by Rubio-Godoy & Tinsley (2002) indicated that this might be the explanation for the greater frequency of early stages of *D. sagittata* on gill 2 and gill 3 in rainbow trout (*Oncorhynchus mykiss*). However, they found that this distribution pattern changes with time, older parasites being more abundant on gill 1 and gill 2. Two possible explanations for these changes in distribution come to mind: (a) differential mortality whereby juveniles survive only in those restricted habitats that are suitable for them and (b) migration of juveniles to the preferred adult sites. Since the numbers of parasites collected by Rubio-Godoy & Tinsley at one, two and three months post-infection were not significantly different, migration of juveniles is the most likely explanation.

In this context, a significant behavioural feature of newly attached polyopisthocotyleans is that they are mobile. Frankland (1955) commented on the leech-like locomotion of freshly attached larvae of *Diclidophora denticulata* and Paling (1969) and Gannicott & Tinsley (1998) noted that young *Discocotyle sagittata* use 'looping movements' to move amongst the gills. Bovet (1967) described two kinds of locomotion in the juvenile stage of diplozoids (known as a diporpa before coupling with another individual – see below), the first being a creeping movement achieved by alternate displacements of the clamps and the second the more typical monogenean leech-like locomotion involving attachment of the haptor and the anterior buccal suckers. In contrast, most adult polyopisthocotyleans are assumed to be immobile (see above).

The idea of post-invasion migration receives further support from observations made on *Diclidophora merlangi* by Macdonald & Caley (1975). They observed that most adults live on the outer hemibranch of the first gill of the whiting (*Merlangius merlangus*) and that most of these adults are attached close together. In their sample, in 73.7% of gills with more than one attached adult, adults were attached to adjacent primary lamellae or were separated by no more than two primary lamellae. In contrast, when a juvenile parasite was located on the same gill as an adult, the parasites occupied adjacent sites in only 31% of cases. This suggests that there is some movement along the gill at the late juvenile or early adult stage.

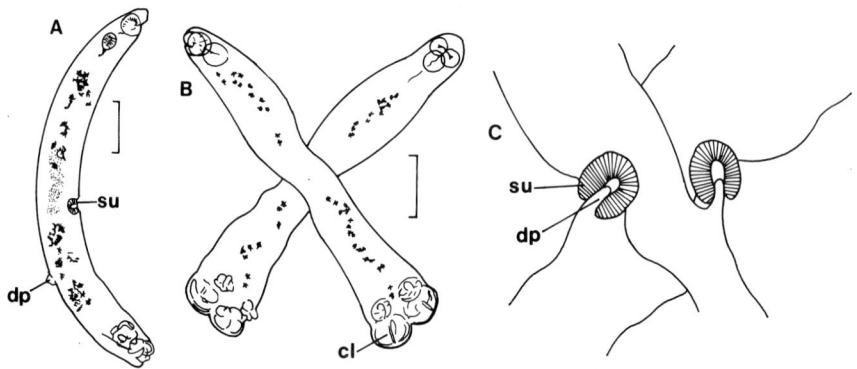

Fig. 6.11. *Diplozoon*. (A) Diporpa with two pairs of clamps (about two weeks old) seen in lateral view. (B) Young couple about one week old. (C) enlarged view of the junction region of a freshly coupled pair of diporpae. cl, Clamp; dp, dorsal papilla; su, sucker on ventral surface. Scale bars for A, B, 100 µm. (A,B) Reproduced from Bovet, 1967. (C) From Khotenovskii, 1985, with permission from Nauka Publishers.

We can judge just how powerful selection pressures are for polyopisthocotyleans to cross-inseminate by a truly remarkable development in diplozoids. After attachment to the host, the diplozoid post-larva or diporpa, which already has a functional pair of clamps, develops a small sucker in the centre of the body ventrally and on the dorsal side a papilla supplied with glands (Figure 6.11A). As noted above, the diporpa is capable of locomotion and if it should meet another diporpa, the ventral sucker of one individual grasps the dorsal papilla of the other. According to Bovet (1967) and Khotenovskii (1985), the bodies of the two parasites then twist so that the ventral sucker that is still free is able to grasp the free dorsal papilla of the other individual (Figure 6.11C). As development proceeds, there follows the complete fusion of the bodies of the two individuals anterior to the haptors, producing a double animal (Figures 6.11B, 6.12), the uniqueness of which is reflected in the generic name *Diplozoon*, meaning double animal.

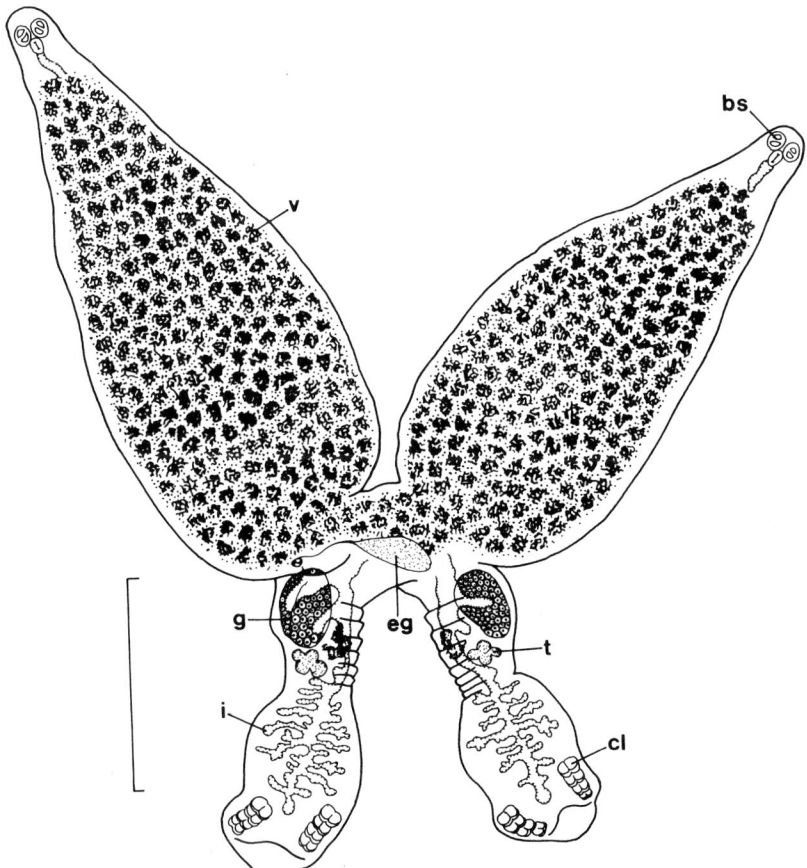

Fig. 6.12. Diplozoon paradoxum, adult couple. bs, Buccal sucker; cl, clamp; eg, egg; g, germarium; i, intestine; t, testis; v, vitellarium. Scale bar: 1 mm. From Khotenovskii, 1985, with permission from Nauka Publishers.

According to Bovet (1967), the intestines of the two parasites become joined, but more importantly the vas deferens of each individual becomes continuous with the vagina of the other, so that sperm from one individual has permanent access to the female reproductive system of its mate. Bovet (1967) found that isolated diporpae that failed to find a mate do not reach sexual maturity and survive as juveniles for no more than five months. Diporpae that find and fuse with a mate and hence achieve maturity live for at least two years.

The two diporpae of a pair are likely to be genetically distant. The permanent grafting together of tissue from genetically distinct individuals is common in plants but has few parallels in the Animal Kingdom.

Mating in other polyopisthocotyleans has been observed rarely. This may be because it is brief, because mating individuals separate when the host's gill-ventilating current ceases or because few parasitologists observe living animals. Macdonald & Caley (1975) observed unilateral mating in *Diclidophora merlangi* and discovered that the process is rather traumatic, since there is no vagina and sperms have to pass through the tegument either via a digested pathway or through wounds made by hooks on the male copulatory organ. *Gastrocotyle trachuri* also has no vagina and although Llewellyn (1983) did not witness mating he surmised that hypodermic impregnation occurs by way of a sclerotised penis tube. Winch (1983) found a vagina in *Atrispinum labracis* but, since it is mid-dorsal and the penis opening mid-ventral, mutual insemination appears to be physically impossible.

7

LEECHES

7.1 INTRODUCTION

In their mode of locomotion and hermaphrodite reproductive arrangement leeches resemble monogeneans and some leeches ventilate their bodies by means of body undulations (Mann, 1962) in a similar way to the monogenean *Entobdella soleae* (Chapter 3). However, this resemblance is superficial and leeches and monogeneans are unrelated and structurally very different.

The body of a leech consists of a series of segments (metameres), each of which contains the same basic components (gut, muscles, nerves, excretory organs). This repetitive segmental arrangement is a feature of the Annelida, the phylum to which the leeches belong. It has long been recognised that leeches are closely related to the oligochaetes, a group of annelids that includes the familiar earthworm *Lumbricus terrestris*. In fact, the Hirudinea (leeches) and the Oligochaeta were grouped together in the same taxon, the Clitellata by Sawyer (1986ab), on the basis that members of both groups possess a specialised annular region or clitellum on the body surface that secretes the cocoon (see below). However, recent molecular studies by Siddall *et al.* (2001) have taken this further, placing the leeches firmly within the Class Oligochaeta. Separate status in the Oligochaeta is indicated for the Hirudinea, the branchiobdellidans (found predominantly on freshwater crustaceans and of no concern to us here) and the acanthobdellidans (leech-like fish parasites regarded as primitive – see below). According to Siddall *et al.* (2001), all three of these groups should be regarded as orders of the Oligochaeta, equal in status to their closest relatives the Lumbriculida. However, these changes have not yet been formally proposed.

Localised modifications of the body segments occur in leeches to accommodate the reproductive system and specialisations of the gut. Groups of adjacent anterior and posterior (caudal) segments are fused to produce attachment suckers. Since new segments are added at the posterior end of the body in annelids, differentiation of the posterior sucker curtails this process and leeches have a constant number of segments. This also means that leeches are unable to regenerate lost or damaged segments. The terminal position of the posterior sucker also demands a dorsal position for the anus (Figure 7.1D).

It is a common misconception that all leeches are bloodsuckers. Many of the leeches that are encountered living freely in ponds and rivers are predators, feeding on small invertebrates such as free-living oligochaete worms, crustaceans or molluscs. Some free-living leeches use their suctorial proboscis to extract body fluids and other contents from their prey, while others ingest whole organisms or part of them. Leeches feeding exclusively on the blood of vertebrates, including fishes, are also likely to be encountered living freely since they may leave their host after feeding and not reattach to a new host until the last meal has been digested.

Leeches are divided into two groups, the Arhynchobdellida and the Rhynchobdellida. As well as blood feeders, both groups contain predatory leeches that do not feed on blood. However, there are fundamental differences in the way in which blood-feeding arhynchobdellidans and rhynchobdellidans extract blood from their hosts. Arhynchobdellidans include the familiar medicinal leech, *Hirudo medicinalis*, which is known to feed occasionally on fishes (see below). *Hirudo* uses jaws to gain access to blood and prevents clotting while feeding by injecting saliva containing a non-enzymatic polypeptide called hirudin, which specifically inhibits the host's clotting enzyme thrombin. The rhynchobdellidans include all other British leeches that feed on fishes and a giant Amazonian leech *Haementeria ghilianii* that readily feeds on humans as well as on cattle. Rhynchobdellidans extract blood via a tubular proboscis lacking jaws and employ different biochemical methods to prevent clotting. *H. ghilianii* does this by secreting an enzyme, hementin, which prevents clots from forming and also dissolves new clots after their formation. According to Sawyer (1986b), these fundamental morphological and biochemical differences between bloodsucking arhynchobdellidans and rhynchobdellidans indicate that the bloodsucking lifestyle has evolved independently in the two groups. However, Apakupakul *et al.* (1999) found evidence to suggest that the common ancestor of all the leeches was a blood feeder and that the blood-feeding habit has been lost at least four times during the course of leech evolution, twice in the rhynchobdellidans and twice in the arhynchobdellidans.

7.2 THE BRITISH FAUNA

7.2.1 *Marine leeches*
Lists of marine leeches parasitic on fishes in North-West Europe have been published by Knight-Jones (1962) and by Hayward & Ryland (2000). Fifteen of the more common species found on British marine fishes are listed in Table 7.1.

It is likely that new records and even new species may be added to this list in the future. *Brumptiana lineata* (Figure 7.1AB) is a leech judged to merit creation of a new genus. It was discovered as recently as 1984 by Llewellyn & Knight-Jones in the oro-nasal groove and mouth of a ray at Plymouth and also on a plaice, *Pleuronectes platessa*, at Port Erin. Other marine leeches may have been missed because their parasitism is seasonal and hosts may have been sampled at times when the leeches are absent (see below). In addition, there are indications from transfer experiments conducted by Hussain & Knight-Jones (1995) that cryptic leech species might also exist. They found that specimens of *Oceanobdella blennii* (Figure 7.1C) removed from the shanny (*Lipophrys pholis*) readily reattached to the shanny. On two occasions the authors succeeded with some difficulty in inducing *O. blennii* to attach to butterfish (*Pholis gunnellus*), but the leeches refused to attach to sea scorpion (*Taurulus bubalis*). When the butterfish carrying *O. blennii* were transferred to an aquarium with other fishes, the leeches on each occasion transferred to shanny within a few days. In contrast, two leeches assumed to be *O. blennii* and collected in Scotland from butterfish refused to attach themselves to shanny, even after starving for a week, raising the possibility that the leeches on butterfish represent a distinct race or separate species.

Some of the marine species on elasmobranch hosts are impressively large. *Branchellion torpedinis* on rays is 3 – 5 cm in length and readily identified by 33 pairs of leaf-shaped, lateral projections or branchiae, which presumably function as gills

(Figure 7.1DE). However, *Branchellion* is outclassed in size by *Pontobdella* spp. on skates and rays (Figure 7.1FG). *P. vosmaeri* reaches eight cm in length, but *P. muricata* is the largest, ranging at rest from 7.5 – 10 cm and measuring when fully extended as much as 20 cm. It is not surprising that such conspicuous parasites are familiar to British fishermen, *Pontobdella* spp. being known as 'skate leeches' or 'skate suckers'.

Table 7.1. Some British leeches parasitic on fishes. Bold type: leeches found in fresh water. From Hayward & Ryland, 2000. Leeches represented by finds of only one or a few specimens not included.

Leech	Host(s)	Locality	Comments
Branchellion borealis	Rays (*Raja* spp.)	South-west	
Branchellion torpedinis	Rays (*Raja* spp.), occasionally teleosts	South-west	
Brumptiana lineata	Rays (*Raja* spp.)	South-west	In mouth and nasal fossae
Calliobdella lophii	Angler (*Lophius piscatorius*)	Norway to Mediterranean	On ventro-lateral body surface
Calliobdella nodulifera	Various, especially gadids	Scotland	
Calliobdella punctata	Sea scorpion (*Taurulus bubalis*)	Isles of Scilly	Transmits *Trypanosoma cotti*
Hemibdella soleae	Common sole (*Solea solea*)	South-west, North Sea	Attached to scales
Hemiclepsis marginata	Various, see text	Widespread	
Heptacyclus myoxocephali	Bull-rout (*Myoxocephalus scorpius*)	Millport, Firth of Clyde	
Hirudo medicinalis	Three-spined stickleback, others?	Romney Marsh, Dungeness	
Oceanobdella blennii	Shanny (*Lipophrys pholis*)	Locally common on east and west coasts	Usually behind pectoral fin, juveniles in gill chamber
Oceanobdella microstoma	Cottids	Plymouth, Anglesey, St. Andrews	Ventrally under head
Oceanobdella sexoculata	Butterfish (*Pholis gunnellus*)	St. Andrews, Northumberland	Behind pectoral fin
Piscicola geometra	Various	Widespread	Also brackish water (plaice, *Pleuronectes platessa*, *Myoxocephalus* in the Baltic)
Platybdella anarrichae	Wolf-fish (*Anarhichas lupus*)	North Sea	On gills and body
Pontobdella muricata	Rays (*Raja* spp.), occasionally plaice (*Pleuronectes platessa*)	North Sea	
Pontobdella vosmaeri	Rays (*Raja* spp.)	Plymouth	
Sanguinothus pinnarum	Sea scorpion (*Taurulus bubalis*)	Firth of Clyde, Berwick-on-Tweed, Anglesey	On fins

The scarcity of the leech *Sanguinothus pinnarum* on its host the sea scorpion (*Taurulus bubalis*) on Anglesey and its apparent absence from this fish further south prompted Hussain & Knight-Jones (1995) to suggest that this is related to the exposed position of the parasite on the fins and the possibility of predation by the corkwing wrasse, *Crenilabrus melops*, a common cleaner fish off the south-west coast of Britain.

Oceanobdella microstoma parasitises the same host but its occurrence extends as far as south Devon. Hussain & Knight-Jones suggested that *O. microstoma* might escape the attentions of cleaner wrasse because it is usually attached beneath the head. Samuelson (1982, in Hussain & Knight-Jones, 1995) recorded removal of *Calliobdella lophii* from the angler (*Lophius piscatorius*) by the rock cook, *Centrolabrus exoletus*.

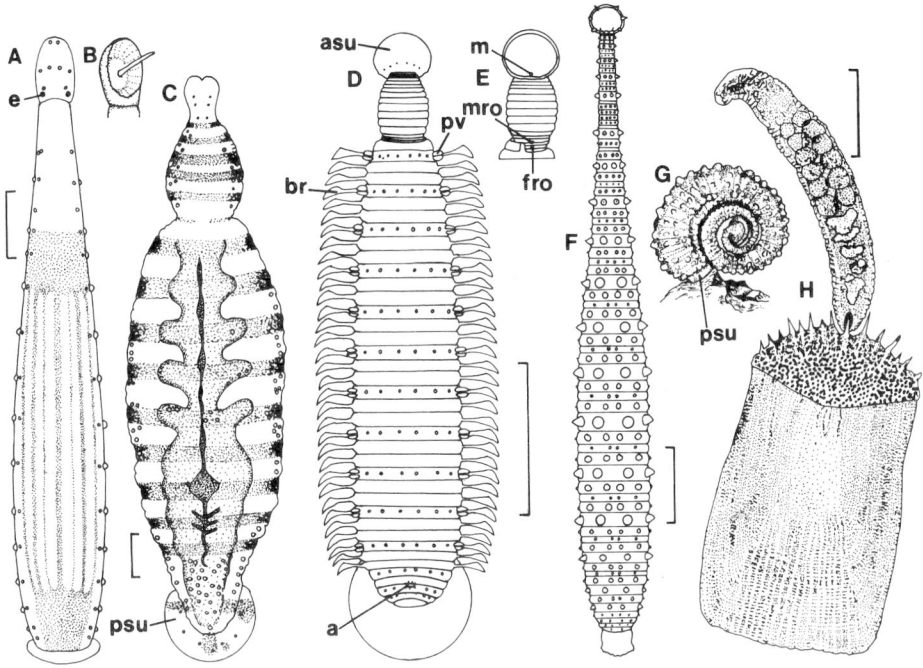

Fig. 7.1. Some marine leeches from British fishes. (A) *Brumptiana* lineata; (B) ventro-lateral view of oral sucker of *B. lineata* with proboscis extruded. (C) *Oceanobdella blennii*; (D) *Branchellion torpedinis*; (E) anterior region of *B. torpedinis* in ventral view; (F) *Pontobdella muricata*; (G) typical posture of *P. muricata* when separated from its host; (H) *Hemibdella soleae*, attached to a scale of the common sole, *Solea solea*. a, Anus; asu, anterior sucker; br, branchia (gill?); e, eye; fro, female reproductive opening; mro, male reproductive opening; psu, posterior sucker; pv, pulsatile vesicle. Scale bars: (ABG) 1 mm; (CDE) 1 cm. Reproduced from the following: (AB) Llewellyn & Knight-Jones, 1984, with permission of Cambridge University Press ; (C) Hayward & Ryland, 2000, by permission of Oxford University Press; (D– G) Harding, 1910, with permission of Cambridge University Press; (H) Llewellyn L., 1965, with permission of the Zoological Society of London.

Hemibdella soleae (Figure 7.1H) is small, with a maximum relaxed length of about 6 mm (Llewellyn L., 1965), but it has some special features that set it apart from other British marine leeches. Like other leeches it uses its posterior sucker for attachment, but this is applied not to a flat surface but to one of the spines projecting from the margin of a ctenoid scale (see Figure 7.1H and below). In British waters *H. soleae* is restricted to the upper surface of the common sole, *Solea solea* (see Frontispiece), but apparently infests other soles in the Mediterranean (Selensky, 1931, in Llewellyn L., 1965). Leeches are frequently found on common soles caught off the coast of Wales and near Plymouth (Llewellyn L., 1965) and I have frequently seen leeches on common soles caught in the North Sea (see Frontispiece). Llewellyn analysed leech populations on common soles caught off the coast of Wales and southern England (Plymouth) and found that only 3% of immature soles (less than 20 cm long and under two years of age) were infected with one or two leeches per fish. On mature fishes the prevalence was about 35%, ranging from 29% to 100% in different localities, with up to 40 parasites per fish. Llewellyn records 70 leeches on one mature sole. Llewellyn found no leeches on the sand sole (*Pegusa lascaris*) or on the solenette (*Buglossidium luteum*) caught off the coast of Wales, and I have not seen *H. soleae* on freshly caught thickback sole (*Microchirus variegatus*) or on solenette caught at Plymouth. The strict host specificity of *H. soleae* is emphasised by experiments performed by Llewellyn (Llewellyn, L., 1965) in which he attempted to establish young leeches on the sand sole. He found that they rarely attached to the sand sole, but the same leeches later attached themselves readily to the common sole.

Another special feature of *H. soleae* is that it spends the whole of its life attached to the host, unlike most other British fish leeches, which are intermittently parasitic, visiting the host to feed and then leaving to digest their meal. Perhaps this shift towards more permanent parasitism in *H. soleae* is related to the host's sandy or muddy substrate, which provides no suitable sites for attachment of the leech after leaving the host. Some estuarine and marine leeches are forced to utilise the surfaces of crustaceans for this purpose (Burreson, 1995).

7.2.2 Freshwater leeches

Three species of leeches feed on British freshwater fishes, but each of these belongs to a different family. Two of them are rhynchobdellidans, namely *Piscicola geometra* belonging to the Piscicolidae and *Hemiclepsis marginata* belonging to the Glossiphoniidae. Surprisingly, the arhynchobdellidan medicinal leech *Hirudo medicinalis*, which belongs to the Hirudinidae, also occasionally feeds on freshwater fishes. Much of our knowledge of how leeches function is derived from work on the obligate fish parasites *P. geometra* and *Hemiclepsis marginata*, and in particular on the former. Consequently these will receive more attention below than the other British fish leeches.

P. geometra and *H. marginata* differ in their environmental preferences. *P. geometra* has a higher oxygen requirement and consequently inhabits faster flowing, well oxygenated and/or cool water. *P. geometra* has also been reported in the brackish waters of the Baltic on plaice (*Pleuronectes platessa*) and on *Myoxocephalus* (see Hayward & Ryland, 2000). *H. marginata* is more or less indifferent to rate of water flow, but is most abundant in ponds or slow moving streams with rich submerged vegetation. Nevertheless, the two species may occur in the same body of water and on the same host fish. They are however readily distinguishable (see Figure 7.2 and colour illustration in Elliott & Mann, 1979).

P. geometra (Figure 7.2A) measures 2 – 3 cm in length and 1.5 – 2 mm in breadth (Harding, 1910). Like other piscicolids each side of the body has segmentally arranged so-called pulsatile vesicles, which contain coelomic fluid and pulsate rhythmically, thereby enhancing the uptake and distribution of oxygen (Mann, 1962). The leech is greenish, yellowish or brownish, with many tiny black or brown stellate pigment cells disposed in more or less regular longitudinal and transverse rows. The dark pigment on the upper surface of the oral sucker is arranged in a cruciform pattern, and the dorsal surface of the posterior sucker bears 14 dark radial rays between which are white eye-like spots. There are also rows of longitudinal white spots on the body. The form and extent of these white spots are variable. There are four eyes near the dorsal surface of the oral sucker. *P. geometra* feeds almost entirely on fish, but is said to feed opportunistically on amphibians and possibly on fish eggs (references in Sawyer, 1986b). It displays little host specificity, having been reported on over 30 species of fresh- and brackish water fishes in Europe. Elliott & Mann (1979) list some of the commonly infested freshwater fishes.

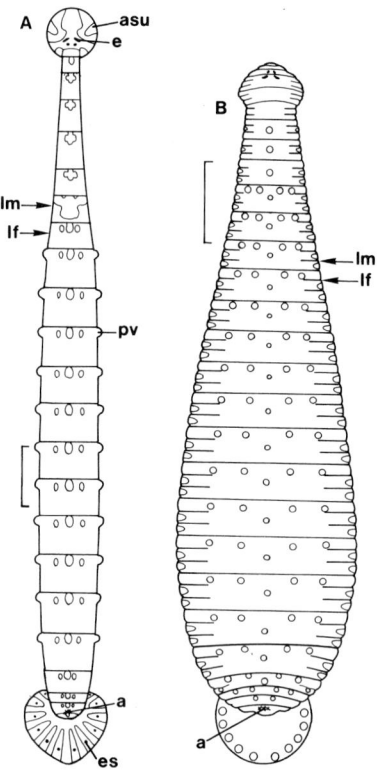

Fig. 7.2. British freshwater fish leeches. (A) *Piscicola geometra* in dorsal view; (B) *Hemiclepsis marginata* in dorsal view. es, Eye-like spot; lf, level of ventral opening of female reproductive system; lm, level of ventral opening of male reproductive system. Other lettering as in Fig. 7.1. Scale bars: 2 mm. Reproduced from Harding, 1910, with permission of Cambridge University Press.

Hemiclepsis marginata (Figure 7.2B) is 16 – 18 mm in length at rest, with a maximum width of 2.5 – 8 mm. It is more or less transparent before feeding, typically with a pale yellow ground colour, variegated above with orange, reddish brown, lemon yellow and intense green (Harding, 1910). There are four eyes in the head region. Like other glossiphoniid leeches, *H. marginata* does not abandon its young in a cocoon, but retains them on its ventral surface and transports them to their first blood meal. Elliott & Mann (1979) list commonly infested fishes. *H. marginata* feeds primarily on coarse fish and also attacks tadpoles but not adult amphibians. It seems to have a predilection for the stone loach (*Noemacheilus barbatulus*), bullhead (*Cottus gobio*) and gudgeon (*Gobio gobio*), which may simply reflect the fact that they are all inactive fishes, living on the bottom where the leeches are also common. Furthermore, the stone loach has reduced scales and the bullhead lacks scales except along the lateral line, so that blood may be easier to reach.

Hirudo medicinalis is uncommon in Britain but persists locally in small numbers (Elliott & Mann, 1979). Generally regarded as a mammalian blood sucker, serological tests and microscopical examination of blood expressed from medicinal leeches collected in the Romney Marsh area revealed that some of them contained fish blood and at Dungeness dead three-spined sticklebacks, *Gasterosteus aculeatus*, were found with fresh bites of this leech (see Ausden *et al.*, 2002).

7.3 ATTACHMENT

The anterior or oral sucker is a muscular and glandular disk typically derived from the ventral half of the first four to five segments, probably the first five segments in piscicolid leeches. An oral sucker with a large diameter, sharply demarcated from the rest of the body, appears to be a feature of many fish leeches, including *Piscicola geometra* and *Hemiclepsis marginata* (Figure 7.2). In *Piscicola* the mouth opening is located in the centre of the disk. In the leeches considered in this book, the posterior sucker is derived from seven segments.

The anterior and posterior attachment organs of leeches are invariably described as suckers, implying that they attach by generating a negative pressure between the organ and the adjacent substrate. In fact, Gradwell (1972) used a transducer to measure the pressure changes involved in the spontaneous attachment of the anterior and posterior suckers of *Placobdella parasitica*, a parasite of freshwater turtles. He achieved this by allowing the leech to attach its sucker over a small hole in a piece of 'Perspex'. When any part of the ventral face of the posterior sucker touched the substrate, the central area of the sucker was immediately everted, creating a convex adhesive surface and facilitating the application of the centre of the sucker against the substrate. Coincident with this, a slight positive pressure was recorded (Figure 7.3). This was followed by the generation of strong suction (usually –2 to –5 cm of water). During prolonged attachment of the posterior sucker, the negative pressure beneath it was not stable, being influenced by movements of the body. For example, when the leech elongated its body an increase in negative pressure (increase in suction) occurred and body shortening led to a decrease in negative pressure. Sub-ambient pressures ranging usually from –2 to –4 cm of water were also recorded between the oral sucker and the substrate.

Adhesive secretions may also make a small contribution to sucker attachment. Van der Lande (1968) reported intense cholinesterase activity in the posterior sucker of *Piscicola geometra* and she suggested that this enzyme might make attachment more effective by digesting and removing the slime on the surface of the host's skin. However, she appreciated that it may have other functions and was at a loss to explain why other blood feeders, including the other fish leeches, lacked such a feature. Gradwell (1972) identified a secretion likely to have a minor adhesive role on the ventral surface of the posterior sucker of *Placobdella parasitica*.

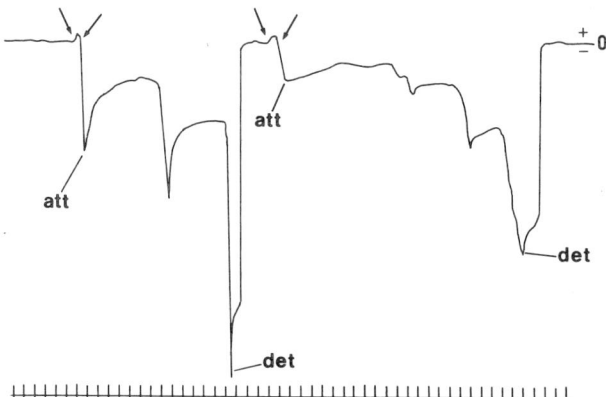

Fig. 7.3. Transducer recording of suctorial pressures during two consecutive attachments of the posterior sucker of the turtle leech *Placobdella parasitica*. Arrows indicate slight positive pressure prior to attachment (att). Note increased pressure prior to sucker detachment (det). Time scale in two-second intervals. Zero indicates ambient pressure. Reproduced from Gradwell, 1972, with permission of NRC Research Press.

A stronger case for the use of cement can be made for *Hemibdella soleae*. The posterior sucker is attached to a single projecting spine of one of the ctenoid scales of its host, the common sole (*Solea solea*) (Figure 7.1H). This seems a most unsuitable surface for the efficient operation of a sucker. However, removal of the leech from the spine reveals the absence of the spine's thin epidermal covering and its replacement by a thick layer of material (Kearn, previously unpublished). It seems likely that the leech secretes this material and that it serves as a bond between the posterior sucker and the spine.

Llewellyn (Llewellyn, L., 1965) noted that, when *H. soleae* is forcibly removed from its host, the posterior sucker expands and becomes disk-shaped. The animal can then move about in a glass dish using the anterior and posterior suckers in typical leech locomotion. Llewellyn also observed that while on the fish the leech could move from one spine to another, changing position in aquaria every few days. Like *E. soleae* (see p. 54) the leeches tend to move in an anterior direction on the upper surface of the host, but unlike *E. soleae*, do not move to the lower surface when they reach the head.

Why *H. soleae* has undergone such a fundamental change in attachment site is an intriguing question. It may be related to the more permanent parasitic life style of *H. soleae*. Perhaps the large suctorial attachment organs of more typical leeches and the damaging blood feeding habits provoke aggressive host defensive skin responses. The posterior sucker of a typical leech is intimately in contact with the host's skin and is likely to be a prime target for these host responses. The temporary parasite may avoid these responses by leaving the host, but this option is not available to *H. soleae* on the sole. By shifting the attachment site to an inert scale spine, *H. soleae* may avoid host attack.

Adhesive secretion appears to be important in the primitive freshwater fish parasite *Acanthobdella peledina*. Sawyer (1986b) regarded this 'living fossil' as a leech (Appendix 2), but observations by Siddall *et al.* (2001) are indicative of an independent status alongside the leeches in the Oligochaeta. The posterior sucker is richly supplied with gland cells secreting a viscous mucoid secretion that sticks the parasite to the substrate (see Sawyer, 1986a). This interesting parasite is widely distributed across the boreal regions of the Northern Hemisphere, but is not found in Britain.

7.4 LOCOMOTION

Behaviour in which leeches use their anterior and posterior suckers for locomotion is generally well known. The specific name of *Piscicola geometra* reflects this movement. Essentially, while initially attached by the posterior sucker the leech extends its body and attaches the oral sucker (Figure 7.4A). The posterior sucker is then detached, drawn forward and reattached near the fixed anterior sucker, which is then released (Figure 7.4B-F). This action is rapid and efficient, so much so that it has been adopted independently by monogenean fish parasites like *Entobdella soleae* (see Chapter 3) and by the caterpillars of geometrid moths (inchworms).

In *P. geometra* there is an extra feature of each locomotory 'step' that is not fully understood. After detachment, the posterior sucker is drawn forward and applied high up on the ventral surface of its own body, producing an exaggerated loop (Figure 7.4C). The posterior sucker then slides down the body until it reaches its usual position behind the oral sucker (Figure 7.4D). Sawyer (1986a) has suggested that this behaviour may serve to clean the ventral surface of the posterior sucker.

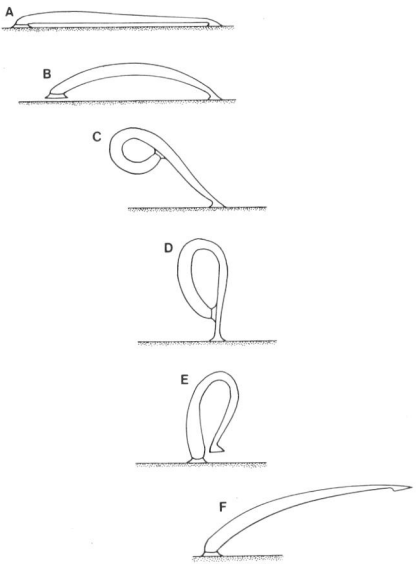

Fig. 7.4. (A-F) Stages in locomotion in *Piscicola geometra*. From Herter, 1968.

P. geometra, but not *H. marginata*, is also able to swim. This is achieved with both suckers detached and the body flattened by contraction of dorso-ventral muscles. Undulations are then propagated along the ribbon-like body in an anterior-posterior direction. In spite of its large size *Pontobdella muricata* is also able to swim by flattening the body (Harding, 1910). Generally, however, *Pontobdella* is sluggish and spends long periods attached to a convenient object, often with the body tightly curled (Figure 7.1 G).

7.5 HOST FINDING

P. geometra is positively geotactic and negatively phototactic, behavioural responses that prompt the leech to seek out the bottom where it rests attached to vegetation or to submerged stones (Sawyer, 1986b). When hungry the leech becomes sensitive to water turbulence and to shadows, both of which are likely to be generated by a passing fish. The leech stretches itself out like a rod and performs 'searching' movements by swaying its body to and fro (Harding, 1910) (Figure 7.5A). These stimuli may also induce inchworm crawling and less commonly swimming. Eye-like spots located on the posterior sucker of *P. geometra* (Figure 7.2A) are the main receptors used to detect shadows cast by a potential fish host (Bielecki, 1999). These spots are absent from the region of the sucker constantly shaded by the leech's body. Immediate host-searching movements were induced in *P. geometra* by a decrease in light intensity of as few as five lux.

Knight-Jones (1940) observed that *Abranchus* (= *Oceanobdella*) *blennii* when separated from the host attached itself by the posterior sucker, flattened the body and usually adopted a zigzag shape. In response to slight water movements or a shadow the leech stretched itself to more than twice its resting length and swung its extended body in all directions "as though attached to the posterior sucker by a universal joint". *Brumptiana lineata* responds to shadowing in a similar way, extending its body to twice its resting length and swinging it about (Llewellyn & Knight-Jones, 1984)

When contact is made with a potential host and the chemical sense of the leech identifies it as appropriate, the oral sucker is attached. According to Harding (1910), this takes place with remarkable speed and precision in *P. geometra*, and release of the posterior sucker permits the leech to be carried off attached to its victim (Figure 7.5B). This is followed by attachment of the posterior sucker to the host and then by movement to a suitable feeding site.

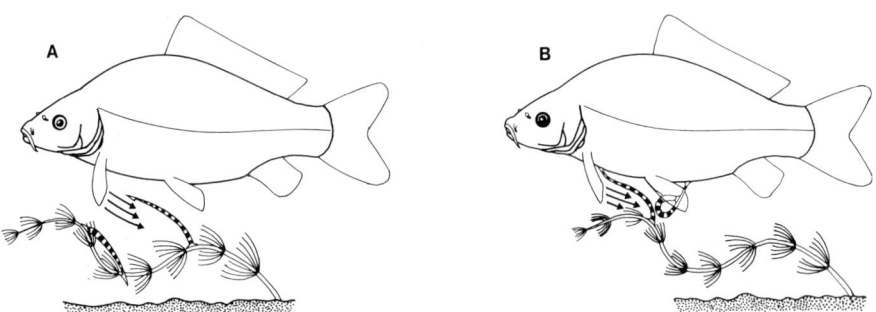

Fig. 7.5. (A) Water currents generated by the fins of a cruising fish stimulate two *Piscicola geometra* to perform 'searching' movements, resulting in (B) contact with the fish and attachment of both individuals to the host. Redrawn from Herter, 1929.

Hemiclepsis marginata behaves in a similar way, performing characteristic oscillatory 'searching' movements of the extended body like a pendulum (see Harant & Grassé, 1959). It is unable to swim but it may adhere to the surface of the water.

7.6 FEEDING

Much of our knowledge of feeding in fish leeches is derived from observations on *Piscicola geometra*, but this leech has some rather unusual features, which suggest that the feeding biology of *P. geometra* may be atypical, with possibly some surprises still in store. There are numerous references to a diet of blood for *P. geometra* (see Harding, 1910, Jennings & van der Lande, 1967, van der Lande, 1968), but Sawyer (1986b) described the food of *P. geometra* as "a clearish substance from the fish rather than blood", suggesting that lymph may be commonly ingested. Unusual features of the digestive process in *P. geometra* will be discussed below.

During feeding, *Piscicola geometra* attaches both suckers, with the oral sucker near the posterior sucker in such a way that its body is markedly curved. The rhynchobdellidan proboscis (Figure 7.1B) lacks jaws, but in some unknown way is able to penetrate the skin.

According to Harding (1910), *P. geometra* feeds mainly from the fins, but feeding has been reported to occur anywhere on the body surface and in the gill chamber (references in Sawyer, 1986b). Feeding may last for over one hour in large leeches and after feeding the leech may remain on the host for some time, attached by the posterior sucker. Further meals may be taken, but usually the leech leaves the host during the next few days. Intervals between feeds may be a few days in young leeches and 8 – 14 days or more in larger specimens. Sated adults can survive without a host for 3 – 4 months at 10 – 15°C and six months at 6°C (for references see Sawyer, 1986b).

In *Hemiclepsis marginata*, feeding lasts for 10 minutes to two hours and the leech leaves the host soon after feeding. Hence it is rarely found on living fishes. According to Sawyer (1986b), *H. marginata* occasionally ingests lymph, but more typically extracts blood. After the first feed, an individual is ready to feed again in about eight days at 20°C and *H. marginata* can live 10 months or more without feeding.

In *Hemibdella soleae* feeding lasts, on average, for 10.5 minutes, during which the pharynx pulsates at a rate of about two per second as it pumps blood into the gut. It seems that a red blood spot is created by each feed. During the first three days or so after finding a host the young leech feeds two to six times, as indicated by the number of blood spots nearby. Thereafter, the leech seems to fast for a period of about 20 days.

Blood-sucking leeches have extensive salivary glands (Figures 7.6, 7.7) to which a variety of functions have been attributed. The salivary secretions of mammalian blood-sucking leeches like *Hirudo medicinalis* are reported to contain an anaesthetic, a spreading factor, which maximises penetration of the secretions around the bite, and a vasodilator, which increases blood flow to the bite area by dilatation of local blood vessels. According to van der Lande (1968), fish hosts of *P. geometra* carry out very violent movements when the leech first inserts the proboscis, suggesting that the secretion lacks anaesthetic properties. There is also evidence that the salivary secretion may contain an agent capable of locally increasing the blood supply (references in van der Lande, 1968). In addition, the salivary gland secretions of *P. geometra* appear to contain an abundant supply of endopeptidases (enzymes that begin the digestion of protein, see below). This feature appears to be unique, not just among fish leeches but also among all leeches that have been studied, and will be considered in more detail below.

Those leeches that pay a short feeding visit to their fish host need to extract as much host blood as possible before abandoning the host. This is achieved by storing the blood in a voluminous lobed sac or crop, branches of which occupy much of the body (Figure 7.6). It is essential that blood in the crop does not clot, since if it were to do so the leech would be unable to extend and contract its body and, because these shape changes are essential for locomotion, would be immobilised. The salivary glands are assumed to be the source of the anticoagulant.

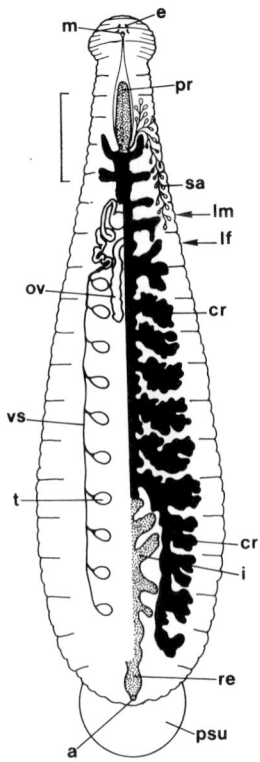

Fig. 7.6. Anatomy of *Hemiclepsis marginata*. Reproductive system shown on the left and the alimentary canal on the right. a, Anus; cr, crop; e, eye; i, intestine; lf, level of ventral opening of female reproductive system; lm, level of ventral opening of male reproductive system; m, mouth; ov, ovary; pr, proboscis inside proboscis sheath; psu, posterior sucker; re, rectum; sa, salivary gland; t, testis; vs, vas deferens. Scale bar: 2 mm. Based on Harding, 1910, and Jaschke, 1933.

A sphincter muscle separates the crop from the intestine (Figure 7.6) and controls passage of food from the former into the latter, but a most surprising finding in *Hirudo medicinalis* is that the complement of digestive enzymes that we would expect to find in the gut is greatly reduced. Graetz & Autrum (in Jennings & van der Lande, 1967) failed to identify proteolytic (protein-digesting) activity in an extract of the gut wall of *Hirudo*. This supported earlier observations by Diwany (in Jennings & van der Lande, 1967), who found that starved *Hirudo* could not digest milk, egg proteins or peptones injected aseptically into the gut. Later, Jennings and van der Lande (1967) used histochemical and other techniques to confirm that endopeptidases are absent in *Hirudo*, as are lipases and carbohydrases (enzymes responsible for the digestion of fats and carbohydrates respectively). In leeches generally, the linings of the crop and intestine are similar and there is no differentiation of these linings at the cellular level into secretory (glandular) and absorptive regions. In fact, glandular structures are absent apart from the specialised salivary glands.

How then do leeches digest their blood meals? Remarkably, *Hirudo* appears to rely almost entirely on a single species of symbiotic bacterium (*Aeromonas hydrophila* = *A. hirudinis*), which is consistently present in the gut. According to Sawyer (1986b), this bacterium secretes an antibiotic that prevents the growth of other bacteria and retards putrefaction so that blood can be stored for long periods. Jennings & van der Lande (1967) have summarised early findings showing that, *in vitro*, the micro-organism is capable of slowly digesting the blood that is the normal food of this leech and that, *in vivo*, the experimental inclusion of antibiotics in the food inhibits digestion. Jennings & van der Lande also demonstrated that the combined digestive capacities of the gut flora of *Hirudo* are sufficient to play an important part in the digestion of proteins, fats and carbohydrates.

In most animals, endopeptidases initiate protein digestion by splitting dietary proteins into shorter chains of amino acids (polypeptides). This permits more effective action by exopeptidases, enzymes with the secondary role of detaching terminal amino acids from the polypeptide chains. The liberated amino acids are then available for absorption. Most leeches have lost the capacity to produce a full range of digestive enzymes and have no endopeptidases. They have compensated partly by emphasising exopeptidases and partly by recruiting symbiotic micro-organisms. Digestion is likely to be slow, since the number of terminal sites for the removal of amino acids by exopeptidases will be limited, but this type of digestive physiology suits the intermittent feeding habits of leeches and the long periods between meals.

Jennings & van der Lande (1967) found that *P. geometra* differs from the other leeches examined in their comparative study, including *Hirudo medicinalis* and *Hemiclepsis marginata*, in the principal site of digestion. In *Hirudo* and *Hemiclepsis*, significant digestion takes place in the intestine, food entering this region a little at a time from the crop. In *P. geometra*, digestion occurs largely in the crop and is much more rapid than in other blood-feeding species. Jennings & van der Lande reported that red blood cells, which are prominent in the crop soon after feeding, are completely digested within 10 days. More intriguing is the discovery by Jennings & van der Lande that endopeptidase, which is universally absent in the other leeches, occurs in the crop contents of recently fed *P. geometra*. They found no endopeptidase activity in the crop lining, but recorded a strong response from the salivary glands and came to the conclusion that the enzyme in the crop lumen came from saliva injected into the host and reingested during feeding.

Like other fish leeches *P. geometra* also has symbiotic micro-organisms, possibly a species of *Pseudomonas*, housed in paired sacs or mycetomes opening into the oesophagus (Figure 7.7). The lining of the mycetomes is similar to that of the crop, but the rod-shaped micro-organisms (0.7 x 3µm) appear to be exclusively extracellular.

The precise role of these micro-organisms in the biology of *Piscicola geometra* is not clear. Jennings & van der Lande (1967) found that the gut flora of *P. geometra* does not have the digestive potential enjoyed by the gut flora of *Hirudo*. They were of the opinion that leeches like *Piscicola* are capable of digesting the protein components of blood using their own endogenous proteolytic enzymes (i.e. independently of their symbiotic micro-organisms). The rather speedy digestion undertaken by *Piscicola* probably reflects the presence of salivary endopeptidase, which mixes with the blood meal and promotes initial proteolysis in the crop, thereby enhancing the activity of exopeptidases perhaps regurgitated into the crop from the intestine and reducing dependency on the symbiotic gut flora. Nevertheless, the prominence of the gut flora in

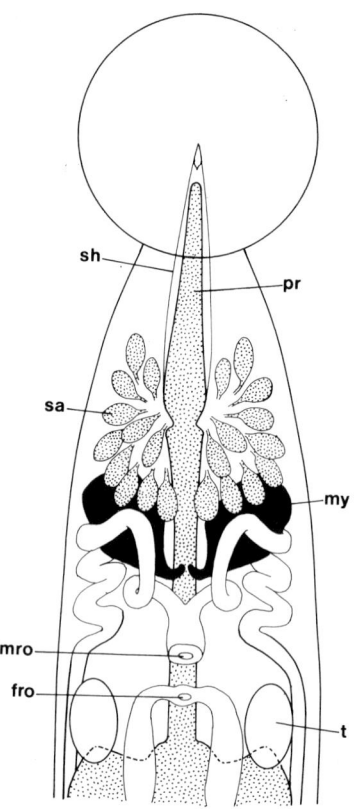

Fig. 7.7. Position of the mycetomes (my) in relation to the proboscis and terminal reproductive ducts of *Piscicola geometra*. fro, Female reproductive opening; mro, male reproductive opening; pr, proboscis; sa, salivary glands; sh, proboscis sheath; t, testis. Redrawn from Jaschke, 1933.

leeches like *Piscicola* indicates that their contribution to leech nutrition is significant. Jennings & van der Lande (1967) suggested that the flora might be concerned with digestion of the limited intake of fats and carbohydrates in the blood meal, since the leech lacks specific enzymes to deal with these substrates. They also pointed out that the diet of blood-feeding animals is poor in vitamins, especially of the B group, and that symbiotic micro-organisms may replenish this shortfall, as they do in blood-feeding arthropods.

If we are right in our conclusion that symbiotic bacteria are important in the nutritional economy of fish leeches, then the leech must ensure that these bacteria are transmitted from one generation of leeches to another. Jaschke (1933, in Sawyer, 1986b) examined the cocoon fluid of *P. geometra* and found many micro-organisms, which varied in number from cocoon to cocoon. How the organisms reach the cocoon fluid is not clear, but there is close proximity between the mycetomes and the reproductive systems in leeches like *Piscicola* (see Figure 7.7). It seems that the young leech becomes infected after breaking the egg membrane by ingesting cocoon fluid. Prior to emergence from the egg membrane the mycetomes of the embryo are devoid of micro-organisms.

7.7 MATING

Like monogeneans, leeches are protandrous hermaphrodites (the male system matures before the female system). In spite of the potentiality for self-insemination, most leeches pair and cross-inseminate. Some leeches have a protrusible penis that is inserted

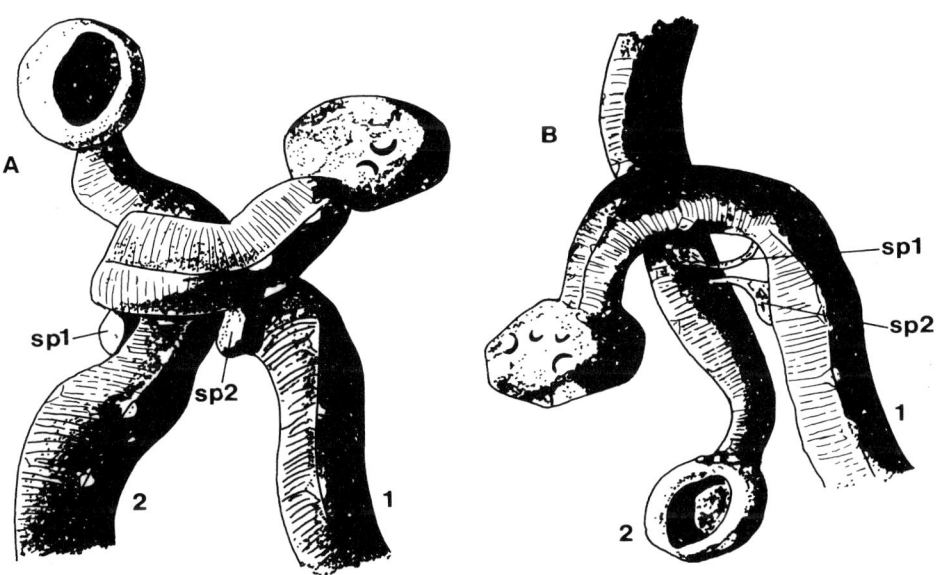

Fig. 7.8. (A) Two individuals (1 and 2) of *Piscicola geometra* mating. (B) The two individuals shown in (A) after treatment with anaesthetic and partial separation. sp1, spermatophore produced by leech 1; sp2, spermatophore produced by leech 2. From Brumpt, 1900a, with kind permission of the Société Zoologique de France.

into the vagina of the mating partner, but the British fish leeches lack a penis and, like the monogenean *Entobdella soleae* (see Chapter 3), exchange sperm in special packets or spermatophores, which are attached externally to the body of the mating partner.

Exchange of spermatophores is often accompanied by intertwining of the bodies of the two mating leeches (Figure 7.8A). Mating may last as long as five or six hours but is generally shorter (Brumpt, 1900a). This is not a haphazard entanglement since it involves mutual exchange and accurate positioning of spermatophores. By anaesthetising mating pairs of *P. geometra*, Brumpt was able to partially separate the co-copulants and observe the stalk–like proximal end of each spermatophore issuing from the male opening and the opposite broad end in contact with a special copulatory reception area of the partner (Figure 7.8B). This reception area lies posterior to the genital pores (Figure 7.9A).

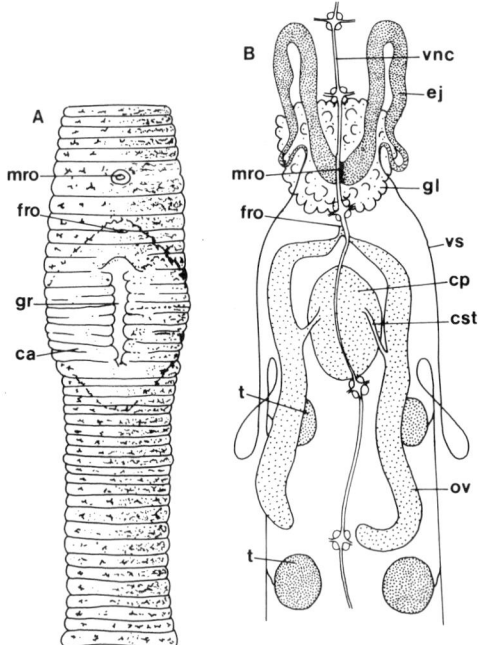

Fig. 7.9. Clitellar region of *Piscicola geometra* . (A) Ventral view of the external surface; (B) internal anatomy of the same region, in dorsal view. ca, Copulatory reception area; cp, 'conducting' pad; cst, 'conducting' strand; ej, ejaculatory duct; fro, female reproductive opening; gl, gland secreting spermatophore; gr, groove in copulatory reception area; mro, male reproductive opening; ov, ovary; t, testis; vnc, ventral nerve cord; vs, vas deferens. From Brumpt, 1900a, (A) with kind permission of the Société Zoologique de France; (B) redrawn.

In the monogenean *Entobdella soleae*, sperm from the externally deposited spermatophore is sucked into a vaginal tube, from which sperm is ultimately transported to the germarium (part of the ovary) where the oocytes are fertilised (Chapter 3). However, there is no convenient ducting to transport the sperms of *Piscicola* directly to the ovarian sac. It is a surprising and remarkable fact that the sperms must somehow penetrate the surface of the leech and find their own way through the leech's tissues to the ovarian sac. How the sperms penetrate the body of the leech is unknown, but this takes place at the site of deposition of the spermatophore, i.e. through the copulatory reception area. Lying immediately beneath this area is a pad of fibrous tissue (Figure 7.9b), which receives the penetrating sperm. Sperms travel readily through this 'conducting' pad and are funnelled into two 'conducting' strands each of which serves to guide the sperms to one of the ovaries (Figure 7.9B).

7.8 COCOON ASSEMBLY

Once the eggs of leeches are fertilised, they are not abandoned individually to the dangers of a free-floating planktonic life as they are in some of their annelid relatives (polychaete worms). The fertilised eggs (or in *P. geometra* and most piscicolids a single fertilised egg) are enclosed in a cocoon, which is cemented not to the fish host but to a suitable substrate (stones or vegetation).

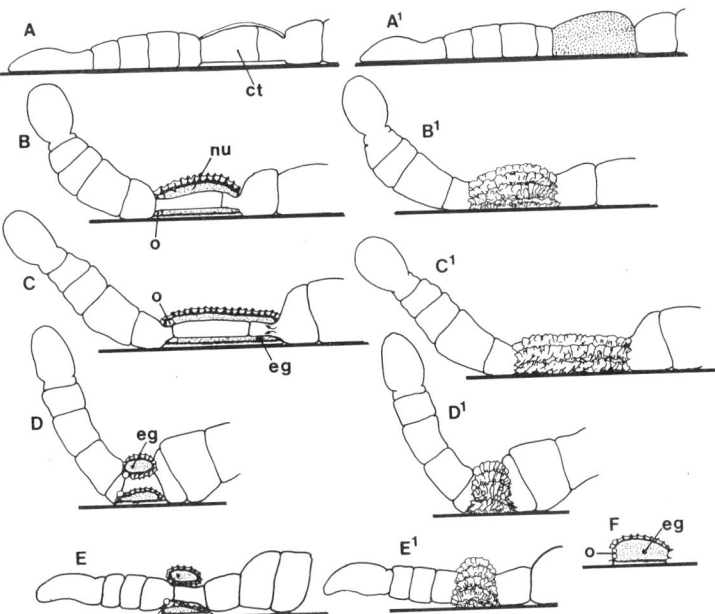

Fig. 7.10. Series of diagrams showing stages of cocoon assembly in *Piscicola geometra*. At each of stages A – F, the clitellar region and cocoon are shown in longitudinal section. $A^1 - E^1$, Lateral views of intact cocoon at assembly stages A – E. (A) Deposition of first secretion around clitellum (ct); (B) rotation of leech and deposition of other cocoon secretions; (C) egg laying and commencement of inrolling (arrows) of cocoon wall; (D) termination of inrolling; (E) commencement of withdrawal from cocoon; (F) completed cocoon after departure of leech. See text for detailed explanation. eg, Egg; nu, nutritive layer lining cocoon wall; o, operculum. From Malecha & Vinckier, 1983.

Cocoons are structurally complex and demanding to assemble, but all of this extra effort is presumably offset by advantages, especially the protection afforded by the thick wall of the cocoon. There is no longer a need to provide each egg with an individual supply of yolk since nutriment is available in the cocoon fluid and can be absorbed as development proceeds.

Cocoon assembly in *Piscicola geometra* (Figure 7.10) was studied by Brumpt (1900ab) and in greater detail by Malecha & Vinckier (1983), who identified no less than five distinct types of gland cell involved in cocoon assembly in *P. geometra*. These glands are unicellular and distributed throughout the length of the body, but their ducts converge on the clitellum, a short, anteriorly situated length of the leech's cylindrical body. Those ducts with cell bodies posteriorly situated may attain a length of a few centimetres in large leeches. The ducts open on the surface of the clitellum in a characteristic pattern in relation to the male and female genital openings, which are located ventrally.

P. geometra begins cocoon assembly by attaching itself to the chosen substrate by its two suckers. One of the sets of glands then secretes a translucent sleeve around the clitellum (Figure 7.10AA[1]). Debris attached to the clitellum is dislodged and transported externally as the secretion emerges. The sleeve adheres to the substrate where it makes contact ventrally and is extensible, providing a sac to hold the other glandular secretions. The leech then disengages its oral sucker (Figure 7.10BB[1]) and rotates about its longitudinal axis for about five minutes, with rest periods interspersed. During this rotating stage the secretions of the other four kinds of gland are released. Three of these contribute layers internally to the wall of the cocoon, the innermost being destined to provide nutrients for the growth of the embryo. The fourth secretion is strictly limited in its deployment, being deposited as a ring of material around the anterior opening of the cocoon. This ring is destined to close the anterior opening of the cocoon and to form a lid or operculum through which the single fully developed offspring will escape.

A single egg cell is then released into the cocoon from the female genital pore (Figure 7.10CC[1]) and this is followed by an anteriorly directed wave of inflation of the post-clitellar region of the body. This serves to invert the posterior border of the cocoon into the cocoon cavity and push it forward until it makes contact with the anterior border, so that the egg and its nutritive medium are enclosed in the ring-shaped compartment so-formed (Figure 7.10DD[1]). This activity temporarily reduces the length of the cocoon by a half.

Finally, the leech withdraws backwards from the cocoon (Figure 7.10EE[1]). In the minute that follows this withdrawal, the posterior part of the wall of the cocoon that is tucked inside the cocoon cavity everts in a posterior direction and the cocoon resumes its original elongated shape (Figure 7.10F).

According to Malecha & Vinckier (1983), this curious inversion creating a ring-shaped compartment effectively seals off the contents of the cocoon from the outside world and prevents contamination of these contents by external micro-organisms as the leech withdraws its anterior end. After withdrawal, the elasticity of the cocoon wall ensures reversion to its original shape. The anterior and posterior apertures of the cocoon are then sealed and the cocoon contents are aseptic, except for symbiotic gut bacteria derived from the parent (see above).

After abandonment by the adult leech, the wall of the cocoon hardens and becomes brown in colour (see also below). At the time of emergence of the young *P. geometra*, the junction between the operculum and the wall of the cocoon is somehow weakened and the leech escapes by pushing off the operculum (Malecha & Vinckier, 1983). Movements of the whole of the body of the young leech seem to be involved since a loop of the body always springs out before the anterior sucker.

The above account of cocoon formation in *P. geometra* is typical of leeches that leave the host and attach their cocoons to the substrate. However, *Branchellion torpedinis* does not leave the host to deposit its cocoon, which, while in situ on the body of the leech, is hidden by a fold of skin like a prepuce covering the clitellum (Figure 7.1DE). The ring-shaped cocoon is shrugged off over the head by an abrupt movement of the anterior region of the body.

7.9 SOME OBSERVATIONS ON LIFE CYCLES

7.9.1 *Marine leeches*

The life cycle of *Oceanobdella blennii* has been described by Gibson & Tong (1969). On the shores of Anglesey in North Wales cocoons are deposited, probably at night in April and May, on the lower surfaces of boulders and on the sides of rock pools. Leeches are absent from their hosts (the shanny, *Lipophrys pholis*) from June to November and, although occasional free-living leeches are encountered in the summer, it is assumed that they eventually die.

The cocoons have an unusually protracted development period, one young leech emerging from each cocoon about 7 – 8 months after deposition, i.e. in December and January. These young leeches attach themselves to their hosts and enter the opercular cavity, where they feed on blood and grow, increasing in length from 3 – 4 mm in December to 12 – 14 mm in May/June. Prior to cocoon deposition most of the growing leeches leave the opercular cavity and take up a position on the body beneath the pectoral fin.

Leeches are more commonly found on male fish and on fish greater than 12 cm in length. Gibson & Tong suggested that these differences in distribution are related to host behavioural differences. They proposed that territorial restriction of movements of males at the time when the leeches are most abundant in the spring increases their vulnerability to leeches and that larger fish might be more susceptible because of their position on the shore.

The scarcity of detached leeches suggests that *O. blennii* may spend most of its short life attached to the host. *Hemibdella soleae* spends the whole of its life attached to a scale spine of the common sole (*Solea solea*). The sole uses powerful body flexing movements when it comes to rest to throw sand onto its upper surface, and this feature of the host's behaviour makes it possible for *H. soleae* living on the upper surface to attach its cocoons to shell fragments or large sand grains without leaving the host. Cocoon assembly has not been described. Each cocoon has a diameter of about 0.6 mm and single cocoons may be found on sand particles just large enough to accommodate them. Llewellyn (Llewellyn, L., 1965) found up to eight cocoons on shell fragments. The cocoons are shaped like a shrapnel helmet (Figure 7.11A) and are attached to the substrate by nearly transparent basal cement. The cocoon wall is pale initially and becomes golden brown with age, like the eggs of *Entobdella soleae*, and, as suggested by Llewellyn, this may indicate the presence of a scleroprotein (tanned protein).

Each cocoon of *H. soleae* has small plugs at opposite ends and contains a single embryo (Figure 7.11B). At temperatures fluctuating between 16 and 17.6°C, the young leeches are ready to emerge from their cocoons after 41 days. Recently hatched leeches are 1.6 to 2.0 mm in length at rest and have a pair of red eyespots, but the rest of the body is almost transparent. Llewellyn (Llewellyn, L., 1965) found that young leeches deprived of a host survived for about two months without growing. When provided with a host they attached immediately, first by the anterior sucker and then by the posterior sucker, and took their first blood meal usually within an hour. Black or brown pigment begins to appear in the body about 20 days after beginning to feed and a day or so later maturity is reached.

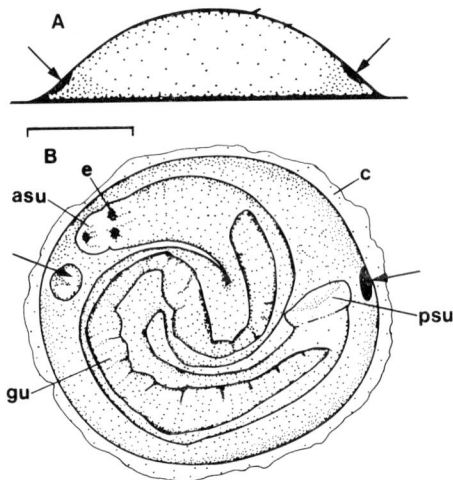

Fig. 7.11. Cocoon of *Hemibdella soleae* viewed from the side (A) and from above (B), the latter containing a 35 day-old embryo. Arrows indicate positions of plugs. asu, Anterior sucker; c, cement; e, eye; gu, gut; psu, posterior sucker. Scale bar: 200μm. From Llewellyn L., 1965, with permission from the Zoological Society of London.

A young leech introduced to a sole 10 days after hatching reached its maximum length of 6 mm about 40 days later at 17°C, but those that had reached lengths of 3.8 mm, about 23 days after finding a host, readily produced cocoons. Thus, the entire life cycle takes less than three months at British summer temperatures.

The common sole's habit of throwing sand onto its upper surface when resting on the sea bottom probably ensures that newly emerged leeches reach the upper surface of the host. Since more of this sand falls on the posterior region of the host's body, leeches may establish themselves more frequently in this region, moving forward as they age and dropping off from the head. Transfer of adult leeches from host to host appears to be a rare event in aquaria, in contrast with *Entobdella soleae* (Chapter 3).

7.9.2 Freshwater leeches
Malecha (1984a) studied the life cycle of *Piscicola geometra*. His observations were made in the north of France where conditions are likely to be similar to those in the south of England. Based on observations in the field and in the laboratory, he found that the leech survives over the winter period and that these overwintering individuals show two growth phases, the first in September/October and the second in January/February.

These overwintering leeches grow to a large size, some reaching weights of 60 – 70 mg in March and April. They do not reproduce until the end of February and die between April and June. The life span of one of these leeches is between seven and nine months.

The old dying leeches are then replaced by their offspring, which begin to emerge at the beginning of May and commence their own reproduction in June when most weigh as little as 4 – 6 mg. Malecha (1984a) estimated that three or four generations of leeches intervene during the summer. This estimate was based on the following figures for development of leeches obtained from laboratory culture: 15 – 20 days for embryonic development, 1 – 2 months for achievement of sexual maturity, an overall survival time of 3 – 4 months. Factors such as light, temperature and nutrition will have a significant influence on the number of summer generations. Reproduction ceases in September and, according to Malecha (1984b), this reproductive hiatus among leeches of the overwintering generation is induced by the daily autumnal photoperiod of 12 hours of light and 12 hours of darkness.

Malecha (1984a) found that the two growth phases of the large overwintering leeches were attributable to different synthetic activities. The first phase in September/October is characterised by accumulation of reserves in adipose (fat filled) cells, while the second phase in January/February coincides with a large increase in the volume and secretory activity of the clitellar glands, which are destined to produce the cocoons (see above). Malecha offered two hypotheses, which are not mutually exclusive, to explain the biological significance of this two-tier population – large overwintering individuals and small summer individuals. He pointed out that the large body size promotes higher reproductive activity, which he regarded as necessary after the harsh winter period. In fact, he was able to show that large *P. geometra* collected in March produce between 30 and 50 cocoons between two consecutive blood meals, while in similar conditions the small summer individuals produce only 5 – 10 cocoons. The second hypothesis concerns the respiration of the leech. It is well known that *P. geometra* has a greater need for oxygen than other leeches (Sawyer, 1986b). Since oxygen availability in the bodies of freshwater inhabited by the leech and its hosts is likely to be at its lowest in the summer, the larger surface area with respect to volume of the smaller leeches may be advantageous.

According to Needham (1969, in Burreson, 1995), *Hemiclepsis marginata* has an annual cycle. Eggs are laid in late spring and early summer when the temperature reaches 15°C. The cocoons are deposited on a substrate, usually protected by the leaf bases of reeds. The parent remains near and on hatching the young leeches attach themselves to the ventral surface of the adult leech, where they remain until transported to a host fish. Here the young leeches leave the parent and take their first blood meal. The sated young then abandon the host and shelter amongst vegetation to digest the blood. They continue to feed intermittently and grow throughout the summer and autumn. Feeding is reduced or ceases during the winter, to be resumed in the spring in anticipation of breeding.

Hirudo medicinalis has been extensively studied in the laboratory, but according to Elliott & Mann (1979) little is known about its life cycle in the wild. The cocoons of *H. medicinalis* are about 1 cm long and are deposited in a damp place just above the water line on the shore or bank (Elliott & Mann, 1979). There may be a delay of 1 – 9 months between copulation and cocoon deposition. In the wild, cocoons have been found to contain 5 – 15 eggs and hatch after 4 – 10 weeks depending on the temperature. The leech takes at least two years to reach breeding condition in the field and slow growing leeches may not breed until they are three to four years old.

7.10 LEECHES AS TRANSMITTERS OF DISEASES

Leeches are rarely pathogenic to their fish hosts. Their effects are usually limited to feeding or attachment sites, and only become serious when infestations are high, as they may become in fish farms. For example, in carp-rearing ponds in eastern Europe, *Piscicola geometra* may produce small bleeding ulcers and the host may become emaciated or may acquire secondary bacterial or fungal infections (see Burreson, 1995). However, the intermittently parasitic habit of many fish leeches lends itself well to the transmission of other parasitic organisms from one fish to another.

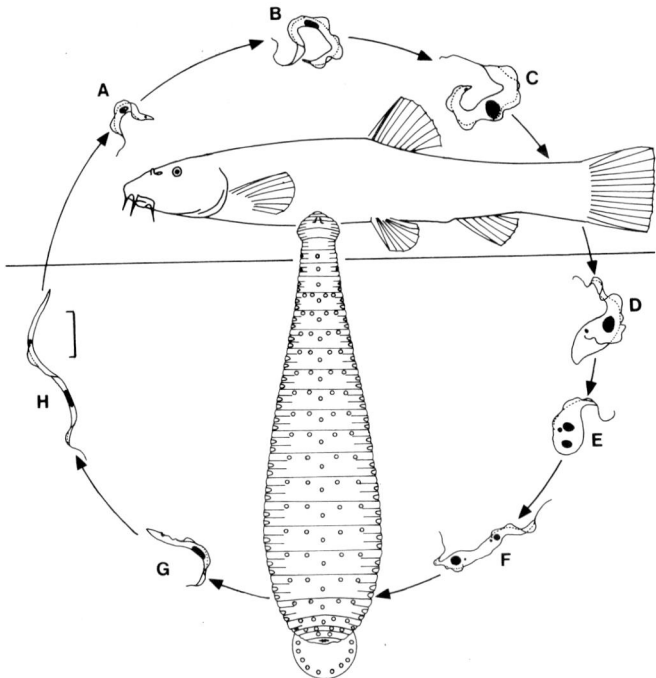

Fig. 7.12. The life cycle of *Trypanosoma cobitis* in the stone loach, *Noemacheilus barbatulus*, and in the leech *Hemiclepsis marginata*. Above the horizontal line, stages in the loach; below the line, stages in the leech. (A-C) Growth in loach blood; (D) early stumpy form in leech crop; (E,F) cell division in crop; (G) elongation and forward migration to proboscis sheath; (H) metacyclic infective stage in proboscis sheath. Scale bar for trypanosomes: 10 µm. Based on Letch, 1980 and Sawyer, 1986b.

It comes as a surprise to most naturalists to learn that beneath the placid surfaces of our ponds and rivers, freshwater fishes may be infected by a blood parasite closely related to the trypanosome protozoans that cause sleeping sickness in humans in Africa. Trypanosomes are elongated, uninucleate protozoans (Figure 7.12A-C), living freely in the blood where they are propelled by a single flagellum. Part of this flagellum is free and part is joined to the trypanosome cell by a membrane, but all of it is motile.

The terrestrial vectors of the human trypanosomes are blood sucking tsetse flies, but in the aquatic environment fish leeches fulfil this role. In 1912, Muriel Robertson established by elegant experimentation that *Hemiclepsis marginata* readily transmits trypanosomes between fishes (Figure 7.12). She found that goldfish, *Carassius auratus*, in a pond in the garden of the Lister Institute at Elstree near London, were infected with a trypanosome and she also found the parasite in the crop and proboscis sheath of *H. marginata*, the only fish leech found in the pond.

The offspring of *H. marginata* are carried about attached to the parent, but since no parasite transmission takes place between parent and her young these broods are a source of uninfected leeches. Robertson (1912) demonstrated that these young leeches became infected after feeding on infected goldfish from the pond. Trypanosomes taken into the crop of the leech undergo a marked change of form and multiply rapidly by asexual cell division (Figure 7.12D-F). Robertson remarked: "it is no infrequent occurrence to find the whole crop simply seething with amazing numbers of the creatures". After a few days, slender forms arise (Figure 7.12G) and at the end of digestion they migrate forwards along the proboscis and enter the proboscis sheath (Figure 7.7). The leech is not infective until the slender forms (Figure 7.12H) are in the sheath, even though the rest of the leech is full of trypanosomes. When the leech takes its next blood meal, these slender forms are inoculated into the fish. Robertson also succeeded in transmitting trypanosomes to uninfected goldfish from another pond and also from perch, *Perca fluviatilis*, to goldfish and from bream, *Abramis brama*, to goldfish.

Robertson (1907, 1927) also contributed to our knowledge of *Trypanosoma raiae*, a relatively large and common blood parasite of rays (*Raja* spp.) transmitted by the large marine leech *Pontobdella muricata*.

Robertson found another haemoflagellate, a trypanoplasm (= *Trypanoplasma*), living in the blood of the goldfish from the Elstree pond. Trypanoplasms have two flagella arising together from a pocket at one end of the organism (see Lom & Dykova, 1992). There is a short free organelle and a long organelle, which runs back down the body and is joined to it to form an undulating membrane. Robertson showed that the trypanoplasm is also transmitted by *H. marginata*. Like the trypanosomes the trypanoplasms taken into the crop of the leech with the blood meal multiply and produce slender infective forms, which pass forward to the proboscis sheath, whence they are injected into another fish host at the next feed.

Leeches are also implicated as vectors of haemogregarines and piroplasms, which are intracellular parasites, living inside white or red blood cells, and, according to Burreson (1995), there is accumulating evidence that leeches can also transmit viruses and bacteria.

8

SIPHONOSTOMATOID COPEPODS: (1)
FISH LICE – CALIGIDS

8.1 INTRODUCTION TO COPEPODS

Members of the Crustacea, which includes the copepods, have a modular construction like leeches, their bodies consisting of a repetitive series of segments. However, crustaceans are arthropods, characterised by possession of an external skeleton and jointed limbs. Each segment possesses one pair of limbs, and in crustaceans each limb basically consists of two branches (biramous). The inner branch is termed the endopod and the outer branch the exopod, and the separate units of a jointed limb are referred to as 'podomeres' or as 'articles' (the former term will be used in this book). No modern crustacean can be found in which individual segments and their biramous limbs can be identified from one end of the body to the other, because fusion of segments has occurred to varying extents, together with modification or loss of segmental limbs. Typically arthropods have segments fused together or regionally grouped to form head, thorax and abdomen, but the divisions between these regions may be obscure. Groups of fused segments are referred to as tagmata (singular: tagma).

Copepoda is a relatively small group of arthropods with currently about 11,500 valid species worldwide (Boxshall & Halsey, 2004). However, in terms of numbers of individuals, free-living copepods are enormously abundant. In marine muds and sands they are a major component of the interstitial fauna living between sedimentary particles, but it is in the marine and freshwater pelagic realms that they dominate. A single sweep with a fine net, either in the sea or in freshwater, is likely to collect large numbers of planktonic copepods. In the shallow waters of the North Sea, for example, they may reach 70,000 per cubic metre and even at depths of 4000 m in the North Atlantic may persist at densities of 100 or so per cubic metre (references in Huys & Boxshall, 1991). In fact, based on his studies of marine plankton, Sir Alistair Hardy (1970) claimed that in terms of absolute numbers of individuals, the copepods are the most numerous multicellular animals on the planet. In his estimation they are likely to outnumber the insects and the nematodes, both of which have been nominated for this position.

Given the density and ubiquity of copepods it is not at all surprising that nearly half of all known species have developed symbiotic relationships with other organisms (Huys & Boxshall, 1991). Copepods engage in commensal, parasitic and other kinds of association with members of virtually every phylum of animals, from sponges to vertebrates. Indeed parasitism of fishes has undoubtedly evolved independently more than once within this group. Most of these fish parasites are ectoparasitic or mesoparasitic and the diversity of their form and life cycles is staggering, so much so that only a small selection of them can be adequately covered in the five chapters devoted to them in this book.

According to Kabata (1979), ancestral bottom-dwelling copepods gave rise to two distinct kinds of planktonic forms, the gymnopleans and the podopleans (the distinction between them need not concern us here). For reasons that are not entirely clear, the gymnoplean anatomical 'blueprint' appears to have had no propensity for parasitism or for any other kind of association with animals. On the other hand the podoplean 'blueprint' appears to have had much wider potential, producing both planktonic and parasitic groups. Kabata (1979) recognised three major groups of podoplean copepods, namely the Siphonostomatoida, Poecilostomatoida[1] and Cyclopoida[1], all of which have produced fish parasites. Kabata (1981) estimated that a massive 75% of all copepods parasitic on fishes were siphonostomatoids, about 20% poecilostomatoids and about 5% cyclopoids.

According to Kabata (1979), 14 families of siphonostomatoid copepods have representatives on British fishes. There is room to deal adequately with only three in this book. This chapter will be devoted to the relatively unspecialised and extensively studied Caligidae, while the Pennellidae will be considered in Chapter 9 and the Lernaeopodidae in Chapter 10. The interested reader is referred to Kabata's work (1979) for information on other British copepod fish parasites not dealt with in this book.

On British fishes Kabata (1979) recorded twenty-two caligid species, mostly belonging to *Caligus* and *Lepeophtheirus*. Two of these species, both belonging to *Lepeophtheirus*, have had a central role in studies of the biology of caligid fish parasites and will figure prominently in this chapter. These are *L. pectoralis*, which is easily accessible and common in the wild on pleuronectid flatfishes (Figure 8.1), and *L. salmonis*, a parasite of salmon (*Salmo salar*) and sea trout (*Salmo trutta*). *L. salmonis*, together with *Caligus elongatus*, have become notorious since the 1960s as serious pests of farmed salmon. For information on other British caligids, the reader is referred again to the work of Kabata (1979).

8.2 GENERAL FEATURES OF CALIGIDS

The flat hard body and crab-like limbs of the caligid copepods readily evoke the image of a louse in the minds of observers. Indeed, the name *Lepeophtheirus* is derived from the Greek words for 'scale' and 'louse'. Thus, 'fish louse', 'sea louse' or 'salmon louse' seem appropriate names for these animals, although, traditionally, the term 'fish louse' refers to the freshwater fish parasite *Argulus*, which is a branchiuran not a copepodan crustacean (see Chapter 13).

Free-living planktonic copepods are moulded by the need for locomotory efficiency. The way such a free-living copepod moves has been described eloquently by Kabata (1979) as follows: "Propelled by the flickering beat of its swimming legs and the sweeps of its first antennae, its fusiform anterior end cleaves the liquid medium, trailing behind it its slender posterior part whose feathery tip acts as a steering oar or vane". In the caligids the antennae are no longer used for propulsion and there has been

---

[1] In the light of new evidence Boxshall & Halsey (2004) have advocated the amalgamation of the Cyclopoida and the Poecilostomatoida, rendering the name 'Poecilostomatoida' redundant. However, for ease of reference to earlier work the term 'Poecilostomatoida' will be retained in this book.

a major change in the shape of the anterior region of the body. The streamlined bullet shape of this region in the planktonic copepod has been transformed into a shallow disk (Figures 8.1, 8.2, 8.9), covered above by a convex dorsal shield (carapace) and concave below (Figure 8.3). This disk is the hallmark of the caligid copepods and has important parasitological significance because it reflects the development of suctorial capacity used for attachment to the fish host. The disk is derived from the whole of the head (cephalon) and the anterior part of the thorax, hence the term 'cephalothorax' used for this tagma. It is probable that as many as nine anterior segments, five from the head and four from the thorax, have been fused together to create it.

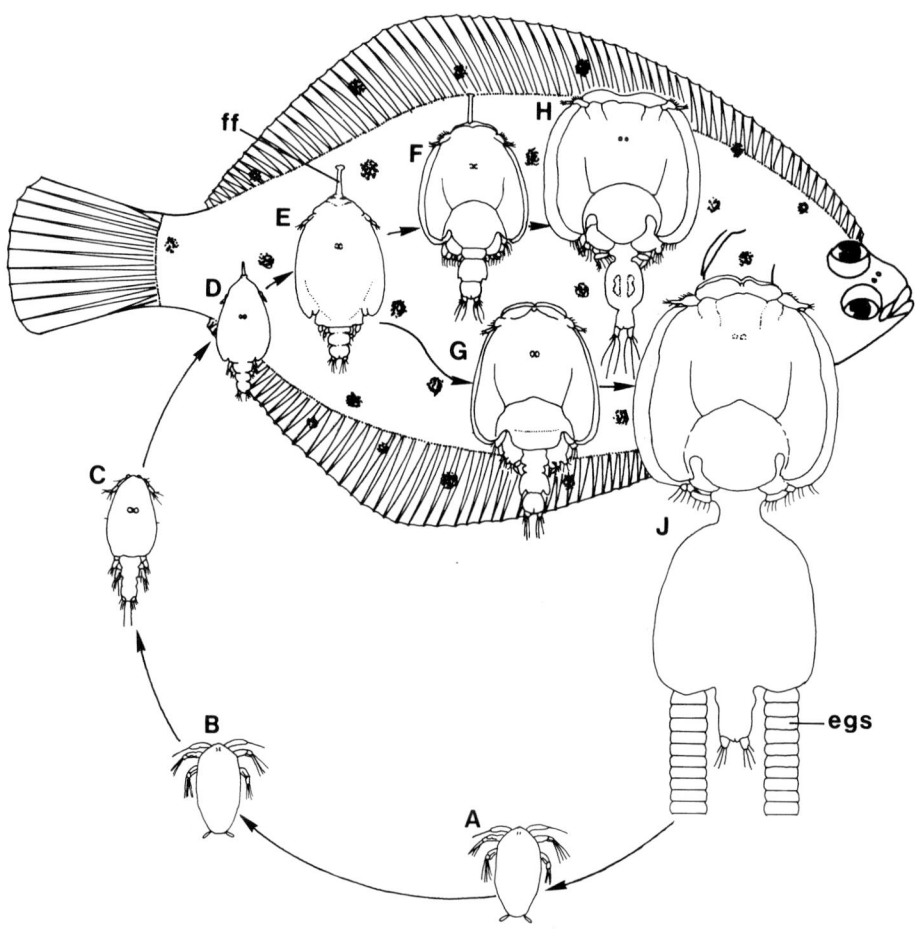

Fig. 8.1. Life cycle of *Lepeophtheirus pectoralis* on plaice, *Pleuronectes platessa*. (A) Nauplius I; (B) nauplius II; (C) copepodid; (D) chalimus I; (E) chalimus IV; (F) first preadult (male); (G) first preadult (female); (H) adult male; (J) adult female. Not all stages are included (see text). ff, Frontal filament; egs, egg sac. Based on Boxshall, 1974 and Kabata, 1979.

Caligids have a second means of attachment to the host, especially when the parasites are young. They are able to tether themselves by means of a filament secreted by glands in the anterior region of the body (Figure 8.1). This remarkable secondary attachment mechanism is a copepod feature with no counterpart in other kinds of fish parasites. It appears to be related to the problems created by the need to moult.

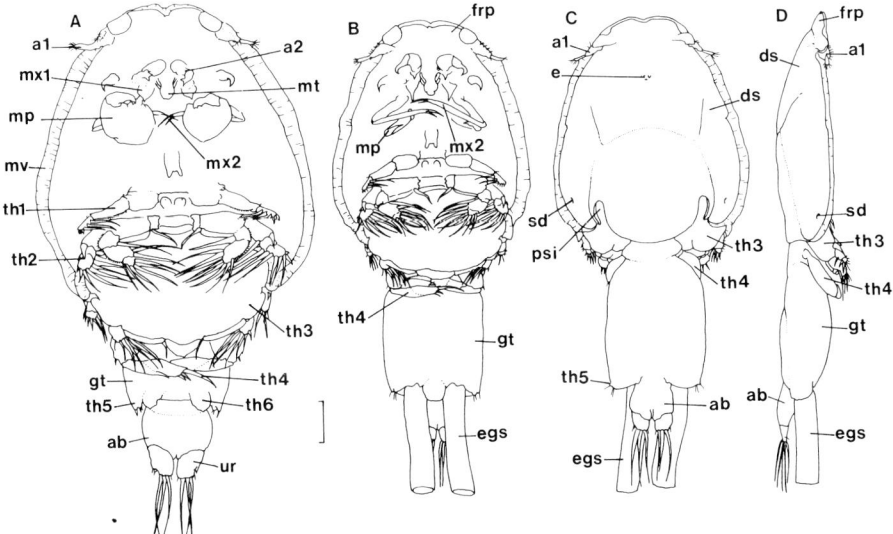

Fig. 8.2. The morphology of a caligid copepod (*Caligus curtus*). (A) Male in ventral view; (B) female in ventral view; (C) female in dorsal view; (D) female in lateral view. a1, First antenna (antennule); a2, second antenna; ab, abdomen; ds, dorsal shield or carapace; e, eye; egs, egg sac; frp, frontal plate; gt, genital tagma; mt, mouth tube; mv, marginal valve; mx1, mx2, first and second maxillae; mp, maxilliped; psi, posterior sinus; sd, sensory depression; th1 – th6, first to sixth thoracopods; ur, uropod. Scale bar: 1 mm. From Parker *et al.*, 1968, reproduced by courtesy of Fisheries and Oceans Canada, with the permission of Her Majesty the Queen in Right of Canada, 2004.

The highly modified, paired limbs belonging to the cephalothoracic segments are to be found on the concave underside of the disk. The anteriormost limbs are the first antennae (or antennules), located on the anterior border of the head (Figure 8.2). Then clustered around the mouth on the underside of the disk are the second antennae, mandibles, first maxillae (or maxillules) and second maxillae. These are followed by the maxillipeds, which are derived from the anteriormost thoracic segment. Three pairs of thoracic limbs (thoracopods) lie beneath the posterior region of the disk.

The cephalothorax is followed by another thoracopod-bearing thoracic segment and then by the genital tagma. The latter is considered to be derived from two thoracic segments and retains vestiges of their limbs (Figure 8.2; th5, th6). The abdomen is small and lacks limbs and there are two terminal uropods bearing hair-like setae. Thus, caligids have four tagmata, namely the cephalothorax, the fourth thoracopod-bearing segment, the genital tagma and the abdomen.

Fish lice have separate males and females (Figures 8.1, 8.2). Individuals with two cylindrical egg sacs (ovisacs) trailing from the genital complex are readily identified as females (Figures 8.1, 8.2, 8.9). Even if these sacs are absent, the female genital tagma is larger than that of the male and there are differences between the maxillipeds in the two sexes.

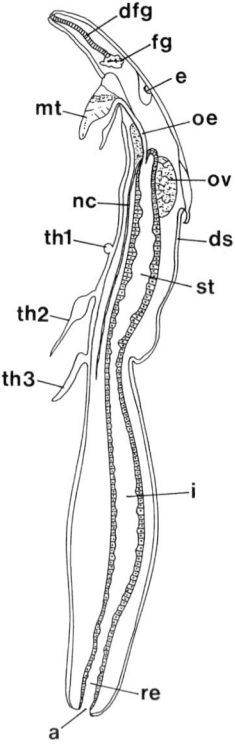

Fig. 8.3. A longitudinal (sagittal) section through an adult female *Lepeophtheirus pectoralis*. a, anus; dfg, duct of frontal gland; ds, dorsal shield (carapace); e, eye; fg, frontal gland; i, intestine; mt, mouth tube; nc, nerve cord; oe, oesophagus; ov, ovary; re, rectum; st, stomach; th1 – th3, first to third thoracopods. From Scott, A., 1901.

The possession of a rigid exoskeleton restricts growth. Consequently, arthropods, including crustaceans, must shed their restrictive armour at intervals by moulting (ecdysis) and replace it with a soft, temporarily extensible, new one. Physical expansion of the body is only possible before this soft exoskeleton fully hardens. The life cycle of caligid copepods (Figure 8.1) typically comprises ten stages separated by moults (Kabata, 1972). The first stage emerging from the egg is nauplius I. This stage is free-living, as is the next stage, nauplius II. This is followed by the copepodid (copepodite), which seeks out and establishes itself on the fish host, and four so-called chalimus stages. Burmeister (1831, in Scott, A. 1901) was sufficiently impressed by the distinctiveness of these attached stages that he placed them in a genus of their own, which he called *Chalimus*. Later they were recognised as larval stages in the life cycle of caligid copepods, but the name 'chalimus' has been retained for these stages (see Scott, A. 1901). Only two chalimus stages (I and IV) are shown in Figure 8.1. The chalimus stages progress at each moult by small steps towards the pre-adult and then the adult condition. According to Schram (1993), it is possible to distinguish male and female chalimus IV stages in *L. salmonis* by small differences mainly in shape and size, but the sexual distinction becomes more obvious in the pre-adult stages I and II.

8.3 THE ADULT CALIGID

8.3.1 *The role of the suctorial cephalothorax in attachment*
A convex dorsal shield (carapace) covers the cephalothorax and extends on its anterior margin to form two frontal plates fused in the mid-line (Figure 8.2CD). The low profile of the flattened cephalothorax (Figure 8.2D) is ideal for attachment to a slippery surface that is often swept by strong water currents (Kabata, 1979). The edge of the shield is sealed by a peripheral flap that acts as a marginal valve when suction is generated, preventing influx of water between the edge of the shield and the host's skin (Figure 8.2). There are two breaches in the marginal seal in the form of notches in each posterolateral margin of the dorsal shield (Figure 8.2C). Each of these so-called posterior sinuses is covered by a flap that permits water to escape from beneath the cephalothorax but prevents influx of water.

The effectiveness of adhesion by suction is illustrated by the following statement by Scott (Scott, A., 1901) referring to *Lepeophtheirus pectoralis*: "By depressing the edges of the carapace (= cephalothorax) and applying them closely to the skin, the parasite can increase its holding power to such an extent that the posterior end can be torn from the anterior part without detaching it". This increase in holding power, so typical of mature females when disturbed (see Scott, A., 1901), presumably reflects a drop in pressure beneath the saucer-shaped tagma, influx of seawater being prevented by the marginal valve. The ability of parasites that have been separated from the host to adhere to the sides and bottom of the holding vessel and to the surface film was described by Scott (Scott, A., 1901) and demonstrates the effectiveness of suction without the assistance of hooked appendages. The importance of intimate contact between the margin of the cephalothorax and the substrate is underlined by an experiment performed by Kabata & Hewitt (1971) using *L. salmonis*. They found that the parasite is unable to attach itself to a piece of coarse sandpaper immersed in seawater.

Kabata & Hewitt (1971) described what they called 'settling' movements immediately after attachment of caligids to stationary or slow moving fishes. These movements involve initial expulsion of water via the posterior sinuses, followed by slight rotating movements of the animal, alternately to the left and right of the longitudinal axis of the body. Kabata & Hewitt came to the conclusion that the rotating movements are effected by the second maxillae, which, by moving alternately, shift first one side of the animal and then the other. These 'settling' movements were interpreted as preparatory movements seating the cephalothorax and sealing its periphery so that suction could be generated, presumably by lifting the ventral surface of the cephalothorax relative to the rigid carapace. According to Kabata & Hewitt (1971), 'settling' is repeated at intervals, perhaps to expel water that has seeped beneath the carapace and to re-establish strong suctorial adhesion, but these movements were rarely observed on vigorously moving fish.

The hook-shaped second antennae and maxillipeds (Figure 8.2AB), described as "strong, rapaciously prehensile appendages" by Kabata & Hewitt (1971), may also play a part in attachment. Kabata & Hewitt speculated that the second antennae and the maxillipeds might serve for attachment when suction is for some reason relinquished.

The special role of frontal filaments for attachment of caligids will be considered below.

8.3.2 *Locomotion*

Movements of the first and second pairs of thoracopods (Figure 8.2AB) propel the animal in open water. A transverse bar links the limbs of each pair, so that the two corresponding limbs work as a unit. In *Caligus clemensi* these units were estimated to beat at a rate of 100 strokes per minute (Kabata & Hewitt, 1971). The first thoracopods are uniramous, but the second thoracopods have a relatively large surface area for propelling water since they retain their primitive biramous condition (see p. 154) and are well endowed with projecting spines and setae.

The relative movements of the first and second thoracopods are such that a jet of water is produced, driving the animal forward and providing lift (see Kabata & Hewitt, 1971, for details). The free-swimming copepod changes direction by using the genital tagma and abdomen as a rudder.

The fact that caligid copepods can swim freely implies that the adults are able to change hosts, but how frequently this occurs is not known. Scott (Scott, A., 1901) found no specimens of *L. pectoralis* in filters used to collect fish eggs from more than 150 parasitised flounders, *Platichthys flesus*, kept in fish tanks, suggesting that, at least in this species, host to host transfer is rare or does not take place. However, he also found that the isolated parasite could live for upwards of six weeks in filtered seawater without visible food. Moreover, the frequent appearance of both male and female caligids in plankton nets at night (Kabata & Hewitt, 1971) is consistent with host changing.

Adult caligids are also able to move rapidly over the surface of their hosts (Kabata & Hewitt, 1971), permitting the parasites to sample fresh feeding sites and to make contact with a mate. According to Kabata & Hewitt, the parasite is propelled forward on the host by jets of water issuing from the posterior sinuses, this water being replaced by water taken in anteriorly beneath the slightly raised frontal plates. These jets are created mainly by the backstrokes of the second thoracopods. Physical contact between the second thoracopods and the host skin during the backstroke may also contribute to the forward glide of the parasite. Kabata & Hewitt also found that when the first and second thoracic limbs are immobilised experimentally by means of an adhesive, the second maxillae (Figure 8.2AB) are capable of dragging the animal forward.

8.3.3 *Sensory capabilities*

The first antennae (Figures 8.2, 8.4) of the parasitic caligid copepods have changed both in structure and in function. They are too short to make a physical contribution to locomotion or to copulation. They consist of only two podomeres, the proximal (basal) podomere often being flattened, so that while the parasite is gliding over the host's surface, they can be tucked out of the way beneath the frontal plate. Their antero-lateral positions and their prominent innervation and abundant projecting setae, underscore

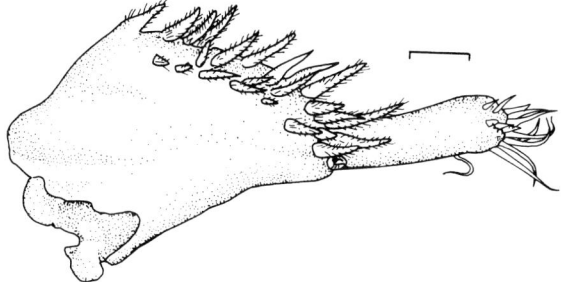

Fig. 8.4. First antenna (antennule) of a caligid copepod in ventral view. Scale bar: 100μm. From Parker *et al.*, 1968, reproduced by courtesy of Fisheries and Oceans Canada, with the permission of Her Majesty the Queen in Right of Canada, 2004.

their importance as sensory probes. They appear to be sites for mechanoreceptors and chemoreceptors, sensory structures responding to physical (touch, vibration, water currents) and chemical cues respectively (Pike & Wadsworth, 1999). Moreover, Kabata (1979) observed that in *Caligus* their tips were repeatedly raised and lowered, apparently palpating the substrate, while the copepod was in motion.

Two eyes lie close together on the dorsal surface of the cephalothorax (Figure 8.2C). According to Scott (Scott, A., 1901), the two eyes of *L. pectoralis* appear as a reddish spot embedded in a mass of reddish-black pigment in the living animal. The nauplius eye is described below.

A depression in each posterolateral region of the carapace of *Caligus curtus* contains many minute, branched, hair-like projections (Figure 8.2CD). According to Parker *et al.* (1968), this depression is sensory in nature, but its specific function is unknown.

8.3.4 *Camouflage*

Scott (Scott, A., 1901) made the interesting observation that specimens of *L. pectoralis* from the dark side of the flounder (*Platichthys flesus*) were deep brown, almost black, in colour, while those on the white side of the host and under the fins were nearly colourless. Thus the parasites seem capable of matching their background by contraction or expansion of their chromatophores. The implication is that these exposed skin parasites may not be entirely free from predation by free-living organisms sharing their aquatic environment (see also p. 177).

8.3.5 *Feeding*

We know remarkably little about feeding in parasitic copepods. The reason for this is that it is not possible to observe what is going on beneath the cephalothorax when feeding is taking place. We do have a detailed picture of the anatomy of the mouth region and the structures associated with it, but published accounts of the feeding mechanism are inevitably speculative.

In *L. pectoralis*, and in related caligids such as species of *Caligus*, the ventral mouth lies in the midline within the concavity of the cephalothorax. It is located at the tip of a short tube created by partial fusion of the upper and lower lips (designated labrum and labium by Kabata, 1974) (Figures 8.2AB, 8.3, 8.5A). This tubular buccal apparatus (or siphon) is a major feature of siphonostomatoids and Kabata (1979) regarded the siphonostomatoid mouth as the key to the success of members of this group as parasites, especially those found on fishes.

The mandibles are rod-shaped (Figure 8.5B) and enter the buccal tube through lateral slits (Figure 8.5A). The distal ends of the mandibles are curved, with teeth along their inner curvature (Figure 8.5C), and lie close to the mouth in association with the labium. Figure 8.6A is a diagrammatic representation of an *en face* view of the buccal opening at the free end of the tube, showing the position of the mandibular teeth. The aperture is surrounded by a marginal flap or valve with a frayed border, and there is a labial fold lying proximal to the marginal valve (Figure 8.6AB). Proximal to the fold is a hard cuticular bar or strigil consisting of two halves joined centrally by a cuticular swelling embedded in the tissue of the labium. The outer edge of the strigil is armed with about 100 small sharp teeth.

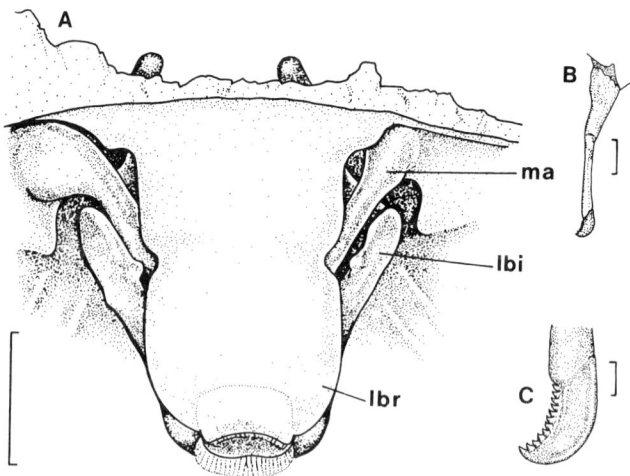

Fig. 8.5. (A) Mouth tube of a caligid copepod with the mandible (ma) *in situ*. (B) Whole mandible and (C) enlarged mandible tip. lbi, Labium; lbr, labrum. Scale bars: (a) 200 µm; (b) 100µm; (c) 25µm. From Parker *et al.*, 1968, reproduced by courtesy of Fisheries and Oceans Canada, with the permission of Her Majesty the Queen in Right of Canada, 2004.

From observations made with an inverted microscope, Kabata (1974) concluded that when moving over the host's surface the buccal tube is folded back along the ventral surface of the body with the mouth pointing in a posterior direction. He assumed that prior to feeding the tube adopted a position perpendicular to the body so that the circular mouth opening could be applied to the skin surface and sealed peripherally by its marginal valve. It is assumed that host skin tissue is abraded by the mechanical action of the mandibles (Parker *et al.*, 1968) and/or the strigil (Kabata, 1974) and that detached tissue is sucked into the gut.

The younger, smaller, parasitic stages appear to abrade and ingest only the epidermis of the host (see below), but the larger individuals (pre-adults and adults) may penetrate the epidermal basement membrane and ingest dermal tissue. According to Scott (Scott, A., 1901), the alimentary canal of an adult *L. pectoralis* rarely displays even the faintest trace of the red material that is to be expected in the gut of a habitual

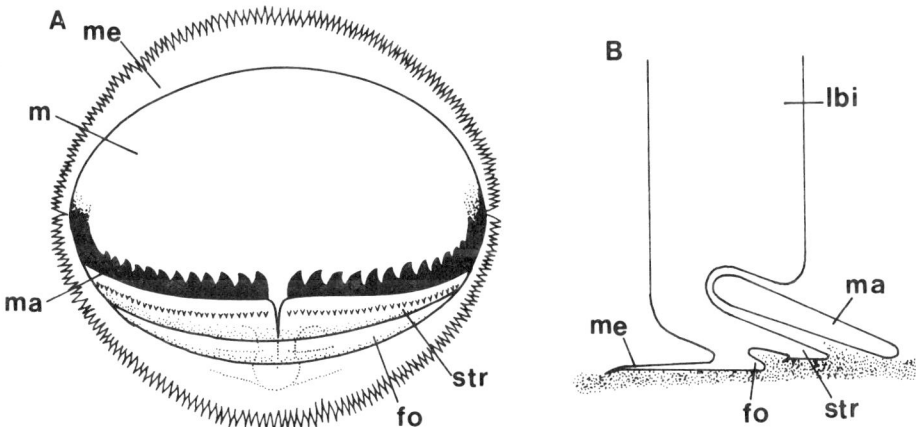

Fig. 8.6. (A) Diagrammatic *en face* view of the opening of the mouth tube of a caligid copepod. (B) Diagrammatic longitudinal (sagittal) section through the distal end of the labium of a caligid copepod. fo, Fold of labium; lbi, labium; m, mouth; ma, mandible; me, membrane; str, strigil. From Kabata, 1974, reproduced by courtesy of Fisheries and Oceans Canada, with the permission of Her Majesty the Queen in Right of Canada, 2004.

blood feeder. However, Scott took the greatest care to remove attached individuals and found that the host skin beneath was lacerated and bleeding, especially when there were several parasites close together. He observed that the pectoral fins of some fishes infested with egg-bearing females were partially destroyed. These observations suggest that host epidermis may provide the bulk of the food for *L. pectoralis* and that this is supplemented by dermal tissue. Penetration of the dermis is likely to lead to rupture of a blood vessel and ingestion of some blood.

It has been demonstrated photometrically that blood is an important component of the diet of *L. salmonis*, especially for adult females, but it is less important for adult males and pre-adults (Brandal *et al.*, 1976). This difference may reflect the larger size of mature females and their correspondingly deeper penetration into the dermis during feeding.

Mobile stages of *L. salmonis* aggregate on the head of salmonids and this is the area where mating occurs (see below). This site preference produces obvious pathology, depicted in many papers on sea lice infestation (see Pike & Wadsworth, 1999). The skin is thin here and grey patches occur on the head where the parasites have been feeding. Damage may be so severe that the skull is exposed or even penetrated if the fish is small (see Wootten *et al.*, 1982).

8.3.6 Mating

Mating involves the transfer of spermatophores from the male to the genital openings of the female, and has been described vividly by Anstensrud (1990) in *L. pectoralis*. Spermatophores may be received by second stage pre-adult females as well as by adult females. Mate guarding, as in other copepods, is a behavioural feature. Adults adopt a pre-copula position (Figure 8.7A) in which the male uses his second antennae to grasp the dorsal surface of the female just anterior to the genital complex. The male may also use his maxillipeds for pre-copula attachment. The pre-copula stage may last from 2 – 4 hours up to six days, during which period the male rarely moves, except for occasional contractions of the genital complex accompanied by a few rapid strokes of the first and

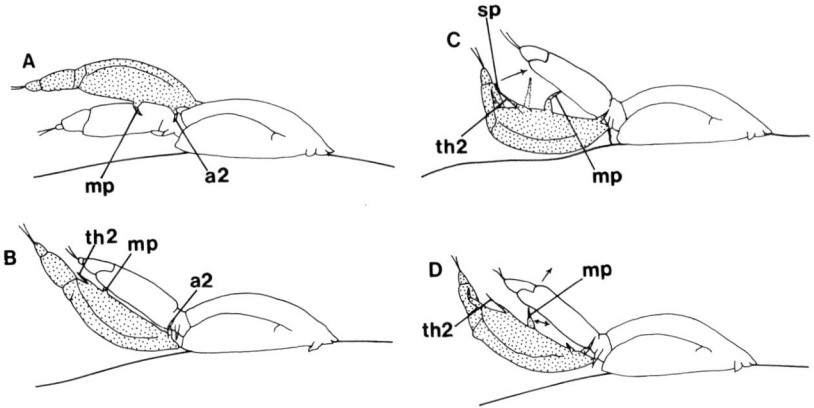

Fig. 8.7. (A-D) Successive positions of partners during mating in *Lepeophtheirus pectoralis*. Male stippled, female unstippled. See text for explanation. a2, Second antenna; mp, maxilliped; sp, spermatophore; th2, thoracopod 2. From Anstensrud, 1990, with permission of Cambridge University Press.

second thoracopods. In most pairs the onset of moulting in the female prompts the male to release his hold and station himself on the host's body surface close to the female. After the female completes moulting, the male returns to the pre-copula position. Anstensrud did not observe insemination of immediately post-moult females, which have a soft exoskeleton.

When the female's exoskeleton is hard the male moves from the upper to the lower surface of the female and adopts the copula position (Figure 8.7B). The male pierces the openings of the female's seminal receptacles with the tips of his maxillipeds and pulls the female's genital complex violently towards him. The male repeatedly stretches and partly deforms this region of the female for periods of up to 20 seconds. During this phase, the male repeatedly flexes his genital complex and abdomen, accompanied by rapid strokes of the first and second thoracopods. The maxillipeds move forwards, thus bending the genital tagma of the female upwards (Figure 8.7C). The male then expels two spermatophores simultaneously, each with a sticky coat that fuses with its neighbour (Figures 8.7C, 8.8A). The spermatophores are picked up from the surface of the male's genital complex by the long setae of his second thoracopods and implanted on the female's genital tagma. The arrow in Figure 8.7C indicates this movement. The transfer of spermatophores takes less than two seconds. The maxillipeds then resume their actions, being used to adjust the position of the spermatophores and then to push the genital complex of the female away from the male (Figure 8.7D, arrows). Anstensrud observed that the spermatophores were stroked three or four times with the second thoracopods at this stage. Within six minutes of transfer the male adopts a post-copula position identical to the pre-copula position on the female's dorsal surface and leaves the female within three hours. The female is motionless during copulation, apart from small movements of the first antennae and maxillae. The reproductive behaviour of *L. salmonis* is similar (see Pike & Wadsworth, 1999).

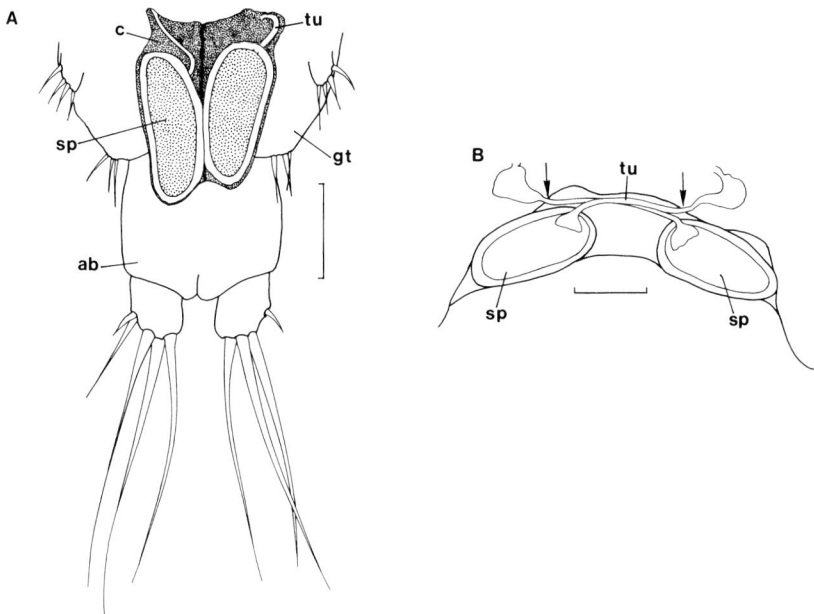

Fig. 8.8. (A) Posterior end of a male *Lepeophtheirus pectoralis* with newly expelled spermatophores (sp). (B) The genital region of a female *Lepeophtheirus pectoralis* with attached spermatophores. Arrows, openings of seminal receptacles. ab, Abdomen; c, cement; gt, genital tagma; tu, spermatophore tube. Scale bar: 100 μm. From Anstensrud, 1990, with permission of Cambridge University Press.

Each spermatophore of *L. pectoralis* has a tubular extension (Figure 8.8) through which the contents of the spermatophore are conveyed to the seminal receptacle of the female. The spermatophore on the left supplies the female's right seminal receptacle and *vice versa* (Figure 8.8B).

Even when many males were present, Anstensrud (1990) observed only one male at a time in pre-copula or copula positions. Extra males sometimes approached a mating pair, but after a brief encounter, the extra males withdrew with no agonistic behaviour observed. Anstensrud also conducted a series of fascinating experiments in which the first males to achieve the pre-copula position were stained with neutral red, a vital dye that has no apparent effect on survival time, or on the vital functions and behaviour of the males. He then challenged these marked males with unstained free males. In matings between second stage pre-adult females and males, unstained males replaced only 7% of stained males in pre-copula. However, in matings between adult females and males, 40% of the male partners changed places. Anstensrud suggested that the opportunity for swapping male partners might occur when the first male is forced to leave the female when she moults.

Anstensrud found no evidence that larger males take over females from smaller males in *L. pectoralis*. When males were introduced simultaneously to virgin mature females, there were no significant differences in the size ranges of the males that succeeded in establishing pre-copula and those that failed. Thus, the first male to locate a female normally establishes the pre-copula position, but if the female undergoes moulting, the original male may lose the female to another male of any size that happens to be nearby.

8.3.7 *Site preference*

Scott (Scott, A., 1901) reported that males of *L. pectoralis* and immature individuals of both sexes occur all over the skin on both sides of the flounder (*Platichthys flesus*), but that mature, egg-bearing females had a preference for the inner surfaces of the pectoral fins. Here, according to Kabata (1979), they aggregate in close ranks with their carapaces in rows and their posterior halves overlapping one another (Figure 8.9). Scott (Scott, A., 1901) reported finding between 20 and 30 mature females under each pectoral fin.

As already stated, mobile stages of *L. salmonis* and especially adult males prefer a dorsal anterior site on the salmon (*Salmo salar*), aggregating mainly just behind the head, and this is where mating pairs appear to be set up (see Pike & Wadsworth, 1999). When adult males are present in this dorsal anterior site, other stages, such as

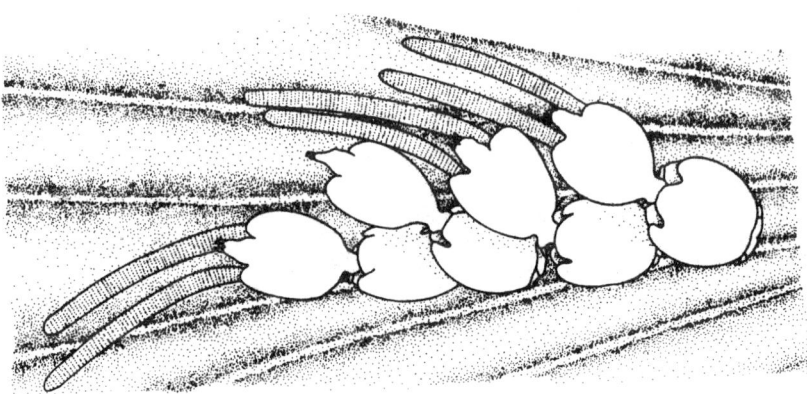

Fig. 8.9. An aggregation of four female *Lepeophtheirus pectoralis* attached to the host's fin. From Kabata, 1979, with kind permission of The Ray Society.

pre-adults of both sexes and females that have already received spermatophores, often occupy other locations. This relocation may avoid the inevitable disturbance created by the activity of new males. According to Ritchie (1993, in Pike & Wadsworth, 1999), mated female *L. salmonis* typically relocate to the dorsal adipose or anal fin areas to produce their egg sacs (ovisacs) – they are capable of producing several pairs of egg sacs without further mating (see below).

8.4 FATE OF THE EGGS

Eggs of caligid copepods are not released into the sea as they are assembled. There is an opening on each side of the genital segment and as each biscuit-shaped egg leaves this opening it is enclosed in a thin tube or ovisac secreted by a gland communicating with the oviduct. Each ovisac extends as newly laid eggs are added to it in a single column, creating a long egg string, which, together with the string on the opposite side of the genital segment, provides a characteristic and easily recognisable feature of female copepods (Figure 8.9). Each ovisac is securely attached to the parent by a hook, which in *L. salmonis* enters a notch, locking the proximal end of the ovisac in position (Schram, 2000). The presence of hook muscles indicates that the female may be able to control release of the egg sac. Up to 700 eggs have been recorded from one pair of egg sacs of *L. salmonis* by Wootten *et al.* (1982).

Development times between egg extrusion and hatching are temperature dependent. In *L. salmonis* Johnson & Albright (1991) recorded times of 17.5, 8.6 and 5.5 days at 5°C, 10°C and 15°C respectively. Hatching of the nauplii begins at the distal end of the egg sacs in *L. salmonis* (see Pike & Wadsworth, 1999). However, retention of eggs until they hatch does not lead to autoinfection as it does in gyrodactylid monogeneans (see Chapter 4), since the released larva or nauplius is not infective and leads a pelagic, independent life (see below).

Johannessen (1978) observed that egg sacs of *L. salmonis* were replaced within 24 hours after release of all their nauplii. A single mated female observed by Ritchie (1993; in Pike & Wadsworth, 1999) produced six sets of egg sacs over a period of 50 days when maintained at 14°C and exposed to a light/dark regime of 6:18 hours.

8.5 NAUPLIUS LARVAE

The newly hatched nauplius larva (nauplius I) of *L. pectoralis* (Figure 8.1A) is about 0.5 mm in length (Boxshall, 1974). It has only four segments and three pairs of limbs on the anterior segments, a uniramous pair and two biramous pairs. These projecting limbs bear setae at the distal ends and provide very effective oars for sculling the animal along. The first and second pairs of limbs are the first and second antennae and, surprisingly, the third pair is the mandibles, at this stage bearing no resemblance to the mandibles of the adult. Nauplius I moults to produce a similar nauplius II (Figure 8.1B). Nauplii are assumed to be unable to feed, subsisting on yolk reserves sequestered during embryonic development (Bron *et al.*, 1993ab).

Nauplii have a median eye. In *L. salmonis* this consists of a pair of dorsal ocelli arranged side by side and a single ventral ocellus (Bron *et al.*, 1993a). Each dorsal ocellus has a large lens projecting anterodorsally through a window in the pigment shield surrounding the eye. Nine retinular cells are arranged in a basket surrounding the lens of each ocellus. The ventral eye has no lens and 10 retinular cells. Each dorsal and ventral ocellus has, like a cat, a tapetum or mirror, which reflects light collected by the lens back into the retinular cells for a second time, thus enhancing the light sensitivity of the eye. According to Bron *et al.* (1993a), this combination of lens, mirror and effective pigment-shielding in the dorsal ocelli indicates that the eye has good light-gathering properties in the anterodorsal field of view, and may have image-forming ability, allowing for precise location of a light or shadow source. The ventral ocellus is in a good position to provide information on the level of illumination from behind and below the animal.

According to Wootten *et al.* (1982), nauplii of *L. salmonis* are photopositive. Active upward swimming is followed by a passive-sinking phase. If they touch the bottom or are mechanically disturbed a new phase of swimming is initiated.

8.6 THE COPEPODID

The next moult involves the most radical change in the whole sequence of moults in the copepod's life. This is the change that denotes the transition from the brief free-living nauplius to the copepodid, the infective stage capable of establishing itself as a parasite.

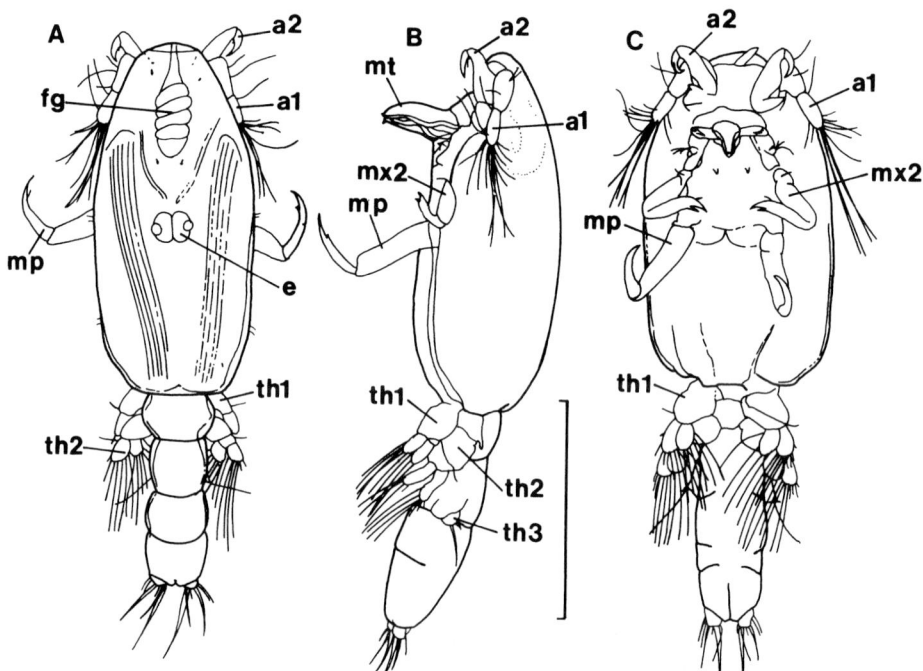

Fig. 8.10. Copepodid of a caligid copepod (*Caligus spinosus*). (A) Dorsal, (B) lateral and (C) ventral views. a1-2, First and second antennae; e, eye; fg, frontal gland; mp, maxilliped; mt, mouth tube; mx2, second maxilla; th, 1-3, thoracopods. Scale bar: 300 μm. From Izawa, 1969, reproduced with permission.

The copepodid has a cephalothorax incorporating the first two thoracic segments and their limbs (maxillipeds and the first pair of thoracopods), followed by four segments, the first two carrying the second and rudimentary third pairs of thoracopods (Figures 8.1C, 8.10).

Like the nauplius, the copepodid of *L. salmonis* is thought to persist on embryonic reserves while free-living (Bron *et al.*, 1993b) and, according to Wootten *et al.* (1982), survives for 4 days on these reserves at 12°C. Boxshall (1976) found that the free-swimming copepodid of *L. pectoralis* has about three days in which to find a host.

8.6.1 Host finding
According to Boxshall (1976), copepodids of *L. pectoralis* are positively phototactic, but after an initial post-moult period of swimming become relatively inactive and sink to the bottom where potential hosts are to be found. Bursts of renewed swimming activity are induced by sudden changes in light intensity and by water turbulence. Boxshall observed that the larva swims against water currents and he believes that this response is crucial for host finding, since copepodids swimming upstream in relation to currents produced by fish breathing movements and locomotion are likely to make contact with the host. In fact, in natural populations of flatfishes from the North Sea, newly attached copepodids were concentrated on the fins and around the gill openings, sites from which water currents originate.

Research on salmon (*Salmo salar*) indicates that they frequent the top three metres of the sea, especially at night (references in Pike & Wadsworth, 1999). On the basis of this, copepodids of *L. salmonis* would be expected to make their way towards the surface at night, but Heuch *et al.* (1995) found the reverse. In six metre deep plastic enclosures suspended in the sea, copepodids ascended during daylight and descended at night, the daylight response being consistent with the photopositive behaviour of copepodids in the laboratory (Wootten *et al.*, 1982; Bron *et al.*, 1993a). Thus, the hosts and the infective stages of the parasite appear to move in opposite directions and, as Heuch *et al.* suggested, opportunities for infection may occur during crossing over, as the fish travel downwards and the parasites upwards and *vice versa*. The presence of infective copepodids at the surface was confirmed by Hevrøy *et al.* (1997, in Pike & Wadsworth, 1999), who found that sentinel fish held at depths of down to four metres acquired substantial infections. However, the discovery that sentinel fishes held at depths of 4 – 8 and 8 – 12 metres acquired few parasites was unexpected, since the indications are that copepodids move to deeper water at night. This clearly requires further investigation.

Responses of copepodids and salmon to stratification produced by sudden changes in salinity with depth (haloclines) may also promote host location. Heuch (1995, 2002) found that copepodids in the sea tend to aggregate near haloclines. Foraging salmonids are also likely to spend time near these discontinuities since chemical cues from their prey organisms spread along them.

Once within range, cues from the host will become important in host location. Copepodids readily infect in the dark, so light-related cues are not essential (Heuch, 2002). Chemical cues are likewise unlikely to be significant since a rubber cast of a salmon head provides no relevant chemical cues and yet copepodids attack the head when it is advanced towards them through the water (Heuch, 2002). On the other hand, the head will provide hydro-mechanical signals and the importance of these has been confirmed by Bron *et al.* (1993a) and by Heuch & Karlsen (1997). The last-named authors established that copepodids of *L. salmonis* respond to low frequency vibrations between one and 10 Hz. Frequencies below 20 Hz are infrasonic, i.e. not detectable by the human ear, but are similar to those propagated within a few centimetres of the advancing front end of a swimming fish. Copepodids react by rapid circular swimming, which probably increases the chances of making contact with the host. It is possible that free-living planktonic copepods may employ similar sensitivity and responsiveness as a means of detecting and avoiding predatory fish and that *L. salmonis* has inherited and modified these traits for host finding.

Boxshall (1976) and Bron *et al.* (1993a) believed that after initial contact with the host, the copepodids of *L. pectoralis* and *L. salmonis* respectively determine whether the fish is a suitable host or not by means of chemoreception. If the substrate is unsuitable the copepodids are capable of resuming swimming.

8.6.2 Settlement and initial attachment to the host
Bron *et al.* (1991) described settlement and attachment of the copepodid of *L. salmonis*. They experimentally infected salmon smolts (mean length 28.7 cm) with copepodids. No more than six copepodids settled and established themselves on each fish, in spite of exposure to over 30 copepodids per fish for one hour. The successful settlers preferred the fins, in particular the ventral surfaces of the paired fins and the entire caudal fin, but settlement also occurred on the ventral flanks, in the buccal cavity and on the operculum and gills (cf. distribution of the mobile stages, pp. 163, 166).

After initial contact with the host, a period of close 'searching' ensued, during which the organism gripped the fish with the prominent maxillipeds (Figure 8.10) whilst moving over a small area and probing the skin with the anterior end of the cephalothorax. In this attitude the first and second antennae were in close contact with the host's surface. At this stage the copepodid was still able to abort the settlement process and leave the host, but if the chosen site were suitable the hooked second antennae (Figure 8.10) were driven into and sometimes through the epidermis, usually with a repeated stabbing action. This action forced the anterior edge of the cephalothoracic shield forwards and downwards, like the shovel of a bulldozer, thereby pushing the epidermis into a heap of compressed cells in front of the animal. The final phase of attachment involved the secretion of cement from a frontal gland (Figure 8.10A). This cement is injected via a single median opening on the head underneath the host epidermis and onto the basement membrane, where it spreads out and sets to form a basal plate. Further secretion produces a stem or filament of different composition from the basal plate and connecting the latter with the anterior end of the copepodid. The attachment plate and filament are covered externally with a sheath that is apparently continuous with the cuticle of the copepodid. In *L. salmonis*, filament production is rapidly followed by the moult to the next stage (chalimus I), since Bron *et al.* (1991) found no copepodids attached by the filament. In other caligids a few days (2 – 3 days in *L. pectoralis*; see Boxshall, 1976) may lapse between settlement and moulting to the first chalimus stage.

In *L. salmonis*, insertion of the frontal filament through the epidermis causes relatively little reaction in the skin and in spite of the invasive nature of the filament it does not provide an uptake route for nutrients. The copepodid feeds on host epidermis. As Bron *et al.* (1993b) have pointed out, the area of host skin accessible to the copepodid is severely limited to the region adjacent to the mouth cone, because the parasite is pinned to the skin by the second antennae.

8.7 CHALIMUS STAGES

The frontal filament provides the only attachment for the chalimus stages (Figure 8.1DE). By the time the second chalimus stage is reached, the disrupted epidermis in the immediate vicinity of the filament is normal or mildly hyperplastic (Jones *et al.*, 1990). Thus the frontal filament appears to provoke little or no reaction from the host and this inert tether relieves the chalimus of any need to expend energy for attachment. All resources can be directed towards growth and development.

It was the view of Gurney (1934) that the filament is formed just once during the progression of fixed chalimus stages, apart from the proximal end (origin) of the filament, which is renewed at each moult. Heegaard (1947) rejected this view, asserting that the whole filament is formed anew at each moult (see account of moulting below). More recent work tends to support the notion that the filament is replaced during development (references in Pike *et al.*, 1993).

The chalimus uses its maxillipeds to scuttle in an almost complete circle around the tethering filament, the length of which determines the radius of the circular patch of skin that is within grazing reach of the parasite's mouthparts. The feeding activity of the chalimus leaves a distinctive ring of erosion around the attachment point. Food material, including host epithelial cells, was frequently found in the gut (Jones *et al.*, 1990; Bron *et al.*, 1993b); blood corpuscles were also recorded. In tissue sections, Jones *et al.* (1990) found all chalimus stages with the mouth cone in contact with the skin, but penetration through the epidermal basement membrane was not seen. Jones *et al.* (1990) raised the possibility that host tissue may be digested externally, as in monogeneans like *Entobdella soleae* (Chapter 3). Digestive secretions may be secreted into the mouth tube of the chalimus and thence onto the skin. However, Jones *et al.* were of the opinion that the epidermal damage appeared more consistent with mechanical rather than chemical action.

8.8 THE PROBLEM OF MOULTING

The tethering frontal filament restricts the area of host skin from which epidermis can be harvested. Larger parasites require the freedom to harvest food from a wider area and excessive local feeding pressure by a large tethered parasite is undesirable. Consequently larger parasites (pre-adults and adults) are no longer tethered, except for the early pre-adult male of *L. salmonis* (see Wootten *et al.*, 1982) and the brief period of moulting in all stages.

Parasitic arthropods like the caligids face a problem that is not shared by other ectoparasites such as the monogeneans and the leeches. In order to grow, arthropods must undergo moulting. This involves sloughing off the whole of its exoskeleton (cuticle), which includes the covering of the limbs. Maintaining secure attachment to the body of an active fish host, while, at the same time, shedding the hard covering of the body and particularly the limbs is difficult enough, but added to this, the new exoskeleton is initially soft, rendering the limbs ineffective for attachment.

Heegaard (1947) gave a detailed and vivid account of moulting in a chalimus, illustrating the importance of the tethering frontal filament in the process. According to Heegaard, the chalimus begins by shrinking and withdrawing backwards slightly within the old cuticle, so that the old and new cuticles separate a little. The old filament connects the front of the old exoskeleton to the fish and this connection is maintained until moulting is complete. The chalimus inside its new cuticle is no longer connected to the old one. Aided by the parasite, the old cuticle splits down each side and the carapace emerges. The chalimus maintains this position, with the posterior part of its body inside the old cuticle until the new cuticle of the carapace and its limbs have hardened somewhat. The parasite then crawls onto the fish's skin, but does not yet withdraw the posterior region of the body from the old cuticle, so that if it should be dislodged it will not be swept away from the host. Only after the parasite has pinned itself securely to the host by means of the newly hardened maxillipeds, does it withdraw its abdomen from the old cuticle, completing the moulting process.

Now the chalimus has only to re-establish its connection with the host via a new tethering frontal filament. Heegaard's sketches of a chalimus of *Caligus curtus* renewing this important link are shown in Figure 8.11. The old empty cuticle remains

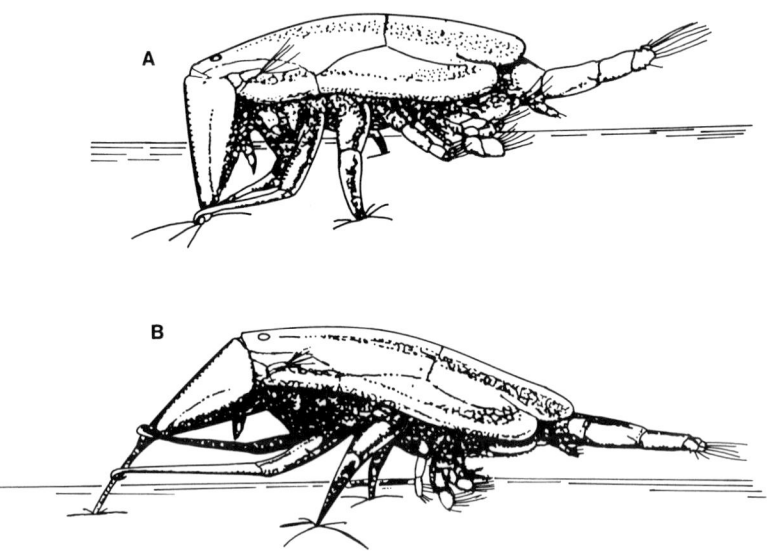

Fig. 8.11. Freshly moulted chalimus of *Caligus curtus*, (A) piercing a hole in the skin of cod, *Gadus morhua*, to reach a fin ray and (B) establishing the tethering filament attached to the fin ray. From Heegaard, 1947.

attached to the host by its original tethering filament. In Heegaard's own words: "If therefore one examines a piece of the fin of a cod in the late summer one can see on its surface, upon a close microscopic examination, the remains of one chalimus cuticle after another".

Pre-adults and adults of *L. pectoralis* are free to roam the surface of their host during intermoult periods but face the same problems as the chalimus stages during moulting. Anstensrud (1990) observed that they solve the problem in a similar way, attaching themselves just before moulting by secreting a temporary frontal filament, which is variable in length, but may reach 0.5 – 1 mm. Lack of reports of this filament in moulting pre-adults and adults of other caligids may be a reflection of the difficulty of detecting this thin and hyaline structure against the host's skin.

According to Anstensrud (1990), after splitting of the old cuticle anteriorly the pre-adult or adult parasite has the freedom to shed it in a posterior direction by undulating body movements. Moulting takes 1 – 2 hours and the parasite remains tethered by the frontal filament for another two or three hours, during which time hardening of the exoskeleton of the body and appendages proceeds. When fully hardened with fully effective appendages, the individual detaches itself from the frontal filament and resumes its free movement over the host.

8.9 HOST SPECIFICITY AND SPECIATION – *LEPEOPHTHEIRUS PECTORALIS*

Scott (Scott, A., 1901) found *L. pectoralis* most frequently on the flounder (*Platichthys flesus*). Boxhall (1976) reported that the most common natural hosts of *L. pectoralis* are flounder, plaice (*Pleuronectes platessa*) and dab (*Limanda limanda*), in the same order of frequency of occurrence. It has also been recorded occasionally from 16 other fish species, including the common sole (*Solea solea*). Boxshall offered single fish species to copepodids derived either from adult parasites on flounder ('*flesi*-copepodids') or from adult parasites on plaice ('*platessae*-copepodids'). Some of his results are shown in Table 8.1 and demonstrate a clear difference between *flesi*- and *platessae*-copepodids in their response to the host offered. *Flesi*-copepodids showed a stronger preference for flounder than for plaice, dab, common sole or turbot (*Scophthalmus maximus*) and *platessae*-copepodids preferred plaice. This pattern of choice was reinforced by experiments in which Boxshall offered the copepodids a choice of two host species. In these experiments flounder was preferred by *flesi*-copepodids to all other fish species and plaice was preferred by *platessae*-copepodids to all other fishes (Table 8.2).

These results suggest that there are two distinct strains of *L. pectoralis*, one (*flesi*) adapted to flounder and the other (*platessae*) adapted to plaice. Boxshall (1976) interpreted this as incipient speciation of *L. pectoralis*. Flounder and plaice are closely related (they hybridise naturally on occasion) and undoubtedly evolved from a common ancestor relatively recently. According to Boxshall, speciation of the parasites has lagged behind that of the hosts – no morphological differences between the parasite strains are evident as yet, but behavioural differences in the form of host preference are already evident. Nevertheless, this host preference is not yet absolute and both strains are still capable of infecting both host species.

Table 8.1. Success of establishment (%) of copepodids of Lepeophtheirus pectoralis *when presented with a single fish species (from Boxshall, 1976).*

Fish species	flesi *copepodids*			platessae *copepodids*		
	No. of expts	Mean	Range	No. of expts	Mean	Range
Flounder	3	69	57 - 87	4	39	20 - 71
Plaice	3	19	8 - 30	4	56	43 - 64
Sole	1	43	-	-	-	-
Turbot	1	3	-	1	6	-
Dab	1	0	-	1	3	-

In the single-host experiments, few copepodids responded to dab (Table 8.1). This is surprising because dab is a commonly infected natural host. It is also a pleuronectid fish like plaice and sometimes hybridises with it in the wild. In the two-choice experiments, dab performed badly when competing with flounder for *flesi*-copepodids, with plaice for *platessae*-copepodids and even with common sole for *platessae*-copepodids (Table 8.2). These observations suggest that there is a third strain of *L. pectoralis* adapted to dab.

In the single-host experiments, common sole was more attractive to *flesi*-copepodids than plaice (Table 8.1) and in the two-choice experiments common sole was preferred in direct competition with dabs for *platessae*-copepodids (Table 8.2). Curiously, the common sole, a member of a different family (Soleidae), is rarely infected in the wild.

These results suggest that ecological factors may be involved in host specificity in natural conditions, as well as factors associated with host relationships. Boxshall (1976) pointed out that flounder, plaice and dab have a predominantly diurnal activity pattern, while the common sole is nocturnal (see also Chapter 3). It is possible that, in the wild, the inactivity and concealment of soles in the sediment during daylight may reduce their chances of infection with copepodids compared with diurnally active flatfishes. There is also evidence that, although *L. pectoralis* copepodids are attracted to common soles in the laboratory, these fishes do not provide a favourable habitat for their further development (references in Boxshall, 1976).

Table 8.2. Success of establishment (%) of copepodids of Lepeophtheirus pectoralis *when presented with a choice of two fish species (from Boxshall, 1976).*

Fish species	No. of expts	Mean	Range	:	Mean	Range
flesi *copepodids*						
Flounder: plaice	3	41	32 - 50	:	0.3	0 - 1
Flounder : dab	1	33		:	0	
Flounder : sole	2	41	34 - 48	:	20	12 - 27
Flounder : turbot	1	51		:	3	
platessae *copepodids*						
Plaice : flounder	3	20	8 - 35	:	1	0 - 3
Plaice : dab	1	67		:	0	
Plaice : sole	1	43		:	3	
Plaice : turbot	2	50	33 - 67	:	1.5	0 - 3
Dab : sole	1	0		:	33	

8.10 'SEA LICE' AND SALMONID FISHES

Much of the following account has been condensed from the extensive review by Pike & Wadsworth (1999), where relevant references will be found.

Two species of 'sea lice' infest salmon (*Salmo salar*) and sea trout (*S. trutta*), namely *Lepeophtheirus salmonis* (also known by the more specific name of salmon louse) and *Caligus elongatus*. *L. salmonis* is a darker brown colour and larger than *C. elongatus*, and is sufficiently conspicuous to receive vernacular names – laxlus, lakselus

– in Scandinavian countries (Berland, 1993). Female *L. salmonis* reach lengths of 18 mm or more (excluding the egg sacs) and males 7 mm, compared with 6mm and 5 mm for *C. elongatus*. *C. elongatus* is generally more active than *L. salmonis* and is more likely to leave the host temporarily. *L. salmonis* is more localised on the host (see above). *C. elongatus* has been reported from over 80 species of fish and has a worldwide distribution. *C. elongatus* was reported to parasitise at least 37 species of British fishes in a review by Wootten *et al.* (1982), and Kabata (1979) regarded it as the most common species of parasitic copepod in British waters. *L. salmonis* is largely restricted to *Salmo*, *Salvelinus* and *Oncorhynchus*.

Salmon and sea trout are anadromous, migrating to sea for varying periods before returning to spawn in their home rivers (see Chapter 1). Some salmon spend considerable periods of their lives at sea and may migrate immense distances. Sea trout tend to remain within the coastal marine environment. The life cycles of sea lice are entirely marine, but, although Atlantic-salmon and sea-trout anglers regard the presence of sea lice as indicating recent entry into freshwater (a fresh-run fish), this may not always be true, since there are records of sea lice persisting on fishes that have been in freshwater for two or three weeks.

Pike & Wadsworth (1999) expressed surprise that in spite of the great distances covered by salmon in the ocean and the low fecundity of sea lice (females produce only a few thousand eggs), infestation of wild salmon is widespread, often with high prevalences and low intensities. On a gill-netted sample of 157 wild fish caught by Johannessen in June 1988, prevalences and mean intensities of 93% and 7.44 respectively were recorded for *L. salmonis* and 17% and 1.42 respectively for *C. elongatus*. Infestations on wild salmon consist almost entirely of adult parasites, suggesting that infection is intermittent and possibly that sea lice are long-lived. There may be two foci for transmission of *L. salmonis* in the wild, one operating among mixed-age populations of salmon and sea trout in coastal areas and the other maintained by long-lived female sea lice on shoaling salmon on the oceanic feeding grounds.

Marine fish farming creates a highly suitable environment for rapid increase in infestation levels. Brandal & Egidius (1977, in Wootten *et al.*, 1982) found over 2000 lice on a single farmed Norwegian salmon and as few as five adults have been known to produce pathology on a small fish, such as newly introduced smolt. Norwegian salmon farmers were the first to experience serious damage to their fish stocks as a result of the activities of sea lice in the 1960s, soon after cage culture of salmon began. By the mid-1970s salmon farms in Scotland were also affected. Annual salmon production has dramatically increased since that time, but this expansion has not been trouble free.

L. salmonis and *C. elongatus* are present on farmed salmon throughout the year (Wootten *et al.*, 1982). The generation time of *L. salmonis*, even at 10°C, is only about six weeks, so that four generations can be completed in Scottish waters between May and October when sea temperatures are 9 – 14 °C. According to Ritchie *et al.* (1993), reproduction continues in the winter months and they discovered distinct seasonal differences in the reproductive output of winter and summer generations of adult females. Winter females were larger, produced significantly longer egg sacs and a greater number of smaller eggs compared with summer females. Ritchie *et al.* (1993) assumed that environmental factors such as temperature and photoperiod influence reproductive output, but the significance of these seasonal differences in terms of reproductive investment remains to be discovered.

8.10.1 *Pathology and host susceptibility*

Most of the damage to salmon is inflicted when the parasite feeds. Erosion of the epidermis by the parasite will affect the water balance of the fish, encouraging leakage of body fluids, with loss of proteins and electrolytes. Heavy infestations of *L. salmonis* may also lead to elevated levels of the hormone cortisol in host blood plasma. The consequence of this is suppression of the host's immune system and failure on the part of the host to limit proliferation and further invasion of other opportunistic pathogenic organisms, which already have access to the host via epidermal lesions (references in Tully & Nolan, 2002). In fact, Mustafa *et al.* (2000) found enhanced susceptibility to the microsporidian *Loma salmonae* in rainbow trout, *Oncorhynchus mykiss*, infected with *L. salmonis*.

Atlantic salmon exposed to *L. salmonis* for the first time fail to mount any significant tissue responses to any of the parasite's developmental stages. In contrast, coho salmon (*Oncorhynchus kisutch*), which are resistant to establishment of *L. salmonis*, mount strong localised tissue responses leading to parasite loss. On the other hand, when this localised response is suppressed by cortisol implantation, coho salmon fail to reject *L. salmonis* (references in Tully & Nolan, 2002). The implication is that the localised inflammation and epithelial hyperplasia generated by coho salmon is instrumental in the rejection of these crustacean parasites. *L. salmonis* releases proteolytic, trypsin-like enzymes, perhaps by regurgitation or in its salivary secretions (Firth *et al.*, 2000) and there is evidence that the enzyme alkaline phosphatase, that hydrolyses phosphoric acid esters, is also secreted (Fast *et al.*, 2003). The obvious role for these enzymes is digestive during feeding, but the possibility that these enzymes serve to suppress the immune system of susceptible hosts has also been raised (Firth *et al.*, 2000; Fast *et al.*, 2003)

There is no unequivocal evidence that sea lice are involved in transmission of other pathogenic organisms.

8.10.2 *Control*

Strategies for the control of sea lice in salmon farms are continually developing and changing. Pike & Wadsworth (1999) traced the development of the salmon farming industry and gave a detailed account of the biology of sea lice, their impact on the industry and the attempts that have been made to control them during the latter part of the last century. However, for a 'snapshot' of the control situation that is up-to-date at the time of writing, the reader is referred to Grant (2002). Grant's account is summarised below.

Five medicines are authorised at present for the control of sea lice in Scotland. These are: (1) cypermethrin (a synthetic pyrethroid that interferes with nerve impulses); (2) azamethiphos (an organophosphate that also interferes with nerve impulses); (3) hydrogen peroxide (mode of action not fully understood); (4) emamectin benzoate (targets peripheral nerves); (5) teflubenzuron (a chitin synthesis inhibitor). Dichlorvos (commercial name 'Aquagard'), another organophosphate, played a major part in reducing sea lice infestations in the latter part of the last century (see Pike & Wadsworth, 1999), but this compound is no longer available.

Cypermethrin is active against both mobile and juvenile stages of *L. salmonis* and *C. elongatus* and emamectin acts against all stages of both species and prevents maturation. Azamethiphos and hydrogen peroxide are effective only against pre-adult and adult stages of the two species, while teflubenzuron, being a chitin synthesis inhibitor, is active only against moulting lice. Three of these compounds, namely cypermethrin, azamethiphos and hydrogen peroxide, are applied topically by enclosing the water body containing the fish and adding the appropriate quantity of the medicine to the water. Emamectin and teflubenzuron are administered in feed. Injection of the fish is a potential method for delivery. All of these application methods have drawbacks. They must take into account the following: potential for chemical contamination of the environment; levels of fish stress; safety and practicability for the operator; the need for accurate dosage; the potential for development of parasite resistance; the danger to the fish consumer of residual pesticide. Grant (2002) discusses these complexities and considers the problems of creating an effective integrated management strategy for sea lice control in Scotland.

Attempts to control sea lice in the fish farms by application of chemicals have provoked often-acrimonious debate, with environmental organisations concerned about the potential effects of these treatments on non-target marine organisms. Also, a steady decline in sea trout populations and premature returns of sea trout to fresh water have been blamed on sea lice infestations originating from salmon in cage culture. Heuch (2002) has proposed that wild salmonids entering coastal regions in the spring from fresh water may be exposed to unusually high densities of infective copepodids from overwintering parasites on caged salmon. This has kindled a prolonged and often heated argument that has not yet been resolved, involving anglers, fish farmers, scientists and politicians (see Pike & Wadsworth, 1999, for further consideration).

The discovery that four species of wrasse, namely goldsinny (*Ctenolabrus rupestris*), rock cook (*Centrolabrus exoletus*), corkwing (*Crenilabrus melops*) and, to a lesser extent, cuckoo wrasse (*Labrus mixtus*), remove lice from salmon, offered an appealing alternative to chemical treatments (see Pike & Wadsworth, 1999). The number of lice consumed by an individual goldsinny was in the range of 26 – 46 lice per day and, by 1994, 30 farm sites in Scotland were using 150,000 wrasse. All of these fishes were wild caught from inshore coastal waters, generating concern about the impact on local stocks. Also, the susceptibility of wrasse to microbial infections and the potential for viral transmission from wild-caught wrasse were further causes for concern that have led to a reduction in their use as cleaner fish in recent years.

9

SIPHONOSTOMATOID COPEPODS: (2) PENNELLIDS

9.1 *LERNAEOCERA*

There cannot be many sea anglers with experience of whiting, cod or bib who have failed to notice as they lifted the operculum and looked into the gill cavity, a bright red swelling, a centimetre or more in length (Figure 9.1). This is a parasite, not as one might suppose some kind of pathological outgrowth from the fish – but it is no ordinary parasite, having more strange features than any other that we have encountered.

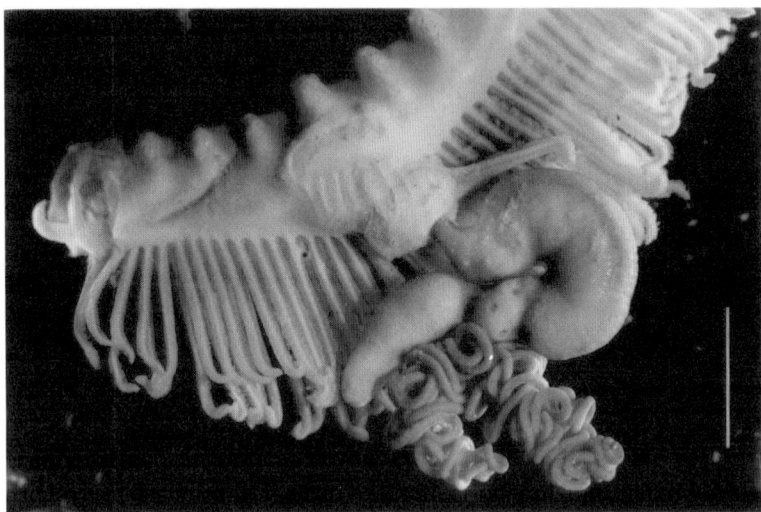

Fig. 9.1. Lernaeocera with the head embedded in the gill arch of a bib (*Trisopterus luscus*). Note coiled egg sacs. Scale bar: 5 mm. Photograph by Sheila Davies.

There are few clues as to what kind of parasite this is. If there is a head it is not visible, because at the fixed end the stalk-like neck joined to the free, swollen, s-shaped body plunges into the flesh of its host and the body itself is inert and smooth, lacking appendages. However, it is this free swollen body that offers the first clue to its identity, because it usually carries a pair of coiled egg sacs, which are the hallmark of the female copepod (Figure 9.1). Any observer with a strong lens or microscope to hand can find another clue –

examination of the neck will reveal the presence of four pairs of exceedingly small jointed limbs (Figure 9.2). Although this parasite is clearly an arthropod and most probably a copepod, we are left wondering how and why the organism has undergone such a transformation, where it came from and why all the individuals are female – no males are to be found.

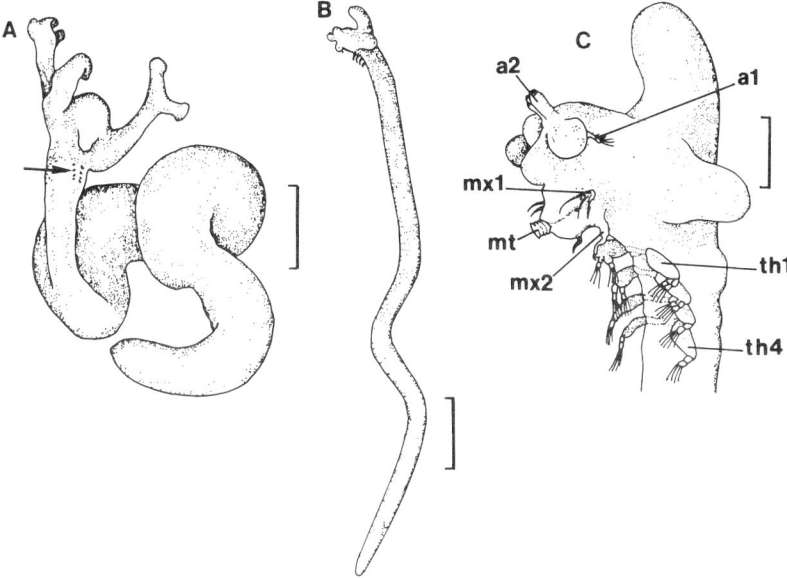

Fig. 9.2. Young adult females of *Lernaeocera* spp., showing reduced limbs. (A) *L. branchialis*; (BC) *L. lusci* – (C) is an enlarged view of the head of (B). Arrow in (A) indicates position of four pairs of swimming limbs (thoracopods). a1, a2, First and second antennae; mt, mouth tube; mx1, mx2, first and second maxillae; th1, th4, first and fourth thoracopods. Maxillipeds are absent in the female. Scale bars: (A) 2mm; (B) 1 mm; (C) 200 µm. From Kabata, 1979, with kind permission of The Ray Society.

The lack of males and of developmental stages of this bizarre creature was one of many similar puzzles for nineteenth century parasitologists and marine biologists. Developmental stages clearly recognisable as parasitic copepods were known to occur on the flounder, *Platichthys flesus*, but Metzger (1868, English translation 1869) recorded that he could not find females with 'egg-threads', in spite of searching for them repeatedly. The realisation that these were two pieces of the same puzzle came when Metzger found some partly transformed females on the gills of a lumpsucker, *Cyclopterus lumpus*. Thus pennellids like *Lernaeocera* spp. are virtually unique among parasitic copepods in possessing a second obligatory fish host. The whiting, *Merlangius merlangus*, is one of the round-bodied hosts of *L. branchialis* (see below for other hosts), but these fishes harbour only the impregnated, egg-laying female, which is transformed to such an extent that it is barely recognisable as a copepod. All other life cycle stages, including copulating males and females, have typical copepod features, but we have to turn to the gills of a flatfish such as the flounder to find them (Figure 9.3).

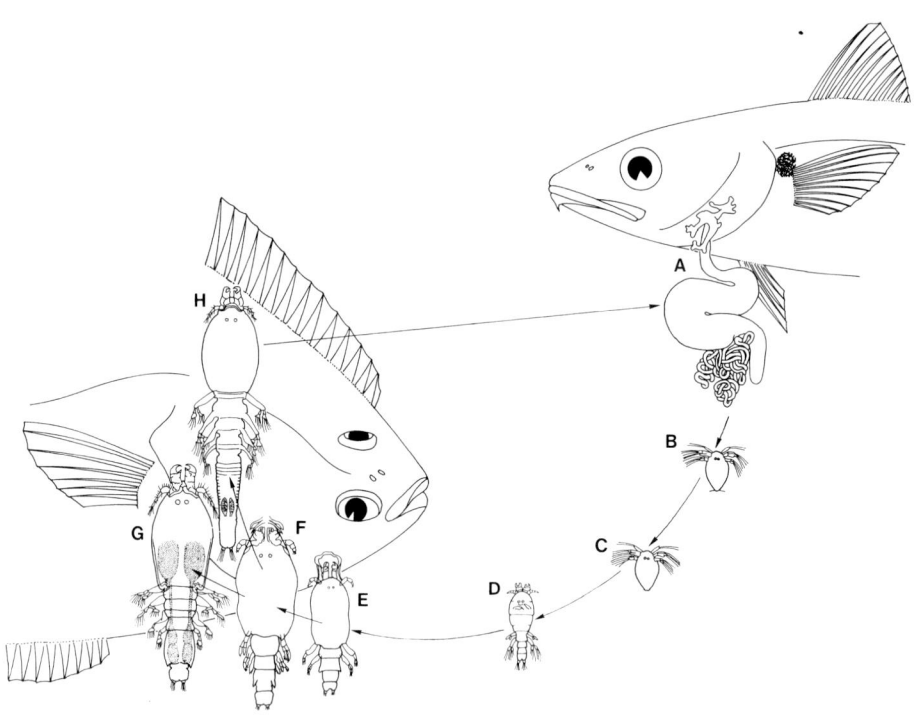

Fig. 9.3. Life cycle of *Lernaeocera branchialis*. Some stages have been omitted from the diagram. (A) Egg-laying female in gill chamber of whiting, *Merlangius merlangus* (B-D) Free-swimming stages. (BC) Nauplii; (D) copepodid. (E-H) Stages on gills of flounder, *Platichthys flesus*. (E) Chalimus I; (F) chalimus III; (G) male; (H) female. Based on Sproston, 1942, and Kabata, 1979.

9.1.1 *Invasion of the first fish host and larval development*
According to Whitfield *et al.* (1988), *L. branchialis* females on whiting can produce more than one set of egg sacs (Figure 9.3A), with a mean number of 1445 eggs per pair of sacs and a maximum of 3000. At 10°C the eggs take about 12.7 days after extrusion to liberate nauplius I larvae (Figure 9.3B). Most eggs from a pair of egg sacs hatch during the first

three days but the rest continue to hatch for 12 days, thereby spreading the nauplius larvae over a wide area. At 10°C the moulting sequence from nauplius I to nauplius II to infective copepodids (Figure 9.3B-D) takes about two days and the non-feeding copepodids have a 50% survival level at 7.5 days.

Field experiments at Lowestoft, England in June 1987, when the sea temperature was about 16°C, revealed that the period between infection of the flounder by the copepodids and copulation on the flounder could be as short as 11 days. Using previously uninfected flounder as sentinel fishes in cages on the sea bed, Whitfield *et al.* (1988) estimated that copepodids were establishing themselves on these fishes at a mean rate of not less than 30 parasites per fish per day.

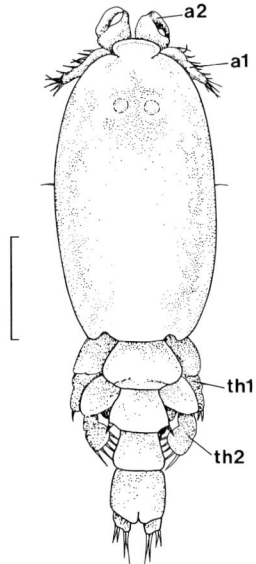

Fig. 9.4. Copepodid of *Lernaeocera branchialis* in dorsal view. a1, a2, First and second antennae; th1, th2, first and second thoracopods. Scale bar: 150 μm. From Kabata, 1979, with kind permission of The Ray Society.

The copepodid has two pairs of seta-bearing swimming legs (thoracopods) and projecting from its anterior extremity a pair of chelate (pincer-like) second antennae (Figures 9.3D, 9.4). According to Sproston (1942), these chelate antennae are remarkably manoeuvrable and are used to grasp the tip of one of the flounder's primary gill lamellae. Before moulting to the first chalimus stage (Figure 9.3E), the parasite tethers itself to the gill by means of a frontal filament, similar to that described in the caligids. According to Sproston (1942), there are four chalimus stages followed by adult males and females (Figure 9.3F-H).

9.1.2 *Mating*

According to Anstensrud (1989), the female *L. branchialis* does not move around in the gill cavity of the flounder although capable of doing so – it is the male that seeks out and guards the female. The pre-copula attitude adopted by males and females on the gills of the flounder has been described by Kabata (1958) and confirmed by Anstensrud (1989). The male is attached to the host's gill by means of its second antennae, which, like those of the copepodid, terminate in pincer-like chelae. The head of the male is positioned adjacent and dorsal to that of an attached immature female (chalimus stage) (Figure 9.5A). The male chelae penetrate the gill tissue into the anchoring frontal filament of the female. The male maintains this position until the female moults to the mature stage. Anstensrud (1989) observed that the male's swimming legs beat rapidly at intervals so that the whole animal vibrates. The male also repeatedly touches the dorsal surface of the female with its genital complex and abdomen.

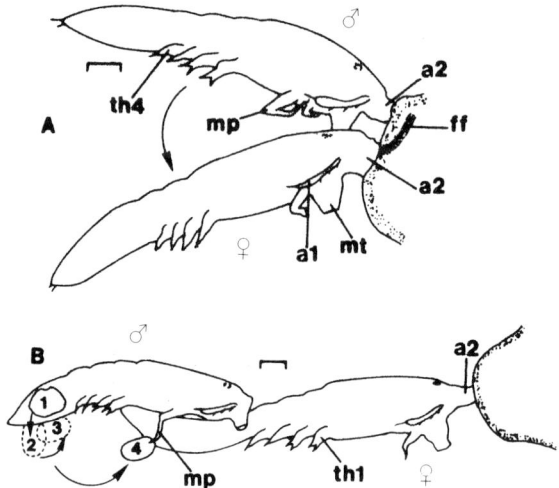

Fig. 9.5. Mating in *Lernaeocera*. (A) Pair in precopula attitude. (B) Pair in copulation. Numbers indicate consecutive positions of the spermatophore: 1, still inside the male; 2, 3, manipulation after extrusion; 4, attached to the female. ♀, Female; ♂, male; a1, a2, first and second antennae; ff, tethering frontal filament; mp, maxilliped; mt, mouth tube; th1, th4, first and fourth thoracopods. Scale bars: 100 μm. From Anstensrud, 1989, with permission of Cambridge University Press; (A) modified according to Kabata, 1958.

While attached in pre-copula attitude, both male and female may browse on host gill epithelium. The mouth opens via a telescopic mouth tube (Figure 9.6A), which is supported by three chitinous rings (Kabata, 1962). The spatulate mandibles, each with a serrated margin (Figure 9.6BC), enter the mouth tube via lateral apertures (Figure 9.6A). In the male, feeding lasts up to a minute and when not feeding the mouth tube rests in contact with the female's dorsal surface (Figure 9.5A).

Moulting in the female begins with the splitting open of the exoskeleton anteriorly. By contractions of the cephalothorax and trunk the female shrugs off the old cuticle in a posterior direction. The maxillipeds of the male (the female has no maxillipeds) increase their activity during moulting of the female and touch her dorsal and ventral surfaces. After moulting of the female is completed, the male travels backwards along the female's dorsal surface using his second antennae. When the copula position is reached, the second antennae of the male grasp the female just behind the last leg-bearing thoracic segment (Figure 9.5B). Unlike the virtually motionless chalimus IV, the mature female performs occasional rapid strokes with the thoracopods. The male also beats his thoracopods repeatedly, bends and stretches the rear of his body and initiates undulating contractions of the genital complex and abdomen.

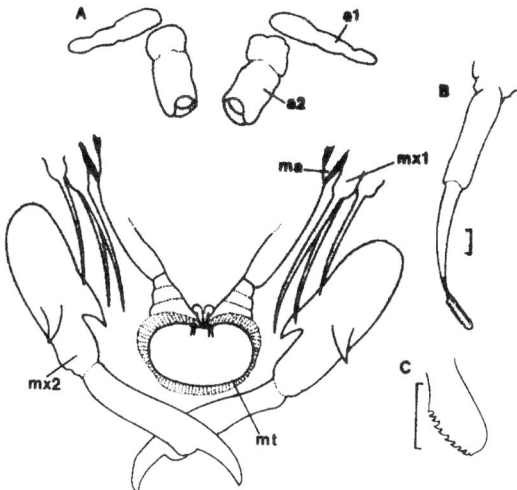

Fig. 9.6. (A) Appendages of the pennellid cephalothorax; (BC) whole mandible and mandible tip respectively of *Lernaeocera branchialis*. a1, a2, First and second antennae; ma, mandible; mt, mouth tube; mx1, mx2, first and second maxillae. From Kabata, 1979, with kind permission of The Ray Society.

Within one hour of establishing the copula position, the male expels two spermatophores, which are pushed forward onto the ventral surface of the genital complex where they are probably retained by rows of small hook-like processes. By bending its posterior region ventrally and anteriorly, the male holds the spermatophores against the vulva region of the female with the long setae of the first thoracopods, while at the same time moving the maxillipeds around the spermatophores and the female's genital openings. The male then withdraws his abdomen, leaving the spermatophores behind on the female (Figure 9.5B).

Only pennellid males have maxillipeds. This feature and the activity of these maxillipeds during mating in *L. branchialis*, suggest that they have a special function in reproduction. Anstensrud (1989) found that males from which maxillipeds had been removed retained the ability to locate a female in the gill cavity of the flounder, perhaps by responding to chemical messages (pheromones) produced by the female, but were usually unable to fertilise the female. When there is a high ratio of males to females, the males tend to congregate in clusters or gather around a pre-copulating or copulating pair. When these extra males disturb a male in the pre-copula position, the resident male grasps the female with his maxillipeds and makes his body vibrate by rapid strokes of the thoracopods. One of the extra males rarely replaces the resident male and in most cases the first male to take up the pre-copula position is the successful male in terms of insemination.

After mating the partners separate and the female leaves the flatfish host. Scott (Scott, A., 1901) reported their occurrence in the plankton and commented on the rarity of planktonic males. According to Scott most males remain on the gills of the flatfish after the females have left and presumably perish.

9.1.3 Establishment and growth of the egg-laying female
The female *L. branchialis* establishes herself inside the gill chamber of a round-bodied host such as whiting or cod (*Gadus morhua*) and there undergoes the remarkable transformation into the grossly swollen and egg-laying female. This involves the penetration of the head region into the tissues of the host and extensive development of the genital region. Any attempt to remove the fully developed female without dissecting out the embedded head will lead to breaking of the neck, leaving the head still embedded in the fish. This is because the bizarre head is deeply anchored in host tissue by three roots, all with short terminal branches (Figure 9.3A). One of these roots is dorsal and the others lateral. The centrally located head appendages persist, but are tiny (Figures 9.2C, 9.6). The mouth tube is also tiny and, amazingly, in *L. branchialis*, usually lies inside the aorta, bulbus arteriosus or even in the ventricle of the heart itself, where it is in a position to directly withdraw blood from the centre of the host's circulatory system.

L. branchialis substantially reduces the haematocrit value (ratio of red blood corpuscles to blood plasma) compared with uninfected hosts. Van Damme *et al.* (1994) found that a single parasite on the whiting reduced this value by 21% and two parasites by 38%. In *L. lusci* from bib, *Trisopterus luscus*, the egg-laying females are commonly found on the gill arch (see also below) and penetrate only as far as the gill arteries. *L. lusci* appears to have less of an impact on its host, as many as three parasites reducing the haematocrit value by about 29%.

The progression of the head into the host's body and the development of the roots are the consequence of activation of a vigorous growth centre located in the cephalic region (Kabata, 1979), perhaps coupled with digestion of host tissue in the path of the growing parasite. Activation of a similar vigorous growth centre in the genital area leads to phenomenal expansion in the length and girth of this region, the different rates of dorsal and ventral growth imposing on it a sigmoid swollen shape. This transformation of the adult female *Lernaeocera* has been described as a metamorphosis (e.g. by Kabata, 1979). However, the phenomenon does not involve the dissolution and reconstitution of organ systems in *Lernaeocera* as it does in many insects and in strigeid digeneans (see Kearn, 1998) and 'differential growth' seems a more appropriate descriptive term.

Kabata (1981) made an interesting point with respect to the sudden and dramatic growth undertaken by female *L. branchialis* after mating. He pointed out that this growth is made possible by an extraordinary physiological change that has gone unnoticed by other researchers. Prior to mating the female undergoes a relatively modest increase in length from about 0.5 to 1.5 mm by way of a series of moults. Then, after mating, moulting ceases but growth continues up to 60 mm, accompanied by a gigantic increase in biomass. The all-important question posed by Kabata is how does the rigid, growth-limiting cuticle (see p. 158) suddenly change its characteristics and become pliable and elastic enough to permit this huge increase in growth? The same questions can be asked of similar changes in other copepod parasites such as the Lernaeopodidae, Lernaeidae and Chondracanthidae (Chapters 10, 11 and 12 respectively).

9.1.4 *Ectoparasite or endoparasite?*
First impressions suggest an ectoparasitic relationship between the egg-laying female of *L. branchialis* and its host, but clearly this description is not appropriate. It is no more appropriate to regard this as an endoparasitic relationship and the useful term 'mesoparasitism' has been coined to describe associations of this kind (see Kabata, 1979). Unfortunately, Euzet (1989) introduced the term 'mesoparasite' at a later date, apparently in ignorance of its usage by Kabata, for a parasite living entirely inside the host but in a cavity with an opening to the outside world (such as a tapeworm inhabiting the intestine). In terms of priority of introduction and usefulness, Kabata's application of the term is recommended and will be used in this book.

9.1.5 *Taxonomic problems and British species*
As Kabata (1979) has pointed out, the egg-producing females of *Lernaeocera* are morphologically plastic. Their anterior regions are moulded by the conditions they meet as they penetrate host tissue, and the exposed genital regions may likewise be moulded and distorted by the physical pressures experienced within the gill cavity. Consequently, a plethora of taxa have been described based on unusually shaped holdfasts or trunks and many of these are unlikely to be valid species.

Kabata (1979) records only three British species, namely *L. branchialis*, *L. lusci* and *L. minuta*, but this apparently simple subdivision conceals a puzzling complexity. Kabata (1957) established a new species, *L. obtusa*, for parasites of the haddock, *Melanogrammus aeglefinus*, but later the species was relegated to sub-specific status as a form of *L. branchialis* (see Kabata, 1979). Kabata (1979) recognised some morphological differences between egg-laying females of *L. branchialis* f. *branchialis* and *L. branchialis* f. *obtusa*, but admitted that these differences are in some measure host-dependent. *L. branchialis branchialis* parasitises cod or whiting and from its location at the ventral end of the gill arch the anterior end penetrates the ventral aorta, bulbus arteriosus or ventricle. On the other hand, *L. branchialis obtusa* often parasitises old and large haddock and has to cover a larger distance to reach the main axis of the circulatory system and seldom penetrates to the heart. The consequence is that *L. branchialis branchialis* has a shorter neck than *L. branchialis obtusa*. Differences between the stages of these two forms on their flatfish hosts are minor.

Host affiliations add a further dimension to the puzzling taxonomy of the group. According to Kabata (1979), the hosts of *L. branchialis* in the southern North Sea are commonly whiting, pollack (*Pollachius pollachius*) and flounder. In the northern North Sea all these fishes are present, but the preferred hosts are haddock and lemon sole (*Microstomus kitt*). In both regions cod is infected with equal frequency. North of British waters in the seas off the Faroes and Iceland, haddock is rarely infected and the life cycle involves cod and lemon sole. Kabata (1979) lists many other occasional hosts for *L. branchialis*, including lumpsucker, *Cyclopterus lumpus*.

Slinn (1970) unexpectedly encountered common soles, *Solea solea*, infected with egg-laying female *Lernaeocera* in a marine fish hatchery in the Isle of Man. His intensive investigations raised the possibility that these egg-laying females were *L. lusci*, commonly found on bib in the Irish Sea. Slinn assumed that the parasite had been introduced to the hatchery as developmental stages on common soles from the Irish Sea and, in the absence of bib in the hatchery, the mated females had completed their development to the egg-laying stage on the gills of the soles, the only hosts available. Slinn found that bib introduced into the tanks containing soles infected with developing stages acquired adult egg-laying females, while whiting, cod and haddock did not become infected. Thus, at least in this artificial environment, the parasite can undergo the whole of its life cycle on a single host species. Slinn found juvenile stages but no egg-laying adults on wild-caught local common soles and about 25% of bib from Manx waters were infected with egg-laying *L. lusci*. He suggested that soles are the natural second host for *L. lusci* in the Irish Sea. A few egg-laying *L. lusci* were found attached to the body surfaces of wild-caught bib and of soles in the hatchery.

9.2 EYE-MAGGOTS – *LERNAEENICUS*

Lernaeocera with its deeply rooted head, blood-sucking habit and swollen blister-like body is repulsive enough, but one of its relatives has an even more unpleasant feature. *Lernaeenicus sprattae* is also a bloodsucker, tapping not the heart or neighbouring arteries but the blood vessels of the choroid coat lying immediately outside the retina of the eye. The really disturbing aspect of this habit is that the egg-producing female parasite or eye-maggot reaches this source of blood from outside the fish's body, by boring through the cornea, passing either through the pupil or the iris and then penetrating the retina (Figure 9.7A). The cephalothorax is embedded in the eye, while the narrow, greatly elongated trunk, comprising thoracic segments, extended genital segment and egg sacs, lies outside the eye, trailing backwards like a pennant along the flank of the fish (Figure 9.7A).

The eye is also the chosen site of the pennellid *Phrixocephalus cincinnatus* (see Kabata, 1969a) and the lernaeopodids *Ommatokoita elongata* and *Lernaeopodina longibrachia* (see Table 10.1 and Kabata 1969b respectively).

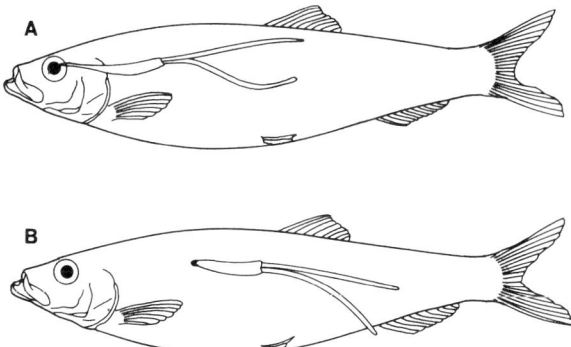

Fig. 9.7. Egg-laying females of (A) the eye-maggot *Lernaeenicus sprattae* and (B) *L. encrasicoli* in their typical locations on a sprat, *Sprattus sprattus*. From Möller & Anders, 1983.

9.2.1 Lernaeenicus *in British waters*

The host of *Lernaeenicus sprattae*, as its name indicates, is the sprat, *Sprattus sprattus*. First year fishes, together with, in some areas, young herring, *Clupea harengus*, are familiar as whitebait. Scott & Scott (1913) examined a sample of 600 sprats captured with a shrimp trawl off Blackpool in 1910 and collected 14 *L. sprattae* (prevalence 2.3%). They reported that the eye of one of the sprats had three parasites fixed to it. Leigh-Sharpe (1935) and Gurney (1947) found similar infection levels: about 3% in samples of 970 and over 1000 fishes respectively taken from Hole's Hole in the river Tamar near Plymouth. Gurney remarked that in no instance was more than one parasite found on each fish. Schram & Anstensrud (1985) reported a similar prevalence of 3.5% on sprats in the Oslofjord, Norway. Anstensrud & Schram (1988) found that 89% of 438 infected sprats caught in the Oslofjord carried a single parasite and 45 fishes carried a double infection. Thirty-seven of these 45 fishes had double infections in the same eye and in 24 of these the holes in the cornea were close together with parasites at different stages of development, indicating infection at different times. Thus, eight sprats were infected in both eyes. Four sprats had triple infections. Two fishes had all three parasites close together in the same eye and the other two fishes had both eyes infected.

Herring has occasionally been reported as a host for *L. sprattae* (see, for example, Kabata, 1979), but Anstensrud & Schram (1988) found no specimens on herring in the Oslofjord (see also below). According to El Gharbi *et al.* (1985) *L. sprattae* parasitises the pilchard, *Sardina pilchardus*, in the Mediterranean.

In an earlier sample of sprats taken off Blackpool in 1906 Scott & Scott (1913) found three *Lernaeenicus* deeply rooted in the tissues at the anterior end of the dorsal fin.

On dissecting out the head of one of these parasites it was found to have penetrated into the visceral cavity. In Leigh-Sharpe's (1935) sample of 970 fishes a single specimen was found in the same location on its host. This is Britain's second species, *L. encrasicoli* (Figure 9.7B), which, according to Kabata (1979) parasitises other clupeid fishes such as the anchovy, *Engraulis encrasicolus*, as well as the sprat, and only occasionally penetrates the eye. The scarcity of *L. encrasicoli* in British waters indicated by Leigh-Sharpe's (1935) figures (prevalence 0.1%), is reflected in similar figures for Norwegian waters (Oslofjord), where Schram & Anstensrud (1985) recorded a prevalence of only 0.3% on 7000 sprat.

9.2.2 *Life cycle*

Gurney in 1947 found a tiny *L. sprattae*, just 3 mm long, embedded in the eye of a sprat and remarked on its correspondence with the free-swimming adult stage of *Lernaeocera branchialis*. He found no earlier stages, in spite of a search of the body, fins and gills and, surprisingly, it was not until 1979 that Schram published an account of the life cycle of the eye-maggot.

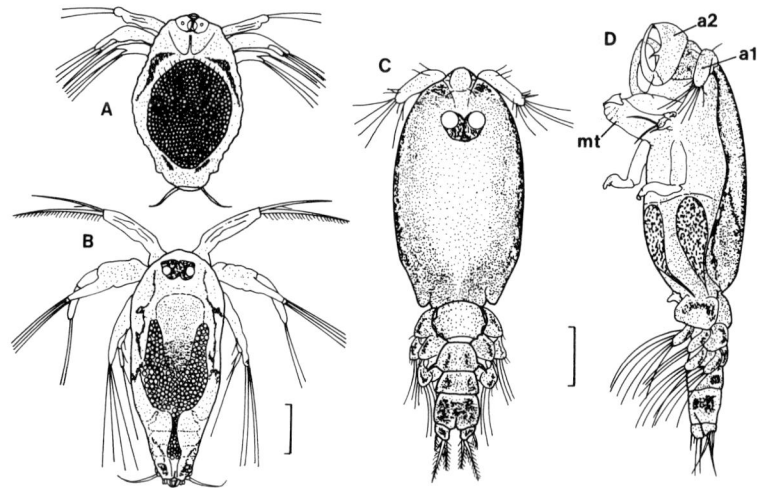

Fig. 9.8. Free-swimming stages of *Lernaeenicus sprattae*. (A) Newly hatched nauplius I; (B) nauplius II; (CD) copepodid in dorsal and ventro-lateral views respectively. a1, a2, First and second antennae; mt, mouth tube. Scale bars: 100 μm. Reproduced from Schram, 1979, with permission of Taylor & Francis AS (www.tandf.no/sarsia).

Schram (1979) established that the life cycle of *Lernaeenicus sprattae* resembles that of *Lernaeocera branchialis*. After hatching, two nauplius stages and a copepodid stage succeed each other (Figure 9.8). Between the copepodid and chalimus I, Schram distinguished a transitional stage, which he called a 'grasping stage', because it was found attached to its host by the powerful chelate second antennae (Figure 9.8D). Schram raised the possibility that the 'grasping' stage may be a second copepodid, but since there is no evidence of a moult prior to the appearance of the 'grasping' larva, any differences may simply reflect a change in the way of life from free-swimming to attachment. Each of the four chalimus stages (chalimus II and IV only are shown in Figure 9.9) secures itself by means of two tethering filaments secreted by the frontal glands. According to Schram (1979), the frontal filaments appear to 'flow through' the second antennae and terminate in two balls embedded in the host. These parasitic early stages are tiny, the female chalimus IV (Figure 9.9B) measuring on average only 1.33 mm in length and the male chalimus IV (Figure 9.9C) 1.45 mm. Bearing in mind their small size and the rarity of these stages in Gurney's (1947) sample, it is not surprising that Gurney failed to find them.

There then follows an important phase in the life cycle during which the males and females relinquish their attachment to the host. The free male (Figure 9.10AB) is now fully mature, but the free female (Figure 9.10C-E) remains immature. This mobility permits meeting between the sexes and mating, followed presumably by death of the male. The mated female needs the resources provided by a fish host to produce her eggs and the sprat again is the source of this nourishment. The question is whether the whole life cycle proceeds on a single host individual or whether the cycle involves two different individual fishes, albeit of the same species. These two scenarios may of course proceed side by side.

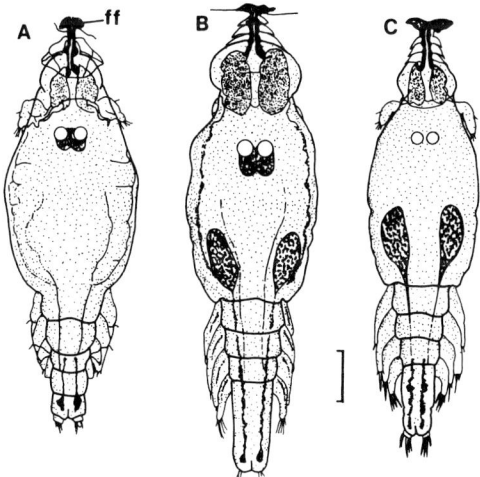

Fig. 9.9. Chalimus stages of *Lernaeenicus sprattae* in dorsal view. (A) Chalimus II; (BC) chalimus IV, female and male respectively. Pigmentation shown in posterior region only. ff, Tethering frontal filament. Scale bar: 200 μm. Reproduced from Schram, 1979, with permission of Taylor & Francis AS (www.tandf.no/sarsia).

Since Schram (1979) collected planktonic females (and males) in Oslofjord, it seems likely that some at least of the females leave the first host fish and, after a period of time of unknown length, infect a new and different sprat. The finding of free-swimming females in aquaria by Anstensrud & Schram (1988) further supports this. Anstensrud & Schram (1988) suggested that the proportion of females leaving the first host might depend on their density on this host.

Work in Norwegian waters has thrown more light on the behaviour of larvae and free-swimming adults of *L. sprattae*. Schram & Anstensrud (1985) found that about 86% of all the pelagic stages (829 specimens) found in the top 90 cm of water in the inner reaches of Oslofjord, were captured at night. Copepodids were particularly abundant – 748 individuals were collected, 87% of them at night. Almost all the free-swimming adults were captured at night, but there was no difference between day and night catches of nauplius II. Schram & Anstensrud came to the conclusion that the higher number of larvae (mostly copepodids) caught at night (approximately 5.5 larvae per m^3 compared with 1.0 larvae per m^3 during the day) can only be explained by assuming that these larvae undertake diurnal vertical migrations.

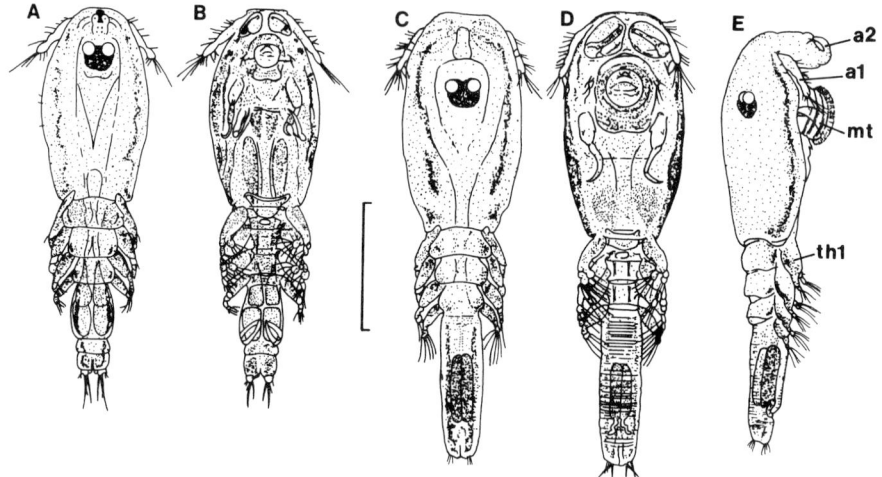

Fig. 9.10. Free-swimming adults of *Lernaeenicus sprattae*. (AB) Male in dorsal and ventral views respectively. (C-E) Female in dorsal, ventral and dorsolateral views respectively. Scale bar: 500 µm. a1, a2, First and second antennae; mt, mouth tube; th1, first thoracopod (swimming limb). Reproduced from Schram, 1979, with permission of Taylor & Francis AS (www.tandf.no/sarsia).

The domination of the plankton hauls by copepodids probably reflects the greater longevity of the copepodid (5 – 6 days) compared with earlier nauplius stages (about 24 hours each). Free-swimming females are more abundant than males in the hauls. In all probability this reflects the demise of the males soon after copulation. Copepodids swim upwards towards the surface in dim light or darkness, while at higher light intensities they are photonegative. Sprat also migrate vertically on a daily basis, coming close to the surface at night, so host and parasite (both copepodids and free-swimming, immature but mated females) are likely to mingle during darkness, offering good opportunities for host infection.

Anstensrud & Schram (1988) exposed anaesthetised sprats and herring to copepodids and discovered that both fish species induced a marked increase in copepodid swimming speed. After making contact with the fish, copepodids were still present on the sprat 24 hours later, but those on herring abandoned the fish within two minutes. In aquaria containing both sprats and herring, 61 – 91% of the sprats became infected but none of the herring. Similarly, no herring became infected when herring alone were exposed to copepodids in aquaria, whereas 68 – 90% of sprats became infected when they were exposed alone to the parasites. In nature Anstensrud & Schram found that fish less than 5 cm long were uninfected, that the highest average prevalence (3.6%) was for length group 8.0 – 9.4 cm and that prevalence was 0.3% for larger sprats. Thus, there is a reduction in prevalence of chalimus infections when fish reach approximately 9.5 cm in length (i.e. about one year old). However, in experimental infections, copepodids readily infected sprats >9.5 cm in length and showed no preference for any specific length group. This suggests that the behaviour of the fish influences recruitment in nature.

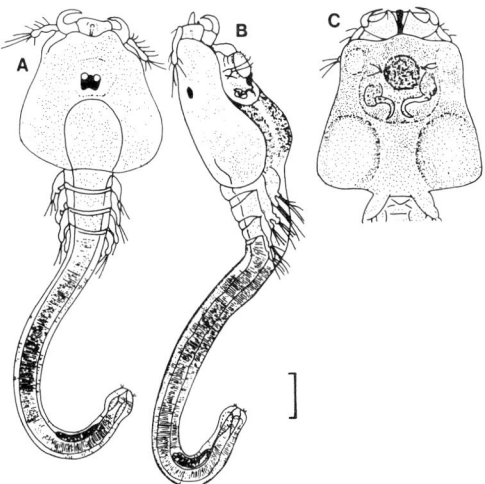

Fig. 9.11. Young specimen of *Lernaeenicus sprattae* from the eye of a sprat, *Sprattus sprattus*. (A) Dorsal view; (B) lateral view; (C) cephalothorax only, in ventral view. Scale bar: 200 μm. Reproduced from Schram, 1979, with permission from Taylor & Francis AS (www.tandf.no/sarsia).

A similar natural pattern of infection of sprats with adult females was recorded. No fishes less than 5 cm long were infected. Infection reached a maximum on 8 – 9.4 cm sprats and then declined to about 1.9% on fish of length 9.5 – 10.9 cm. Sprats longer than 5 cm in length can harbour both chalimus stages and adult females.

Copepodids appear to settle randomly on their hosts, but during the first 24 hours there is a shift towards the fins, especially the pectoral and dorsal fins.

Anstensrud & Schram (1988) found that mature females are initially scattered over the host's surface, but those adults that go on to produce egg sacs are invariably rooted in the eyes. When the developing females first penetrate into the eye they are about 3 mm in length and hard to detect (Schram, 1979). The hole in the cornea is only 0.2 mm in diameter at this stage (slightly greater than the abdomen width) and only 0.2 – 1.0 mm of the posterior end of the parasite protrudes from the cornea. The cephalothorax becomes trapezoidal in shape (Figure 9.11AC) and lies between the retina and the black-pigmented, vascularised choroid layer. The mouth tube points somewhat anteriorly and the mouth opening is in contact with the choroid. Because of the position of the cephalothorax between

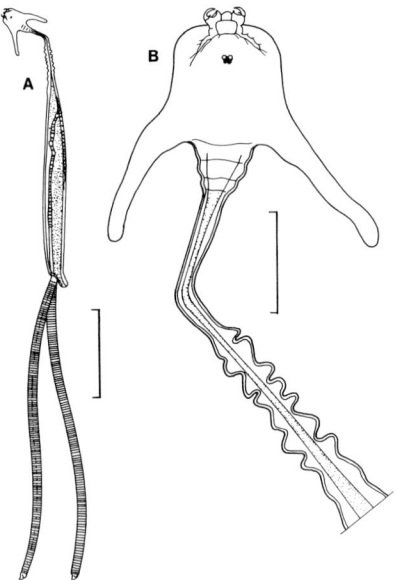

Fig. 9.12. Egg-laying female of *Lernaeenicus sprattae* dissected out of the eye of a sprat, *Sprattus sprattus*. (A) Whole animal; (B) enlarged view of cephalothorax. Scale bars: (A) 5 mm; (B) 1 mm. Reproduced from Schram, 1979, with permission from Taylor & Francis AS (www.tandf.no/sarsia).

the retina and the choroid and the need for the genital region to communicate with the outside world via the cornea, the parasite is bent into an S-shape as seen from the side (Figure 9.11B). Later, a pair of postero-lateral cephalothoracic projections ('horns') develops (Figure 9.12). These eventually reach about 1.2 mm in length but are rarely symmetrical. The 'neck' is derived from the posterior part of the cephalothorax and has a sharp bend where it enters the cornea (Figure 9.12). Further development produces a greatly elongated trunk, comprising part of the thorax, the genital segment containing the ovary and giving rise to the egg sacs, and a terminal relatively small abdominal segment (Figure 9.12A).

According to Schram (1979), *L. sprattae* is a blood feeder and may also derive nourishment from the host's coelomic and tissue fluids. The intestine, which is narrow as it passes through the neck, widens in the trunk and is slightly black-pigmented and filled with light red contents. Suspended in these contents are grey and black particles (from the choroid?).

A survey of 491 rooted adult females revealed no preference for right or left eye. However, the parasites do show a strong preference (88 – 89%) for the upper rear quadrant of both eyes (Figure 9.13).

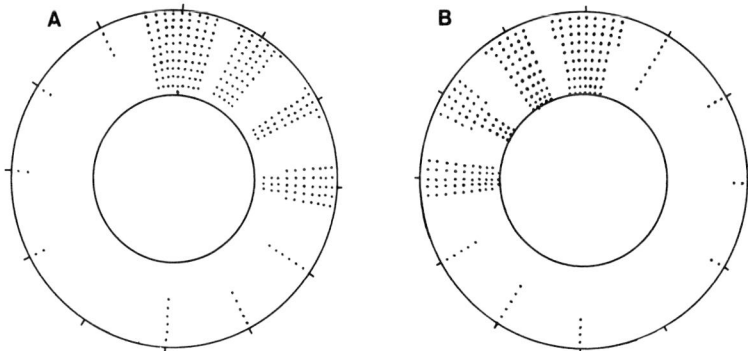

Fig. 9.13. Distribution of the anchored heads of 491 eye-maggots, *Lernaeenicus sprattae*, in (A) the left eye and (B) the right eye of the sprat, *Sprattus sprattus*. From Anstensrud & Schram, 1988 (Copyright © 1988), with kind permission of Kluwer Academic Publishers.

9.2.3 *The question of pathogenicity*

According to El Gharbi *et al.* (1985), *L. sprattae* infections inflict considerable damage on their pilchard hosts (*Sardina pilchardus*) in the Mediterranean. Loss of blood to the parasite leads to anaemia and enlargement of the spleen. They reported local damage to the cornea, retina and choroid and claimed that the vascularisation of the eye may be affected, depriving the organ of adequate oxygen. According to Leigh-Sharpe (1935), *L. sprattae* causes partial or total blindness and El Gharbi *et al.* (1985), observed that pilchards infected with *L. sprattae* lagged behind healthy fishes in shoals. El Gharbi *et al.* also found that the only fishes taken in nets were the less agile infected fishes. They attributed these behavioural differences to poorer vision. On the other hand, Anstensrud & Schram (1988), based on their study of infected sprats (*Sprattus sprattus*) in nature and in aquaria, found no indication that the parasite causes unilateral blindness or had any significant effect on the swimming ability of the host. Rauk (in Anstensrud & Schram, 1988) believed that bilateral infection by adult *L. sprattae* would most probably lead to immediate death, but there are reports of apparently healthy fishes with bilateral infections (Anstensrud & Schram, 1988). The location of the parasite in the posterior upper quadrant of the eye and its trailing attitude might obscure little of the visual field of the fish. However, the effects of the parasite on the vision of the host have not been assessed. If they were minimal this would be truly remarkable, in view of the invasive nature of the parasite.

10

SIPHONOSTOMATOID COPEPODS: (3) LERNAEOPODIDS

10.1 INTRODUCTION – CHANGES IN ATTACHMENT

Crustaceans seem eminently suitable for attachment to fishes. Their jointed limbs are sufficiently plastic in the evolutionary sense to produce formidable claws or pincer-like chelae with a vice-like grip. Nevertheless, moulting, a fundamental feature of crustacean biology, temporarily renders attachment by such limbs ineffective (see Chapter 8). During moulting the animal must slough off the whole of its exoskeleton, and this can only be achieved by relinquishing the grip of each limb temporarily, so that its old covering can be discarded. Moreover, the new exoskeleton enclosing each limb takes time to harden and until this has taken place secure attachment is compromised. Some copepod parasites have solved this problem by deploying a unique frontal filament or thread secreted by a frontal gland in the head. This filament is cemented at its free end to a stable structure in the host, such as a fin ray or basement membrane, while the other end is attached to the head of the parasite. Thus the parasite is securely tethered to the host during the dangerous process of moulting. The frontal filament has the added advantage that it is non-living and may be less provocative to the host's immune system than living tissue. However, whether it is provocative or not, it is likely to be more resilient to host attack.

This leads us to wonder why this remarkable and special attachment system has not been more widely adopted by fish parasites. One possible reason is that tethering severely reduces mobility and accessibility of food. Like a tethered horse, grazing is restricted to a circular area with a radius equal to the length of the tether. Reduction in mobility also reduces opportunities for meeting mates. Another problem is that the filament needs to be slender to permit the limited mobility that it provides. Stresses and strains on the filament will increase as parasites increase in size and the increase in filament thickness needed to prevent large parasites breaking free will further reduce mobility. So filament attachment is ideal for juvenile copepods that moult frequently, are small enough to satisfy their nutritional requirements from a relatively small area of host skin and place relatively weak strains on the filament.

Some adult female lernaeopodids have successfully exploited this attachment system by changing the way in which the inert tethering material is deployed. Prior to attachment the frontal glands of the female secrete filament material in the shape of a button or 'bulla', which is implanted in a hole excavated by the parasite in host tissue. The parasite's mode of attachment to the bulla is then changed. The parasite detaches the bulla from the cephalothorax and establishes a new hold on the bulla with the extremities of the two elongated second maxillae (Figure 10.1). Thus the female relies on a deeply embedded button of secreted material rather than on a dangerously thin filament to tether it to the host.

This inevitably has meant a more sedentary way of life, but mobility is not entirely lost since many female lernaeopodids such as *Lernaeopodina* spp. have exceptionally long second maxillae which permit the now free mouth region to roam over a large area. The long second maxillae of *L. longimana* are shown in Figure 10.9D. However, this is a relatively modest elongation compared with *L. longibrachia*, in which the second maxillae reach the extraordinary length of about 10 cm while the body excluding the egg sacs is a mere 14 mm in length (Kabata, 1969b). If blood feeding is adopted little mobility is demanded of the parasite for feeding, especially on the gills where blood is widely available and easy to access. Retaining the mobile male (see below) facilitates sexual interactions.

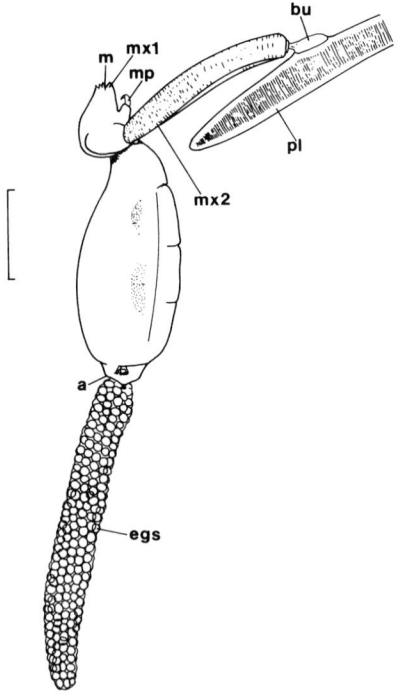

Fig. 10.1. Egg-laying adult female gill maggot (*Salmincola salmoneus*) attached in its typical location near the tip of a primary gill lamella (pl) of a salmon (*Salmo salar*). a, Anus; bu, bulla; egs, egg sac; m, mouth; mp, maxilliped; mx1, mx2, first and second maxillae. Only the left egg sac is shown. Scale bar: 2 mm. From Friend, 1941, reproduced by permission of the Royal Society of Edinburgh from *Transactions of the Royal Society of Edinburgh*, volume 60 (1939-42), pp. 503-541.

10.2 THE SALMON GILL MAGGOT

10.2.1 *Occurrence and site of attachment*

In Britain the lernaeopodid *Salmincola salmoneus* is a common and widespread parasite of the Atlantic salmon, *Salmo salar*. The so-called salmon gill maggot may be up to 8mm long (egg sacs excluded). Its counterpart on Pacific salmonids is *Salmincola californiensis* (see Kabata & Cousens, 1973). Friend (1941) examined the attachment sites of more than 4000 *S. salmoneus* and found that all of them were attached to the efferent edge of the primary gill lamella and, with only one exception, within 1 cm of the free distal end of the primary lamella (Figure 10.1). Friend (1941) described the mature female attached to a primary gill lamella as resembling a gymnast hanging from a vertical bar. Every one of the conspicuous gill parasites examined by Friend was an adult female, and although the female had been recorded by Linnaeus in 1758 as *Lernaea salmonea* (see Berland & Margolis, 1983) and reported from Scotland as long ago as 1766 (see Friend, 1941), it was not until the 1930s that the other stages of the parasite were recognised and the events of the life cycle unravelled.

Friend (1941) highlighted an important difference between the salmon gill maggot *S. salmoneus* and sea lice (*Lepeophtheirus salmonis*; see Chapter 8). Sea lice are essentially parasites of salmonids during their time at sea and are lost sooner or later in fresh water. Gill maggots are present on salmon both in the sea and in fresh water but they reproduce only in fresh water.

10.2.2 Feeding in the adult female

Friend (1941) observed no gill maggot (*S. salmoneus*) in the act of feeding or in the attitude of feeding, i.e. with the mouth tube in contact with the gill. However, he reported finding several individuals in which the contents of the midgut were red anteriorly, shading to brownish black in a posterior direction. This is consistent with a diet of blood and accumulation of haematin residues. However, Friend reported small areas of missing gill tissue in the vicinity of attached females, raising the possibility that some gill tissue in addition to blood may be ingested.

Dedie (1940) described a new species which he called *Salmincola mattheyi* from the body surface and fins, but not the gills, of Arctic charr, *Salvelinus alpinus*. He claimed that the parasite gnaws the skin and described homogeneous stomach contents containing cell debris. Later Kabata (1969c) came to the conclusion that *Salmincola mattheyi* is synonymous with *S. edwardsii*, a parasite of brook charr, *Salvelinus fontinalis*, and this seemed at odds with a statement by Fasten (1921) that *Salmincola edwardsii* removes "enormous quantities of blood" from its host. However, Black (1982) showed that *S. edwardsii* is found on the body surface, fins and opercula of small brook charr (<250 mm long), while on larger hosts the gills were infected. *S. edwardsii* has been recorded in Britain (see below).

Kabata & Cousens (1977) regarded *Salmincola californiensis* as essentially a browser on epithelial cells of sockeye salmon, *Oncorhynchus nerka*, but pointed out that where blood vessels lie close to the surface (e.g. on the gills and inner surface of the operculum), blood is readily taken. They also observed that the gills of adult sockeye salmon are usually chosen as attachment sites by adult parasites, while on young fishes the body surface is preferred.

10.2.3. Hatching and invasion of the host

Hatching is preceded in *Salmincola californiensis* by osmotic swelling of the eggs and splitting of the egg sac (Kabata & Cousens, 1973). This is closely followed by bursting of the eggs and the simultaneous moulting of the nauplius to release a rather crumpled copepodid, the infective stage. The copepodid spends the first 30 minutes after hatching on the bottom, where it grooms and extends the new cuticle. The copepodid then becomes active, progressing in a jerky manner by simultaneous backward strokes of the two pairs of swimming legs (thoracopods) (Figure 10.2). One or more swimming strokes are followed by a brief inactive period during which the copepodid sinks. On contact with the bottom, the copepodid usually resumes swimming, flexing the hind body dorsally so as to direct the animal upwards. Poulin *et al.* (1990) found that the copepodids of *S. edwardsii* greatly increase the rate and length of upward swimming bursts when exposed to passing shadows or subjected to shock waves by tapping the vessel containing the larvae. Such stimuli are likely to be generated by potential fish hosts. Kabata & Cousens (1977) reported evidence to suggest that similar stimuli were operative in the activation of copepodids of *S. californiensis*.

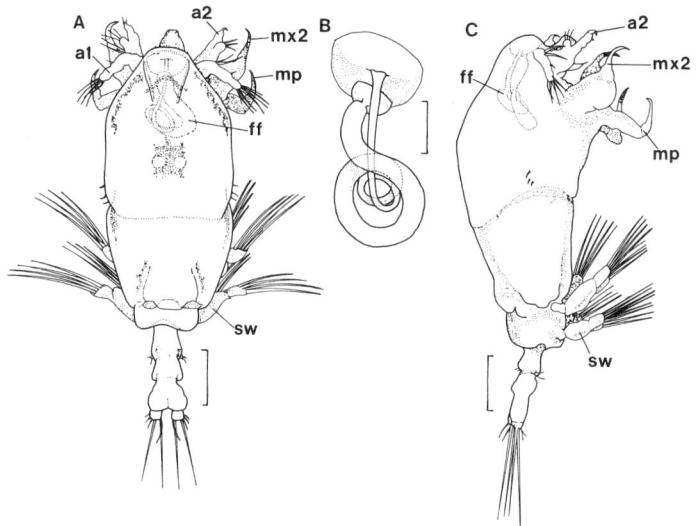

Fig. 10.2. The copepodid of *Salmincola californiensis*. (A) Dorsal view; (B) coiled frontal filament; (C) lateral view. a1, a2, First and second antennae; ff, frontal filament; mp, maxilliped; mx2, second maxilla; sw, swimming leg. Scale bars: (AC) 100 µm; (B) 5µm. From Kabata & Cousens, 1973, reproduced by courtesy of Fisheries and Oceans Canada, with the permission of Her Majesty the Queen in Right of Canada, 2004.

Friend (1941) found that 210 *S. salmoneus* from six salmon (*Salmo salar*), were distributed as follows in relation to the gill slits: 28, 59, 59, 54, and 10. He observed that the areas of gill available for colonisation in each gill slit were in the following proportions: ½:1:1:1:¼. So close are the parasite figures to these proportions that the indication is that the larvae are carried into the gill slits by the gill ventilating current and that the establishment of the parasites is related to the area available for colonisation.

10.2.4 *Development on the host*
The most detailed account of development in *Salmincola* is that of Kabata & Cousens (1973) on *S. californiensis*. It is basically similar to that of *S. salmoneus* will be summarised here.

The copepodid of *S. calforniensis* uses the hooks on its second antennae for initial attachment to the host, but soon brings the hooked second maxillae into play

(Figure 10.2). The copepodid then shifts from place to place on the host until the time comes to implant its frontal filament. The animal is able to move backwards and forwards by shifting first one of its second antennae then the other, then one of its second maxillae then the other. The hooked maxillipeds may also take part in this shifting sequence, but are the last to change position. The swimming legs remain immobile during these movements.

The frontal filament is attached to a dermal support of some kind, such as a fin ray, the supporting rod of a primary gill lamella or a scale. These supports are reached using the maxillipeds to excavate a cavity in the skin between the attached points of the second maxillae. The copepodid takes up to 90 minutes to prepare this cavity and may start as many as three separate excavations. When the cavity is large enough it is tested with the mouth tube and with the anterior margin of the cephalothorax. The latter is then pushed into it. The large terminal plug on the already formed frontal filament (Figure 10.2) is then inserted into the cavity and glued to the skeletal support with rapid hardening cement produced by the frontal gland. The parasite then walks backwards, pulling out the frontal filament from its accommodation in the cephalothorax as it goes. For the moment, the smaller terminal plug at the proximal end of the filament (Figure 10.2B) remains attached to the anterior margin of the body.

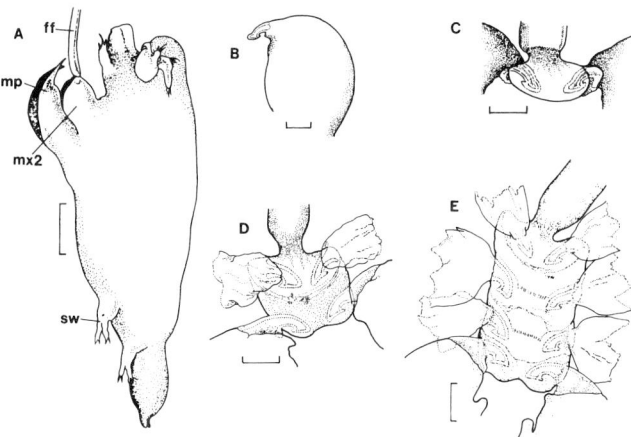

Fig. 10.3. Chalimus larvae of *Salmincola californiensis*. (A) First chalimus in lateral view; (B) second maxilla of first chalimus; (C) first chalimus, base of frontal filament with embedded tips of second maxillae; (D) second chalimus, base of frontal filament and shed tips of second maxillae; (E) fourth chalimus male, base of frontal filament with shed tips of second maxillae. ff, Frontal filament; mp, maxillipeds; mx2, second maxilla; sw, swimming leg. Scale bars (A) 100 µm; (B-E) 20 µm. From Kabata & Cousens, 1973, reproduced by courtesy of Fisheries and Oceans Canada, with the permission of Her Majesty the Queen in Right of Canada, 2004.

It is interesting to note that the larval stages of the related lernaeopodid *Vanbenedenia kroeyeri* attach their frontal filaments not to the host but to the surface of the adult female, where they remain until capable of establishing themselves on the host (Kabata, 1979). Females of *V. kroeyeri* attach themselves to the dorsal spine of the ratfish, *Chimaera monstrosa*.

In *Salmincola californiensis* moulting then occurs and the first chalimus emerges (Figure 10.3A). During the moult the second maxillae exchange their terminal claws for barbed spikes (Figure 10.3B). The end of the frontal filament attached to the parasite is then transferred from the anterior margin of the body to the tips of the new second maxillae (as in Figure 10.3C). The barbed spikes are embedded in the small plug at the free end of the frontal filament, aided by the secretion of more cement from the frontal gland.

Within about 36 hours from the time of initial contact with the host, the parasite moults to the second chalimus stage. During this process the second maxillae are withdrawn from their old cuticle and cemented to the free end of the frontal filament. This involves considerable contortion because the free end of the filament, while still attached to the old cuticle by the spikes of the first chalimus, must be manoeuvred over the mouth tube so that new cement can be added to the end of the filament from the frontal gland. The spikes of the new second maxillae are then implanted in this cement, which dries quickly. The second maxillae are then manoeuvred back over the mouth tube to their normal position. After the second chalimus has shuffled out of the old cuticle, the latter remains suspended from the base of the frontal filament by the old second maxillary spikes. The newly moulted second chalimus then cuts away the old cuticle using its mandibles, usually leaving behind the old second maxillary spikes embedded in the filament (Figure 10.3D).

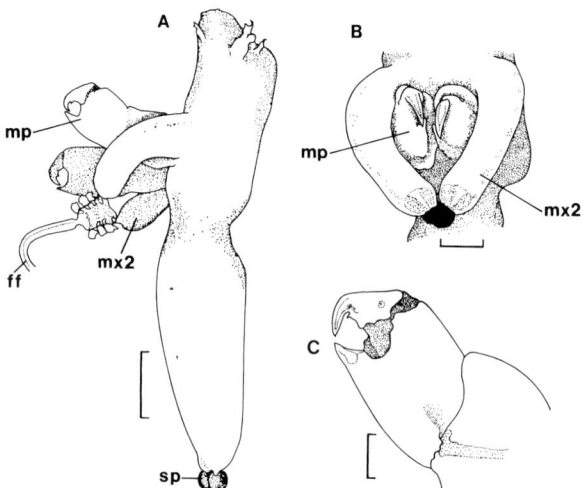

Fig. 10.4. Female fourth chalimus of *Salmincola californiensis*. (A) Lateral view of whole animal; (B) ventral view of limbs of cephalothorax; (C) enlarged view of maxilliped. ff, Frontal filament; mp, maxillipeds; mx2, second maxilla; sp, spermatophore. Scale bars: (A) 200 µm; (B) 250 µm; (C) 100 µm. From Kabata & Cousens, 1973, reproduced by courtesy of Fisheries and Oceans Canada, with the permission of Her Majesty the Queen in Right of Canada, 2004.

The second chalimus is followed by the third and fourth chalimus. All stages of development, namely the copepodid, four chalimus stages and adult males and females, occur on the same host fish. Each intervening moult involves the same contortions described above, ensuring the cementation of each new pair of second maxillary spikes to the proximal end of the frontal filament and the abandonment anterior to it of the old spikes of the preceding second maxillae. The fourth chalimus male for example may retain the three abandoned pairs of spikes from the three preceding stages (Figure 10.3E).

During the fourth chalimus stage important events overtake the female. First, the parasite reverses the position of the second maxillae and the maxillipeds, so that the second maxillae come to lie posterior to the maxillipeds (Figure 10.4A). The female begins to move, with the consequence that the frontal filament is stretched taut and ultimately breaks, usually at the junction of the last drop of secretion with its proximal end. This leaves the tips of the second maxillae still linked by the secretion droplet (Figure 10.4B). The free female then seeks a site of permanent attachment and excavates a cavity, which will eventually accommodate the bulla. This task is prolonged because the female must rely on the buccal appendages and to a lesser extent on the second antennae to excavate the hole, because the second maxillae are immobilised by the frontal secretion binding their tips and the pincer-like maxillipeds are concerned with attachment to the host (Figure 10.4C).

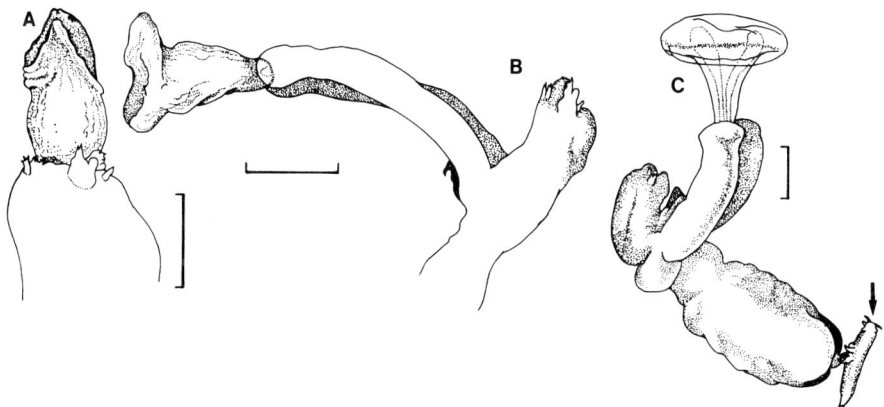

Fig. 10.5. Deployment of the bulla in young female *Salmincola californiensis*. (A) Early stage of bulla eversion; (B) second maxillae newly attached to the bulla; (C) fully developed female with bulla inflated and male attached (arrow). Scale bars: (A) 200 µm; (BC) 500 µm. From Kabata & Cousens, 1973, reproduced by courtesy of Fisheries and Oceans Canada, with the permission of Her Majesty the Queen in Right of Canada, 2004.

The moult that transforms the fourth chalimus female into a young adult presents some problems because the parasite is attached to the host at this stage by the maxillipeds and second antennae and not by the frontal filament. Moulting begins with a rupture of the cuticle in the frontal area and withdrawal of the second antennae from the cuticle. These take over as attachment organs while the maxillipeds are withdrawn from the old cuticle. The remaining frontal secretion bridging the tips of the second maxillae acts as a restraint, permitting withdrawal of one second maxilla and allowing the old cuticle to slide down and free the other. In the few females in which this bridge is missing it proves impossible to extract the second maxillae and these individuals presumably die.

During or immediately after moulting into an adult, the female begins to push out the crumpled and inverted bulla (Figure 10.5A), which is still housed in the cephalothoracic cavity previously occupied by the frontal filament. The bulla is everted and implanted in the prepared cavity and contact with the cephalothorax is lost. The second maxillae make contact with the stalk or manubrium of the bulla (Figure 10.5B) and the tip of each limb enters the opening of a duct running down into the bulla and ending blindly inside it. Secretion from the maxillary glands cements the second maxillae permanently in place and liquid from these glands also enters the reservoirs in the bulla, leading to its inflation (Figure 10.5C). When successfully attached (Figure 10.1) the female continues to grow and many females at this stage have males fixed to the genital area (Figure 10.5C; see also Figure 10.9E). Kabata & Cousens (1977) reported that the gills show the most vigorous response to the bulla, which becomes enclosed by the host in a capsule welding the bulla to the gill. Thus the response of the host is exploited by the parasite to ensure secure attachment.

Kabata & Cousens (1977) discovered that the burrowing habits undertaken by the female to excavate a cavity for the bulla could be fatal for fry of sockeye salmon. The female ceases to excavate when the cavity is large enough and when the female reaches tissue that is sufficiently hard. Should the female fail to reach suitable supporting tissue, burrowing will continue and the female may penetrate into the visceral cavity of the fish and pass right through the viscera with potentially fatal consequences.

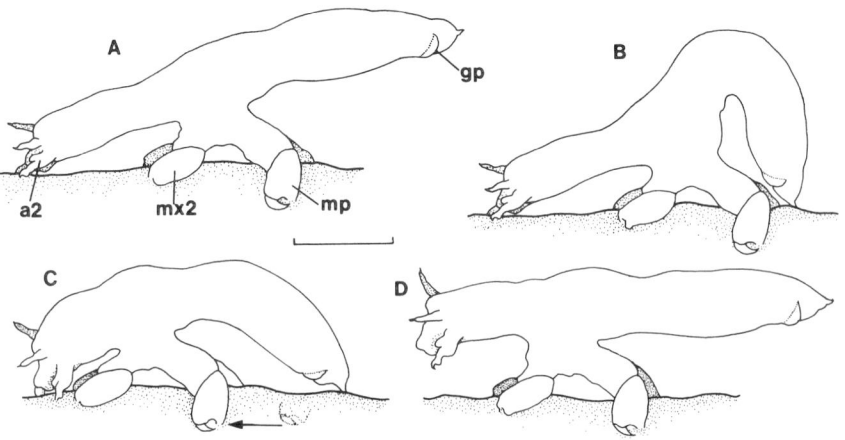

Fig. 10.6. Salmincola californiensis, successive stages in locomotion of the male (diagrammatic). See text for explanation. a2, Second antenna; gp, plate covering genital opening; mp, maxilliped; mx2, second maxilla. Scale bar: 250 μm. From Kabata & Cousens, 1973, reproduced by courtesy of Fisheries and Oceans Canada, with the permission of Her Majesty the Queen in Right of Canada, 2004.

The fourth chalimus male moults to produce the adult male, which now relinquishes its grip on the frontal filament and maintains attachment to the host using the second maxillae and the maxillipeds (Figure 10.6A). This change in function of the second maxillae is reflected in a fundamental change in their structure, the limbs being transformed into pincer-like chelae, similar in size and structure to the adjacent maxillipeds (Figure 10.6).

10.2.5 *Mating*

The male now seeks a female, travelling across the skin using a caterpillar-like gait in the following way. With the second maxillae and the maxillipeds attached, the anterior part of the body is extended and the anterior extremity secured to the skin by the second antennae (Figure 10.6A). The posterior extremity of the body is then brought forward close to the attached maxillipeds and the posterior end is fixed by ramming the projecting spikes on the tail diagonally into the skin (Figure 10.6B). Finally the second maxillae and maxillipeds are released and shifted forward close to the buccal region where they are reattached (Figure 10.6C; note arrow). The second antennae then release their hold (Figure 10.6D) and the process is repeated.

Having located a female, the male attaches itself to the genital region with its maxillipeds (Figure 10.5C) and proceeds to implant a spermatophore in each vaginal opening in the following way. The posterior tip of the male is brought forward under the trunk and inserted between the male's two attached maxillipeds. This brings the two male genital openings adjacent to the two vaginal openings of the female. Each spermatophore consists of a pyriform bulb and a narrow tube (Figure 10.7A). The distal (closed) third of the bulb contains a transparent liquid, the middle third contains cement and the proximal third, adjacent to the origin of the tube, contains sperm. The tube is implanted in the vagina and then the sperm packet followed by the cement is forced down the tube (Figure 10.7B), probably by osmotic pressure generated by the transparent liquid. The sperm enters the body of the female and the cement permanently

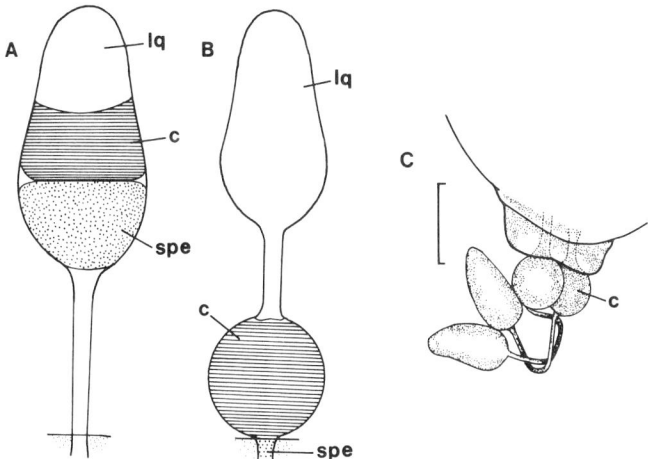

Fig. 10.7. Spermatophore of *Salmincola californiensis*. (AB) Diagrammatic representation of a spermatophore before and after discharge respectively; (C) genital region of female with discharged spermatophores. c, Cement; lq, liquid; spe, sperm. Scale bar for C: 100 μm. From Kabata & Cousens, 1973, reproduced by courtesy of Fisheries and Oceans Canada, with the permission of Her Majesty the Queen in Right of Canada, 2004.

seals off the vaginal openings preventing further matings (Figure 10.7C). At least 75% of females are inseminated while still at the chalimus stage (Figure 10.4A). The remains of the spermatophores eventually become detached and egg sacs are present 28 to 32 days after first contact is made with the host by the copepodid.

It is worth mentioning here an interesting finding by Caillet & Raibaut (1979) concerning host selection by the infective copepodids of another lernaeopodid, namely *Clavellodes macrotrachelus* (= *Alella pagelli*). This parasite is found on the gills of a range of sparid fishes in British waters (see Kabata, 1979). Caillet & Raibaut showed

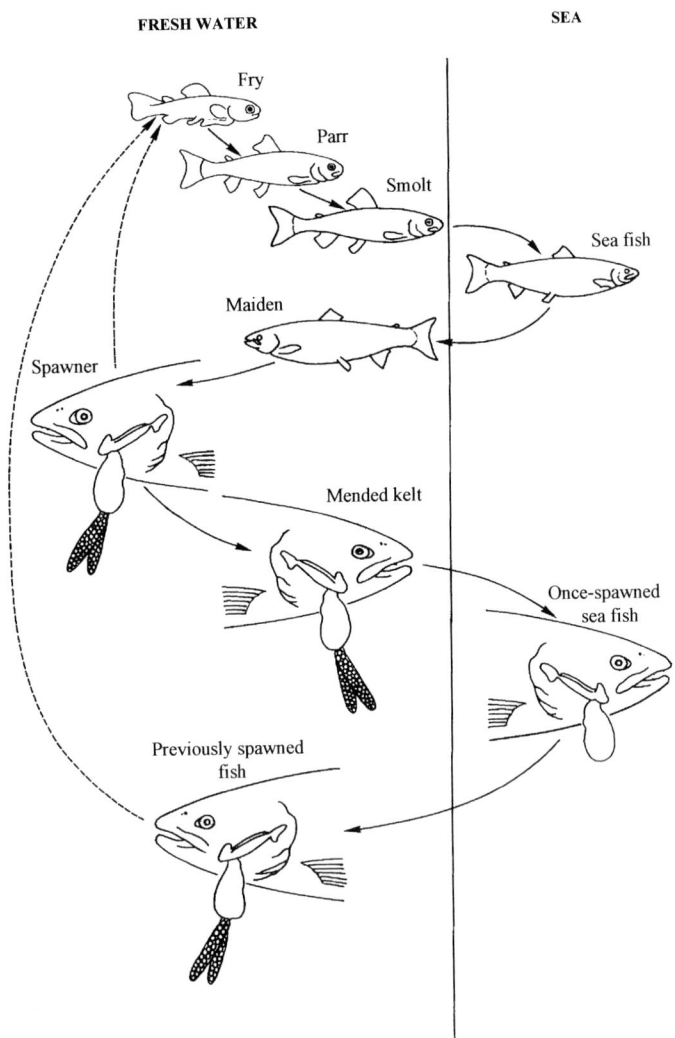

Fig. 10.8. The life cycle of the salmon, *Salmo salar*, in relation to parasitism by the salmon gill maggot, *Salmincola salmoneus*. Broken arrow indicates spawning. See text for explanation. Based on diagrams by Friend (1941) and by Fryer (1982).

experimentally that copepodids destined to become males fasten themselves only to fishes already infected by an adult female, while copepodids destined to become females attach themselves to any fish (they used the Mediterranean fish *Diplodus sargus*). This remarkable adaptation, perhaps based on some kind of pheromonal (chemical) attraction by the mature female, increases the chances of sexual encounter.

10.2.6 *Relationship with the salmon life cycle*

Friend (1941) made a study of salmon (*Salmo salar*), caught by Scottish commercial fisheries and determined the relationship between its life cycle and parasitism by the salmon gill maggot (*Salmincola salmoneus*). A short but clear summary of this relationship will be found in Fryer (1982). In fresh water, smolts (see Chapter 1) and earlier stages are not infected (Figure 10.8). This is because the infective copepodid larvae are too large to pass between the gill arches of these small fishes. Copepodids also fail to reach the gills by the alternative route via the opercular opening because of the strength of the gill ventilating current. Consequently, Atlantic salmon returning from their first sojourn at sea, prior to their spawning run in the rivers (so-called maiden fishes), are uninfected. They acquire their first parasites in fresh water from parasitised, older, previously-spawned fishes and within five to six months these parasites are themselves capable of releasing eggs and infecting new hosts.

Fishes that survive the hazards of their first spawning return to the sea as mended kelts. The parasites on these fishes survive while their hosts are at sea but do not breed. Soon after entering the rivers for the second time, their parasites produce egg sacs. Some factor associated with the transition from marine to fresh water as the fishes head for their spawning grounds appears to stimulate the extrusion of egg sacs. The process then continues as before, and is limited only by the number of times the fish is able to return to spawn.

10.3 OTHER BRITISH LERNAEOPODIDS

In 1952, Fryer (1981) had the opportunity to examine a lernaeopodid copepod about 4 mm long from an Arctic charr (*Salvelinus alpinus*) from Ennerdale Water, Cumbria. He recognised the parasite as belonging to the genus *Salmincola*, but could not assign it to any species previously recorded in the British fauna. He went on to identify the copepod as *Salmincola edwardsii*, a parasite of northern circumpolar regions.

The distribution of this parasite in Britain is intriguing. It occurs in no other lake in Cumbria, although the host is present in these lakes, but, curiously, it also occurs in four Scottish lochs, namely Stack (Sutherland), Lee (Angus), Tay (Perthshire) and Doon (Ayrshire) (Fryer, 1982). In Ennerdale Water the parasite is restricted to the fins. In Scotland the fins are infected, but it occurs also under the opercula and, in Loch Tay, adjacent to, but not on, the gills, even though it occurs on the gills outside Britain (see above). This indicates that there may be genetic differences between these isolated populations.

Even more curious is the fact that Ennerdale Water is also the home to two rather special free-living crustaceans: the mysid *Mysis relicta*, found in Ireland but in no other lake in Britain, and the copepod *Limnocalanus macrurus*, which is unknown in the British Isles outside Ennerdale Water (Fryer, 1982). Why should Ennerdale Water alone among English lakes have these three crustaceans and why does *S. edwardsii* occur in Scottish lochs but not in most lakes of the English Lake District? For a possible answer to these intriguing questions see Fryer, 1981.

Salmincola gordoni appears to be confined to the trout, *Salmo trutta*, both brown trout and sea trout being parasitised (Fryer, 1982). Parasites of the brown trout spend their entire lives in freshwater, while those on sea trout presumably tolerate fresh and salt water. *Salmincola gordoni* is much smaller than *S. salmoneus*, reaching just over 4 mm but often much less, and it prefers the walls of the gill chamber, including the inner faces of the operculum. According to Fryer (1982) it has a northerly distribution, having been reported at his time of writing from Scotland, Yorkshire and Port Erin Bay, Isle of Man.

Table 10.1. Some common lernaeopodid copepods from British marine fishes selected to illustrate the range of hosts and microhabitat (from Kabata, 1979).

Parasite	Host(s)	Microhabitat	Comments
Ommatokoita elongata	Sharks	Cornea	
Charopinus dalmanni	Thornback ray (*Raja clavata*)	Gills, nasal cavities, spiracles	
Pseudocharopinus bicaudatus	Spurdog (*Squalus acanthias*)	Spiracles	Figure 10.9C
Lernaeopodina longimana	Rays (*Raja* spp.)	Gill arches, buccal cavity	Figure 10.9D
Lernaeopoda galei	Dogfish (*Scyliorhinus canicula*), smooth hound (*Mustelus mustelus*), tope (*Galeorhinus galeus*)	Cloacal region, less commonly male claspers	Figure 10.9E
Albionella globosa	Dogfish (*Scyliorhinus canicula*)	Nasal cavities	
Alella pagelli	Red sea-bream (*Pagellus bogaraveo*), black sea-bream (*Spondyliosoma cantharus*), pandora (*Pagellus erythrinus*)	Gills	
Neobrachiella merluccii	Hake (*Merluccius merluccius*)	Near tips of gill rakers	Figure 10.9F
Clavella adunca*	Cod (*Gadus morhua*), Whiting (*Merlangius merlangus*)	Gills, gill arches Gill arches, gill rakers	Figure 10.9G
	Haddock (*Melanogrammus aeglefinus*) Other gadids	Gills, gill arches	Forms tumour of attachment
Clavella stellata	Hake (*Merluccius merluccius*)	Mostly on narrow strip of body skin parallel with edge of operculum	Bulla cemented to scale

*Kabata (1960) claimed that antagonism existed between this parasite and the monogenean *Diclidophora merlangi* (see Chapter 6) on the gills of whiting, but Smith (1969) denied this.

Salmincola thymalli is also small, up to about 3 mm in length, and as its name suggests is specific to the grayling, *Thymallus thymallus*, in Britain. At the time of Fryer's publication (1982) it had been found only in the River Ouse system in Yorkshire.

Two other lernaeopodids are likely to be encountered on British freshwater fishes, namely *Achtheres percarum* (Figure 10.9A) in the buccal cavity of perch, *Perca fluviatilis*, and *Tracheliastes polycolpus* (Figure 10.9B) on the fins of cyprinid fishes.

Many lernaeopodids have been recorded from British marine fishes. *Allela pagelli* is mentioned above. A selection of the more common species is given in Table 10.1, illustrating the range of hosts and microhabitats occupied by these parasites. Some of the parasites listed in the Table are illustrated in Figure 10.9. The reader is referred for further details to the comprehensive survey of British lernaeopodids in Kabata (1979). This publication also includes a useful host/parasite checklist (pp. 417 – 420).

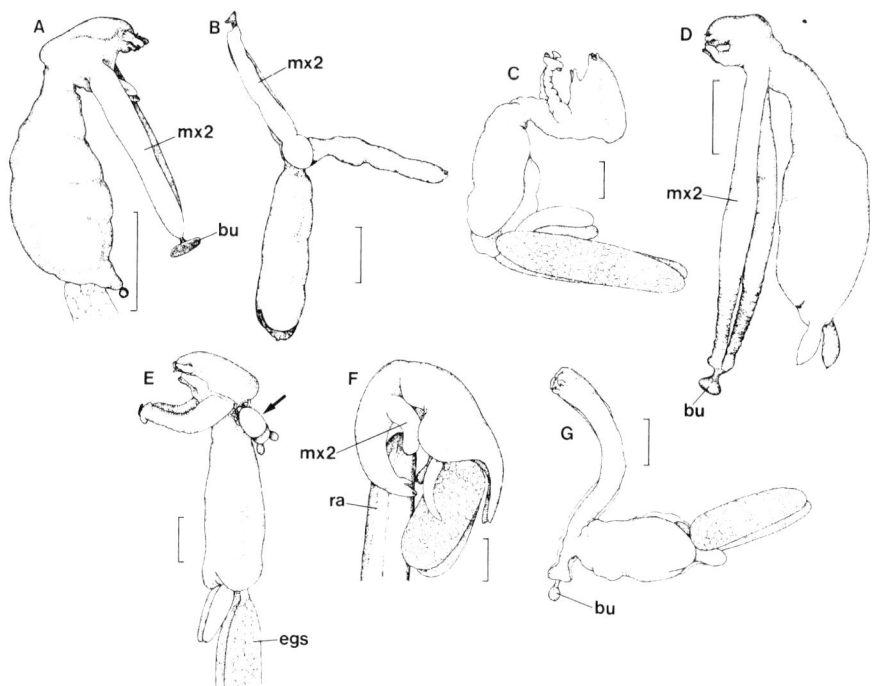

Fig. 10.9. Some common lernaeopodids (mature females) from British freshwater and marine fishes. (A) *Achtheres percarum*; (B) *Tracheliastes polycolpus*; (C) *Pseudocharopinus bicaudatus*; (D) *Lernaeopodina longimana*; (E) *Lernaeopoda galei*, female with male (arrow) attached; (F) *Neobrachiella merluccii*; (G) *Clavella adunca*. bu, Bulla; egs, egg sac; mx2, second maxilla; ra, gill raker. Scale bars: 1 mm. From Kabata, 1979, with kind permission of The Ray Society.

11

CYCLOPOID COPEPODS – THE ANCHOR WORM

11.1 INTRODUCTION

It is easy to be misled by a casual glance at the adult female anchor worm, *Lernaea cyprinacea* (Figure 11.1A). The head is buried in the flesh of its fish host (freshwater fishes such as crucian carp, *Carassius carassius*) and its elongated body trails backwards across the surface of the fish (Figure 11.2P). These features are reminiscent of pennellids (see Chapter 9) and this resemblance is strongly reinforced when the head is dissected out of the fish, because, like the pennellid *Lernaeocera*, the head is elaborated to produce a holdfast or anchor consisting of extensive, sometimes branched roots (Figures 11.1AB, 11.2N). However, in spite of appearances, *Lernaea* is not a pennellid and these common features have arisen by convergent evolution. *Lernaea* and its relatives on the one hand and the pennellids on the other are rather remotely related groups of copepods that have adopted similar life styles as fish parasites and have come to resemble each other as a result of the influence of similar selection pressures.

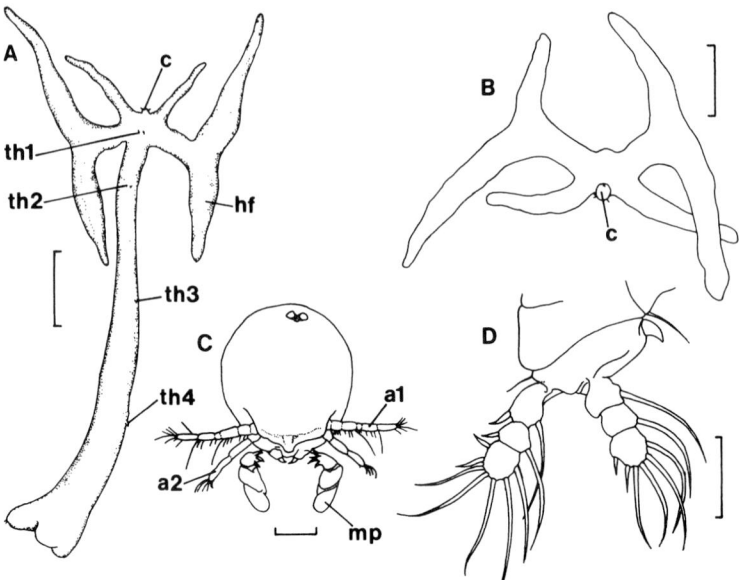

Fig. 11.1. The adult female anchor worm, *Lernaea cyprinacea*. (A) Whole parasite without egg sacs; (B) holdfast; (C) cephalothorax; (D) first thoracopod. a1, a2, First and second antennae; c, cephalothorax; hf, holdfast; mp, maxillipeds; th1- 4, first to fourth thoracopods. Scale bars: (AB) 2 mm; (C) 100 µm; (D) 50 µm. (ACD) From Kabata, 1979, with kind permission of The Ray Society; (B) from Grabda, 1963, with permission of the Institute of Parasitology, Polish Academy of Sciences.

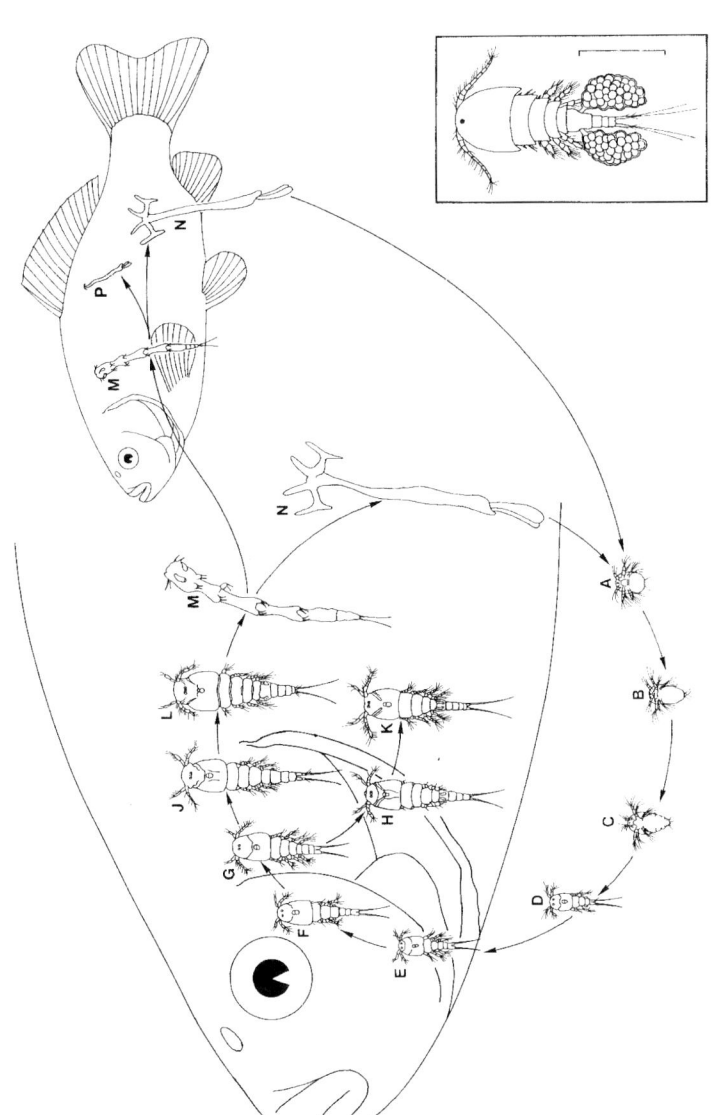

Figure 11.2. Life cycle of the anchor worm, *Lernaea cyrinacea* on crucian carp (*Carassius carassius*). (ABC) Nauplii I, II and III; (D) infective copepodid I; (EFG) copepodids II, III and IV; (H) ovigerous mesoparasitic female, (P) *in situ*. Copepodid I larvae usually settle on the gills, but may settle elsewhere; the parasite leaves the gills at the copepodid female stage. The life cycle may be completed on the same host fish (left) or on a different fish of the same (upper right) or a different species. Parasite stages not drawn to scale. Based on Grabda, 1963, and Fryer, 1982. Inset: a free-living freshwater cyclopoid copepod; female with egg sacs. Scale bar: 0.5 mm.

In the light of this, it will come as no surprise to learn that the early days of systematics were plagued by confusion about the relationships of *Lernaea* and *Lernaeocera*. The following is an abbreviated account of these events as documented by Kabata (1979), where the references will be found. Linnaeus first described *Lernaea cyprinacea* in 1758 in the 10th edition of his Systema Naturae. In the 12th edition (1766 – 1768) he added '*Lernaea*' *branchialis*, which was later moved to a new genus *Lerneocera* by Blainville in 1822 (much later amended to *Lernaeocera*). Unfortunately, Blainville also moved *Lernaea cyprinacea* to *Lerneocera*. In 1840, an inadvertent switching by Edwards of the names *Lernaea* and *Lerneocera*, which was not disentangled until 1917 by Wilson, further compounded this confusion. Even at this late date Wilson continued to regard *Lernaea* and *Lerneocera* as belonging to the same family.

Parasitism then can lead to morphological changes in crustaceans that greatly obscure their phylogenetic affinities, but these changes usually occur late in life so we still have a living record of their ancestry preserved in the earlier stages of the life cycle. The eggs of *Lernaea cyprinacea* liberate typical copepod nauplii, which moult twice to produce similar but progressively larger nauplii II and III (Figure 11.2 ABC). It is the next series of moults that reveals the ancestry of *Lernaea*. Following nauplius III are five copepodid stages (Figure 11.2 D-G plus H and J, copepodid V male and copepodid V female respectively) resembling the copepodids of free-living members of the Cyclopoida. The free-living cyclopoids are hugely abundant in freshwater and anyone with a good eye or a hand lens will have no difficulty in recognising *Cyclops* (see Figure 11.2, inset) and similar tiny pelagic relatives. The evolutionary progression from such a free-living organism to a highly specialised mesoparasite is truly remarkable.

11.2 LIFE CYCLE

Grabda (1963) has described the life cycle of *Lernaea cyprinacea* (Figure 11.2). It is the first copepodid stage (Figure 11.2D) that makes initial contact with a fish, settling on the gills or skin sometimes as early as during the first day of life. Grabda found that copepodid I can survive without a host for six days but fails to develop further. Each successive copepodid stage acquires an extra free segment and at the copepodid V stage the sex of each individual is recognisable (Figure 11.2 HJ). Grabda reported that copepodid V males are already sexually mature, with a spermatophore visible inside the body on each side, but she was of the opinion that copulation did not take place until after the next moult, at the cyclopoid stage (Figure 11.3), since it was not until this stage had been reached that spermatophores were found attached to the genital segment of the female (Figure 11.3B). However, Bird (1968) claimed to have witnessed copulation regularly between what she identified as copepodid V individuals. She observed that this act was followed by another moult in the female; then penetration of host tissue took place and transformation into the ovigerous mesoparasitic female. Bird also showed experimentally that the events of host penetration and transformation of the female do not take place if copulation does not occur. Isolated unmated copepodid V females or females reared only with other females do not penetrate or transform.

The cyclopoid males and females (Figure 11.2 KL) are capable of swimming freely. After mating the male makes no further contribution and subsequently dies. The mated female then becomes sedentary, burrowing beneath a host scale and passing through the skin to the muscles (Grabda, 1963). Shariff & Roberts (1989) observed that females of *Lernaea polymorpha* penetrate the bighead carp, *Aristichthys nobilis*, at an angle, an attitude dictated by the need to slide between the host's scales.

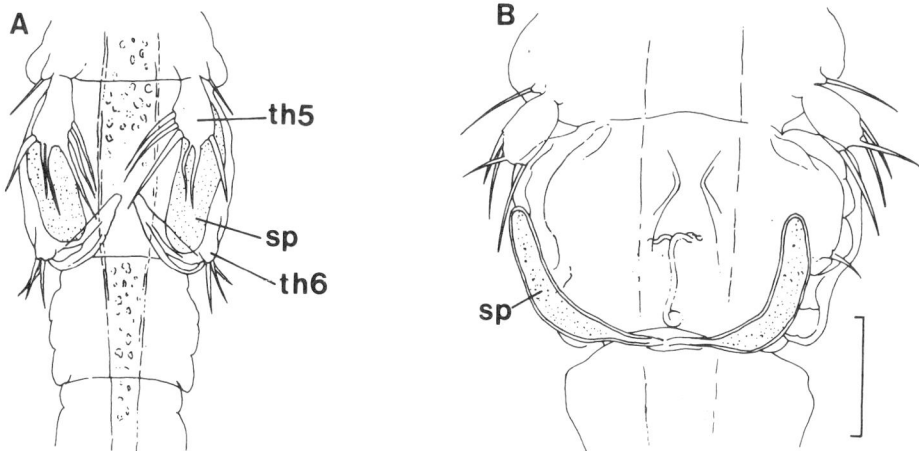

Fig. 11.3. Genital segments of the cyclopoid stage of *Lernaea cyprinacea.* (A) Male; (B) female. sp, Spermatophore; th5, 6, fifth and sixth thoracopods. Scale bar: 50 µm. From Grabda, 1963, with permission of the Institute of Parasitology, Polish Academy of Sciences.

Transformation of *L. cyprinacea* to a mesoparasitic ovigerous female is dramatic and involves rapid differential growth (see Grabda, 1963). The most vigorous development is the extension of the thorax (Figure 11.2M), with ultimately a loss of external evidence of segmentation. The thoracic appendages (thoracopods) are not caught up in this growth phase and, size-wise, persist exactly as they are in the copepodid stages (Figures 11.1AD, 11.2M). The abdomen also undergoes modest extension, some increase in girth and loss of external segmentation. Grabda (1963) reported that, between the cyclopoid female stage (body length 0.9 mm) and the ovigerous female (body length 12 mm), the free thoracic segments increased in length by between 17.8 and 40 times, while the cephalothorax and the abdomen increased in length by a mere 1.9 and 3.9 times respectively.

In parallel with these changes the branches of the anchoring holdfast begin to grow outwards from the posterior region of the cephalothorax (Figure 11.4), but the rest of the cephalothorax changes little and is dwarfed by the enormous development of the anchor (Figure 11.1ABC).

Little information is available on the nature of the food of *Lernaea* spp. Fryer (1968a) found the subject puzzling and observed that most species of *Lernaea* have no red blood in the gut, even though the head of the parasite is completely buried in host tissue. He pointed out that all of the head and the anchor roots are often encased in an envelope, moulded to the shape of the parasite and presumably produced by the host. In addition there may be a tough fibrous capsule firmly attached to the head and roots and Fryer expressed the view that this would preclude operation of the mouthparts and restrict or prevent access to host fluids. However, the production of large numbers of yolky eggs by such a parasite, demonstrates that nutrients are obtained in abundance. According to Fryer (1968a), *Afrolernaea* differs from most of its relatives by feeding on blood, which must be obtained with the aid of the mouthparts, but the possibility that nutrients are absorbed by the anchor roots, either as the only means of obtaining sustenance or as a supplement to ingested material, seems worth investigating. According to Kabata (1981) deep-seated copepods like lernaeids feed on tissue debris.

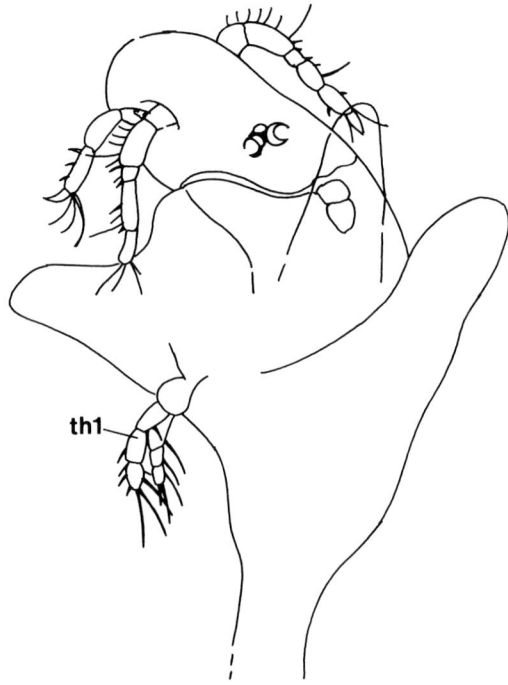

Fig. 11.4. Cephalothorax of young female anchor worm, *Lernaea cyprinacea*, with budding roots of the holdfast. Dissected out from embedded position beneath a host scale. th1, first thoracopod. From Grabda, 1963, with permission of the Institute of Parasitology, Polish Academy of Sciences.

Opinions differ as to whether *Lernaea cyprinacea* has an obligatory second host in the life cycle like *Lernaeocera branchialis* (see Chapter 9). Grabda (1963) came to the conclusion that "no intermediate host *sensu stricto*" occurs in *Lernaea cyprinacea*. This was based on her finding that crucian carp (*Carassius carassius*) infected experimentally with parasites at the copepodid I stage developed fully ovigerous females. However, we do not know how often cyclopoid females changed hosts in her experiments. It is possible that all cyclopoid females of *L. cyprinacea* are obliged to swim away from the first host and attach to a new one before their development can continue. This second host could be the same host individual on which larval development occurred, a different host individual of the same species (as shown in Figure 11.2), or even an individual of a different host species. The life cycle of *L. cyprinacea* published by Fryer (1982, fig. 95) illustrates the second of these three options, i.e. penetration and transformation to the ovigerous stage on a different individual of the same carp species.

11.3 PATHOLOGY AND RESISTANCE

Pathology produced by infection of the Asian bighead carp, *Aristichthys nobilis*, with *Lernaea polymorpha* has been described by Shariff & Roberts (1989). Punctate haemorrhages develop at the sites of penetration of the female copepods and each parasite causes extensive disruption of tissues, necrosis (tissue death) and haemorrhage along its path of entry. This is followed by an acute inflammatory response, succeeded by encapsulation of the holdfast roots by collagen fibres (see also above). Of special interest is the finding that adult parasites fail to develop on fishes with previous experience of infection with *L. polymorpha*, suggesting that the immune system of the fish is able to resist renewed invasion. Such immune (?) hosts developed haemorrhages that were much larger than those associated with penetrating females on naïve hosts, but adult females were not found in these enlarged lesions. Rejection of adult female *L. cyprinacea* by goldfish, *Carassius auratus*, has been described by Shields & Goode (1978).

The potential threat to freshwater fishes is emphasised by the comments of a Mrs A.J. Davies of Milford Haven, Pembrokeshire, quoted by Fryer (1968b). She described some of her infected goldfish from her garden pond with up to 20 parasites per fish as "grotesque" and "speared at every conceivable angle". Many died. *L. cyprinacea* may be rapidly fatal for small fishes.

11.4 THE ANCHOR WORM IN BRITAIN

Fryer (1968b, 1982) pointed out that *Lernaea cyprinacea* is probably not a native of Britain. It was first reported by Fryer (1968b) in garden ponds in the London area and was most probably imported into the country on one or more consignments of goldfish. It now seems to be well established, having been recorded from England, Wales and Scotland on common carp (*Cyprinus carpio*), ide (*Leuciscus idus*), roach (*Rutilus rutilus*) and three-spined stickleback (*Gasterosteus aculeatus*) (see Kabata, 1979). Fryer (1982) thought it likely that the parasite can utilise any British freshwater fish. Kabata (1979) likewise claimed that over 100 species, including 46 cyprinids, could act as hosts, and there are reports from tadpoles and adult amphibians.

Grabda (1963) found ovigerous females of *Lernaea cyprinacea* only on crucian carp under natural conditions in Poland. In contrast, she found copepodid stages developing on a variety of fishes such as sticklebacks, tench (*Tinca tinca*), bitterling (*Rhodeus sericeus*) and bleak (*Alburnus alburnus*), as well as on crucian carp. Thus, in her experience in Poland, the ovigerous female is strictly host specific, while the larval stages show no obvious host specificity.

12

POECILOSTOMATOID[1] COPEPODS

12.1 INTRODUCTION

According to Huys & Boxshall (1991), the origin of the Poecilostomatoidea is linked with a shift in their mate-grasping behaviour. During copulation the male holds the female with the maxillipeds not with the antennules, and this has resulted in marked sexual dimorphism in the maxillipeds in all known members of the group. Virtually all poecilostomatoids have parasitic or other associations with animals and most of them are marine. Many are associated with invertebrate hosts, but about ten families are parasites of fishes. Huys & Boxshall (1991) regard the group as probably the most diverse order of copepods in terms of gross body morphology. This diversity is apparent even among the subset of poecilostomatoids parasitising British fishes. These range from the ergasilids (including the only freshwater parasitic poecilostomatoids), bomolochids and taeniacanthids, still readily recognisable as copepods, through the bizarre chondracanthids to the virtually endoparasitic *Philichthys xiphiae*.

12.2 ERGASILIDS

Parasites such as *Lernaeocera branchialis* are impressive because of the extent of their adaptations to parasitism, so extensive in fact that they are not easy to recognise as crustaceans. The ergasilid poecilostomatoids have great appeal too, but for another reason – they demonstrate that crustaceans can be successful parasites without major anatomical or other biological changes. We might expect this to be the case since it is generally assumed that parasites have evolved from free-living antecedents by gradual evolutionary steps, the first of which demanded a minimum of change from the free-living ancestor. Ergasilids, therefore, generate two thoughts. First they may provide us with an image of what crustaceans looked like when they took that first step on the road to parasitism. Secondly, ergasilids may themselves have embraced parasitism relatively recently – they may have been parasites for a much shorter length of time than most of their parasitic relatives.

After hatching, an ergasilid passes through a surprising number of stages separated by moults (Figure 12.1). In *Ergasilus sieboldi* and *E. briani*, there are six nauplius stages, five copepodid stages and the adult male and female (Abdelhalim *et al*., 1991, Alston *et al*., 1996, respectively). Basically, all copepods have six naupliar stages so there has been no reduction in nauplii in these ergasilids like there has been in other parasitic copepods. In ergasilids, only one stage, the mature female, is parasitic on the gills of the fish host, and then only after mating has taken place. All the other stages including the male and the unmated female live a free, non-parasitic life. The most

---

[1] Boxshall & Halsey (2004) regard poecilostomatoids as cyclopoids. See footnote on p. 155.

striking feature of the parasitic female is the enlarged and claw-like second antennae (Figures 12.1K, 12.2, 12.3, 12.11A), used for attachment to the host.

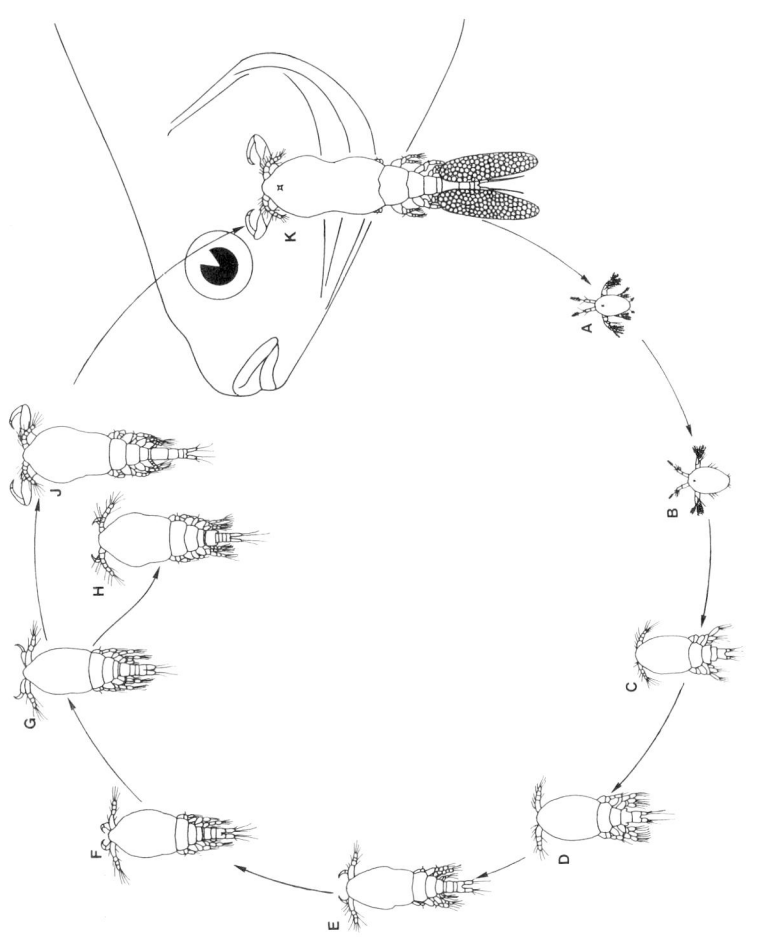

Figure 12.1. Life cycle of *Ergasilus briani*. (AB) Nauplii. Only two of six are shown. (C-G) First to fifth copepodid; (H) adult male; (J) adult female; (K) parasitic adult female. Stages not drawn to scale. Based on Alston *et al.*, 1993, 1996.

12.2.1 *British ergasilids*

There are six ergasilid species belonging to three genera in the British fauna, namely *Ergasilus sieboldi* (freshwater), *E. briani* (freshwater), *E. gibbus* (brackish water), *E.*

lizae (= *E.nanus*)[2] (marine records only in Britain), *Neoergasilus japonicus* (freshwater) and *Thersitina gasterostei* (brackish water) (see Kabata, 1992).

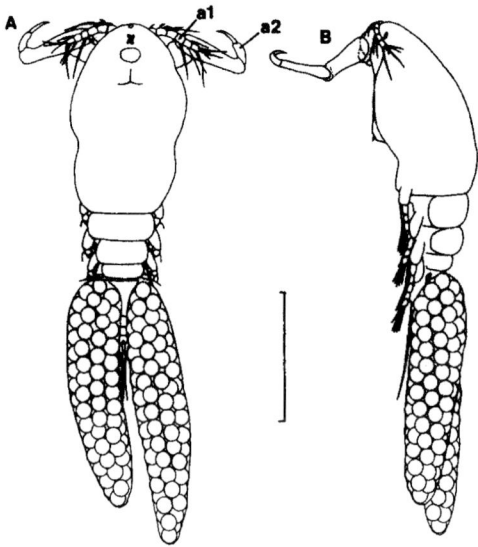

Fig. 12.2. Adult female *Ergasilus sieboldi.* (A) Dorsal view; (B) lateral view. Scale bar: 0.5 mm. a1, First antenna (antennule); a2, second antenna. Reproduced from Fryer, 1982, with permission of The Freshwater Biological Association and the author.

Ergasilus sieboldi (Figure 12.2) was first found in Britain in 1967 on trout (*Salmo trutta*) from Howbrook Reservoir, between Sheffield and Barnsley in Yorkshire (Fryer, 1969). The parasite was probably introduced into Britain accidentally (Fryer, 1982). By 1982 there were further records in Yorkshire as well as in Lancashire, Essex and (on aquarium fishes) in London (Fryer, 1982).

Parasitic female *E. sieboldi* are 1 – 2 mm in length, excluding their hair-like caudal setae, and occur in the northern Palaearctic on the gills of a wide range of fishes, including in Britain the following: brown trout (*Salmo trutta*), bream (*Abramis brama*), roach (*Rutilus rutilus*), rudd (*Scardinius erythrophthalmus*), carp (*Cyprinus carpio*), crucian carp (*Carassius carassius*) and tench (*Tinca tinca*). Infestation with this parasite is, according to Fryer (1969, 1982) potentially damaging. Fryer (1982) gives as an example a 34 cm long tench with 5400 parasites. This fish was only half its normal weight and its emaciated condition is likely to be the result of ergasilid parasitism.

Ergasilus briani was first discovered on the gills of bream, *Abramis brama*, in a pond at Gawthorpe near Dewsbury, Yorkshire in 1982 (Fryer & Andrews, 1983). It is also, in all probability, an accidental introduction into Britain, but there is no way of

---

[2] Kabata (1979) regarded *E. lizae* and *E. nanus* as separate species, but later Kabata (1992) accepted evidence that they are a single species, for which the chronologically earlier name *E. lizae* is correct.

determining when this occurred and it may have been here undetected for some time (Fryer & Andrews, 1983). Its small size (usually less than 1 mm) and its location between the hemibranchs of the gill rather than on the outside of the gill like *E. sieboldi*, make it easy to miss. It has now been found in other sites in Britain (Surrey, Hampshire, Buckinghamshire; see Alston *et al.*, 1993) but undoubtedly occurs elsewhere. *E. briani* is widespread in Eurasia on many cyprinid fishes, and in Britain has been recorded on bream, roach, rudd, tench and gudgeon (*Gobio gobio*).

Ergasilus gibbus lives on the gills of the eel, *Anguilla anguilla*, and in spite of the wide salinity tolerance of the host the parasite appears to be confined to brackish water. Fryer (1982) mentions a single record by Canning *et al.* (1973) (attributed to R. Wootten) from Slapton Ley in Devon. It is interesting that eels exported to Hungary from Ireland were infected and the parasite's occurrence in Ireland has been confirmed (see Fryer, 1982). The parasitic female may reach 2 mm in length.

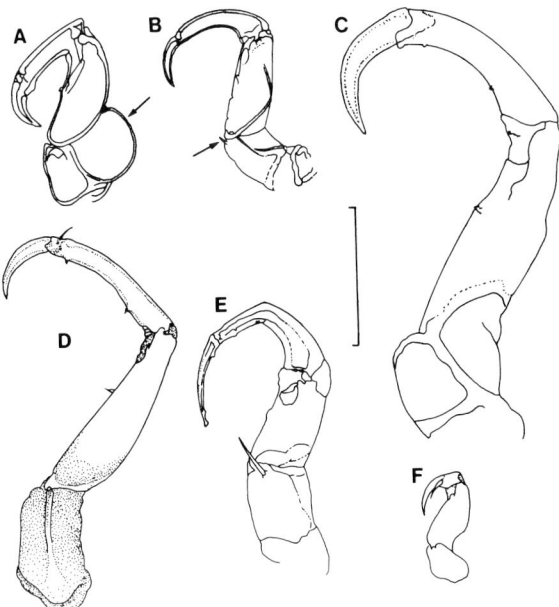

Fig. 12.3. Second antennae of British ergasilids, all drawn to the same scale (Scale bar: 150 μm). (A) *Ergasilus gibbus* (arrow indicates swelling on basal podomere); (B) *E. briani* (arrow indicates peg on basal podomere); (C) *E. sieboldi*; (D) *E. lizae* (= *E. nanus*); (E) *Neoergasilus japonicus*; (F) *Thersitina gasterostei*. (AB) Reproduced from Fryer, 1982, with permission of The Freshwater Biological Association and the author; (CDF) from Kabata, 1979, with kind permission of The Ray Society; (E) from Mugridge *et al.*, 1982, with kind permission of the Zoological Society of London.

Records of *Ergasilus lizae* in Britain are few and exclusively marine. Scott & Scott (1913) referred to a record by A. M. Norman at Swan Pool, Falmouth in 1884. Scott (Scott, T., 1901) also recorded it as *E. nanus* on *Mugil chelo* (= *Chelon labrosus*) near Aberdeen and Kabata (1979) collected it from an undetermined species of mullet in the same locality. According to Kabata (1979), this parasite is able to tolerate the transition from a marine to a freshwater environment on anadromous hosts and is able to infect completely freshwater species. The female ranges from 0.8 to 1.2 mm in length.

In addition to features already mentioned, the second antennae (Figure 12.3) and fifth thoracopods (Figure 12.4) are useful for identifying the British species of *Ergasilus*. *E. gibbus* has a prominent swelling on the basal podomere of the antenna (Figure 12.3A), *E. briani* has a sensory peg on the same podomere (Figure 12.3B) and the fifth limb is vestigial bearing a single seta (Figure 12.4A). In *E. sieboldi* the fifth limb has two podomeres with two setae on the distal podomere (Figure 12.4B). The fifth limb of *E. lizae* apparently has one podomere with four setae (Figure 12.4C).

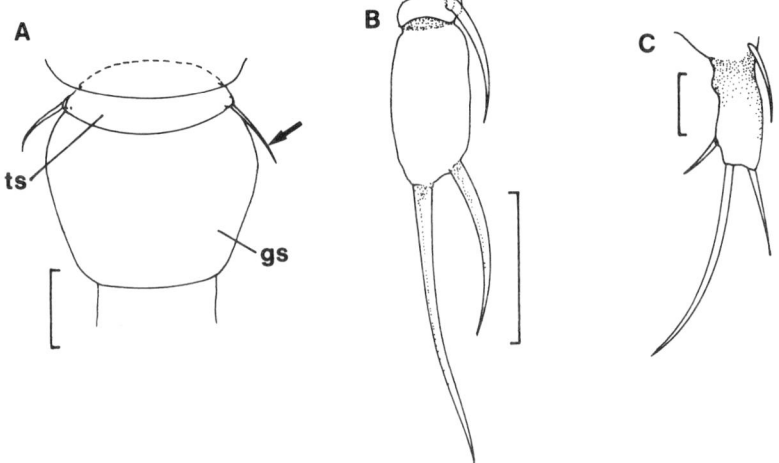

Fig. 12.4. Fifth thoracopods of (A) *Ergasilus briani*. Vestigial fifth limb represented by hair-like seta (arrow) attached to fifth thoracic segment (ts); (B) *E. sieboldi*, whole fifth limb; (C) *E. lizae* (= *E. nanus*), whole fifth limb. gs, Genital segment. Scale bars: (AB) 50 μm; (C) 10 μm. (A) Reproduced from Fryer, 1982, with permission of The Freshwater Biological Association and the author. (BC) From Kabata, 1979, with kind permission of The Ray Society.

Neoergasilus japonicus is a skin parasite, unlike the other ergasilids found in Britain. It is another alien, being first recorded in Britain in two small ponds in West Sussex in 1980 (Mugridge *et al.*, 1982). One view is that it originated in the Far East and has extended its range slowly westwards, reaching Europe (Czechoslovakia and Hungary) in the 1960s (Fryer, 1982). It was recorded in the Paris area in 1979 by Lescher-Moutoué. Fryer (1982) also expressed the opinion that it might be indigenous to Europe. The adult females reach 0.65 – 0.85 mm in length and are most commonly found attached to the fins and to scale-less areas at the base of the fins, although individuals are occasionally found on the gills. In Britain the parasite has been recorded from bream, roach, rudd, common carp and tench, but not on perch (*Perca fluviatilis*) from the same sites (Mugridge *et al.*, 1982). Elsewhere in its range it parasitises perciform fishes and catfishes (Fryer, 1982). Modifications of the first thoracopod distinguish the genus *Neoergasilus*. The outer branch (exopod) carries an extra, flattened appendage and there are curious scraper-like spines on the inner and outer branches (exopod and endopod) (Figure 12.5). The function of this interesting specialised limb is unknown. The basal podomere of the second antenna also has a prominent spine (Figure 12.3E) (see below).

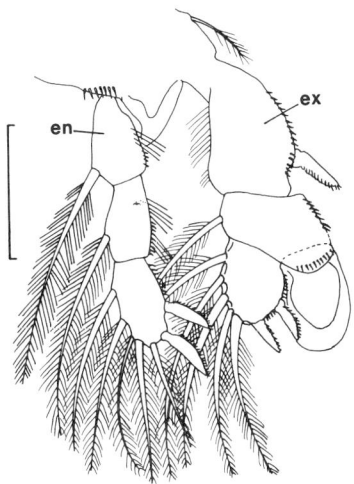

Fig. 12.5. First thoracopod of *Neoergasilus japonicus*. en, Endopod; ex, exopod. Scale bar: 50 µm. From Lescher-Moutoué, 1979.

Thersitina gasterostei (Figure 12.6) is a native species, commonly found on the three-spined and nine-spined sticklebacks (*Gasterosteus aculeatus* and *Pungitius pungitius*) (see Fryer, 1982). It has a distinctively circular cephalic region when viewed from above or from below and relatively small antennae each with a spine near the base of the distal podomere (Figure 12.3F). The female is small (0.6 mm – 1mm) and usually found attached to the inner surface of the operculum. However, in heavy infections females may be found on the gills and elsewhere. When given the choice of sticklebacks the parasite prefers the three-spined stickleback (Gurney, 1913; Walkey *et al*., 1970) and Walkey *et al*. suggested that this might relate to the more voluminous gill chamber in this fish. Walkey *et al*. (1970) working in Norfolk found that, unlike its host, *Thersitina* does not tolerate water more dilute than 2% seawater. Consequently, the parasite is restricted to sticklebacks inhabiting the lower reaches of rivers and has an essentially coastal distribution in Britain (but see below). In addition to Norfolk, it has been recorded from the following localities: Aberdeen, Loch Etive, the lower reaches of the River Forth, the Wirral, Yorkshire (see below), Northumberland, the Thames Estuary, Sussex and probably also the Severn estuary. Its salinity tolerance has permitted it to reach Barra in the Outer Hebrides (Fryer, 1982). Elsewhere *T. gasterostei* appears to have a circumpolar distribution in the Northern Hemisphere (British Columbia, Greenland, Newfoundland, as well as northern Asia). (See also Chapter 13, p. 249).

An inland record of *Thersitina* on three-spined and nine-spined sticklebacks in Yorkshire is of special interest. Fryer (1978) found the parasite in two lagoons or flashes in the Lower Aire Valley near Mickletown. These lagoons originated relatively recently as a result of mining subsidence and contained distinctly brackish water, presumably as a result of leaching of mineral salts from colliery waste dumped nearby. Fryer found that the high salt content was, in fact, rather different in chemical composition from coastal brackish water, being richer in calcium and magnesium in relation to sodium than seawater. This unique habitat attracted a variety of brackish water plants and animals, including *T. gasterostei*. How the parasitised sticklebacks

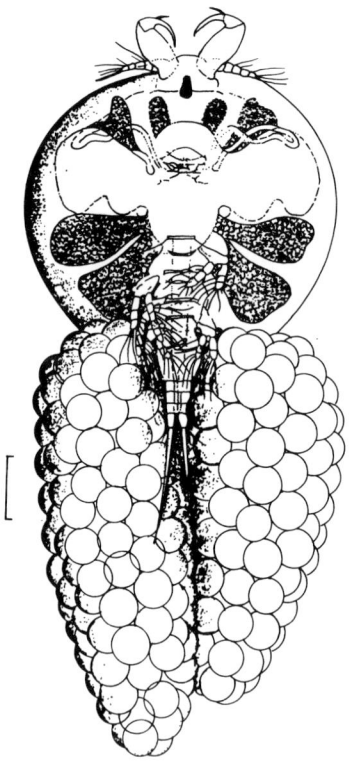

Fig. 12.6. Mature female *Thersitina gasterostei*, in ventral view. Scale bar: 100 µm. Some appendages removed for clarity. From Gurney, 1913, with permission from Taylor & Francis Ltd. (http://www.tandf.co.uk/journals).

reached this inland site is unknown, but Fryer speculated that high salinity in the adjacent River Aire, presumably resulting from similar mining contamination, may have permitted survival of the parasite during inland penetration from coastal brackish areas. Fryer (1978) referred to a second earlier report of *Thersitina* from an inland site, namely at Wintersett about 13k south of Mickletown. Fryer suggested that the parasite might have reached this site via canals linking the Aire/Calder and Dearne/Don river systems.

12.2.2 Development and attachment of the female
The development of *Ergasilus briani* has been followed by Alston *et al.* (1993, 1996) and is summarised here. The development of *E. sieboldi* is similar (see Abdelhalim *et al.*, 1991).

Nauplius I of *E. briani* (Figure 12.1A and shown enlarged in ventral view in Figure 12.7) is about 120 μm long with a red-pigmented median nauplius eye and a ventral mouth overhung by a plate-like upper lip or labrum. The animal is free-living at this stage and capable of feeding, unlike the nauplius of the caligid *Lepeophtheirus*

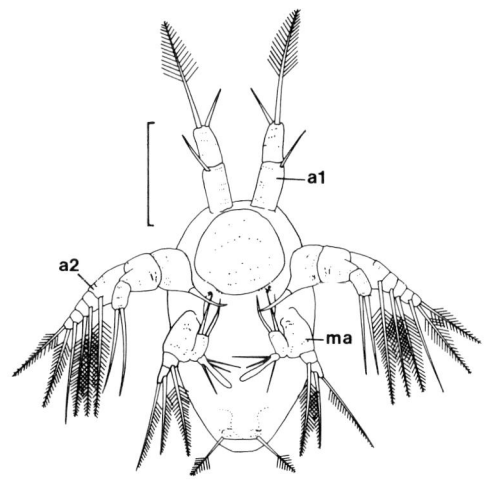

Fig. 12.7. Nauplius I of *Ergasilus briani*, in ventral view. a1, a2, First and second antennae; ma, mandible. Scale bar: 50 μm. From Alston *et al.*, 1996, (Copyright © 1996), with kind permission of Kluwer Academic Publishers.

pectoralis, which is thought to be unable to feed (Chapter 8). Nauplius I of *E. briani* has only three pairs of limbs, namely first antennae (antennules), second antennae and mandibles, the two last-named pairs of appendages being biramous (Figure 12.7). At this stage the mandibles bear little resemblance to the crushing/biting appendages of adult arthropods. The antennules, antennae and mandibles of Nauplius I are equipped with feather-like (plumose) setae which seem well equipped for propulsion (Figure 12.7). However, the second antennae and mandibles are multifunctional, the stiff non-plumose setae projecting medially from their bases being used to hold and manipulate food particles taken from the plankton. In fact, the feeding apparatus changes little through the many naupliar stages and is typical for plankton-feeding copepod nauplii.

Nauplius VI (Figure 12.1B) is twice the size of nauplius I and possesses in addition to the three pairs of limbs of nauplius I, two pairs of medial processes on the ventral surface just anterior to the anal slit. These are the beginnings of the first two pairs of swimming limbs (thoracopods).

Features of copepodid I (Figure 12.1C) foreshadow those of the adult. The second antennae are no longer concerned with feeding. The antennal endopod terminates in a prominent curved claw, although the presence of a vestigial exopod betrays the limb's biramous origin (Figure 12.8A). A significant change has taken place in the mouthparts. The feeding apparatus of the copepodid and of all subsequent stages comprises the mandibles and two new pairs of mouthparts, the first maxillae (maxillules) and the second maxillae (Figure 12.11A). The mandibles are now unjointed and barely recognisable, each one having lost its plumose setae and developed three terminal blades (two long and one very short) armed mainly along their posterior margins with pointed teeth (Figure 12.8B). The first maxillae are small, but the second maxillae are large, each one terminating in a spatula-like structure again armed with cutting teeth (Figure 12.8B).

The body of copepodid I comprises a cephalothorax and three segments bearing swimming legs (thoracopods). Only the first two pairs of swimming legs are biramous and armed with plumose setae (Figure 12.8C); each third swimming leg is represented at this stage only by a bilobed process (Figure 12.8D). A pair of lobes indicates the sites of the developing fourth swimming legs (Figure 12.8D).

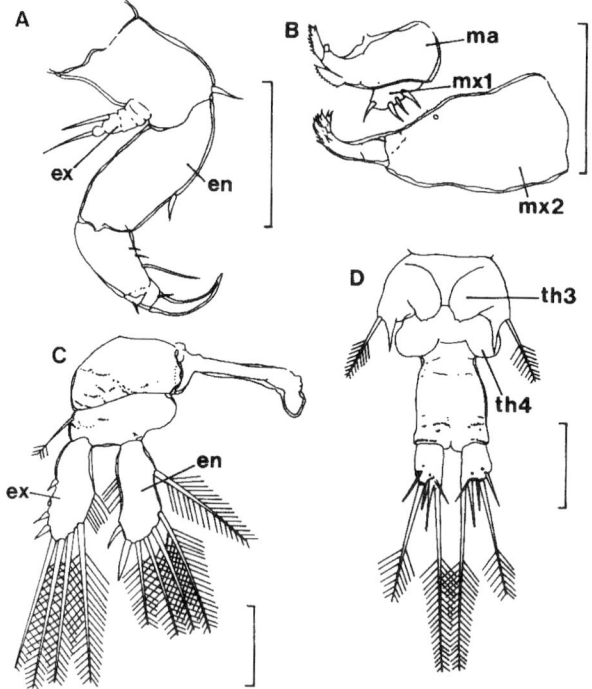

Fig. 12.8. Features of copepodid I of *Ergasilus briani*. (A) Second antenna; (B) mouthparts; (C) first swimming leg (thoracopod); (D) posterior end of body. en, Endopod; ex, exopod; ma, mandible; mx1, mx2, first and second maxillae; th3, th4, third and fourth thoracopods. Scale bars: 40 μm. From Alston *et al.*, 1996, (Copyright © 1996), with kind permission of Kluwer Academic Publishers.

In copepodid II the presence of a lobe (Figure 12.9A) on each side just posterior to the second maxilla distinguishes potential males from females, since these lobes are the developing maxillipeds, which are present only in males. In other male ergasilids the maxillipeds put in their appearance later in development. In the adult male *Ergasilus briani* the maxillipeds are long appendages each of five podomeres (Figure 12.9B). The maxillipeds are sub-chelate, i.e. they are capable of a grasping function, the terminal claw folding back against the proximal shaft of the limb. The male also possesses an extra body segment behind the leg-bearing segments and is a little smaller than the free-swimming female (Figure 12.1HJ).

The maxillipeds are absent in the female. When she reaches maturity (Figure 12.1K) the maxillipedal segment is fused with the cephalon. Technically, the tagma so-formed is a cephalosome, but the first leg-bearing segment is indistinctly separated from the cephalosome, creating an incipient cephalothorax. According to Kabata (1979), ergasilids can be arranged in a series displaying the phenomenon of cephalisation, in which the leg-bearing segments progressively fuse with the cephalosome.

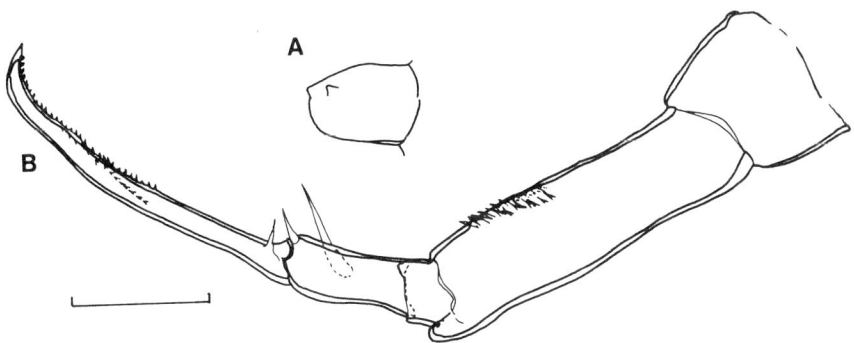

Fig. 12.9. Male maxillipeds of *Ergasilus briani*. (A) Second copepodid; (B) adult. Scale bar: 20 μm. From Alston *et al.*, 1996, (Copyright © 1996), with kind permission of Kluwer Academic Publishers.

There are four pairs of biramous thoracopods, the fifth pair being reduced (Figure 12.4A). A genital double segment (comprising fused genital and first abdominal segments) and three free abdominal segments follow the leg-bearing segments. The paired genital openings are longitudinal slits in the dorso-lateral surface of the genital double segment. The last abdominal (anal) segment is deeply incised medially from its posterior margin and the anus opens medially on its dorsal side. The dorsal positions of genital openings and anus presumably serve to direct genital products and faeces away from the gill and into the exhalent gill ventilating current. The anal segment gives rise to two cylindrical caudal processes armed with large and small setae.

It is the second antennae (Figures 12.2, 12.3) that are used by the female ergasilids to attach herself to the host, in a manner appositely described by Kabata (1982) as a "staple" type of attachment. Each antenna has four podomeres (see for example Fig. 12.3B), the two distal ones being fused together to form a rigid claw. There is a hinge between the claw and the next podomere, so that the claw can close against this podomere (sub-chelate condition). Thus, with the claws extended as in Figure 12.2 A, the parasite resembles a crab spider and it uses the two claws to embrace and grip one of the host's primary gill lamellae, often constricting the lamella (see fig. 13 A in El-Rashidy & Boxshall, 1999). The two embracing claws may overlap and in a South American ergasilid, *Acusicola spinuloderma*, the tip of the claw of each antenna is located in a cavity in the opposite antenna (see fig. 13 CD in El-Rashidy & Boxshall, 1999).

The question of how *Neoergasilus japonicus* uses similar antennae to attach itself to a very different substrate – fish skin and fins – does not appear to have been answered. However, Mugridge et al. (1982) illustrated an antenna with a blunt claw and a strong and sharp spine on the proximal podomere (Figure 12.3E). Fryer (1982) suggested that this spine is concerned with attachment to fins, but its precise role remains to be discovered.

Another change coinciding with the adoption of a parasitic life style and reflecting the loss of mobility that this change entails, is the progressive loss of all the setules that give the setae on the thoracopods their feather-like appearance and propulsive capability (as in Figure 12.5).

The female does not moult after mating but does grow bigger after attachment to the host. In *Ergasilus briani* the mean body length and breadth increase by about 13.5% and 2.3% respectively. This is because eggs are produced and stored in the cephalothorax. A region of flexible cuticle between the dorsal cephalic shield and the tergum (dorsal plate) of the first leg-bearing segment presumably permits this volume increase. After extrusion of stored eggs in the form of egg sacs (Figures 12.1K, 12.2 AB), the cephalothorax returns to its original smaller size, expanding again to accommodate the next batch of eggs. However, ergasilids are generally small, the parasitic female of *E. briani* ranging in length from about 680 to about 910 µm.

12.2.3 *Food, feeding and colour*

The food and feeding behaviour of the free-living stages are unknown (Abdelhalim et al., 1991), but it is very likely that all these stages, including the unmated females, are able to subsist on a diet of free-living planktonic organisms. Thus it is highly likely that the ergasilid mouthparts and digestive system have to cope with a drastic change of diet when the mated female takes up residence on its fish host. There are no significant changes in the ergasilid mouthparts during copepodid development. The mandibles and second maxillae of the adult female have toothed blades like copepodid I (Figure 12.8B). These mouthparts are presumably used to capture planktonic organisms during the non-parasitic phase of the ergasilid's life and the same mouthparts must serve the parasitic female. This remarkable flexibility permits ergasilids to combine and exploit in a single lifetime a free-living and a parasitic existence.

It is particularly interesting that, with the onset of egg production, the ergasilid female foregoes the uncertainties and vagaries of plankton as a food source and adopts the accessible, nutritious and virtually unlimited supply of food provided by the host. In other words the ergasilid maximises egg output by switching to the more reliable resource provided by a fish host.

There is one interesting clue to the nature of the food of the free-living, non-parasitic stages. These stages are frequently brightly coloured and some of these pigments may be derived from ingested food. In *Ergasilus briani* the ventral cuticle between the mandibles of nauplius VI and the primordia of the second legs is orange. This coloration persists through the various copepodid stages to the free-living males and females. The pigment extends forward to the anterior margin and into the antennules, mouthparts, first and second swimming legs and, in the males, the maxillipeds. Alston et al. (1996) were of the opinion that this orange pigment in *E. briani* is almost certainly a carotenoid assimilated from food ingested by the free-living stages. They pointed out that red or orange pigments are present in many free-living copepods. These copepods assimilate β-carotene from ingested algae and synthesise carotenoids such as β-cryptoxanthin and astaxanthin, which are deposited in fat globules in the gut wall and in the epidermis underlying the cuticle. Once attached to the host the pigmentation of *E. briani* females is gradually lost, and ovary development in the cephalothorax causes the parasites to appear white.

The accumulation of these carotenoids by pelagic copepods may not be entirely accidental. It has been suggested that they may have a photo-protective role (see Alston et al., 1996, for references). Highly pigmented morphs of free-living copepods of the genera *Heterocope* and *Diaptomus* survive better in bright light than less pigmented morphs. The interpretation is that these pigments absorb light and prevent the photo-oxidation of vital compounds. It is especially interesting in this respect to note that Alston et al. (1996) found the planktonic stages of *Ergasilus briani* most commonly in the top few centimetres of their culture tank, even when illuminated by intense sunlight. *E. sieboldi* behaves similarly in aquaria and planktonic stages of *Paraergasilus rylovi* were reported to be most abundant in the upper 100 cm of a Russian lake (references in Alston et al., 1996).

Orange is not the only pigment in *E. briani*. Cobalt blue pigmentation in the form of minute droplets is already present in the cells of the gut wall of nauplius I on hatching. This pigment intensifies in colour to blue black in the cephalothorax of the copepodid stages. Posterior to the cephalothorax the gut of copepodids and free-living adults displays olive green and cobalt blue zones, as well as zones lacking colour. Whether some or all of this gut-associated pigment is derived from the food and whether it serves a function other than photo-protection remains to be determined. Its presence in the freshly hatched nauplius I suggests that some of it is synthesised by the ergasilid.

Alston et al. (1996) noted that the strongly pigmented planktonic stages of *E. briani* are easily visible with the naked eye and commented, as others before them, that predators might find them equally conspicuous. They assumed that this disadvantage is offset by the photo-protection provided by the pigment. Alston et al. also made the interesting suggestion that the conspicuousness imparted by the pigmentation might be advantageous to the free-swimming mated female. They suggested that the fish host might be sufficiently interested in the brightly coloured planktonic female to ingest it. Inside the mouth the ergasilid could prevent itself from being swallowed by attachment to the fish's gill rakers using its claw-like antennae. Later it could move from the gill rakers to the gills. Ingestion of the earlier and equally conspicuous naupliar and copepodid stages could, of course, lead to their destruction.

Einszporn (1965) explored the nature of the food of the parasitic female. She preserved gills of tench (*Tinca tinca*) infected with *Ergasilus sieboldi*, embedded them in paraffin wax and cut sections through the parasites and gills. Host epithelial cells at various stages of digestion were identified and were present in abundance in the gut of the parasite, indicating that the gill epithelium provides the bulk of the parasite's diet. Not surprisingly, host mucous cells, which are located in the gill epithelium, were also identified in the gut, but Einszporn also found many fish erythrocytes and leucocytes and came to the conclusion that blood is a regular and important component of the parasite's food. Einszporn rejected an earlier claim by Halisch that digestion takes place outside the body as a result of regurgitation of digestive enzymes, and proposed that the mouthparts bite off and ingest gill tissue. She also believed that gill tissue dislodged by the movements of the swimming legs might be collected by the setae on the first swimming legs and transferred to the mouth.

12.3 BOMOLOCHIDS AND TAENIACANTHIDS

There are two groups of poecilostomatoids that are closely related to the ergasilids, so close in fact that some specialists have included them in the family Ergasilidae. However, these groups are now recognised as independent families, the Bomolochidae and the Taeniacanthidae (see Kabata, 1992; Boxshall & Halsey, 2004).

The bomolochids have two representatives in Britain, *Bomolochus bellones* (Figure 12.10A) in the gill chamber of the garfish (*Belone belone*) and *Holobomolochus confusus* (Figure 12.10B) in the nasal cavities of cod (*Gadus morhua*) and other gadid fishes. A third species, *B. soleae*, occurs on the common sole (*Solea solea*) along the Atlantic seaboard of Europe and in the Mediterranean Sea, but has so far not been recorded in Britain.

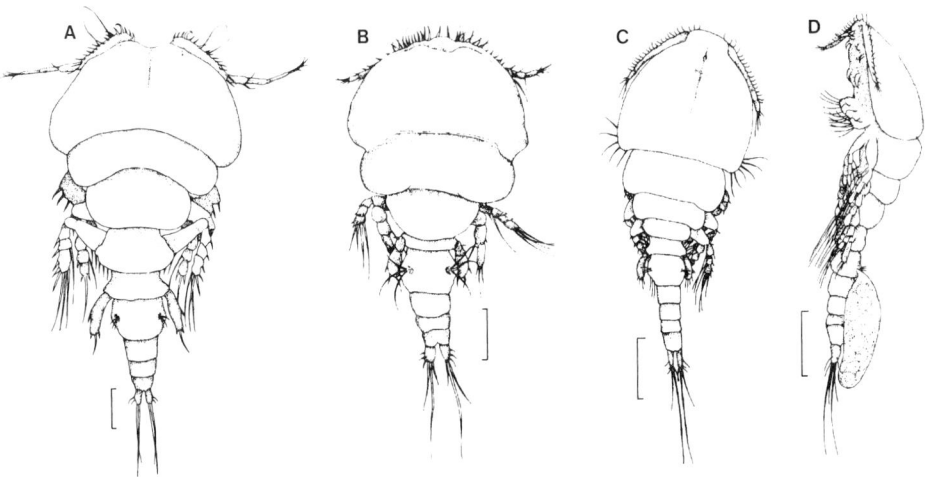

Fig. 12.10. British bomolochid and taeniacanthid copepods. (A) *Bomolochus bellones*, in dorsal view; (B) *Holobomolochus confusus*, in dorsal view; (CD) *Taeniacanthus onosi*, in dorsal and lateral views respectively. Scale bars: 200 μm. From Kabata, 1979, with kind permission of the Ray Society.

Taeniacanthids are represented in Britain by four species of *Taeniacanthus*. All except one, *T. wilsoni*, were formerly assigned to the genus *Anchistrotos* (see Kabata, 1979, 1992). *T. onosi* (Figure 12.10CD) attaches itself to the inner surfaces of the opercula of the rocklings *Enchelyopus cimbrius* and *Ciliata mustela* and *T. zeugopteri* to the upper surface of the topknot, *Zeugopterus punctatus*. *T. wilsoni* is a parasite of elasmobranch fishes, occurring in the gill cavities of rays. *T. laqueus* has been recorded only twice on the gills of comber, *Serranus cabrilla*, which is uncommon in British waters.

Information on the biology of bomolochids and taeniacanthids is scanty, but what little we do know indicates that there is much of interest to discover. Like ergasilids, they have made few concessions to the parasitic way of life and strongly resemble their free-living poecilostomatoid relatives. According to Kabata (1979) their hallmark is their clearly preserved segmentation, these copepods having sacrificed few of the original segmental boundaries to the demands of parasitism.

Bomolochids and taeniacanthids do not attach themselves in the same way as ergasilids, since the second antennae are not enlarged and hooked as they are in ergasilids (Figure 12.11). The dome-shaped cephalothorax of bomolochids and taeniacanthids, with its slightly concave ventral surface and transparent 'membrane' along each lateral margin (Figures 12.10D, 12.11BC), resembles that of caligid copepods and is evidence of a remarkable evolutionary convergence. Poecilostomatoid (bomolochid and taeniacanthid) and siphonostomatoid (caligid) fish parasites (see Chapter 8) appear to have independently solved the problem of attachment to the fish host's flat surfaces by using the whole of the cephalothorax as a sucker. A further point of resemblance between these two distantly related groups is the employment of a specially flattened pair of swimming legs (first pair of thoracopods in the bomolochids and taeniacanthids; third pair in the caligids) to seal off the posterior border of the cephalothoracic sucker. In the bomolochids and taeniacanthids the marginal 'membranes' act as valves sealing the lateral borders of the sucker. Höglund & Thulin (1988) observed specimens of *Holobomolochus confusus* attached by the ventral side of the cephalothorax to the walls of a Petri dish.

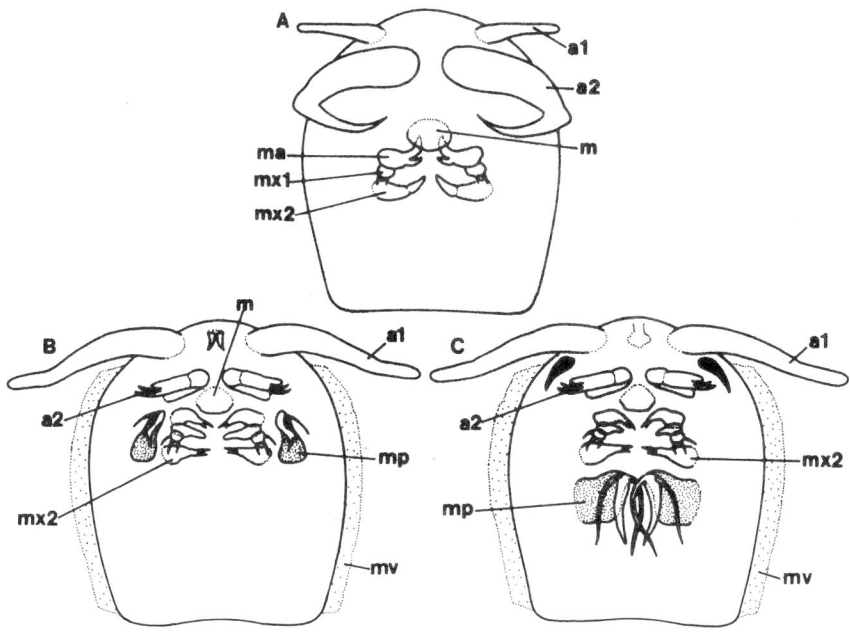

Fig. 12.11. Diagrammatic comparison of the anterior cephalothoracic appendages of (A) an ergasilid, (B) a bomolochid and (C) a taeniacanthid. a1, a2, First and second antennae; m, mouth; ma, mandible; mp, maxillipeds; mv, marginal valve; mx1, mx2, first and second maxillae. From Kabata, 1979, with kind permission of The Ray Society.

Modern literature contains no references to the life cycles and development of bomolochids and taeniacanthids. We have to go back to references from the beginning of the last century for indications that the commitment of the life-cycle stages to the parasitic way of life may be more extensive than in ergasilids. Scott & Scott (1913) collected *Holobomolochus confusus* (undoubtedly identified incorrectly as *Bomolochus soleae*) from the nasal cavities of cod (*Gadus morhua*) captured in the Moray Firth and recorded the presence of males and juveniles at all stages of development, as well as females some of which had egg sacs. The implication is that males and juveniles may be parasitic. Höglund & Thulin (1988) observed that living specimens of *H. confusus* are fast and lively swimmers when disturbed and believed that the parasites, especially the males, may be able to change hosts.

12.4 THE CHONDRACANTHIDS

The chondracanthids with their bizarre shapes (Figure 12.12) bear little resemblance to the relatively unmodified ergasilids and it is hard to believe that they are related. Nevertheless, the mouthparts of chondracanthids (Figure 12.13) are typically poecilostomatoid and bear witness to the common ancestry of the ergasilids and the chondracanthids. The difference is that the chondracanthids have responded to the evolutionary pressures and opportunities of the parasitic way of life, while the ergasilids have made few concessions to parasitism. Several species belonging to four genera, namely *Chondracanthus*, *Acanthochondria*, *Acanthochondrites* and *Lernentoma*, have been recorded from British marine fishes (see Kabata, 1979). Chondracanthids mainly

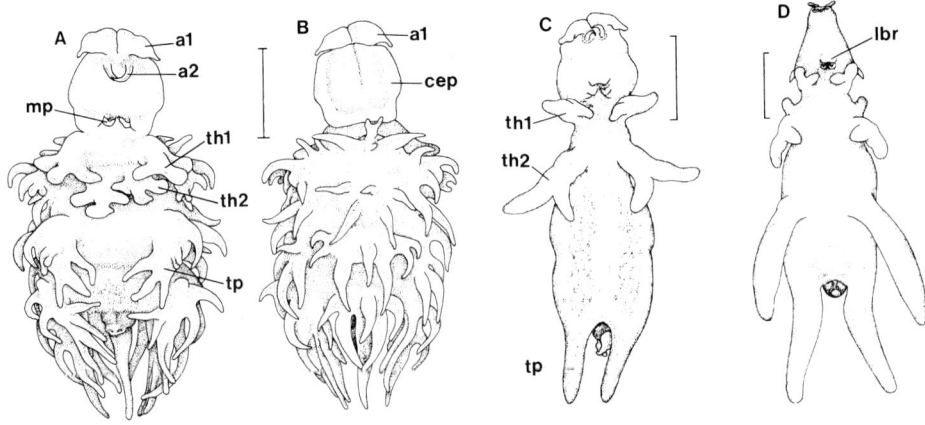

Fig. 12.12. Some British chondracanthids (females). (AB) *Chondracanthus zei*, in ventral and dorsal view respectively; (C) *Acanthochondria soleae* in ventral view. (D) *Chondracanthus merluccii*, ventral view. a1, a2, First and second antennae; cep, cephalosome; lbr, labrum covering mouth; mp, maxilliped; th1, th2, first and second thoracopods; tp, trunk process. Scale bars: 2mm. From Kabata, 1979, with kind permission of The Ray Society.

inhabit relatively sheltered corners of the gill cavities of bony fishes (Kabata, 1982) and the presence of egg sacs on these strange parasites (Figure 12.15A) reveals that they are female.

12.4.1 *The adult female*
The British female chondracanthids have a cephalosome incorporating the first thoracic maxillipedal segment (retaining the maxillipeds unlike the ergasilid females) (Figure 12.13). A thick dorsal shield usually covers the cephalosome (e.g. Figure 12.12B). The first antenna is fleshy, but the second antenna is a heavily sclerotised hook (Figure 12.12A). The latter is the only means of attachment for the female, the parasite using both second antennae to staple itself firmly inside the gill chamber.

According to Kabata (1982) chondracanthids are sessile and this is reflected in the reduction of the legs of the British species to two pairs of lobed structures (Figure 12.12ACD). The mouth, on the ventral side of the cephalosome, is partly covered by the

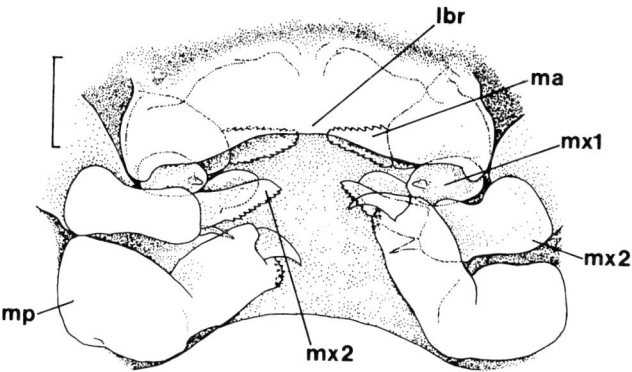

Fig. 12.13. Mouth region of female *Chondracanthus zei*. lbr, Labrum covering mouth; ma, mandible; mx1, mx2, first and second maxillae; mp, maxilliped. Scale bar: 100 μm. From Kabata, 1979, with kind permission of The Ray Society.

upper lip (labrum) with the mouthparts clustered around it (Figure 12.13). The mandibles have a sickle-shaped distal blade with teeth along both margins and the second maxillae have a distal blade with teeth along one margin (Figure 12.13). The first maxillae are small lobes lying between the mandibles and the second maxillae, and there is an extra lobe on each side called a paragnath. The maxillipeds are prominent, immediately posterior to the second maxillae. Each maxilliped is a jointed limb armed terminally with a stout claw.

It is the thoracic region bearing the legs, referred to by Kabata (1979) as the trunk, that has undergone substantial changes in the chondracanthids (Figure 12.14), and it is these changes that have obscured the relationship between the chondracanthids and other parasitic poecilostomatoids. In addition to the loss of legs and modification of those remaining, thoracic segmentation has disappeared and the trunk is greatly enlarged to accommodate the gonads. In addition, the trunks of many chondracanthids have developed finger-like or branched processes. In *Acanthochondria soleae* from the common sole, there are only two of these projecting from the posterior end of the trunk (Figure 12.12C). In *Chondracanthus merluccii* on hake, *Merluccius merluccius*, there are two pairs of these processes (Figure 12.12D), but in *C. zei* from the dory, *Zeus faber*, the whole of the trunk bristles with them (Figure 12.12AB). The function of these processes is unknown, although the bulky trunk is the site of the metabolically demanding process of egg production and the increased surface area provided by the processes may be a reflection of a regional need for enhanced gaseous exchange. At the posterior end of the trunk is a much-reduced genito-abdomen.

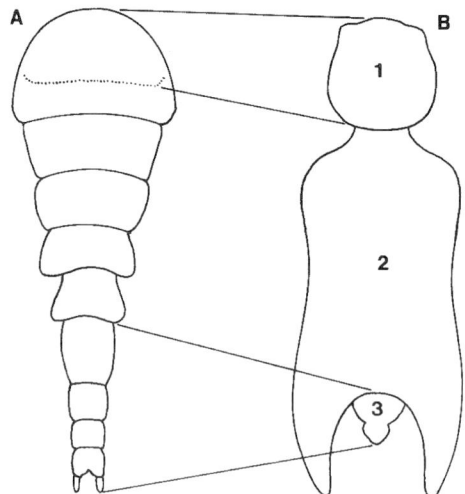

Fig. 12.14. Diagram illustrating changes in body proportions required to transform a primitive podoplean copepod (A) into a chondracanthid with three tagmata (B). Tagma 1, cephalosome or cephalothorax, depending on number of thoracic segments fused to cephalon; tagma 2, 'trunk' derived from fused and expanded thoracic segments; tagma 3, genito-abdomen. Modified from Kabata, 1979.

Very little is known about food and feeding of chondracanthids. Kabata (1982) refers to pre-digestion by the secretion of digestive fluids, but no further details or observations are available, except to note that hypertrophy of host tissue around the feeding/attachment site is a common feature. At the attachment site of *Chondracanthus zei* in the gill cavity of the dory, a so-called tumour of attachment completely encapsulates the cephalosome, presumably rendering the antennal hooks redundant and preventing any change of location by the parasite.

12.4.2 *Where is the male?*

In ergasilids, matings take place between free-living males and females and only the mated females become parasitic, the males presumably dying. Since all chondracanthids attached to the host are females, it is natural to expect them to have a similar life cycle to that of their ergasilid relatives. However, it is only by taking a close look at the parasitic female that we discover that there is a fundamental difference between ergasilid and chondracanthid reproductive biology.

It is easy to overlook the small swelling in the genital region of the female chondracanthid or to dismiss it as an anatomical feature of the female. This tiny, almost spherical structure is in fact the male chondracanthid permanently attached to the female (Figure 12.15). This strange creature is virtually all head, since the cephalosome is enlarged and attached to a substantially reduced trunk (Figure 12.15B). Typical chondracanthid appendages are recognisable, comprising first and second antennae, mandibles, first and second maxillae, maxillipeds and two pairs of reduced legs.

There is a similarity here with oceanic bathypelagic angler fishes such as *Ceratias holboelli*, which is occasionally caught on the deeper fishing grounds of the north-east Atlantic (Wheeler, 1978). Female anglers reach 120 cm in length, but the dwarf males are only 4 – 6 cm in length and attach themselves as sub-adults by their jaws to the female. According to Wheeler (1969, 1978) the vascular systems of the females and males unite and the male's gut degenerates, so that the male is nourished by the female and is truly parasitic. Male chondracanthids have a gut, but this is closed at the posterior end (Turner & Wilson, 1862; El Saby, 1930; Rousset *et al.*, 1978). The dwarf male is not fused to the body of the female. Turner & Wilson (1862) observed that the diminutive male of *Chondracanthus lophii* is attached to the female by its hooked second antennae. They never found more than one male attached to the female and noted that the male is attached to one of two specialised papillae located on the genital segment of the female close to the two separate, lateral, slit-like genital apertures. They observed that each of the nipple-like papillae consists of soft tissue lacking the general cuticular covering of the female and speculated that the male may extract nutriment from these papillae. More than a hundred years later, Rousset *et al.* (1978) made similar observations on *C. angustatus*. They found only one male per female attached to one of two papillae on the genital segment of the female.

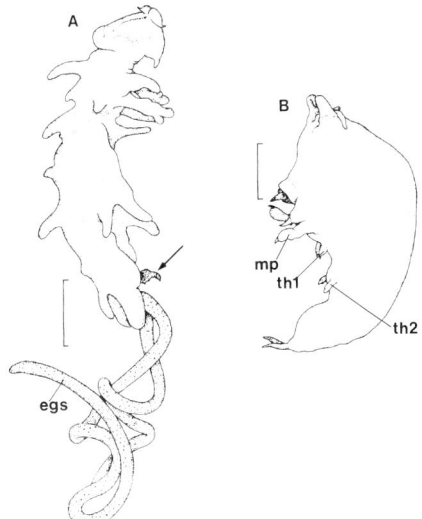

Fig. 12.15. Fully mature female (A) and male (B) of *Chondracanthus lophii*, both in lateral view. Arrow, male attached to female. egs, Egg sac; mp, maxilliped; th1, th2, first and second thoracopods. Scale bars: (A) 1 mm; (B) 200 µm. From Kabata, 1979, with kind permission of The Ray Society.

According to Rousset *et al.* (1978), each of the two testes of *C. angustatus* communicates via a vas deferens with a seminal vesicle. Each of the two seminal vesicles has a separate genital opening and provides temporary storage for a spermatophore. The two genital openings of the female lead to a single median seminal receptacle where sperms are held until required for fertilisation. Turner & Wilson (1862) thought it likely that the male could change its position from one side to the other and impregnate the female via both of her genital openings, but this is hard to confirm.

12.4.3 *The life cycle*

We have a little information about the life history of *Acanthochondria cornuta* from the work of Heegaard (1947). He collected parasites from the gill cavity of the long rough dab, *Hippoglossoides platessoides*, and hatched eggs from these parasites. The nauplius was found to be about 180 μm in length with an internal mass of yolk but no alimentary canal. Heegaard obtained only two copepodids from huge numbers of cultured nauplii, but he observed many nauplii containing apparently fully developed copepodids that failed to escape from the n025pliar skin. This led him to believe that *A. cornuta* has only one nauplius stage.

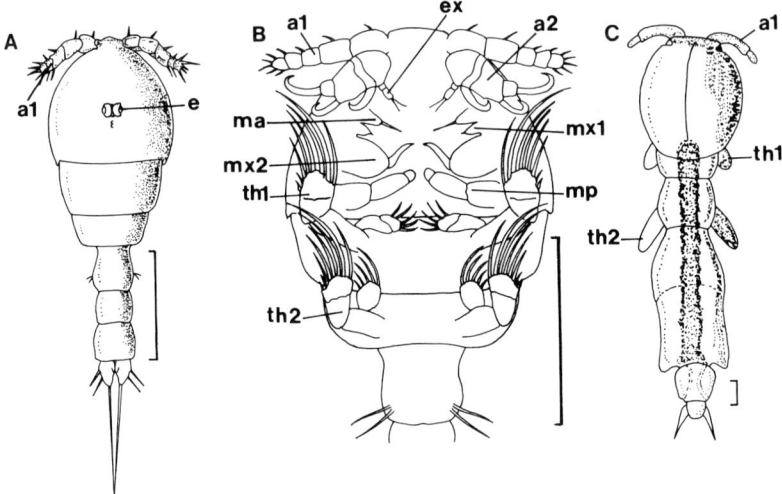

Fig. 12.16. Developmental stages of *Acanthochondria cornuta*. Copepodid I, (A) in dorsal view, (B) anterior region in ventral view. (C) Young female at the stage when male becomes attached. a1, a2, First and second antennae; e, eye; ex, exopod; ma, mandible; mp, maxilliped; mx1, mx2, first and second maxillae; th1, th2, first and second thoracopods. Scale bars: 100 μm. From Heegaard, 1947; (B) modified.

Heegaard's sketches of the first copepodid are shown with some modifications in Figure 12.16AB. He noted the well-developed first and second pairs of thoracopods equipped with plumose setae, which served to drive the copepodid through the water so rapidly that Heegaard found the movement difficult to follow with the eye. He also noted a major change in the second antenna. In the nauplius this appendage is biramous with swimming setae. In the first copepodid Heegaard described a greatly elaborated endopod, with three formidable hooks (Figure 12.16B) and a relatively small exopod, comprising two podomeres. Heegaard claimed that the exopod is lost during subsequent development, as are the two proximal hooks on the endopod. In his view it is the terminal hook on the distal podomere that persists and continues to grow, becoming the adult antennal hook. According to Kabata (1979) this interpretation needs confirming.

Heegaard (1947) claimed that most of the yolk had been used up at this stage and, since he believed that the mouthparts were "already shaped as for the later parasitic life", this led him to believe that the first copepodid, well equipped for attachment by means of the multiple hooks on the second antennae, was the infective stage. However, Heegaard failed to find female parasites at this early stage in the gill cavities of the long rough dab. The smallest female that he found on the fish was already 1 mm in length (length of first copepodid about 300 µm), but Heegaard pointed out that smaller females would be extremely difficult to find in the gill cavity of a large fish.

The smallest males found by Heegaard were attached not to the fish but to young parasitic females measuring about 1.7 mm in length (Figure 12.16C). He regarded these males as second copepodids. They were little different from free-swimming first copepodids, an observation compatible with the notion that the first copepodid is the infective stage. Thus the male in this chondracanthid may have no contact with the fish at any time during its life.

In spite of the association of males at the second copepodid stage with females, Heegaard was of the opinion that insemination of the females did not take place until the male had passed through four more copepodid stages to reach adulthood.

The copepodid stages are, of course, punctuated by moults and Heegaard (1947) appreciated the dangers that beset the male during the moulting process. The male must at all times maintain secure attachment to the female, because a detached male would be rapidly swept out of the gill chamber by the strong gill-ventilating current. The male uses the hooks on its second antennae to pin itself to the genital segment of the female. Moulting begins with a dorsal, longitudinal split in the carapace, permitting the anterior region of the male to fall backwards out of the old skin. However, the rest of the animal remains within the old skin, which is still attached to the female by the now empty hooks of the second antennae. According to Heegaard, the male now remains still, suspended in this way, for about 30 minutes, during which time the new skin hardens. When the hooks of the second antennae are hard and the muscles are again able to function, the male begins cautious movement until its hooks make contact with the female. Once its hooks are buried in the female, the male is in a position to slough off the rest of the old skin, thereby completing the moult without, at any time, abandoning its tight hold on the female. If because of its contortions the male reattaches to a site some distance from the genital segment of the female, the male is capable of returning to its original site using cautious movements of the second antennae and the maxillipeds (Heegaard, 1947).

Many other questions remain unanswered about the reproductive biology of chondracanthids. For example: how many spermatophores does the female require to maximise her output of viable eggs; can the male contribute more than two spermatophores to the female; can a new male replace a spent male?

12.4.4 *A mesoparasitic chondracanthid* – Lernentoma asellina

Lernentoma asellina is an uncommon parasite in the gill chambers of fishes of the gurnard family (Triglidae) in British waters. In this parasite anchorage in host tissue has developed to an astonishing degree. The middle of the cephalosome has become elongated (Figure 12.17) and the isolated antennal region is deeply embedded in host tissue. Strangely, the oral region and associated mouthparts remain just outside the host's tissue, close to the host's epithelial surface. It is hard to imagine what evolutionary advantages drove this extraordinary development. It seems unnecessarily elaborate for attachment, and the superficial position of the mouth denies access to deep-seated blood vessels and other tissues. However, the greatly elongate 'neck' has a

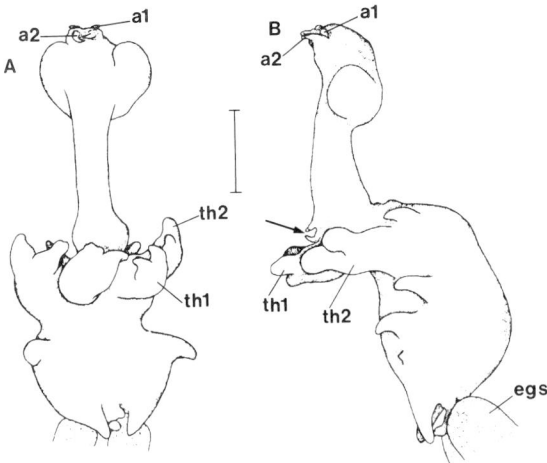

Fig. 12.17. The mesoparasitic chondracanthid *Lernentoma asellina*, (A) ventral view and (B) lateral view. Arrow, position of mouth and mouthparts. a1, a2, First and second antennae; egs, egg sac; th1, th2, first and second thoracopods. Scale bar: 1 mm. From Kabata, 1979, with kind permission of The Ray Society.

relatively large surface area in contact with host tissue and if the integument covering this 'neck' is permeable, this unique mesoparasite may be able to absorb nutrients directly from the host.

12.5 PHILICHTHYIDS

The inclusion of the philichthyids in an account of external parasites of fishes needs some justification because of the virtually endoparasitic life style of these unique copepods. The maggot-like females accompanied by similar but smaller males inhabit mucus ducts or lateral line canals (see Chapter 1) near the surfaces of their fish hosts. There are three reasons to mention them briefly here: (a) they are usually detectable externally because they produce local surface swellings; (b) unlike other endoparasites, they permanently retain a connection with the outside world via the pore through which they entered the fish; (c) they represent an evolutionary extreme in the poecilostomatoids, with the unspecialised ergasilids at the other end of the spectrum, and to omit them would give an incomplete and unsatisfactory picture of poecilostomatoid evolution.

There is one record of *Philichthys xiphiae* in Britain. Its host, the swordfish *Xiphias gladius*, is rarely encountered, but according to Kabata (1979) two parasites were collected by Dr S.F. Harmer from the frontal bones of a swordfish taken off Lowestoft in 1892 and deposited in the Cambridge Museum (Harmer & Shipley, 1909). These specimens were the subject of a study by Scott & Scott (1913). However, Kabata (1979) proposed that British waters might be a profitable hunting ground for other philichthyids. He pointed out that the ballan wrasse, *Labrus bergylta*, which elsewhere is infected with the philichthyid *Colobomatus bergyltae*, occurs around the British coast, as does also the corkwing wrasse, *Crenilabrus melops*, which has been reported to harbour the philichthyid *Leposphilus labrei* (see below) off the coast of Donegal in Ireland (Donnelly & Reynolds, 1994). Philichthyids belonging to *Sarcotaces* have also been reported from two deep-sea bottom-dwelling fishes, the North Atlantic codling, *Lepidion eques*, and the swordsnout grenadier, *Coelorinchus occa*, frequenting the continental slope to the west of the British Isles (Bullock *et al.*, 1986).

The female *P. xiphiae* is 9 – 12 mm long and was described by Scott & Scott (1913) as pale purple in colour. The male is much smaller, being about 2.5 mm in length, and keeps the female company inside the burrow. The female is elongated and indistinctly segmented, with numerous projecting processes on the body (Figure 12.18AB). Limbs are reduced or absent. The mouth is of poecilostomatoid type (see Kabata, 1979), but details of the mouthparts are lacking. The female is recognisable as a copepod when ovigerous, because it possesses a pair of typical egg sacs. However, an atypical feature is that the egg sacs are attached to the genital orifices at mid-length and do not project beyond the posterior end of the animal (Figure 12.18AB).

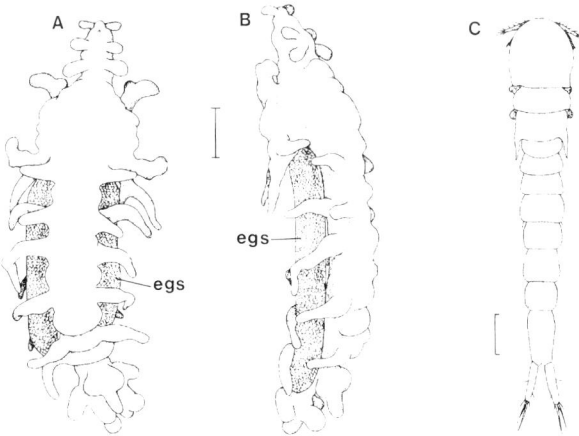

Fig. 12.18. The endoparasitic chondracanthid *Philichthys xiphiae*. (AB) Female in ventral and lateral views respectively. (C) Male in dorsal view. egs, Egg sac. Scale bars: (AB) 1 mm; (C) 250 µm. From Kabata, 1979, with kind permission of The Ray Society.

The small male (Figure 12.18C) is not highly modified like the female, having distinct segmentation and recognisable limbs, including first and second antennae and two pairs of biramous swimming legs. The second leg-bearing segment has two spines, each of which projects from the postero-lateral corner of the segment in a postero-dorsal direction (Figure 12.18C). According to Kabata (1979), these spines may aid the male in maintaining its position in the burrow alongside the female.

The female of *Leposphilus labrei* is about 2 mm long. It does not possess body processes and resembles a maggot much more closely (Figure 12.19A). Typically, the parasite inhabits the lateral line canals and has been recorded from a variety of labrids off the coast of France (see Quignard, 1968), as well as from corkwing wrasse in

Ireland. Donnelly & Reynolds (1994) found that over 30% of Irish corkwing wrasse were infected, the parasites betraying their presence by creating swellings. Almost all of their large sample of nearly 2000 infected fishes had only one female parasite, on either the left or the right side. On just two occasions they found two parasites, but these occurred one on each side of the fish, not in the same burrow.

Quignard (1968) found *Leposphilus labrei* in the lateral line canals of the head of the rock cook (*Centrolabrus exoletus*) and rainbow wrasse (*Coris julis*). This discovery was important since the parasitological cause of the swellings on the head of these fishes (Figure 12.19B) had, up to that time, not been suspected, the protuberances being regarded as normal features of older fishes or features associated with the sexual cycle of the fish.

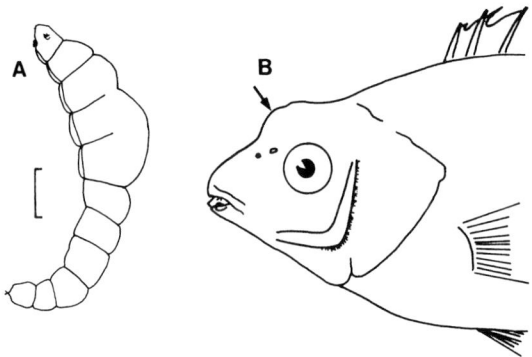

Fig. 12.19. The endoparasitic chondracanthid *Leposphilus labrei*. (A) Whole female from the cephalic canals of the rock cook. (B) Head of rock cook with frontal swelling (arrow) produced by *L. labrei*. Scale bar for (A) 1 mm. From Quignard, 1968.

Sarcotaces sp. inhabits cyst-like swellings beneath the skin. Bullock *et al.* (1986) found that the swellings were generally on the snout of the North Atlantic codling and on the flank near the vent in the swordsnout grenadier. The swollen females contained inky black fluid, probably derived from a diet of host blood, hence the name 'iodine worms' for these organisms (Boxshall & Halsey, 2004). Males were not found.

Izawa (1973) has made some observations on the life history of the Japanese philichthyid *Sarcotaces pacificus* from the frogfish *Antennarius tridens* (= *A. striatus*). Izawa found many eggs and first nauplii inside the burrow. He assumed that the nauplii escape into the water where they pass through a further four nauplius stages to reach the first copepodid stage. According to Izawa, the nauplii are unable to feed, since their appendages have no projecting medial surfaces (gnathobases) that could be used for this purpose and no masticatory setae. Presumably the nauplii rely on yolk from the egg for their development and survival. The first copepodid is said to have no mouthparts but can persist on contained yolk for more than a week. The first copepodid is able to swim and was regarded by Izawa as the infective stage, although the results of his attempt to infect a host with these copepodids were by no means conclusive. If Izawa is correct then, at least in *Sarcotaces*, none of the free-living stages feed and all of their development and activity is fuelled by yolk supplied by the parent to the egg.

13

THE COMMON FISH LOUSE – *ARGULUS*

"The *Argulus foliaceus* is an exceedingly pretty and graceful little animal; and as it can leave the fish on which it feeds, and swim freely in the water, there are many opportunities for watching its gambols through its native element".

W. Baird (1850)

13.1 INTRODUCTION

Argulus is perhaps one of the most interesting and enigmatic fish ectoparasites that we are likely to encounter (Figure 13.1). It is common on the skin and may be found on flat non-respiratory surfaces inside the gill cavity. *Argulus foliaceus* is large enough to be a familiar sight for most freshwater fishermen and has been the subject of many studies since Linnaeus named it as *Monoculus foliaceus* in 1758 (see Baird, 1850). In spite of all this attention, it has proved extremely difficult to obtain answers to the following simple and basic questions about its biology: "how does the parasite feed?"; "what does it feed on?"; "how is sperm transfer achieved?"; "what are its nearest relatives?".

It is safe to say that there is still no consensus about the nature of its food and an added complication is uncertainty about the role of a unique stylet, like a hypodermic needle, located anterior to the mouth but independent of the gut (Figure 13.1). The absence of a penis and spermatophores in the male is a challenge that has led to very different ideas on the mechanics of mating, culminating in recent discoveries that are surprising and possibly without precedent in animals.

Baird (1850) regarded *Argulus* as a copepod and Scott & Scott (1913) included it, although with less confidence, in their book on the British parasitic Copepoda. Certainly at a casual glance the flattened disk-shaped body of *Argulus* (Figure 13.1) looks very much like that of a caligid copepod and there is a resemblance too between the mouth tubes of these parasites. However, if we take the trouble to look a little closer, many differences become apparent. These have led to the elevation of *Argulus* and its relatives to the status of a sub-class, known as the Branchiura, which, along with the sub-class Copepoda is currently included in the crustacean class Maxillopoda (see Appendix 2 and also below, p. 263).

The disk-shaped carapace makes little if any suctorial contribution to attachment, unlike the carapace of the caligids. Nevertheless, suction is important for *Argulus* and its most conspicuous and unique feature, not shared with any copepod, is a pair of powerful stalked suckers located beneath the carapace, one on each side of the body (Figure 13.1BD). According to Fryer (1982), these may be the most elaborate suckers in the Animal Kingdom. It is hard to believe that these sophisticated organs are modified first maxillae (maxillules), but a study of the developmental stages of the parasite provides convincing evidence that this is so (see below).

Unlike copepods *Argulus* does not retain its fertilised eggs in strings or sacs attached externally to the body. *Argulus* leaves the host after insemination and attaches its eggs in groups to hard surfaces such as stones.

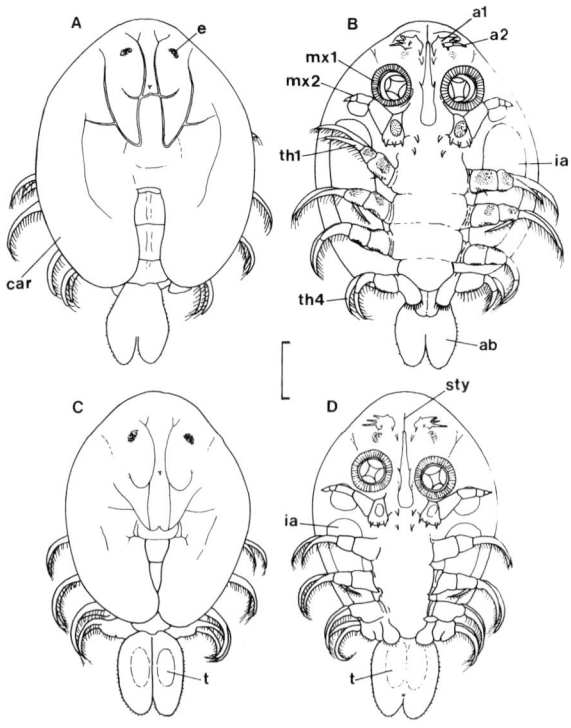

Fig. 13.1. The common fish louse, *Argulus foliaceus*. (AB) Female, in dorsal and ventral view respectively. (CD) Male, in dorsal and ventral view respectively. a1, a2, First and second antennae; ab, abdomen; car, carapace; e, eye; ia, inner ventral area of carapace lacking spines; mx1, mx2, first and second maxillae; sty, stylet; t, testis; th1, th4, first and fourth thoracopods. Variations in carapace length may leave the bases of the fourth pair of thoracopods exposed and visible from above. Scale bar: 1 mm. Reproduced from Fryer, 1982, with permission of The Freshwater Biological Association and the author.

Males are generally smaller than females and this led Baker (quoted by Jurine, 1806) to regard the small and large individuals as belonging to two different species. However, unlike copepods, branchiurans moult frequently after attaining sexual maturity and, because of this, mature individuals of any species range widely in size. Males are readily identified by their second, third and fourth pairs of swimming legs (thoracopods) which have basal modifications used in mating and also by the outlines of the testes in the abdominal lobes (Figure 13.1CD).

13.2 GENERAL MORPHOLOGY

The body of *Argulus* comprises three regions. The first of these, the cephalon, gives rise to the shield-shaped carapace and bears the following appendages: relatively small first and second antennae; mandibles enclosed in the mouth tube; the first maxillae (attachment suckers); leg-like second maxillae (Figure 13.1BD). The second region is the thorax, consisting of four separate segments, the first of which is partly fused with the cephalon. Each thoracic segment bears a single pair of biramous swimming legs (thoracopods), the first two pairs of which in both sexes have a backwardly projecting process or flabellum. The third region, the abdomen, projects in a posterior direction beyond the carapace, is small, bilobed, unsegmented and lacks appendages.

The carapace is roughly circular in shape but has a deep median notch in its posterior margin, revealing the segmented thorax (Figure 13.1AC). The carapace is partly fused dorsally to the first thoracic segment (Martin, 1932) and extends laterally to form thin 'wings' (Figure 13.2). In *A. appendiculatus*, Sutherland & Wittrock (1986) used the scanning electron microscope to survey the ventral surfaces of these 'wings' and observed that their ventral margins have backwardly directed spines (see also below). These spiny areas are sharply demarcated from the non-spiny inner areas (Figure 13.1BD), to which a respiratory function has been attributed. Evidence in support of a respiratory role for these non-spiny areas is the presence of a dendritic pattern of blood sinuses separated from the external aqueous environment by only a single layer of cells (Martin, 1932; Sutherland & Wittrock, 1986). Martin demonstrated that, after treatment with gold chloride and formic acid, reduced gold penetrated these non-spiny areas, showing that they are thinner and more permeable than other regions of the surface.

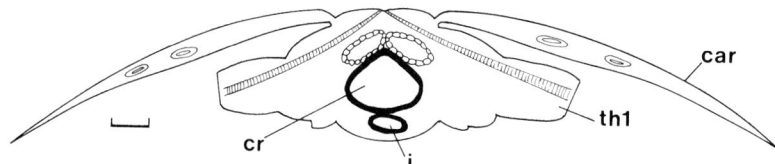

Fig. 13.2. Transverse section through the thorax of *Argulus*. car, Carapace; cr, crop; i, intestine; th1, first thoracopod. Scale bar: 1 mm. Redrawn from Martin, 1932.

13.3 BRITISH BRANCHIURANS

In Britain we have three common freshwater species of *Argulus*. Two of these, namely *A. foliaceus* (Figure 13.1) and *A. coregoni* are native and the other, *A. japonicus*, has been introduced. Fryer (1982) included *A. japonicus* in his booklet on the parasitic Copepoda and Branchiura of British freshwater fishes, even though this parasite had not been reported in Britain at the time of printing. Fryer included the species because it had already spread through continental Europe from the Far East, probably originating from carp and goldfish in China. He suspected that it would appear sooner or later in Britain and this prediction proved correct. Kennedy (1975b) refers to a report of the North American *A. appendiculatus* on chub (*Leuciscus cephalus*). This parasite was probably introduced on large-mouth bass (*Micropterus salmoides*).

The notes that follow on the British freshwater argulids are from Campbell (1971) and from Fryer (1982). The three common species can be distinguished by differences in the sexual ornamentation on the swimming legs (thoracopods) of the males, but there are easier ways to tell them apart. *A. coregoni* can be distinguished from its British relatives by the shape of the abdominal lobes, which are pointed not rounded (Figure 13.3B). *A. japonicus* has rounded abdominal lobes like *A. foliaceus*, but differs from the latter in the depth of the incision between the abdominal lobes of the female (Figure 13.3C). This incision is more than half the abdominal length in *A. japonicus* females, while in *A. foliaceus* the incision is less than half the abdominal length in both sexes (Figure 13.3A). Also *A. foliaceus*, but not *A. japonicus*, has darkly pigmented areas on the first, second and third thoracopods.

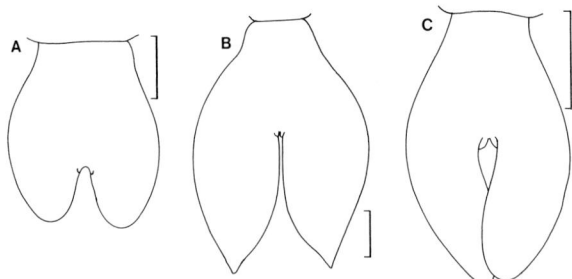

Fig. 13.3. Abdomens of female argulids. (A) *Argulus foliaceus*; (B) *A. coregoni*; (C) *A. japonicus*. Scale bars: approximately 0.5 mm. Reproduced from Fryer, 1982, with permission of The Freshwater Biological Association and the author.

A. foliaceus can parasitise any member of our freshwater fish fauna and is widely distributed throughout the British Isles south of the Central Highlands of Scotland and excluding the Scottish islands. Records indicate that it prefers lowland habitats and 'richer' waters rather than clear rivers and oligotrophic lakes (lakes with a poor nutrient supply), though it is known from weakly acidic waters. It was not reported in the Lake District of England until 1971 when it appeared in Esthwaite Water. It was recorded in Blelham Tarn in 1973 and was first seen in Windermere in 1975. Early in 1976 it was found on spawning Arctic charr, *Salvelinus alpinus*. This fish had not previously been parasitised, but later in 1976 fishermen found it to be numerous, with 18 out of 20 fishes caught by one fisherman being infected. It was also reported on perch in 1976. Presumably its appearance in the Lake District was the result of introduction of infected fishes and/or the liberation of live bait, and its establishment may have been favoured by the fact that the lakes mentioned above are three of the richer (more eutrophic) lakes in this district.

Female *A. foliaceus* are up to 10 mm in length and males up to 8 mm. Fryer (1982) reported that *A. foliaceus* can be found anywhere on the body surface and even inside the mouth; there seems to be a preference for fins and fin bases of scaly fishes and the flanks of smooth-skinned fishes. It is a pest in fishponds and there are reports of death and damage to hosts. Fryer mentions a European report of 4250 parasites collected from a single 28-cm tench (*Tinca tinca*). *A. foliaceus* is common and widespread, extending through much of Europe and across central Asia to Siberia.

Argulus coregoni females reach 13 mm in length and males 12 mm. It is less common and has been reported from fewer hosts than *A. foliaceus*. There are scattered reports of the parasite throughout England, Wales and Scotland, south of the Central Highlands, with an Irish record from Lough Erne. There are several records from brown trout (*Salmo trutta*) and grayling (*Thymallus thymallus*), indicating a liking for clear and sometimes briskly flowing water, but this is by no means a hard and fast rule and Fryer mentions its occurrence on bream (*Abramis brama*) in the River Yare in Norfolk. According to Fryer, the presence of the parasite in the oligotrophic Llyn Tegid (Bala Lake) in Wales may emphasise the differences between the ecological preferences of *Argulus coregoni* and those of *A. foliaceus*. *A. coregoni* is widespread in northern Europe and extends eastwards to the Amur basin.

A. japonicus females reach about 9 mm in length and males about 7 mm. According to Fryer (1982) its distribution in Europe hints at a preference for warmer and perhaps more alkaline waters than those favoured by *A. foliaceus*. *A. japonicus* was probably accidentally imported into Britain on koi (*Cyprinus carpio*) used to stock ornamental ponds and lakes. It was first recorded in February 1990 on koi in a private pond in Kent and at the same time on mirror carp (a variety of *Cyprinus carpio*) in Hereford (Rushton-Mellor, 1992). In April 1991 *A. japonicus* was collected from koi from a private pond in Hampshire and from native fish populations, including rudd (*Scardinius erythrophthalmus*), roach (*Rutilus rutilus*) and common carp (*Cyprinus carpio*) in Kent and in Dorset.

In addition to the three British freshwater species of *Argulus*, Rushton-Mellor (1992) reported a marine species, namely *Argulus arcassonensis* Cuénot 1912, but she did not mention the host or the locality where the parasite was found. However, Professor G.A. Boxshall (personal communication) has provided additional information about this interesting record. The specimen of *A. arcassonensis* was found by an amateur collector on a large triggerfish, *Balistes carolinensis*, caught in a lobster pot at Falmouth Bay, Cornwall in September 1985. The specimen was sent to Professor Boxshall for identification and then returned to the collector. These extra details were not published and there have been no further records of *A. arcassonensis* in British waters.

A. arcassonensis was first described by Cuénot (1912) in the Bay of Arcachon on the Atlantic coast of France. Parasites were collected from several fishes including the red gurnard (*Chelidonichthys cuculus*) and the grey triggerfish (*B. carolinensis*). While comparing Cuénot's type specimens of *A. arcassonensis* with *Argulus* collected off the west coast of Africa, Monod (1928) found two scales of a fish identified as red mullet, *Mullus barbatus*, in a tube with the type specimens. These scales carried large numbers of eggs, presumed to be from *A. arcassonensis*, raising the possibility that this parasite attaches its eggs directly to the host fish. However, these eggs were on the internal surfaces of the scales and concentrated at the proximal ends. Since this region of the scale is buried in the dermis when the scale is *in situ*, it seems likely that these eggs were attached after the fish shed the scale.

13.4 ATTACHMENT

The conspicuous ventrally directed suckers (modified first maxillae) are the principal organs of attachment in *Argulus*. Wilson (1902) likened them to the suckers on the arms of octopus and squid and he described their alternate use in a kind of walking motion, which enabled the parasite to scuttle rapidly over the host's skin, so long as it remained moist. Their effectiveness as attachment organs is emphasised by Bower-Shore (1940) who observed that a fast-running stream of water in the laboratory did not dislodge the parasites.

Wilson was aware that the suckers were not the only means of attachment, being supplemented by a variety of hooks and spines. In *A. foliaceus* each first antenna (antennule) has a prominent hook (Figure 13.4A). There are two pairs of large spines pointing in a posterior direction near the bases of the first and second antennae (Figure 13.4A) and two other pairs of spines with a similar orientation lie between the bases of the second maxillae (Figure 13.1B). Each leg-like second maxilla has hooks at the tip, spiny papillae along the side and three stout, backwardly directed spines on the basal joint (Figures 13.1BD, 13.4B). The undersides of the anterior margin and edges of the carapace are studded with short, backwardly directed spines.

Wilson (1902) noted that *Argulus* attaches itself with its longitudinal axis parallel with that of its host and with its head anterior with respect to the host. He pointed out that parasites would be subjected to strong water currents, especially in the gill cavity and on the surface of a rapidly swimming fish. By facing into the current, the parasite ensures that the backwardly directed spines and hooks engage in the host's skin and prevent any sliding backward. However, Wilson believed that the main function of the three basal spines of the second maxilla (Figure 13.4B) was to provide a firm brace when the stylet was thrust into the host's skin, the antennular hooks at the same time pinning down the anterior edge of the carapace.

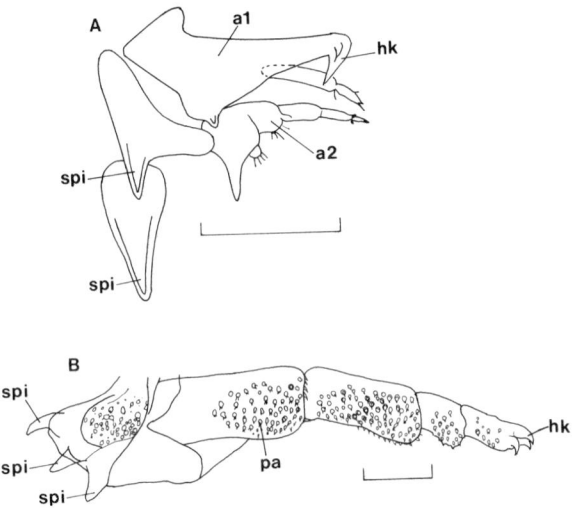

Fig. 13.4. (A) The first and second antennae and (B) the second maxilla of *Argulus foliaceus*. Appendages of the left side seen in ventral view. a1, a2, First and second antennae; hk, hook; pa, spiny papillae; spi, spine. Scale bars: 200 µm. Redrawn from Stammer, 1959.

Gresty *et al.* (1993) studied the musculature and mode of action of the suckers of *A. japonicus*. Each sucker is a relatively deep cup supported by a stalk (Figure 13.5). Distally the cup has a flattened rim, which is seen to taper in cross-section creating a marginal flap, which undoubtedly functions as a valve to prevent the influx of water when suction is generated. This marginal valve is supported by closely spaced, radial, rib-like sclerites. Thin cuticle covers the outside of the stalk and the cup, but the cuticular lining of the cup is thicker and sclerotised, providing the rigidity without which the cup would collapse when suction is generated. This thick inner cuticular lining consists of two separate hoops, which are joined together by a ring of thin cuticle. The hoops overlap slightly as shown in Figure 13.5 inset. This arrangement permits a little telescopic movement between the distal hoop with its attached marginal valve and the proximal hoop.

The similarities between the suckers of *Argulus* and the suction clamps of polyopisthocotylean monogeneans like *Diclidophora* spp. (see Chapter 6) provide a striking example of convergent evolution. The floor of the suction cup of *Argulus* has a central plate of thickened cuticle suspended by a surrounding area of thin flexible cuticle (Figure 13.5). Five powerful extrinsic muscles are attached to this plate and at the other end are inserted on the rigid exoskeleton of the animal. The course of one of these muscles, which has an elongated origin on a thickened dorsal ridge on the inner surface of the carapace, is shown in Figure 13.5. Contraction of these muscles will lift the central plate in the floor of the cup and, provided that the rim of the cup is watertight and sealed by the marginal valve, the pressure within the cup will fall, i.e. strong suction will be generated.

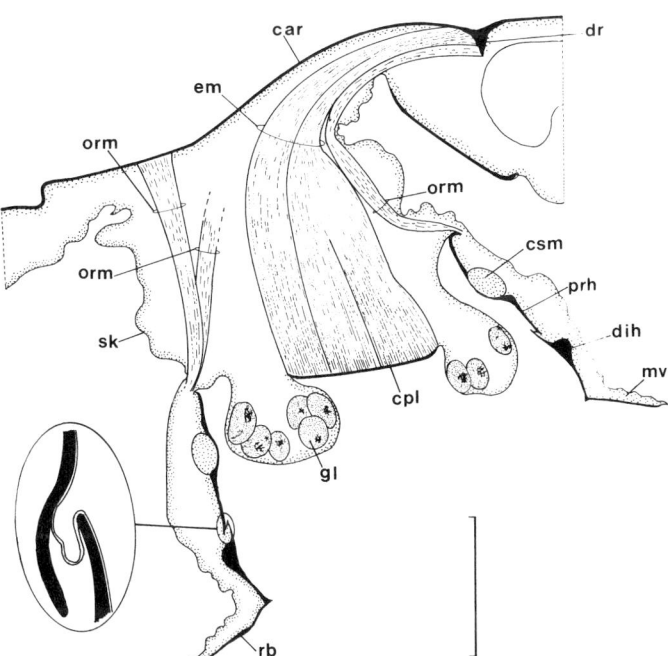

Fig. 13.5. Diagrammatic transverse section through a maxillary sucker of *Argulus japonicus*. Thickened cuticle is indicated by solid black. Inset: detail of region of overlap between proximal and distal hoops of thickened cuticle lining the wall of the suction cup. car, Carapace; cpl, central plate; csm, circular sucker muscle; dih, distal hoop of sucker; dr, dorsal cuticular ridge; em, extrinsic adductor muscle; gl, gland cell; mv, marginal valve; orm, orientation muscle; prh, proximal hoop of sucker; rb, supporting rib; sk, sucker stalk. Scale bar: 100 µm. Redrawn from Gresty *et al.*, 1993.

There are some further refinements. A second set of muscles originating proximally on the exoskeleton is attached distally around the circumference of the base of the sucker or in the sucker wall (Figure 13.5). Differential contraction of these muscles serves to orientate the sucker. Each sucker is able to move independently of the other and by co-ordinating the movements of the two suckers the parasite is able to move relatively quickly by a kind of walking locomotion over the host's body (Bower-Shore, 1940; Gresty *et al.*, 1993). By contracting all these orientation muscles at the same time the sucker stalks shorten and the attached parasite pulls itself down onto the surface of the host. The flexibility and limited telescopic movement of the stalk permit the sucker rim to make small adjustments to irregularities in the host surface, ensuring a watertight fit.

Gresty *et al.* also discovered abundant gland cells, situated close to the thin cuticular floor of the suction cup. Strong staining for mucopolysaccharide using the Alcian blue – PAS technique, indicated that they produce mucus. This mucus may enhance the seal between the marginal valve and the host's skin, but this has not yet been demonstrated and Gresty *et al.* failed to find gland openings through which the secretion could be released.

The sucker may be freed partly by relaxation of the extrinsic suction muscles and partly by equalising the internal and external pressures by contraction of a circular muscle running round the base of the suction cup (Figure 13.5).

13.5 HOST FINDING

Argulus is a temporary ectoparasite attaching its eggs not to the fish but to the substrate and visiting a fish when food is required. Parasites can survive without a host for less than two days at the larval stage and usually for no more than one week when adult (Mikheev *et al.*, 2000). Thus, finding a suitable host quickly and efficiently may be crucial for the survival and successful reproduction of the organism. Completely at odds with this concept is the conclusion reached by Herter (1927, in Mikheev *et al.*, 1998), that *A. foliaceus* encounters a host fish by random movement. He claimed that neither optical nor chemical stimulus elicits any directed response from *Argulus* towards a distant swimming fish, although at close range he believed that the parasite responded to water movements generated by the fish and to tactile cues.

In spite of the fact that the lateral eyes are the most highly developed sense organs in *Argulus* (see Madsen, 1964), Herter (1927, in Mikheev *et al.*, 1998) attributed to them the relatively minor role of seeking out suitably illuminated areas. However, Kollatsch (1959) claimed that *A. foliaceus* responds to sudden changes of illumination by moving rapidly, but he regarded the movements as non-directional and not related to host finding. In contrast, Bohn (1910, in Mikheev *et al.*, 1998) reported a directional response to a fish as a result of stimulation by the fish's shadow.

This rather confused and untidy picture of host finding in *Argulus* persisted until 1998 when Mikheev *et al.* conducted a thorough re-investigation. Their findings are of exceptional interest, not only because they revealed that, far from being random, host finding in *Argulus* relies on a sophisticated and flexible behaviour pattern, but they also shed some light on how earlier observers may have been misled.

The breakthrough in understanding host finding in *Argulus* came as a result of pilot studies by Mikheev *et al.* (1998) in which they introduced various materials into aquaria containing parasites. They observed a strong positive response on the part of *A. foliaceus* to highly reflective objects plunged into the aquarium, such as the blade of a stainless steel knife. This response took the form of a rapid swimming movement towards the reflective object from distances up to 10 – 15 cm. This led them to question the validity of observations made on *Argulus* by the earlier workers, who kept their animals in standard glass aquaria where they would be subjected to reflections from extraneous sources and from the internal surfaces of the aquarium, especially in the corners. Mikheev *et al.* suspected that *Argulus* might perceive these changes in illumination as spurious visual targets, leading to confusion, sensory overload and loss of host-finding efficiency.

Mikheev *et al.* (1998) confirmed these suspicions experimentally. They found that juvenile perch (*Perca fluviatilis*) and roach (*Rutilus rutilus*) acquired fewer parasites in a highly reflective glass aquarium than in an aquarium lined with black plastic to reduce reflectivity (Figure 13.6AB). They also compared the activity of single specimens of *A. foliaceus* in the absence of potential hosts in highly reflective glass aquaria and lined tanks with low reflectivity. The behaviour of the parasite in these two situations was strikingly different. In the black tank with reduced reflectivity the parasite was most frequently observed hovering in the water column, gliding slowly or

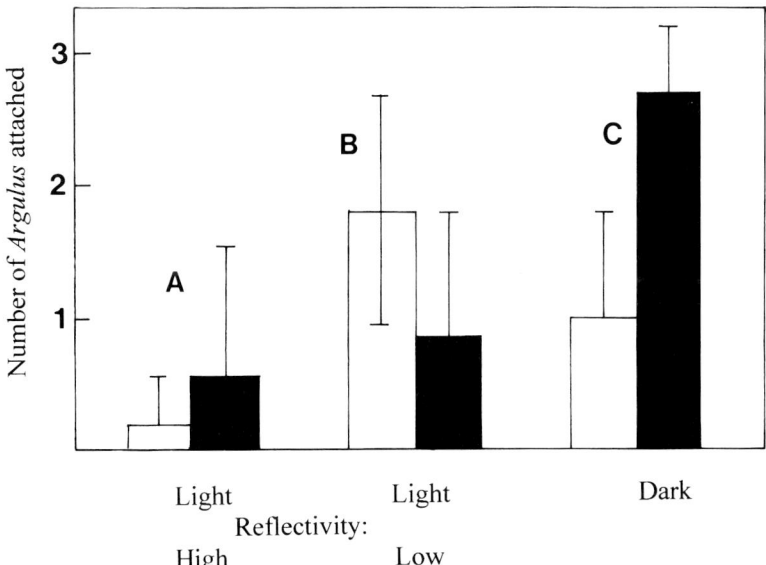

Fig. 13.6. Vulnerability of roach, *Rutilus rutilus* (white columns) and perch, *Perca fluviatilis* (black columns) to *Argulus foliaceus* in different optical conditions. Redrawn from Mikheev *et al.*, 1998.

with no obvious movement. This behaviour persisted for periods ranging from 10 seconds to several minutes, followed by movement to a new hover station. In the reflective glass tank hovering was rare. The average swimming speed in the reflective tank was 4.4 times faster on average than in the black tank and, compared with those in the black tank, parasites in the reflective tank were 5.6 times more likely to be found close to the tank wall. In the reflective tank parasites swam erratically close to the walls and especially the corners and occasionally struck the glass surface.

If *Argulus* finds its host by random swimming as Herter suggested then we would expect host search success by *A. foliaceus* to be positively correlated with average swimming speed. Mikheev et al. (1998) found just the opposite. Hunting sorties from the hovering (almost motionless) attitude were most successful. In fact, actively moving parasites were not observed to advance towards a potential host even when as close as 2 – 3 cm.

In aquaria with low reflectivity, juvenile roach and perch not only acquired more parasites than in the highly reflective aquarium (cf. Figure 13.6A and B) but roach were noticeably more attractive than perch. Moreover, the parasites preferred the silvery sides and belly of the roach and the head area (anterior to the pectoral fins) of the perch, that is they preferred the more reflective parts of the fish's body. Thus, in the light, host seeking *A. foliaceus* respond to flashes of light from the bodies of passing fishes and launch an attack in the direction of the light flash from a hovering position in the water column. Consequently, they are more likely to respond to the silvery bodies of roach than to the duller bodies of perch. In fact, an attack rate four times greater than that for perch was recorded for the brighter more reflective roach juveniles.

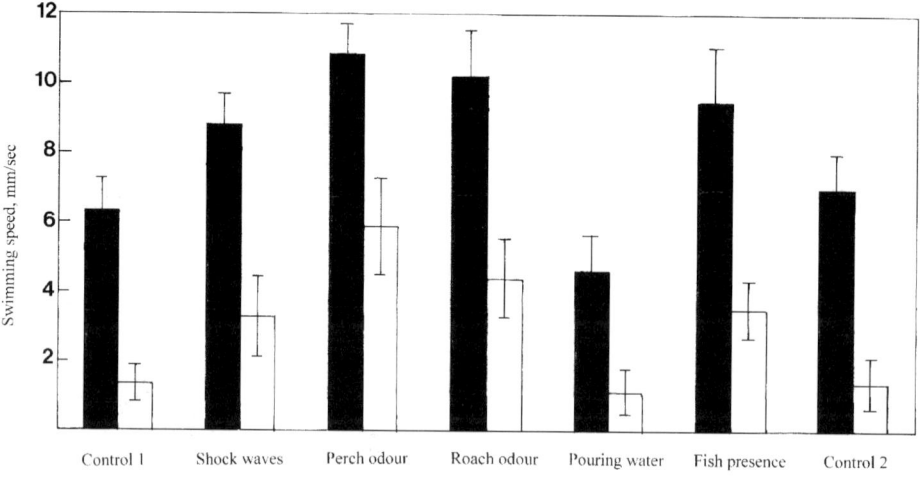

Fig. 13.7. Effects of external stimuli on the swimming activity of *Argulus foliaceus*. Black columns in darkness, white columns in the light. Shock waves were created with semi-rigid plastic tubing. Fish odours were collected by immersing four juvenile fishes for 30 minutes in 60 ml of water, which were then poured into the test aquarium. Note that pouring water without fish odour produced no significant change in swimming speed. Redrawn from Mikheev et al., 2000.

Relevant to this behaviour is the observation of Poulin & FitzGerald (1989a) that juvenile three-spined sticklebacks (*Gasterosteus aculeatus*) infected with *Argulus funduli* (incorrectly identified as *A. canadensis*; see Poulin, 1999) are more likely to acquire additional fish lice than uninfected sticklebacks. Since infected sticklebacks swim erratically, bright reflections from their bodies may be more frequent than when swimming normally, providing a greater attraction for hovering parasites and leading to an aggregated distribution of parasites among the fish population. The silvery sides of fish like the roach are thought to be an effective means of camouflage against predators in open water, provided that the body and its in-built side mirrors are vertical (see Denton, 1971), but deviations from the vertical will produce a flickering effect which may greatly increase the conspicuousness of the fish.

There is another twist to this story because Valtonen *et al.* (1997) found that perch in Finnish lakes were more heavily infected with *A. foliaceus* than roach. This is at first a surprising observation since it appears to contradict the apparent preference of *Argulus* for roach described above. However, Mikheev *et al.* (1998) found that this adaptable parasite is able to find its hosts in the dark. In fact, their highest infection levels were obtained when perch were exposed to the parasites in the dark (Figure 13.6C). These infection levels were significantly higher than for roach under the same conditions.

The elegant experiments of Mikheev *et al.* (2000) have provided a fascinating explanation for these observations and have revealed a surprising level of sophistication in the host finding behaviour of *A. foliaceus*. They discovered that *A. foliaceus* switches to a quite different host searching strategy in the dark. In the light, the parasite's behaviour can be described as a sit-and-wait (ambush) strategy, although the parasite hovers or slowly glides rather than attaching itself to the substrate. In the dark, the parasite switches to an active, widely foraging (cruising) strategy, spending most of its time actively swimming, often in long straight lines. Its swimming speed is, on average, 4.7 times higher in the dark than in the light (Figure 13.7). The only times when no significant difference was found between swimming speeds in the light and in the dark were when the parasite was fully fed (freshly detached from the host) and under conditions of starvation (five or six days after detachment from the host).

Chemical stimuli (washings from perch and roach; presence of juvenile perch) and hydrodynamic stimuli (shock waves) significantly increased swimming speeds in *A. foliaceus* both in the light and in the dark (Figure 13.7). Chemical stimuli (fish smells) are especially effective, and Galarowicz & Cochran (1991) also recorded enhanced swimming speeds when *A. japonicus* was tested with washings from a common carp (*Cyprinus carpio*).

Thus, host searching in *A. foliaceus* is a 24-hour occupation, but involves a strategic change from day to night. The sit-and-wait daytime and the active foraging night-time strategies exploit chemical and mechanical clues from the host, but visual clues are also used in the daytime. This means that during the hours of daylight the highly reflective silvery roach is more likely to attract *Argulus* than the duller perch. Why then does *Argulus* change its foraging strategy in the dark? Visual clues are of course no longer available, but Mikheev *et al.* (2000) also found that the swimming speed of juvenile perch was significantly lower in the dark than in the light and that their swimming was more intermittent. Roach did not change their swimming behaviour significantly in the dark, so at night the less active perch are easier targets than roach, provided that the parasites are prepared to invest energy in an active foraging strategy. This is exactly what *A. foliaceus* is prepared to do. In fact, the parasite has to risk high energy losses in the dark because this is the most favourable time for host finding. As Mikheev *et al.* (2000) have pointed out, the swimming activities of host and parasite complement each other on a diurnal basis – when the host is active the parasite chooses a sit-and-wait strategy and *vice versa* at night. Should a potential host advertise its presence during the day by visual, chemical or mechanical means the parasite is capable of responding by a burst of activity that may lead to successful host invasion. At night visual stimulation is not an option but chemical and mechanical stimuli from potential hosts are still available and probably make their contribution to host finding at least by inducing an increase in swimming speed.

Thus, not only variation in search strategy induced by light but also the interplay between host and parasite behaviour, determine the success of host-finding in *Argulus foliaceus* and also which host fishes are likely to be infected by the parasite.

In laboratory tests, Poulin & FitzGerald (1989bc) working in Canada studied interactions between *Argulus funduli* (incorrectly identified as *A. canadensis*) and juvenile three-spined and black-spotted sticklebacks (*Gasterosteus aculeatus* and *G. wheatlandi* respectively). *A. funduli* is potentially lethal to small sticklebacks and may also transmit bacterial diseases (see below). The fishes were 20 – 25 mm long and the parasites 3 – 5 mm long. Bearing in mind the possible constraints of experiments in tanks revealed by the work of Mikheev *et al.* (1998), the work of Poulin & FitzGerald reveals a strong behavioural interplay between this parasite and its hosts. Poulin & FitzGerald (1989b) found that the prevalence and intensity of infection of the two fish species were greater in vegetated microhabitats than in open water. *A. funduli* swam near the bottom and in the vegetation. Without parasites, both species of fish preferred to swim amongst vegetation. With parasites present, the distribution of the three-spined stickleback changed little, but the black-spotted sticklebacks spread from the vegetated microhabitats into open water. In the absence of parasites both species of fish swam near the bottom, but when parasites were present the fish swam near the surface. These observations suggest that juvenile sticklebacks are aware of the parasites and may reduce the risk of predation by changing their microhabitat. Dugatkin *et al.* (1994) also found evidence to suggest that juvenile three-spined sticklebacks avoid shoals of conspecifics in which individuals are parasitised by *A. funduli*. Parasites alone are not avoided and Dugatkin *et al.* were of the opinion that avoidance may be a response to some feature of infected fishes such as a change in their behaviour.

Poulin & FitzGerald (1989c) suspected that juvenile fishes of both species might lower their risk of parasitism by shoaling. The fish form shoals, sometimes of mixed species, when they are less than two weeks old. At this age there is little danger from predators because they are too small to be worth attacking, but they are vulnerable to the attentions of *A. funduli*. Poulin & FitzGerald (1989c) set up tanks containing parasites along with one, five, 10 or 20 juvenile sticklebacks. As shoal size increased, the number of attacks by parasites on individual fishes decreased, but attack rate and attack success by the parasites were not reduced by increased shoal size. Poulin & FitzGerald also set up two plastic pools each containing 30 sticklebacks and then added 40 parasites to one of them. They found that the fishes in the pool with the parasites formed bigger shoals. In addition, the risk faced by the parasite of being counter-attacked and eaten by the host increased with shoal size in the three-spined stickleback. Parasites were captured and eaten only by fish in shoals, not by isolated fishes. In the black-spotted stickleback the risk faced by parasites was not related to shoal size and the numbers of parasites eaten showed no evidence of an effect of shoal size.

Thus, these laboratory experiments support the notion that parasitism by *Argulus* may have been a selective force in the evolution of social behaviour of juvenile sticklebacks. The indication is that individual shoaling sticklebacks are less likely to be attacked because the parasite has a multitude of targets and because shoaling fishes are more likely to eat parasites. However, in field studies Poulin (1999) failed to find a significant tendency for decrease in abundance of *A. funduli* with increasing shoal size and was forced to conclude that selection by *A. funduli* appears to be negligible in this context. In contrast, parallel studies conducted by Poulin on the interaction between sticklebacks and the copepod parasite *Thersitina gasterostei* (see Chapter 12) revealed that fish in large shoals carry more parasites than fish in small shoals.

13.6 FOOD AND FEEDING

In spite of the attention that *Argulus* has received it is hard to find any clear statement in the literature on the nature of the parasite's food. Published accounts of how feeding takes place are also conflicting. Part of the problem concerns the presence of a unique anatomical feature, which takes the form of a single, slender stylet, resembling a hypodermic needle, located in the mid-line anterior to the mouth between the maxillary suckers (Figure 13.1BD). This stylet is not derived from the mouthparts or appendages, is not found in any other group of parasitic crustaceans and, indeed, is not universal among the branchiurans, being limited to species of *Argulus* and *Dipteropeltis*. Various authors have regarded it as a sting, as a tactile sensory structure and as a suctorial proboscis (references in Swanepoel & Avenant-Oldewage, 1992).

The stylet of *A. japonicus* can be protruded or withdrawn. When protruded much of the stylet is exposed (Figure 13.8A), but on withdrawal it is retracted into a sheath, which is simply a tubular covering that inverts (turns outside in) as retraction proceeds (Figure 13.8B). It is thought that protrusion is brought about by increased hydrostatic pressure and withdrawal by contraction of muscles running from the proximal end of the stylet to points of anchorage on the exoskeleton (Swanepoel & Avenant-Oldewage, 1992; Gresty *et al.*, 1993). In a small adult about 2 mm in body length the fully everted stylet and sheath is about 600 µm long and the slightly swollen tip of the stylet (Figure 13.8C) is 1.5 – 2 µm in diameter (Gresty *et al.*, 1993).

It is the belief of some authors that this stylet functions like the proboscis of a mosquito, enabling the parasite to penetrate the skin of the host fish and extract either blood or lymph (van Duijn, 1956 and Kabata, 1970 respectively). The problem with this idea is that, although the stylet is hollow with a subterminal opening 0.5 to 1 µm in diameter (Figure 13.8C), this opening is not the mouth (Gresty *et al.*, 1993). In fact, there is no communication between the stylet and the gut (Figure 13.8; Madsen, 1964). In the adult, the mouth is located at the distal end of a tube or proboscis, superficially similar to the mouth tube of a siphonostomatoid copepod; this tube lies posterior to the stylet (Figure 13.8).

Scott & Scott (1913) described the mouth as siphon-like but mistakenly placed the stylet inside the siphon. Another misleading statement in a textbook by Parker & Haswell (1961) places the mandibles and maxillae inside this mouth tube and describes these mouthparts as 'piercing organs'. However, Gresty *et al.* (1993) found only the mandibles near the mouth opening at the end of the tube (Figure 13.8). These mouthparts seemed better suited for scraping than piercing (Figure 13.8D).

Gresty *et al.* (1993) demonstrated that the stylet is used to inject fluid into the host rather than to withdraw it. A duct conveys the secretion from a large gland just posterior to the base of the stylet (Figure 13.8A) to the hollow lumen of the stylet. The slightly rounded tip of the stylet and the sub-terminal position of its distal opening (Figure 13.8C) discourage clogging when the stylet is inserted.

Another misleading idea that is well established in the literature is that the stylet is some kind of venomous sting. Lapage (1958) stated that fish lice "possess a kind of sting". Noble & Noble (1971) mentioned a "poison spine" and Schmidt (1992) referred to "a pre-oral sting". Although there may be adverse reactions to substances injected into the host by *Argulus* – Bauer (1959) mentioned rapid killing of young fish by one or two parasites – it seems unlikely that the primary function of the stylet secretion is to paralyse or kill the fish.

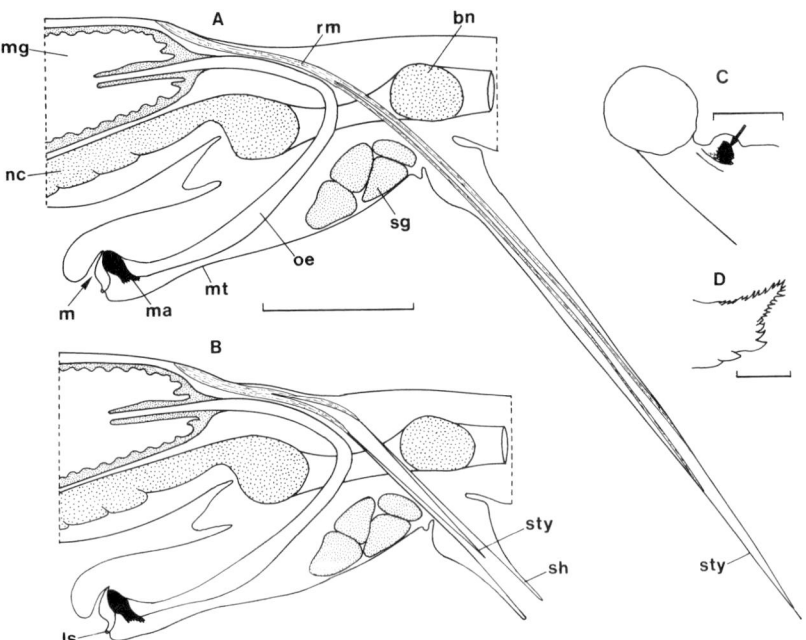

Fig. 13.8. The stylet and mouth tube of *Argulus japonicus*. (AB) Median (sagittal) sections through the stylet/mouth-tube region, with the stylet (A) fully extended and (B) fully retracted; (C) enlarged blunt tip of stylet with sub-terminal aperture (arrow); (D) enlarged tip of mandible. bn, 'Brain'; ls, labial spine; m, mouth; ma, mandible; mg, mid-gut; mt, mouth tube; nc, nerve cord; oe, oesophagus; rm, retractor muscle of stylet; sh, stylet sheath; sg, stylet gland; sty, stylet. Scale bars: (AB) 200 µm; (C) 2 µm; (D) 25µm. Modified from Gresty *et al.*, 1993. (C) Redrawn from a scanning electron micrograph in Gresty *et al.*, 1993.

There seems little doubt that the stylet plays some part in feeding, but the exact nature of its role remains obscure. According to Gresty *et al.* (1993 and references therein), the stylet is thrust repeatedly into the skin during feeding and creates splits and fissures in the host epithelium by mechanical action of the stylet and/or by possible lytic properties of the stylet secretion. Curiously, the stylet is directed anteroventrally, while the mouth tube, when not in use, is directed posteroventrally. Even when the mouth tube is erected into the feeding position and at right angles to the fish's skin, the parasite needs to shift forwards to place its mouth opening in contact with any lesion produced by injected stylet secretion (see Figure 13.8). According to Gresty *et al.* (1993), this mobile type of feeding can be observed on captive fishes, but at other times the evidence of association with a well-defined lesion suggests that the parasite has been feeding at the same place for some time.

Gland cells associated with the mouth tube must also make some unknown contribution to the feeding process. Some of these glands open into the oesophageal lumen and others open to the exterior via two labial spines at the distal end of the mouth tube (Figure 13.8) or elsewhere in the labial region of the mouth tube (see Swanepoel & Avenant-Oldewage, 1992).

Many authors regard *Argulus* as a bloodsucker (e.g. Wilson, 1902; van Duijn, 1956; Lapage, 1958; Bauer, 1959; Pavlovskii, 1962; Schmitt, 1965), but according to Kabata (1970) red blood cells have not been recorded in the gut of the parasite. Clark (1902) described the food of *A. foliaceus* as blood plasma. Minchin (1909) starved *Argulus* for a day or two and then gave them access to fishes for varying periods of time. He reported that none of these parasites contained red blood cells and no change of colour indicative of ingestion of blood was detected. Minchin wrote the following: "I very much doubt, therefore, if *Argulus* feeds on blood, or at least on blood-corpuscles". In contrast, Bower-Shore (1940) claimed to have observed tiny blood clots in large female *A. foliaceus*. He took this further and applied three separate tests for haemoglobin and claimed that all gave positive results. According to Cressey (1983) the stylet is used to inject digestive secretions into host flesh; the mouth tube then ingests blood and 'host juices' oozing from the stylet-produced wound. Shimura & Inoue (1984) injected an extract made from the mouth tubes, stylets and associated glands into the muscles of rainbow trout (*Oncorhynchus mykiss*) and found that the extract produced a haemorrhagic response but had no haemolytic or cytotoxic effects. They believed that this haemorrhagic response would facilitate sucking of host blood and that haemorrhagic spots in areas of the host inhabited by *A. coregoni* represented feeding sites of the parasite.

The study of Gresty *et al.* (1993) has advanced our understanding of the functional morphology of the stylet and the mouth tube, but we are no closer to an appreciation of how the stylet, the mandibles and the mouth tube interact during the feeding process. Perhaps *Argulus* combines the ability to ingest both blood and epidermis from its host. The penetrating ability of the stylet, perhaps coupled with vasodilatory and/or anticoagulant properties of the stylet secretion, may promote the release of blood from dermal blood vessels, while the mandibles have the ability to bite off pieces of superficial epidermis. Such a versatile dual capability may explain why *Argulus* is sometimes observed indulging in the mobile type of feeding behaviour while at other times the parasite seems to be stationary and engaged in feeding at a single site (see above).

13.7 REPRODUCTIVE SYSTEMS AND MATING

Jurine, as long ago as 1806, gave a vivid account of the amorous activities of the male "Argule foliacé" (*Argulus foliaceus*). He described the males as very ardent, cruising rapidly over the surface of the fish in search of females. According to Kollatsch (1959), parasites do not commence copulation in open water, requiring attachment of the female to the surface of the host to establish sexual relations. Jurine (1806) referred to the male stimulating the female with his legs or with convulsive movements of the extremities of the wing-like lateral expansions of the carapace. The responses of the female are subdued and involve only slight movements of the 'wings', without leaving her place of attachment. After this interplay the male mounts on the back of the female. According to Jurine (1806), it is at this stage that the female may put up some resistance, either by bending her abdomen downwards away from the male or by raising her 'wings' to which the male is attached by means of his suckers. This apparently increases rather than decreases the determination of the male. If the searching male fails to find a female, it detaches itself from the body of the fish and searches elsewhere. According to Jurine, the male is not selective, coupling with the first female it meets. In fact, Jurine

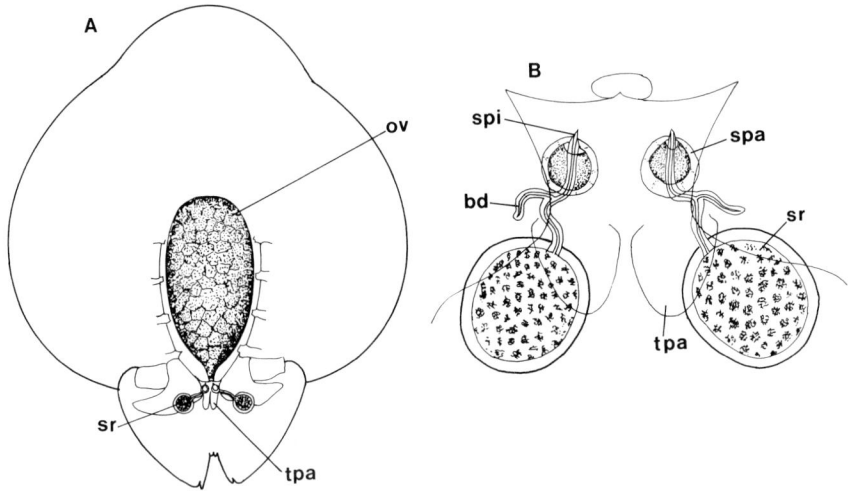

Fig. 13.9. Female sexual organs of *Argulus* (A) Organs *in situ* in relation to the body. (B) Region of seminal receptacles and papillae enlarged. bd, Blind diverticulum; ov, ovary; spa, seminal papilla; spi, seminal spine; sr, seminal receptacle; tpa, tactile papilla. From Wilson, 1902.

saw a male coupling with a female whose body was so full of eggs that she could barely move and another with the mutilated body of a female recently killed. If a male should meet another male, it fails to recognise the sex of the other and may briefly attempt to mate with it.

Jurine (1806) commented on the strong adhesion between male and female *A. foliaceus* during mating. He recalled pouring a mixture of alcohol and water and then spirits of wine on a mating pair without inducing separation. Even after death the pair remained coupled and Jurine detected some resistance when he separated the thoracopods of the male and female with two needles. During mating Jurine (1806) described convulsive movements of the legs of the male, especially those concerned with copulation, succeeded by periods of rest. The female made no special movements, although she often detached herself from the fish carrying the male with her. According to Jurine (1806) mating may sometimes continue for several hours, but Stammer (1959) claimed that it scarcely lasts longer than one hour.

In spite of Jurine's detailed account of mating, it is only comparatively recently (almost 200 years later) that we have begun to appreciate what may be happening between two mating individuals and to understand the anatomy of the female and male reproductive systems.

The female and male reproductive systems of *Argulus* are full of anatomical surprises. *Argulus* has a single, medianly situated, sac-like ovary, located in the thorax dorsal to the alimentary canal (Figure 13.9A), but a cross-section through the ovary of

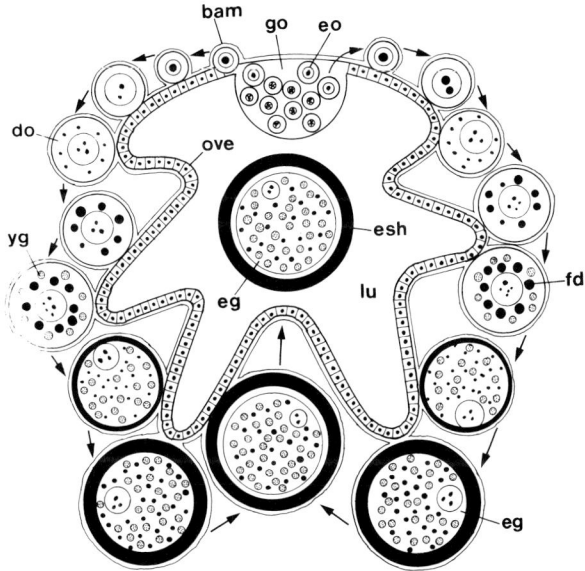

Fig. 13.10. Diagrammatic transverse section through the ovary of *Argulus japonicus*. Arrows indicate direction of movement of oocytes. bam, Basement membrane of ovarian epithelium; do, developing oocyte; eg, fully developed egg; eo, early oocyte; esh, egg shell; fd, fat droplet; go, germinative region of ovary; lu, lumen of ovary; ove, ovarian epithelium; yg, yolk granule. Modified from Ikuta & Makioka, 1997.

A. japonicus reveals an unexpected and unusual arrangement shown diagrammatically in Figure 13.10. First, contrary to expectations, developing eggs are located not inside the ovarian lumen but attached to the outside of the ovary, adjacent to the haemocoel. Secondly, the germinative region of the ovary is embedded in the dorsal ovarian wall (at the top of the picture in Figure 13.10) and runs along the median longitudinal line of the ovary. According to Ikuta and Makioka (1997), this germinative zone produces oocytes (egg cells) that leave the ovary dorso-laterally, moving out onto the outer surface of the ovary wall where they are enclosed externally by the basement membrane of the ovarian epithelium and internally by the ovarian epithelium itself. These early oocytes continue to migrate down the lateral walls of the ovary towards its ventral surface, growing as they go and undergoing vitellogenesis (deposition of yolk granules in the cytoplasm) and formation of the external envelope of the egg. By the time the eggs reach the ventral surface of the ovary they have reached maximum size and enter the ovarian lumen for the first time.

A single short median oviduct extends forwards from the anterodorsal end of the ovary and communicates on each side with a long and wide lateral oviduct (Figure 13.11). One of these lateral oviducts extends backwards and connects via a genital atrium to a single median female genital opening or gonopore. The other lateral oviduct ends blindly without connecting to the gonopore. The large oocytes from the ovarian lumen enter the median oviduct and then the functional lateral oviduct where they mature and await oviposition. The blind non-functional lateral oviduct remains empty and flattened.

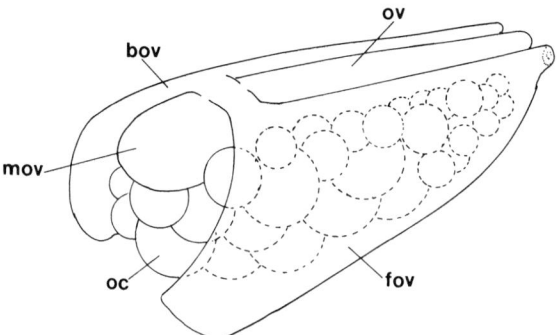

Fig. 13.11. Diagrammatic three-dimensional view of the ovary and oviducts of *Argulus japonicus*. bov, Blind lateral oviduct; fov, functional lateral oviduct; mov, median oviduct; oc, oocyte; ov, ovary. Modified from Ikuta & Makioka, 1997.

Completely separate from this remarkable ovary and posterior to the median gonopore in the abdomen are two spherical seminal receptacles where spermatozoa from the male are stored (Figure 13.9B). From each receptacle a spermathecal duct runs in an anterior direction towards the gonopore and terminates in a conical seminal papilla. According to Wilson (1902), the tip of this semen-carrying duct is hardened to form a sharp chitinous spine that can be protruded from the papilla or withdrawn into it. Wilson also described a blind diverticulum of the spermathecal duct (Figure 13.9B).

In the male there are two testes, but unlike the ovary these are located in the abdomen (Figures 13.1CD, 13.12). From the anterior border of each testis a sperm duct runs in an anterior direction and enters a transversely orientated seminal vesicle. Each lateral region of the seminal vesicle gives rise to a vas deferens, which travels in a posterior direction. Each vas deferens enters a short ejaculatory duct together with a duct from a large accessory gland (function unknown) situated anterior to the seminal vesicle in the thorax. The ejaculatory ducts from each side converge on the median genital atrium and genital pore (gonopore), which is located in the same place as the female gonopore. We might expect to find a penis in this position, but there is no such organ.

We must turn now to the thoracopods to complete our picture of the sexual equipment of the male. As already mentioned, the second, third and fourth pairs of thoracopods of the males have sexual embellishments that are absent in the female and the shapes of these embellishments can be useful for distinguishing one species of

Argulus from another. In *A. japonicus*, for example, there is a prominent peg on the anterior (pre-axial) margin of the basal podomere of the fourth (last) thoracopod and a corresponding socket on the posterior (post-axial) margin of the third thoracopod (Figure 13.13). There are additional spiny areas on the second and third thoracopods.

In the absence of a penis in the position where one would expect to find such an organ and with no evidence that spermatophores are produced and exchanged in *Argulus*, two completely different functional interpretations of the male sexual equipment have emerged. Jurine (1806) regarded the peg on the fourth thoracopod of the male as a penis and described the socket on the third thoracopod as a vesicle, which he claimed was filled with sperm. Claus (1875, in Martin, 1932) observed that the peg

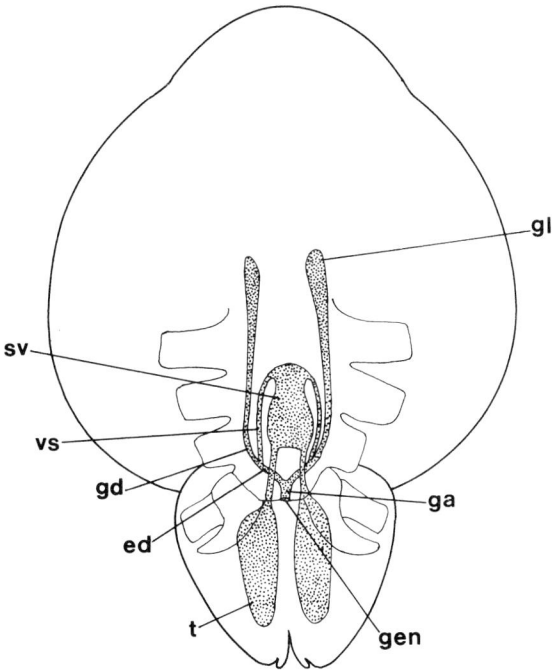

Fig. 13.12. Male sexual organs of *Argulus*. ed, Ejaculatory duct; ga, genital atrium; gd, duct from male accessory gland; gen, genital opening; gl, male accessory gland; sv, seminal vesicle; t, testis; vs, vas deferens. From Wilson, 1902.

fitted into the socket and believed that sperm was transferred to the socket in this way. As late as 1959 Stammer described the socket as a 'Samenkapsel' (seminal capsule). Further confusion was introduced by Jurine's suggestion that the 'penis' enters the female sexual organ and that a small hook near the base of the 'penis' holds the male and the female together during copulation. The possession of two penises, one on each side, neither of which have any connection with the testes, stretches credibility, as Martin (1932) has pointed out.

Martin (1932), working with parasites identified as *Argulus viridis* from fishes in the Brighton Aquarium, deserves credit for bringing some clarity to this situation. She pointed out that no traces of sperm were found associated with the peg or with the socket and that the male genital opening is single, median and on the ventral side of the last thoracic segment. During male/female interactions she observed that the last two pairs of thoracopods of the male are clasped around the last (fourth) thoracopods of the female from the dorsal side. The peg and socket arrangement on the legs of the male (Figure 13.13) then clips the legs together like a press-stud around the leg of the female. Martin likened the socket to a patch pocket of a coat, with the top turned in loosely and with pleats that allow the pocket to expand. The peg fits into the pocket and is held in position like 'Velcro' by interlocking spines on the peg and inside the pocket. Spiny areas on the second and third thoracic limbs of the male (Figure 13.13) are also important, enabling the limbs to grasp the third thoracic limb of the female. Thus, according to Martin (1932), with the male and female attached together and stabilised, the abdomen of the male could twist under that of the female, first to one side and then to the other. This would bring the male genital opening directly into contact with the opening of each of the female seminal receptacles.

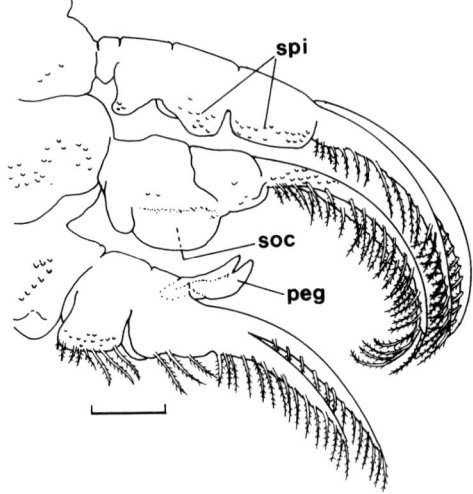

Fig. 13.13. The last three thoracopods of the male *Argulus japonicus*, showing structural modifications used in mating. peg, Peg; soc, socket; spi, spines. Scale bar: 200 μm. Redrawn from Avenant-Oldewage & Swanepoel, 1993.

However, this is not the end of the story. *Argulus* has yet another surprise for us because Avenant-Oldewage & Swanepoel (1993) claimed that the two ejaculatory ducts of the male *A. japonicus* have no communication with the genital atrium, ending blindly. They envisaged only one possible explanation. The wall of the ejaculatory duct must be punctured in some way in order for sperm to be released for transfer to the female. The obvious candidate for the task of perforation is the seminal spine, the protrusible spine-like termination of the duct from each of the two seminal receptacles of the female (Figure 13.9B). Avenant-Oldewage & Swanepoel (1993) suggested that during copulation the muscular walls of the vas deferens and the accessory gland duct contract, forcing sperm and accessory gland secretion into the blind-ending ejaculatory duct. Sphincter muscles at the proximal ends of the two ejaculatory ducts then close, maintaining a high internal pressure in each sperm-filled ejaculatory duct. Insertion of the female seminal papilla into the genital atrium of the male and protrusion of her seminal spine would puncture the blind end of one of the two ejaculatory ducts. The pressurised sperm within would be forced out of the ejaculatory duct down the duct from the papilla and into the seminal receptacle (Figure 13.9B).

According to Fryer (1982), enough sperm is received during one mating to fertilise all the eggs that the female produces during her lifetime. Nevertheless, more than one mating may occur (Kollatsch, 1959). Jurine (1806) noted that "la matrice" (translates as 'uterus' but = ovary) of unmated females of *A. foliaceus* remained small, with the transparent eggs hard to see, but soon after mating this enlarges.

13.8 EGGS

According to Jurine (1806), the female carries the eggs for 13 – 19 days from the time of insemination, then leaves the fish and seeks out a suitable substrate for laying, usually a stone. Thus, unlike parasitic copepods, female argulids do not retain their fertilised eggs in external strings or sacs until they hatch.

The female *Argulus foliaceus* attaches herself to the substrate by means of her suckers and expresses the first egg, which is attached to the substrate by an adhesive coat. According to Wilson (1902), each egg is fertilised as it is laid. He observed that, in one of his American argulids, the bases of the posterior legs grasped each egg as it was laid. As the egg was moved backward by these legs it was likely to make contact with the seminal spines, perhaps with sufficient force to perforate the egg envelope. Sperm is presumably delivered to the egg by this means and Wilson observed sperm swarming around the jelly envelope of a freshly deposited egg. Stammer (1959) also described fertilisation of each egg as it is laid by means of a prick delivered by the papilla located near the female gonopore.

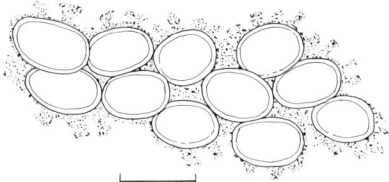

Fig. 13.14. Sketch by Martin (1932) of freshly laid eggs of a parasite identified as Argulus viridis. Scale bar: 5 mm. With kind permission of the Zoological Society of London.

After each egg is laid and attached, the female takes a step forward, such that the body advances obliquely. Consequently, the next egg to be laid is attached in front of and to the side of the last. Next follows another step, such that the female is in a position to lay the next egg in the same orientation as the first. By alternately stepping forward and from side to side and releasing one egg in each position, the female lays eggs usually in two rows, one layer thick (Figure 13.14). Jurine (1806) also recorded that sometimes the eggs were laid in three, four or five rows. Some females interrupt their laying and may eject them in several groups in different places.

The number of eggs per clutch varies widely. Fryer (1982) refers to a female *A. foliaceus* 8.5 mm in length that produced 1181 eggs in 15 clutches over a period of 38 days. Clutch size ranged from 18 to 232. According to Kollatsch (1959) a female 9 mm in length produced a similar number of eggs (1192) in only four clutches, ranging from

71 to 481 eggs, over a period of 15 days. According to Fryer (1982), the largest clutches of eggs are laid by the largest females, which may contain as many as 600 ready-to-lay eggs. Gravid females may sometimes be unable to attach themselves to the host.

According to Shafir & van As (1986), *A. japonicus* produces clutches of eggs in a similar way, the female laying at a rate of one egg every 3.5 – 10 seconds. Gravid females lay between one and nine clutches of eggs arranged in one to six rows. All the eggs of the female (a maximum of 226 was recorded) may be laid in one clutch, but Shafir & van As recorded one egg clutch of only five eggs. The time interval between laying of successive clutches ranged from 30 seconds to a few days. During these longer intervals the female often returned to the fish where feeding occasionally occurred.

A. foliaceus lays eggs in shallow water, while *A. coregoni* prefers deeper water for oviposition (Hakalahti *et al.*, 2004). Mikheev *et al.* (2001) believe that this habitat preference in Finland reflects the preferred hosts of the two parasites. The main hosts of *A. foliaceus* are cyprinids and percids that remain in shallow habitats during the summer, while the salmonid hosts of *A. coregoni* inhabit deeper, well-oxygenated water. According to Hakalahti *et al.* (2004), dark substrates are preferred for egg laying by *A. coregoni*, with some preference for rough rather than smooth surfaces.

In Finland, *A. coregoni* survives the harsh winter period only as eggs. *A. foliaceus* also overwinters as eggs, but may also survive as adults (references in Hakalahti & Valtonen, 2003). *A. coregoni* has only one generation annually in Finland, but the extended period of warm weather in Japan appears to favour the production of a second generation (see Hakalahti & Valtonen, 2003).

13.9 HATCHING AND DEVELOPMENT

Jurine (1806) found that the eggs of *A. foliaceus* began to hatch about 35 days after laying (temperature not given), the larva emerging through a longitudinal split in the shell. Shafir & van As (1986) observed that eggs of *A. japonicus* incubated at 35°C began to hatch after about 10 days, but at 15°C hatching took 61 days. Overwintering eggs of *A. coregoni* are known to hatch over an extended period in the laboratory. If this phenomenon occurs in the wild it may enhance the chances of larvae locating and infecting a host (Mikheev *et al.*, 2001). Clark (1902) observed that the newly hatched *A. foliaceus* immediately seeks a host, attaches itself and begins feeding.

The account that follows of the development of *Argulus foliaceus* is taken from Rushton-Mellor & Boxshall (1994). On hatching, the young *Argulus foliaceus* is between 630 and 820 µm in length. It resembles the adult when viewed from the dorsal side (Figure 13.15A), but in ventral view (Figure 13.15B) several important differences are evident and the animal can be regarded as an advanced nauplius (= metanauplius). Conspicuous by their absence in the metanauplius are the suckers, which are so characteristic of the adult. The metanauplius is propelled through the water by the exopods of the second antennae and by a mandibular palp (Figure 13.15B). Propulsion is aided by the presence of long plumose setae at the free ends of these structures. The first pair of biramous thoracic swimming legs is moderately developed with two plumose setae at the end of the exopod, but the other three pairs are rudimentary and non-functional.

The frontal margin of the carapace of the metanauplius bears a fringe of short processes, amongst which are longer sensilla (Figure 13.15). On the dorsal surface there are two large compound eyes and, just behind and between them, is a median nauplius eye (Figure 13.15A). Areas that are thought to have a respiratory function (see above) are already identifiable in the metanauplius. There are two such areas on each side on the ventral surface, a large posterior and a much smaller anterior area (Figure 13.15B).

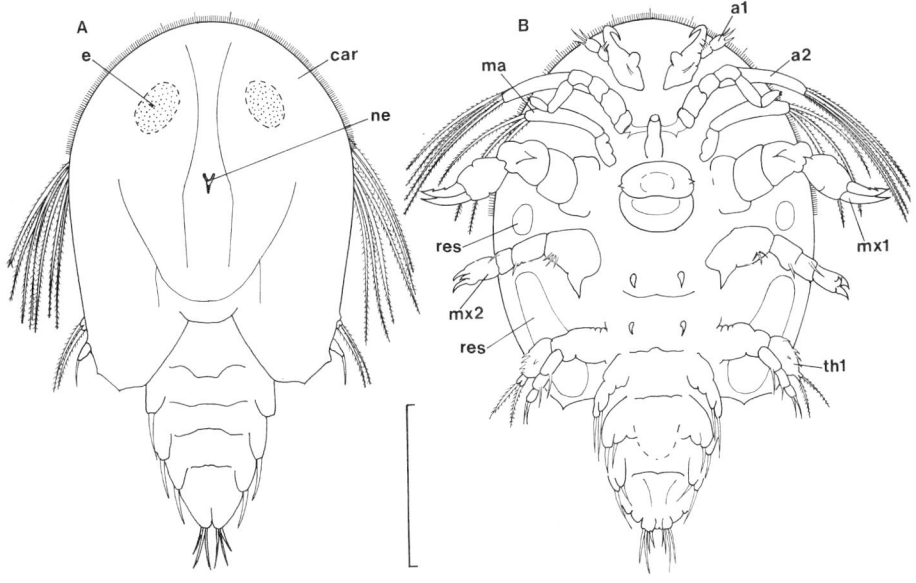

Fig. 13.15 Metanauplius stage of *Argulus foliaceus*, (A) in dorsal view and (B) in ventral view. a1, a2, First and second antennae; car, carapace; e, compound eye; ma, mandible; mx1, mx2, first and second maxillae; ne, nauplius eye; res, respiratory (?) area; th1, first thoracopod (swimming leg). Scale bar: 250 μm. Redrawn from Rushton-Mellor & Boxshall, 1994.

The first moult marks a fundamental change in the method of propulsion. The antennal exopod and the mandibular palp are reduced in length and lose their terminal setae (Figure 13.16). Their role in locomotion is lost and the task switches to the swimming legs (thoracopods), all of which are well developed with plumose terminal setae. These thoracic legs retain their swimming role in all subsequent stages of development and in the adults. Because of this shift towards adult features, Rushton-Mellor & Boxshall (1994) regarded 'juvenile' as an appropriate name for this second developmental stage.

The next major development must rank as one of the most remarkable transformations among fish parasites, if not in the Animal Kingdom. This is the transformation of the proximal podomere of the first maxilla (maxillule) into a structurally complex and highly efficient sucker (see above). The process takes place gradually, beginning at the second stage and continuing with successive moults until the sucker is functional at the sixth stage (Figure 13.17A-D).

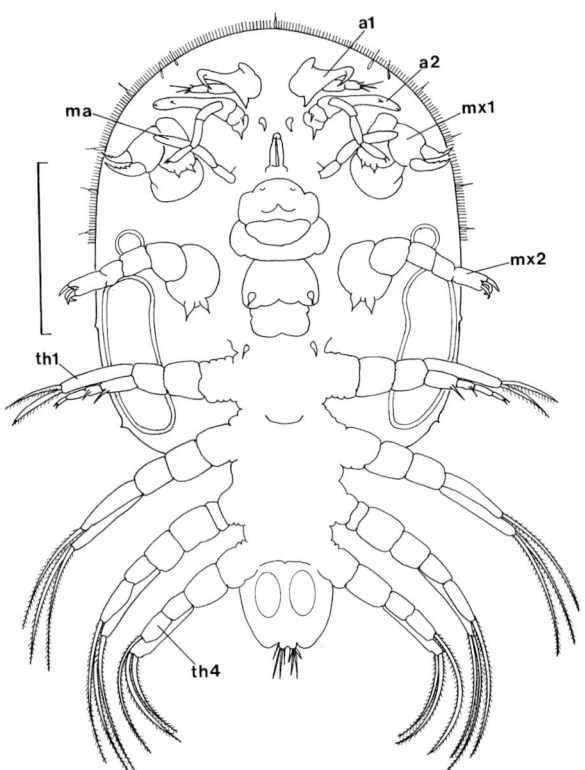

Fig. 13.16. 'Juvenile' (second stage male) *Argulus foliaceus* in ventral view. a1, a2, first and second antennae; ma, mandibular palp; mx1, mx2, first and second maxilla; th1, th4, first and fourth thoracopods. Scale bar: 250 μm. Redrawn from Rushton-Mellor & Boxshall, 1994.

While the suckers are developing and non-functional, the parasite must rely on its hooked appendages to cling to its fish host. There is a hook on each of the first antennae, two barbed recurved spines at the end of each first maxilla and three recurved spines at the end of each second maxilla (Figure 13.16). It is at stage six (Figure 13.17D) that the suckers at the base of the first maxilla functionally replace the barbed spines at the free ends of the same appendages. However, the distal jointed region of the first maxilla persists, albeit in a degenerate state, throughout the rest of the animal's life, and the presence of this small protrusion projecting from each sucker is a reminder of the sucker's origin and the amazing transformation that has created it. Although the maxillary suckers are functional at the 6th stage, the parasite is not yet adult and additional radially arranged supporting sclerites are added around the margin of each sucker during successive moults.

In the 6th stage male the accessory copulatory structures on the thoracopods are not yet fully developed. Rushton-Mellor & Boxshall (1994) reported that the seminal receptacles were fully developed in 7th stage females, but eggs were not detectable in the thorax of the female until after the 10th moult. Thus, there are apparently 10 developmental stages before adulthood is achieved. The interval of time between successive moults from the 1st to the 7th stage was found to be remarkably uniform.

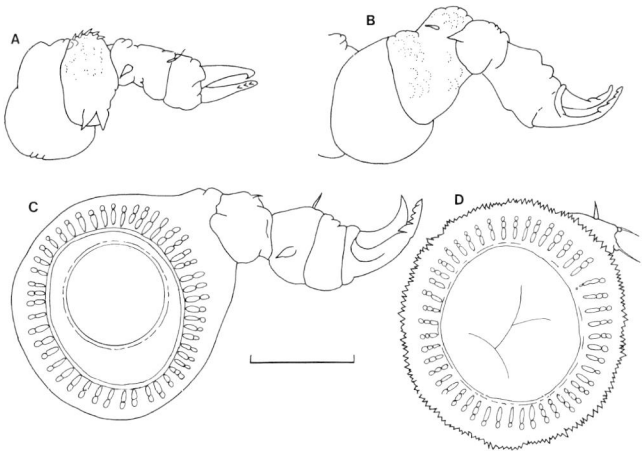

Fig. 13.17. Stages in the transformation of the first maxilla to a sucker. (A) Third stage female; (B) fourth stage male; (C) fifth stage female; (D) sixth stage male. All drawn to the same scale. Scale bar: 100 µm. Redrawn from Rushton-Mellor & Boxshall, 1994.

13.10 INTERACTIONS WITH OTHER ORGANISMS

Fish hosts may not be entirely at the mercy of *Argulus*. Clark (1902) observed that three-spined sticklebacks (*Gasterosteus aculeatus*) sometimes swallow the parasite (presumably free-swimming *Argulus*), but he noted that generally they are avoided or ejected from the mouth. Mikheev *et al.* (2000) came to the conclusion that juvenile roach (*Rutilus rutilus*) and perch (*Perca fluviatilis*) usually avoid free-swimming *A. foliaceus*, except after a long period of starvation. They observed that parasites taken into the mouth were usually spat out immediately and suggested that their unpalatability may be connected with their spiny bodies and limbs.

Poulin & FitzGerald (1989d) recorded predation on *Argulus funduli* (incorrectly identified as *A. canadensis*) by juvenile three-spined sticklebacks in aquaria. They sought to investigate the possibility that selective predation on female *Argulus* by sticklebacks accounted for the larger numbers of male *Argulus* on these fishes during the summer breeding season of the parasites. Since sticklebacks are visual predators, responding to cues such as prey size and movement, they hypothesised that female *Argulus* would be more attractive as prey because they are larger than males and because they leave the fish to deposit eggs and therefore spend more time swimming freely. This appeared to be supported because more females were eaten than males, but statistically this difference had no significant effect on the sex ratio. They concluded that the male-biased sex ratio on juvenile sticklebacks is a simple reflection of the female's habit of leaving the host to deposit eggs.

Poulin & FitzGerald (1989d) also surveyed the stomach contents of juvenile sticklebacks from natural tide pools on Canada's east coast and found that *A. funduli* accounted for less than 1% of their diet. They suggested that predation on *Argulus* may be reduced in the natural environment because the parasite can rely on camouflage or other forms of protection unavailable in aquaria.

Clark (1902) observed that *A. foliaceus* kept in confinement is readily attacked by water boatmen (*Notonecta*) and speedily killed.

Jackson & Marcogliese (1995) described a quite different relationship. They examined a total of 637 free-living pelagic mysid crustaceans (*Mysis stenolepis*) from Canadian Atlantic waters and collected from them 30 branchiurans identified as *Argulus alosae*. These were juveniles and located ventrally in the marsupium or brood pouch of the mysid, near the junction of the cephalothorax and the abdomen. *A. alosae* adults have been reported from a variety of brackish water and marine fishes. The authors observed that the preoral stylet of the branchiurans had not pierced the bodies of the mysids and interpreted this as meaning that the relationship was phoretic (see also pp. 70, 112, 334) rather than parasitic, i.e. that the branchiurans were using the mysids merely as carriers to increase their dispersal capability. Mysids are active swimmers and undergo migrations, so they are potentially capable of spreading the parasite over large distances.

As mentioned elsewhere (Chapters 7, 15) British fishes are the victims of blood-dwelling protozoan parasites. Because leeches, gnathiid larvae and branchiurans move from host to host between their meals of blood or tissue fluids, they have come under scrutiny as potential transmitters of diseases. This is a role analogous to that of mosquitoes, tsetse flies and other biting arthropods in the transmission of diseases like malaria and sleeping sickness between land vertebrates. It transpires that leeches and gnathiid larvae are guilty as charged, but there is no evidence that branchiurans transmit protozoan fish parasites. Minchin (1909), working at the Sutton Broad Laboratory in Norfolk (a privately run laboratory closed before World War I and referred to briefly by Ellis, 1965, p. 4), observed that almost every fish caught in the broad was infected with *Argulus*. He kept these branchiurans without food for a day or two and then offered them various potential host fishes, all of which were infected with blood protozoans. The branchiurans readily attached to their hosts and after a suitable interval were removed and dissected. Minchin found no parasites in any of these crustaceans and, as already mentioned above, failed also to find blood corpuscles, leading him to doubt that *Argulus* feeds on blood.

Argulus has been implicated in the transmission of other organisms. Ahne (1985) showed that *A. foliaceus* is able to transmit spring viraemia of carp virus mechanically from carp to carp (*Cyprinus carpio*). No multiplication of the virus was detected in the vector.

Argulus also transmits skrjabillanid and probably other dracunculoid nematodes (see Moravec, 1994). The vector here acts as an intermediate host and permits development of the nematode. The adult nematodes are widespread parasites of freshwater fishes in Europe, but, as far as I am aware, they have not yet been recorded from Britain. *Skrjabillanus scardinii* from the rudd (*Scardinius erythrophthalmus*) will serve as an example (see Moravec, 1994). *Skrjabillanus scardinii* is long and threadlike, the males with bodies $3.2 - 6$ mm in length but only $20 - 28$ µm in width and the females $8 - 16.5$ mm long and $49 - 70$ µm wide. Males and gravid females are most commonly found beneath the membrane or serosa covering the posterior part of the fish's swimbladder, but may also be found beneath the serosal covering of the kidneys and gonads. Female nematodes, including gravid ones, also occur in the abdominal cavity, kidneys, bladder, orbits and inside the eye. The gravid females periodically release living larvae into the surrounding tissues of the fish host. These first-stage larvae are only 140 – 180 μm long and 5 – 6 μm wide and their shape and small size enables them to move freely through the fish's blood vessels. Travelling via the bloodstream they enter the muscles and then accumulate in the superficial layers of the skin where they develop no further. It was shown experimentally by Tikhomirova (in Moravec, 1994) that these larvae are taken up during feeding by *Argulus foliaceus* and *A. coregoni*. The larvae quickly penetrate the intestine wall of the branchiuran, enter its body cavity and then accumulate in the thoracopods, where they undergo two successive moults. Immediately after the second moult the third-stage larva is 440 – 460 μm long. The larvae then leave the thoracopods and concentrate in the suckers where they become infective. When *Argulus* takes its next meal these larvae enter the skin of the fish host by an unknown route. The larvae migrate to the fish's internal organs and having undergone two more moults become young adults. The females begin to release larvae about two months after entering the fish host.

Rudd populations in Europe may be heavily infected with the nematode *Skrjabillanus scardinii*, with 50 – 80% of hosts infected (100% in older groups of fishes) and more than 100 nematodes per fish. Moreover, a close relative with a similar life cycle, namely *Molnaria intestinalis*, often accompanies *S. scardinii*. *Daniconema anguillae* from the eel (*Anguilla anguilla*) and *Lucionema balatonense* from the zander (*Stizostedion lucioperca*), are also thought to have life cycles involving *Argulus* as an intermediate host (Molnár & Székely, 1998), but these worms belong to different nematode families, namely the Daniconematidae and Lucionematidae respectively.

13.11 RELATIONSHIPS

As this chapter draws to a close, it will be evident to the reader that the branchiurans have earned their independent status as a subclass of the Maxillopoda (also including as independent sub-classes the ostracods, copepods, and barnacles). Within the Maxillopoda the relationship between the constituent sub-classes is obscure and, indeed, the relationship between maxillopodans and other crustaceans is still debated (see Ikuta & Makioka, 1997). However, there is one fascinating link to the branchiurans that has emerged recently. This is the finding that they are most probably related to the so-called tongue worms or pentastomids.

At first glance this seems an unlikely alliance, because although the pentastomids are parasites, they are worm-like organisms, most of which are found in the lungs of carnivorous reptiles (snakes, lizards and crocodilians), with a few species in the respiratory tracts of birds and mammals (see Roberts & Janovy, 2000). The carnivorous vertebrate is usually infected by eating a cold-blooded or warm-blooded vertebrate intermediate host, containing quiescent larvae or nymphs in their tissues. In the carnivore the activated nymph bores its way out of the intestine into the lung (or air sacs of birds and nasal cavities of mammals), where it reaches adulthood and feeds on blood and tissue fluids. Intermediate hosts are infected by eating eggs of the parasite. It has been suggested that the currently extant 100 or so species of pentastomids represent only a small surviving fragment of a once abundant and widespread group that reached its zenith in the Mesozoic 'Age of Reptiles' (Roberts & Janovy, 2000).

Pentastomids have been linked with annelid worms (like the earthworm), but most modern zoologists regard them as arthropods. The remarkable similarity between the spermatozoa of pentastomids and those of *Argulus* led Wingstrand (1972) to propose a close relationship between pentastomids and branchiurans. Riley *et al*. (1978) gave their general approval to this proposal, based on a comparison of embryonic development, the structure of the surface layer or integument and the development of their sex cells. Ikuta & Makioka (1997) noted the following similarities between the pentastomid and branchiuran ovaries: both are single, median and sac-like, with many lateral and ventral folds of the ovary wall; both hang down from the dorso-median region of the body wall; both possess a single germinative region arranged longitudinally in the dorso-median ovarian wall; in both groups the oocytes remain on the outer surface of the ovarian wall during most of their development. Comparison of nucleotide sequences of 18S ribosomal RNA lends further support to this strange alliance (see references in Ikuta & Makioka, 1997). Riley *et al*. (1978) suggested that the ancestral pentastomid was originally a fish parasite that adopted an endoparasitic life-style in aquatic reptiles through predation.

We are left with the most intriguing questions of all, which currently are without answers. Where do we look among the free-living crustaceans for the ancestors of the branchiuran/pentastomid lineage and what were the evolutionary events that thrust the ancestors of these two groups of parasites apart and set them on such very different courses?

14

A MESOPARASITIC BARNACLE – *ANELASMA*

It would be hard to find anyone who is unaware of the contribution made by Charles Darwin to our understanding of evolution and the origin of species, but his many other contributions to biology are less well known. He was a keen collector of beetles and published books on topics as diverse as earthworms, orchids and barnacles. Surprisingly, it is this last-named topic that provides a parasitological connection.

It is not easy to appreciate that the barnacles (cirripedes) we find on a rocky shore are living organisms and even less easy to accept that they are crustaceans. Their heavy calcareous shells are often hard to distinguish from the rocky substrate to which they are tightly cemented, and early zoologists, although aware that they were living organisms, regarded them as molluscs. There is little sign of life when the tide is out and the barnacles exposed, but if we patiently observe them as they are covered by the incoming tide some important clues are revealed. There is a palisade of four or six calcareous plates, with the space enclosed by these plates roofed over by smaller plates (Figure 14.1A). These smaller plates open like doors, and the six pairs of biramous, jointed appendages that emerge have the hallmark of crustaceans (Figure 14.1A). In fact, these appendages are the jointed biramous thoracic limbs (thoracopods), each with an endopodite and an exopodite, and are abundantly supplied with hair-like setae. As we watch, the newly emerged limbs perform repeated sweeping movements through the water. Planktonic organisms and particles of debris combed from the water by this activity provide most of the food for the sedentary barnacle, although some barnacles are known to be capable of grasping and ingesting relatively large prey items offered in the laboratory (Pearse *et al.*, 1987).

Inside the space enclosed by the calcareous plates the body of the barnacle provides few obvious anatomical clues to its crustacean affinities. However, strong confirmation is provided by the emergence of a nauplius larva from the barnacle egg. After several moults a so-called 'cypris' or 'cyprid' larva is produced, which retains obvious crustacean features such as first antennae and six pairs of biramous thoracic appendages. The animal is enclosed by a carapace consisting of two valves, like the shell of a bivalve mollusc (see Chapter 16). Each first antenna terminates in an adhesive disk, which is used by the cypris for attachment to a suitable substrate, usually near other barnacles. After attachment the carapace is lost and the soft body of the organism, which is now attached virtually upside down, undergoes major reorganisation to produce the adult barnacle. Like many other sessile animals and unlike most of its mobile crustacean relatives, most barnacles are hermaphrodite. However, they retain a propensity for cross insemination, this being achieved by the often enormously extensible penis, which has the ability to reach out between the gaping plates, sometimes over a distance of several centimetres, to a neighbour in the barnacle colony.

Barnacles are entirely marine and spend their adult lives cemented to an inert substrate such as a rock or to man-made structures like ships and piers. They are also extremely abundant, not in terms of numbers of species (Pearse *et al.*, 1987, quote 950 living species), but in terms of individuals (as many as 107,000 per m^2 in the densest part of a barnacle zone, which may extend for many kilometres on an English rocky shore). In view of this abundance, it is not surprising that many barnacles use other animals as substrates, not only those with hard non-living surfaces like molluscs or crabs, but also animals with more flexible and living surfaces. What is surprising is that the parasitic way of life has been adopted by so few of these epizoic barnacles. One of the most spectacular parasitic barnacles, *Sacculina*, has adopted a host with a hard exoskeleton (Pearse, 1987). Its cypris enters the body of a crab and dedifferentiates into

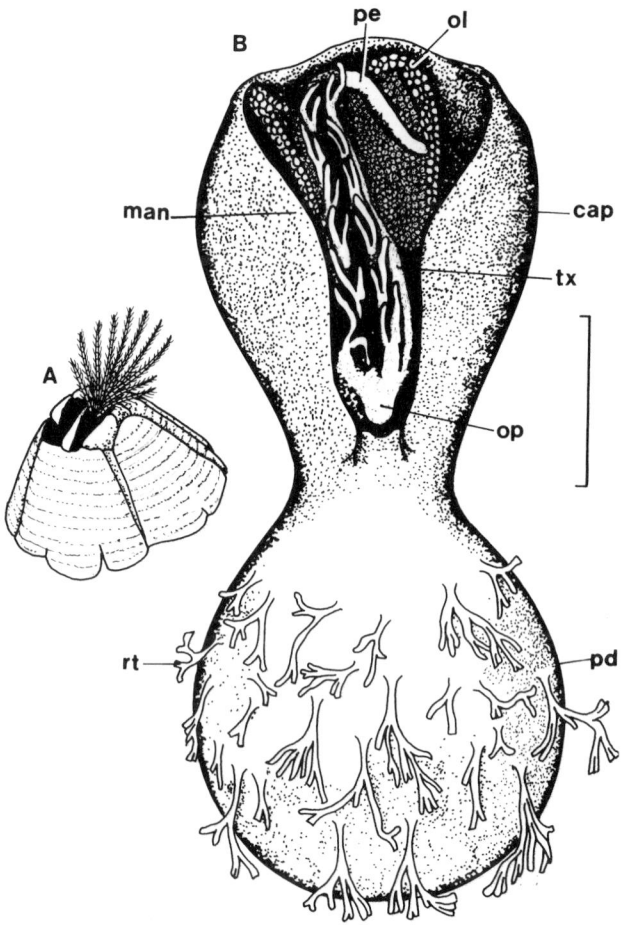

Fig. 14.1. (A) A free-living barnacle (*Balanus*). (B) The barnacle *Anelasma squalicola* removed from its cavity of implantation in the skin of the shark *Etmopterus spinax* and viewed from the ventral aspect. cap, Capitulum; man, mantle; ol, ovigerous lamella; op, oral proboscis; pd, peduncle; pe, penis; rt, root; tx, thorax. Roots represented rather diagrammatically. Scale bar: 5 mm. From Johnstone & Frost, 1927.

a tumour-like branching structure. The opportunities for symbiotic relationships to develop between barnacles and fishes must have been frequent, given the likelihood of contacts between cypris larvae and fishes, but only one association seems to have taken the road towards parasitism.

Lovén (1844 in Darwin, 1851) discovered a barnacle (Figure 14.1B), which he called *Alepas squalicola*, attached to the body of a deep-water shark, now called *Etmopterus spinax* (the velvet-belly). This is one of the smallest sharks, females reaching a length of 60 cm and males 50 cm, and one of very few sharks known to possess light-organs, small luminous raised pores on the skin (Wheeler, 1978). It is relatively common in the north-east Atlantic. The barnacle found on this shark became the subject of an anatomical study by Charles Darwin (1851). The special features of the organism led him to transfer it to a new genus, *Anelasma*. Johnstone & Frost extended this study in 1927, pointing out that Darwin's description was a masterpiece of anatomical research carried out entirely by dissection of a single individual in a poor condition. Hickling (1963) examined large numbers of sharks caught between 1927 and 1944 on the hake fishing grounds to the west of the British Isles. These fishes were taken at depths of 120 – 380 fathoms and provide information on prevalence, intensity and the effect of *Anelasma* on the velvet-belly.

The notion that *Anelasma* is confined to the north-east Atlantic has been dispelled recently by the finding of *Anelasma* on the combtooth dogfish, *Centroscyllium nigrum*, caught off the coast of southern Chile (Long & Waggoner, 1993). The host is related to *Etmopterus*, but the availability of insufficient material prevented the authors from determining whether this barnacle is *A. squalicola* or a new species.

Anelasma squalicola is highly localised on the velvet-belly. Johnstone & Frost (1927) examined several dozen infested fishes and found that most of them harboured two barnacles close together, just in front of, or alongside, the first dorsal fin. Sometimes the barnacles were in similar positions in relation to the second dorsal fin. Hickling's findings (1963) were similar. In a sample of 79 fishes with barnacles, 61 (77%) had *Anelasma* in front of or alongside the first dorsal fin and 13 (16%) in front of or alongside the second dorsal fin. Four of the remaining five fishes had barnacles above the pelvic fins and one above the pectoral fin. Thus attachment at positions other than near the first dorsal fin is exceptional.

The velvet-belly has a long and very sharp spine in front of each dorsal fin and, before birth (the shark gives birth to living young), each spine is enclosed in a sheath. This sheath ruptures after birth, leaving a circular scar around the base of the exposed spine. This scar seems to offer the most favourable substrate for attachment.

It is rare to find only one barnacle on a fish. In a sample of fishes studied by Hickling (1963) eleven fishes (13%) had a single *Anelasma*, 70 fishes (81%) each had two barnacles, four fishes (5%) each had three barnacles and one fish (1%) had four barnacles (although these were grouped in two pairs). Johnstone & Frost (1927) and Hickling (1963) agreed that the barnacles on any given fish tend to be alike in size and we will return to this phenomenon later.

Figures for the prevalence of *Anelasma* on fishes of different lengths are derived from Hickling (1963) and are given in Table 14.1. Nearly 10% of sharks measuring 25 – 29 cm in length had barnacles, but in fishes longer than 40 cm, the prevalence fell to 2% or less. Hickling found no significant difference between male and female fishes in barnacle incidence and no seasonal variation in levels of infection. He claimed that *Anelasma* breeds throughout the year and has a life span of more than one year.

Apart from having a soft body with no enclosing calcareous plates, *Anelasma* is readily recognisable as a barnacle. The part of the body (capitulum) protruding from

Table 14.1. Prevalence of Anelasma squalicola *on* Etmopterus spinax *(velvet-belly) of different sizes*

Length of fish (cm)	Number of fish examined	Number of fish parasitised	Prevalence of parasite
15 - 19	2	-	-
20 - 24	28	-	-
25 - 29	102	10	9.8
30 - 34	250	12	4.7
35 - 39	968	44	4.5
40 - 44	1031	21	2.0
45 - 49	125	2	1.6

the fish measures about 1.25 cm in height on average and is deep purple brown in colour. It consists of a fleshy mantle, which like an overcoat wraps around the thoracic region of the animal, with a vertical gap on what will be regarded as the ventral side where the two lobes of the mantle approach each other (Figure 14.1B). The main features of the thoracic region correspond with those of a typical shore barnacle and can

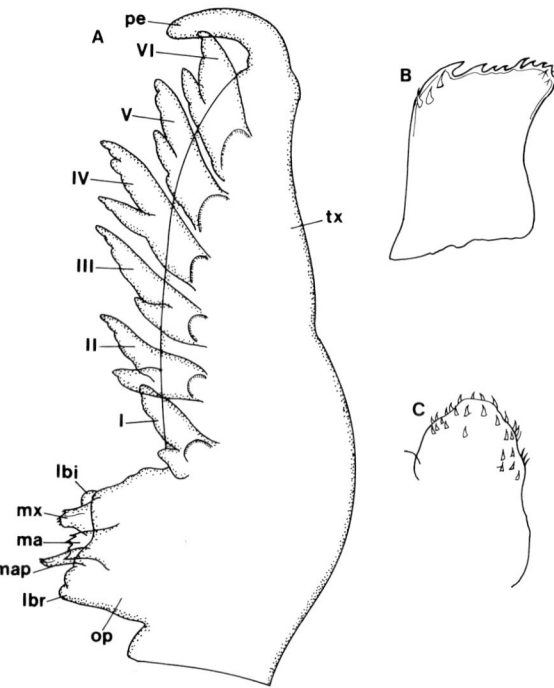

Fig. 14.2. (A) Thorax and oral proboscis of a small *Anelasma squalicola*. Mantle removed and mouthparts everted. lbi, Labium; lbr, labrum; ma, mandible; map, mandibular palp; mx, maxilla. I – VI, Biramous thoracic appendages. Other labelling as in Fig. 14.1B. (B) Mandible, enlarged; (C) maxilla, enlarged. From Johnstone & Frost, 1927.

be seen through this mantle opening. Projecting in a ventral direction are six pairs of thoracic limbs (thoracopods) (Figures 14.1B, 14.2A). These are soft cylindrical biramous appendages, with no signs of a jointed structure except in young individuals, and bearing no spikes or setae. Proximal to these appendages and also projecting ventrally near the bases of the mantle lobes is a cone-shaped structure, which Johnstone & Frost (1927) called the 'oral proboscis', bearing the mouth and greatly reduced mouthparts (Figure 14.2A).

Johnstone & Frost (1927) identified two pairs of stubby mouthparts: a pair of mandibles and a pair of maxillae (referred to by Johnstone & Frost as "maxillae" and as "maxillulae"), all roughly similar in size. According to Darwin (1851), the mandible is about 16 thousandths of an inch (about 0.4 mm) in length, measured in its longest direction. Each mandible has a strongly toothed edge (Figure 14.2B) and each maxilla is armed with spines (Figure 14.2C). The mouthparts are enclosed between an upper and a lower lip (labrum and labium) (Figure 14.2A). There is a pair of stubby palps associated with the mandibles and another small pair associated with the labrum. The anus opens distally on the dorsal side (Figure 14.3A).

This brings us to the interesting question of how *Anelasma* is attached to the fish. When the organism is *in situ* we see only about one half of its body (the capitulum), the other half (the peduncle) being embedded in the fish, creating an obvious swelling. When this swelling is cut open, the buried half of the barnacle is revealed and comprises a roughly spherical peduncle joined to the exposed half of the body by a narrow waist (Figure 14.1B). In spite of this surgical operation the peduncle cannot be easily withdrawn from the implantation cavity because numerous highly branched roots originating from the peduncular surface (Figure 14.1B) penetrate in all directions into the dorsal musculature of the fish, sometimes to a depth of over 1 cm. These roots must be broken to free the barnacle.

Thus, *Anelasma* has two potential routes for the intake of food: via the gut, which is confined to the exposed capitulum (Figure 14.3), and through the integument of the embedded peduncular half of the animal, especially the integument covering the peduncular roots. Darwin (1851) made no reference to the possibility of peduncular uptake of nutrients, but he did describe *Anelasma* as "Parasitic on Squalus". On the other hand, he expressed the belief that the barnacle is capable of catching and swallowing small living organisms that crawl within reach of the mouth. It has already been noted (see above) that some free-living, filter-feeding barnacles are capable of ingesting relatively large prey and *Anelasma* may use its thoracopods and mouthparts in a similar way. Johnstone & Frost (1927) found no traces of ingested food in serial sections of the gut of *Anelasma*, but they reported that "Mr Scott is satisfied that one specimen dissected by him had food debris in the oesophagus and adjacent parts of the alimentary canal". In addition, the extensive development of the digestive gland (Figure 14.3) and the obviously secretory nature of the cellular lining of the digestive tubules, are not features of a vestigial and redundant digestive system. In addition, Johnstone & Frost were convinced that the peduncular roots were capable of withdrawing nutrients from the host because the roots have ducts that communicate with a central lacunar system in the body (Figure 14.4) and their tips have cells that appear to be metabolically active. They doubted that the roots serve only for attachment because when *Anelasma* is enclosed within the intact implantation cavity, the host tissue is tightly constricted around the barnacle's waist and the organism cannot be withdrawn from the fish without rupture.

Stronger evidence in support of a parasitic way of life comes from Hickling's comparative study (1963) of the effects of *Anelasma* on the velvet-belly. Hickling pointed out that this shark has a remarkably large liver. Hickling found that the livers of small sharks appeared to be unaffected by the presence of *Anelasma*, but the livers of large, barnacle-infested fishes were smaller than those of similar-sized fishes lacking barnacles. This suggests that *Anelasma* may become a drain on the resources of the host. Since large fatty livers are thought to increase buoyancy in many sharks (Denton, 1961), large infected sharks may have to work harder to maintain their position in the water.

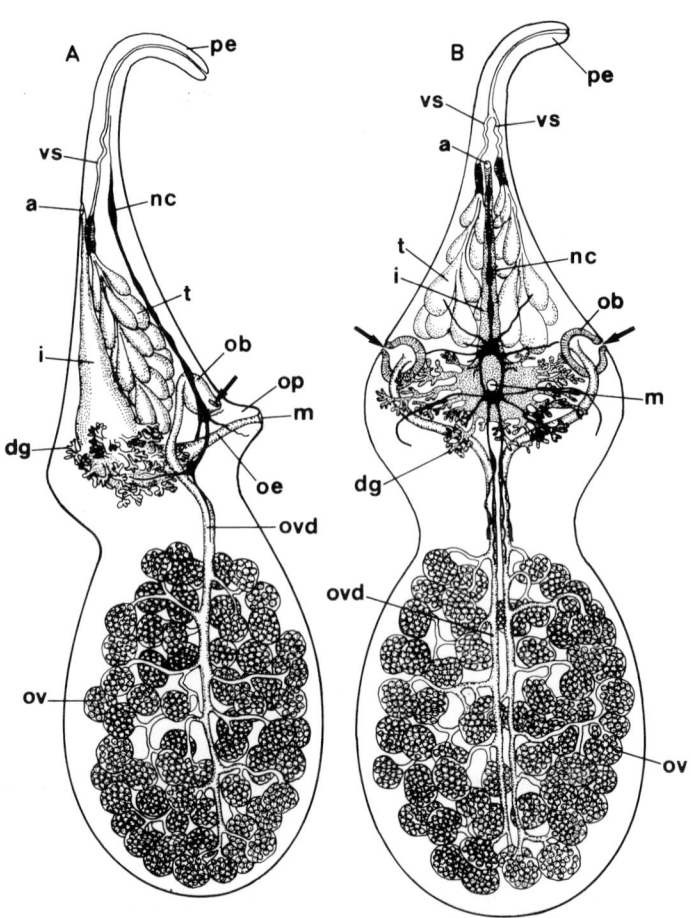

Fig. 14.3. Diagram illustrating the main features of the anatomy of *Anelasma squalicola*, in (A) lateral view and (B) ventral view. Mantle folds not shown. a, Anus; dg, digestive gland; i, intestine; m, mouth; nc, ventral nerve cord; ob, oviducal bulb; oe, oesophagus; op, oral proboscis; ov, ovary; ovd, oviduct; pe, penis; t, testis; vs, vas deferens. Arrows indicate openings of the oviduct. From Johnstone & Frost, 1927.

A further indication that these barnacles are parasitic is the observation of Hickling (1963) that barnacle-infested female sharks were never found to be pregnant and, with one exception, displayed no enlargement of the eggs in the ovary. The testes of infested males were also markedly reduced in size. However, spermatozoa were observed in some of the larger testes from infested hosts, suggesting that *Anelasma* does not go as far as its relative *Sacculina*, which is known to castrate its crab host.

Thus, *Anelasma* appears to be a drain on the resources of older hosts and significantly affects their fertility, leaving little doubt that *Anelasma* is a committed mesoparasite. Whether *Anelasma* supplements its parasitic diet by ingesting material from the surrounding water like its free-living relatives remains to be seen, but, if it does so, such a semi-parasitic life style would be unique amongst animals.

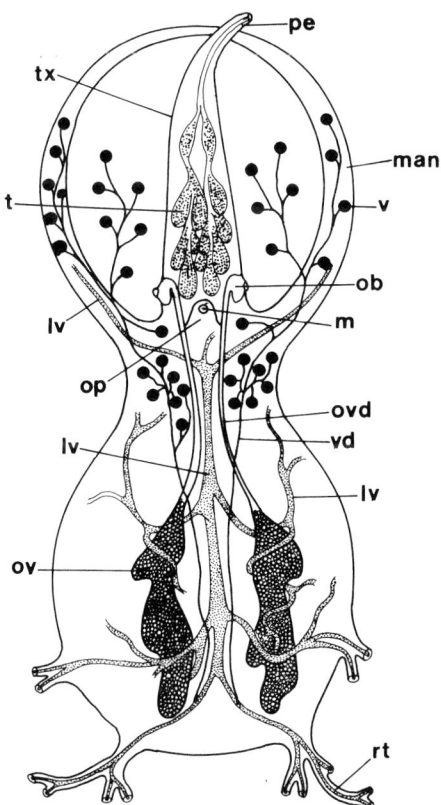

Fig. 14.4. Diagram showing the location of the reproductive organs in *Anelasma squalicola*. lv, Lacunar vessel; v, 'vitelline' (?) gland; vd, 'vitelline' (?) duct. Other labelling as in Figs 14.1B and 14.3. From Johnstone & Frost, 1927.

It is probably also significant that the most metabolically active organs in the body are the two ovaries and that, in the adult, the ovarian follicles with their yolky eggs fill the peduncle and are intimately associated with the lacunar system (Figure 14.4). This lacunar system has blind-ending, capillary-like branches ramifying among the ovarian follicles and penetrating into the roots as far as their tips (Figure 14.4). The 'capillaries' spread throughout the entire body and are especially well marked in the mantle. Thus, the ovarian follicles are in a position to receive nutrients via the lacunar system from the digestion of occasional prey, but they are even better placed to receive a continuous supply of nutrients directly from the host via the peduncular roots and their lacunar 'capillaries'.

Each of the two oviducts of *Anelasma* communicates with an oviducal bulb (Figures 14.3, 14.4), which opens into the mantle cavity via a slit-like aperture at the base of the thorax. Darwin (1851) described these bulbs as "acoustic (?) sacks", but Johnstone & Frost (1927) believed that they serve as seminal receptacles. Like free-living barnacles, *Anelasma* is hermaphrodite with the testes housed in the thoracic region (Figures 14.3, 14.4). Most individuals have a neighbour (see above) and cross-insemination using the long penis (Figures 14.3, 14.4) is theoretically possible. The eggs of *Anelasma* are retained inside the mantle cavity, where they are cemented together to form two pairs of roughly quadrangular plates or ovigerous lamellae, the edges of which are usually visible through the gap between the mantle lobes (Figure 14.1B). Johnstone & Frost (1927) observed that the ova in these lamellae are at various stages of development. Some embryos are close to hatching and dispel any lingering doubts about the crustacean affinities of the animal since the newly hatched larva is a nauplius (Figure 14.5).

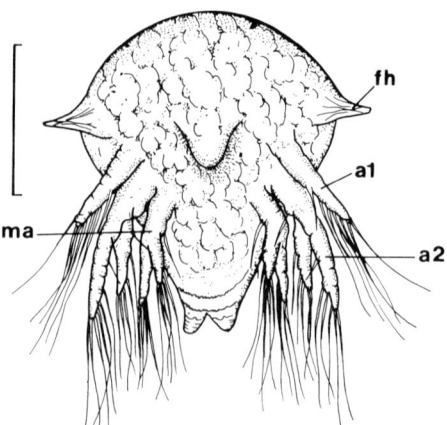

Fig. 14.5. The nauplius of *Anelasma squalicola* viewed from an antero-ventral direction. a1, a2, First and second antennae; fh, fronto-lateral horns; ma, mandible. Scale bar: 250 µm. From Frost, 1928, with permission of Cambridge University Press.

Johnstone & Frost (1927) found that those barnacles less than 10 mm in length were immature, while those over 20 mm generally had eggs in the mantle cavity. According to Hickling (1963), the swelling of the peduncle as the ovaries enlarge within it indicates maturity and this occurs at a total length of about 16 mm. No very young individuals were found accompanying barnacles more than 18 mm long and some unknown factor (or factors) associated with maturity and swelling of the peduncle must deter settlement of young barnacles adjacent to established adults.

There is one more anatomical puzzle concerning a system of so-called cementary glands. This system comprises numerous, apparently syncytial, glandular follicles, each one drained by a collecting tubule. The follicles are scattered throughout the proximal regions of the mantle and the waist region between the capitulum and the peduncle (Figure 14.4v). Johnstone & Frost (1927) suspected that the secretion from these follicles might contribute to the yolk reserves of the eggs and called the glands "vitelline glands" and their ducts "vitelline ducts". However, this interpretation seems unlikely, since the two main ducts into which all the tubules open are said to communicate with the lacunar system, not the female reproductive system.

15

ISOPODS

15.1 INTRODUCTION

The word 'isopod' may not be familiar, but woodlice are within the experience of most people. Their flattened, jointed bodies, covered with a row of overlapping armoured plates or terga, are about a centimetre long and they are frequently encountered in the garden beneath stones, plant pots and the like. Two long antennae protrude from the head and the animal is propelled by seven pairs of similar legs (hence the name 'isopod'), usually called pereopods or pereiopods[1], which tell us immediately that this is a crustacean and not an insect. In fact, it is not always easy to get a prolonged view of the underside of a woodlouse because of the animal's inclination when disturbed to roll

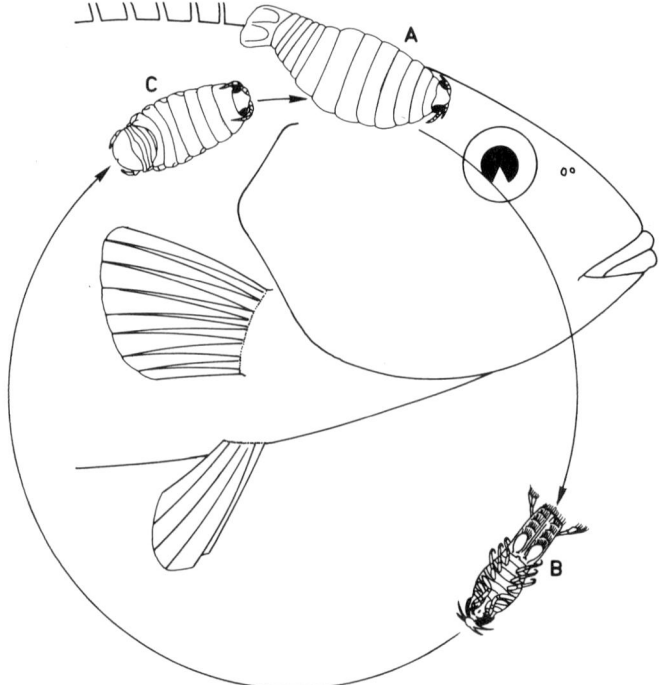

Fig. 15.1. The life cycle of a cymothoid such as *Anilocra* sp. (A) Permanently attached and sessile female; (B) manca larva (shown enlarged in Fig. 15.6); (C) mobile male. See text for explanation. Based on Adlard & Lester, 1994, 1995.

---

[1] 'Pereopods' and 'pereon' are alternative names for 'thoracopods' and 'thorax' (see Glossary) and are preferred here since they are in use in the literature on isopods.

up into a ball (hence the name 'pill bug'). Most crustaceans are aquatic, but woodlice are truly terrestrial and although they prefer moist microenvironments, some are able to tolerate remarkably dry places, such as crevices at floor level in my house.

Although we refer to these animals as 'lice' they are not of course parasites, feeding for the most part on decaying vegetable and sometimes animal material (Edney, 1954). Nevertheless, there are some 'fish lice' in the true parasitological sense of the expression that we would readily recognise as isopods based on our experience of the free-living inhabitants of our gardens. These are the cymothoids. Their life cycle is uncomplicated (Figure 15.1). Morphologically they have undergone remarkably little change as a result of the demands of parasitism, although this conservatism is not matched by their life styles and behaviour, as we will see shortly. The only conspicuous morphological modification concerns the terminations of the many legs, which form formidable grappling hooks for attachment. An enduring and rather disturbing image is of a single, comparatively enormous, blood-feeding parasite like *Anilocra physodes* crouching like a jockey on the top of the head of its host (Figure 15.1), but even more disturbing is the thought that other relatively large species inhabit the mouth cavities of their hosts.

A much more widespread and abundant group of isopod fish parasites are the gnathiids. They are of exceptional interest to naturalists, having several truly remarkable features. They depart a little from the 'woodlouse' image since they are more cylindrical in body shape with swollen central body segments, but, nevertheless, are unmistakably isopods. However, their life cycle is unique (Figure 15.2) and differs in fundamental ways from that of cymothoids. Like *Anodonta* and other unionacean molluscs (Chapter 16) they are parasitic only as larvae, but they differ from the parasitic molluscs because they spend three separate larval periods as parasites, ingesting fish blood, taking no food of any kind between these periods or as adults.

15.2 BRITISH ISOPOD FISH PARASITES

In the 1957 edition of the Plymouth Marine Fauna (Marine Biological Association, 1957) cymothoids are barely mentioned. The Fauna records the finding on two different occasions of single specimens of *Anilocra physodes* (the first records for the British Isles, excluding the Channel Islands). One specimen was described as a "Large male (33 mm) attached to a Red mullet" (striped red mullet, *Mullus surmuletus*). In addition there is a record of a single specimen of *Nerocila neapolitana*, presumably not attached to a host. However, interest in British cymothoids has received a boost from the finding by Horton (2000, 2001) of *Ceratothoa steindachneri*, a parasite of the buccal cavity of the lesser weever, *Echiichthys vipera*. She has discovered established colonies of this parasite in Whitsand Bay, Sennan Cove and Ligger/Perran Bay, Cornwall.

Gnathia is represented by four species in the British fauna (Naylor, 1972). The adults of *G. maxillaris* and *G. dentata* are the commonest intertidal forms, where they may be encountered in rock crevices or in shelters such as empty barnacle skeletons, sponges and holdfasts of the seaweed *Laminaria*. Davies (1982) found that more than 50% of shanny, *Lipophrys pholis*, caught in rock pools at Aberystwyth were infected with from one to seven larvae of *G. maxillaris*. In the same pools, five-bearded rockling (*Ciliata mustela*), sea scorpion (*Taurulus bubalis*) and corkwing wrasse (*Crenilabrus melops*) were also parasitised. Two other species, namely *G. oxyuraea* and *G. vorax* (both sometimes confused with *G. maxillaris*) are more typically found offshore but may occasionally be found in the intertidal zone. According to Potts (1973), *G. maxillaris* has established itself in the fish tanks of the Plymouth Marine Aquarium (see also below) and blood-feeding larvae of the same isopod have been observed occasionally to be pests in marine aquaria in the UK (A. J. Davies, personal communication).

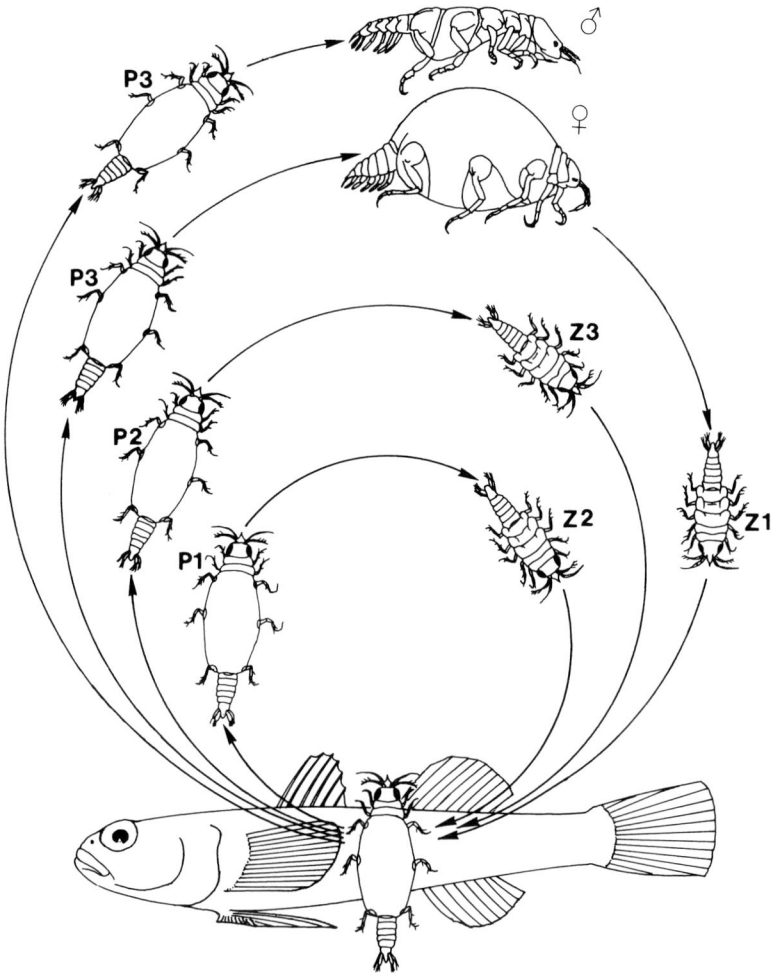

Fig. 15.2. The life cycle of *Paragnathia formica*. The larva spends three periods of its life as a fish parasite; the adults are free-living. The adult female (♀) liberates zuphea larvae (Z1). Z1 ingest blood of the common goby, *Pomatoschistus microps*. Fully fed praniza 1 larvae (P1) leave the host, digest their blood meals and moult to zuphea 2 larvae (Z2). Z2 then return to a host, take a blood meal and leave the host as P2 larvae. P2 moult to zuphea 3 larvae (Z3). Z3 return to a host and take a blood meal. When fully fed the parasites leave the host as praniza 3 larvae (P3) and do not return. P3 larvae moult, producing free-living and non-feeding males (♂) or females (♀). Based on Monod, 1926 and Upton, 1987a.

Paragnathia is represented by only one species in Britain, namely *Paragnathia formica*. The life cycle of *P. formica* takes place in the drainage creeks of salt marshes, such as those on the north coast of Norfolk where it has been studied extensively by Upton (1987ab) and by Tinsley & Reilly (2002) (see below). The common goby, *Pomatoschistus microps*, acts as a frequent host for the larvae.

15.3 CYMOTHOIDS

Much of our knowledge of the biology of cymothoids stems from observations made on species of *Anilocra*. An important contribution is that of Legrand (1952) on *A. physodes*

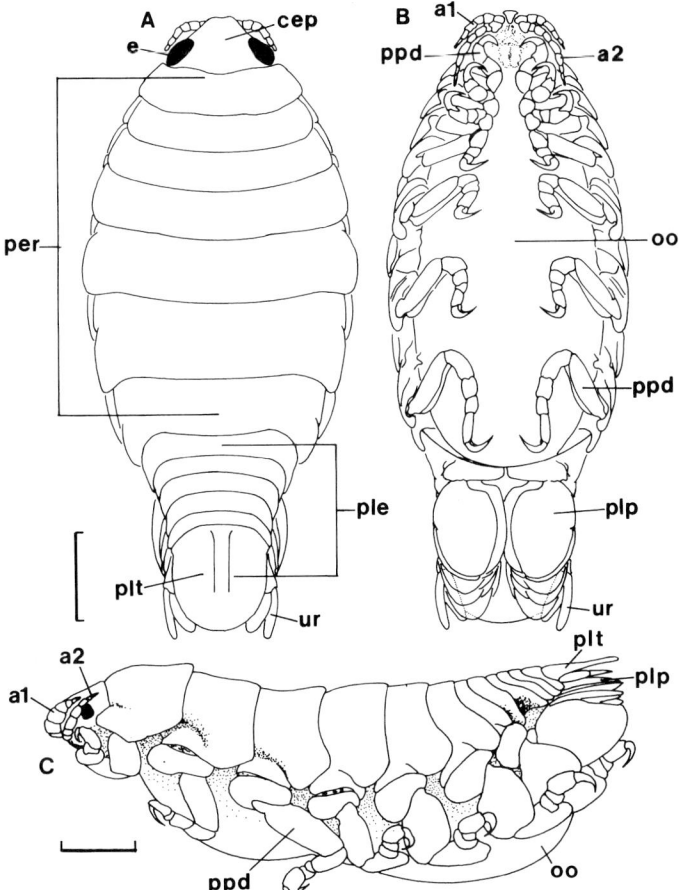

Fig. 15.3. The morphology of female cymothoid isopods. (AB) dorsal and ventral views respectively of *Anilocra pomacentri* from the coral reef fish *Chromis nitida*. (C) Lateral view of *Ceratothoa guttata* from the flying fish *Parexocoetus brachypterus*. a1, a2, first and second antennae; cep, cephalon; e, eye; oo, oostegite; per, pereon; ple, pleon; plp, pleopod; plt, pleotelson; ppd, pereopod; ur, uropod. Scale bars: 2 mm. (AB) From Adlard & Lester, 1995, reproduced with kind permission of CSIRO Publishing (http://www.publish.csiro.au/journals/ajz). (C) From Bruce & Bowman, 1989.

from corkwing wrasse, *Crenilabrus melops*, and ballan wrasse, *Labrus bergylta*, collected in the English Channel. Additional information on the life cycle and development of cymothoids is provided by the studies of Adlard & Lester (1994, 1995) on the Australian parasite *A. pomacentri* from the Barrier Reef chromis, *Chromis nitida*.

Wherever *Anilocra* alights on the host, the parasite usually makes its way to a position on the head (Figure 15.1). It attaches itself just posterior and dorsal to the fish's eye on either the right or the left of the midline, with the parasite's head directed anteriorly with respect to the fish.

15.3.1 *General morphology*

The main features of the morphology of a cymothoid isopod are shown in Figures 15.3 and 15.4. There is no carapace and the clearly segmented body is divisible into three regions. There is a head or cephalon (strictly speaking this tagma is a cephalosome), a thorax or pereon (peraeon) and an abdomen or pleon (Figure 15.3). The cephalon bears two pairs of relatively short jointed antennae (first and second antennae = antennules and antennae) together with a pair of eyes. Neatly packed beneath the cephalon are the mouthparts (Figure 15.4), which are the tiny modified limbs of the cephalic segments, plus a similarly miniaturised and specialised pair of limbs from the first pereon segment, which is fused with the cephalon. The anteriormost mouthparts are the mandibles, each one with a short palp, followed by a pair of maxillules (= first maxillae)

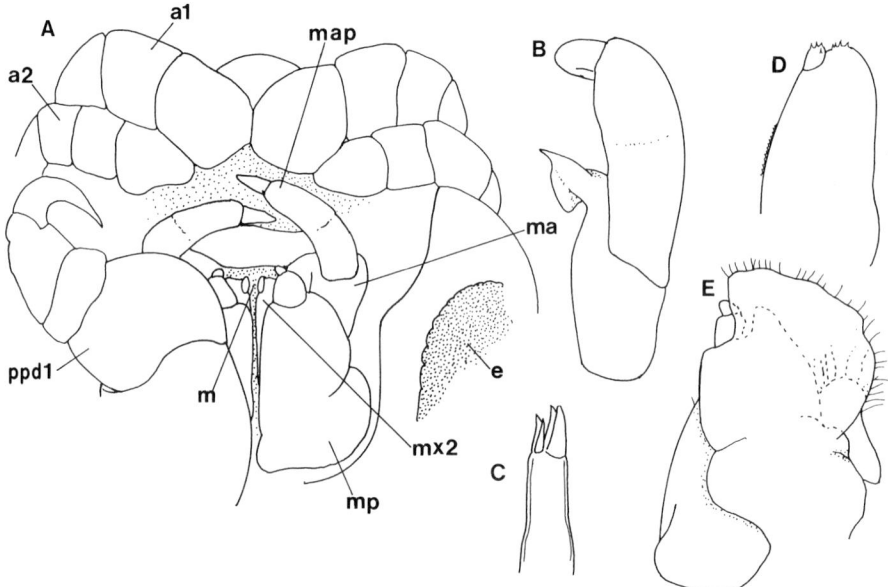

Fig. 15.4. Mouthparts of a cymothoid isopod. (A) Ventral view of mouth region with first pereopod and part of the maxilliped removed on the left side to reveal mouthparts beneath. a1, a2, first and second antennae; e, eye; m, mouth; ma, mandible; map, mandibular palp; mp, maxilliped; mx2, second maxilla; ppd1, first pereopod. (B-E) mandible, first maxilla, second maxilla and maxilliped respectively. From Bruce & Bowman, 1989.

and a pair of maxillae (= second maxillae). All of these are shielded ventrally by a pair of maxillipeds, which are the limbs derived from the first pereon segment.

Each of the succeeding seven segments of the pereon bears a pair of large uniramous limbs (pereopods), each terminating in a powerful claw (Figure 15.3B). The claws of the anterior pereopods are forwardly directed while those of the rear pereopods are directed backwards. In the case of externally located parasites like *Anilocra*, this opposing arrangement of grappling hooks resists any forces threatening to unseat the parasite, whether they are generated by acceleration, deceleration or turning movements of the fish. These hooks are likely to provide equally effective anchorage for parasites like *Ceratothoa*, living in the current-swept gill chamber.

The pereon is followed by five pleon segments (Figure 15.3A), each bearing a pair of highly flattened, biramous limbs or pleopods (Figure 15.3B), well suited for propelling the organism through the water. A sixth pleon segment is fused with the terminal telson to form a so-called pleotelson and this bears a single pair of biramous uropods.

A few distinctive features enable us to distinguish ovigerous female individuals from males. The females do not release their eggs into the sea. Large plate-like lamellae or oostegites project inwards from the bases of some of the anterior pereopods, creating a space or marsupium between the oostegites and the plates (sternites) covering the animal's ventral surface (Figure 15.3BC). This acts as a brood chamber, which receives the fertilised eggs from two gonopores near the bases of the pereopods on the sixth pereon segment. Developing embryos remain in the marsupium until the 'manca' larva hatches.

Males are generally smaller than females (Figure 15.1) and have a pair of penises near the midline between the pereopods of the eighth pereon segment. The pleopod of the second pleon segment bears a process called the appendix masculinum of uncertain function. It will be obvious to the reader that mating (see below) must take place before the oostegites of the female develop, because after this has taken place the female gonopores are covered and inaccessible.

15.3.2 Sex change and mobility
Taken at face value these sexual distinctions between individuals point to the familiar reproductive pattern in which the organism spends its entire lifetime either as a female or as a male (gonochorism). However, in cymothoids, these apparently straightforward features conceal a surprisingly different and interesting reproductive pattern. The immature cymothoid first becomes a mature male, but later in life undergoes a sex change and becomes a female.

Specialists on isopods have referred to this sex change phenomenon as protandrous hermaphroditism. This is true in the sense that, within the lifetime of an individual, both functional male and functional female organs are present, the male organs becoming functional before the female organs. However, it is different from the protandrous hermaphroditism displayed by other parasites such as the monogeneans (Chapter 3) in which the male organs mature first and the female organs later, with the male organs continuing to function so that in later life the male and female systems are operational side by side.

Legrand (1952) highlighted an intriguing feature of the population distribution of *Anilocra physodes*. Out of 511 parasites collected from July to October 1949, 346

(68%) occurred singly on their host. These isolated individuals ranged from very young males to very large females. He also reached the conclusion that only young parasites and males are mobile, attaching themselves temporarily to the host and after taking a meal moving to a new host. He described the young stages of *A. physodes* as agile swimmers, able to perform somersaults that permit them to make contact with and attach to any fish in their path. Legrand confirmed earlier observations that males may detach spontaneously from the host and that young parasites abandon dead hosts. He observed that the simple act of removing a fish from the water might induce a parasite to leave its host.

Individuals lose their ability to leave the fish when they change sex. Individuals between 14 and 20 mm in length appear to be in this transitional sexual stage and isolated individuals longer than 14 mm are usually sedentary. A positive correlation between the sizes of these isolated parasites and the sizes of their hosts lends support to this conclusion, as does the more or less complete disappearance of the setae on the swimming appendages (pleopods). Sedentary individuals also acquire asymmetrical body pigmentation because of the asymmetrical position they adopt on the head of the fish (Figure 15.1). The dorsal region of the fish is dark, but moving down the flank of the fish the colour lightens. The parasite's longitudinal axis is shifted either

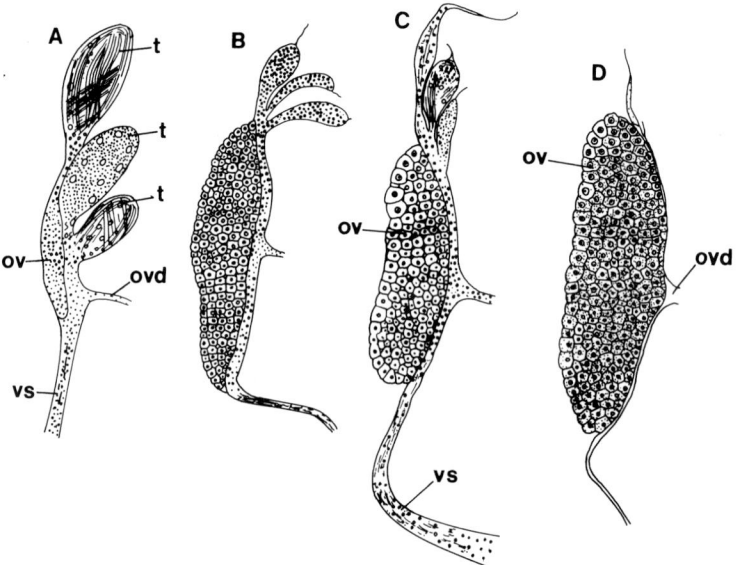

Fig. 15.5. The gonads of specimens of the cymothoid *Anilocra physodes* of various sizes. (A) male, length 10 mm; (B) transitional stage, isolated individual length 17 mm; (C) large paired male, length 26.5 mm; (D) paired female, length 26 mm. ov, Ovary; ovd, oviduct; t, testis; vs, vas deferens. From Legrand, 1952.

to the left or to the right of the fish's longitudinal axis. The chromatophores of the parasite contract or expand to match the local background of the fish's skin. The part of the parasite's body lying close to the dorsal apex on the fish thus becomes darker than the rest of the parasite. Parasites that retain their mobility and regularly change hosts are assumed not to remain long enough in one asymmetrical position for this pigment asymmetry to develop. Among Legrand's sample of 346 isolated parasites, about half of the 86 young parasites/males between 8.5 and 14 mm in length had asymmetrical pigmentation indicating that they were already sedentary.

The male features decline in prominence in the transitional individuals. In particular, a progressive enlargement of the thorax transforms the slim contours of the male and a reduction takes place in the appendix masculina (see above) and in the genitalia. Internally the ovary increases in size and the testis regresses (Figure 15.5). The oviduct develops rapidly, but the female genital openings are not present in individuals smaller than 20 mm, the size at which female maturity was first detected. It is these larger individuals, above 20 mm in length, that are likely to be accompanied by another individual. The maximum size of *A. physodes* recorded by Legrand (1952) at Roscoff was 41 mm.

Legrand's observations indicate that the presence of a mature female greatly retards progression of an accompanying male parasite towards the female phase, but does not retard its growth. Thus, although most males in couples are between 8.5 and 14 mm in length, a few may reach 27.5 mm. However, the transition from the male to the female phase is not arrested indefinitely. Should one of these large males lose its female partner, it does not remain in the male phase for the rest of its life but transforms to the female phase.

Legrand (1952) demonstrated experimentally the influence of the mature female on the development of her partner. He collected fishes each with a single couple of *A. physodes*. The female partners were all more than 22 mm in length and the males between 13 and 16 mm. Some of these couples were left undisturbed and maintained in the laboratory for two to eight months. The females were removed from each of the other couples, leaving the males in isolation on their hosts. Those males that retained their females remained in a fully functional male state, whereas those deprived of their females underwent a substantial shift towards the female condition. This evidence suggests that the mature female prolongs in some way the male status of its partner.

Further evidence presented by Legrand also indicated that a young male might accelerate the transition of a partner towards the female phase. Feminisation of parasites at an intermediate sexual stage made greater progress when a young male was present than when the parasite in the intermediate stage was in isolation.

Legrand described an unusual triple association of parasites on one fish, involving a large female (37 mm), a young male (9 mm) and a parasite of intermediate size (26.5 mm). The last-named would be expected to display full male status maintained by the influence of the female, but, in fact, the parasite was assessed as having an intersexual status. Legrand hypothesised that the intermediate status of this parasite was the consequence of antagonism, the influence of feminisation from the young male counterbalancing the influence of masculinisation from the large female.

15.3.3 Mating
Legrand described a behavioural interaction between members of a couple that he assumed to be copulation. From an immobile position, side by side, the partners became active, indulging in scraping movements with the pereopods. The male moved slowly towards the rear of its companion. The female then raised her abdomen permitting the male to pass beneath, where he turned upside down so that the ventral surfaces of the

partners were in contact. The female raised her abdomen until it was almost perpendicular to the surface of the fish and the two individuals remained in contact for 5 – 10 minutes and then returned to their original positions. Legrand noted that this behaviour took place independently of the moult of the female.

15.3.4 *Egg incubation and hatching*

Legrand (1952) observed that egg laying by the female *A. physodes* is preceded by a moult in the course of which the oostegites appear. As we saw earlier (Chapter 8), moulting is a dangerous business for a parasitic crustacean because of the need for appendages to relinquish their grip while their exoskeleton is shed and also the need for the new cuticle to harden sufficiently for the appendages to regain their usefulness for attachment. This is not a problem for *A. physodes* since moulting occurs in two stages as in woodlice. First, the exoskeleton of the posterior part of the body is renewed and then, five days later, that of the anterior part of the body. The anterior pereopods ensure

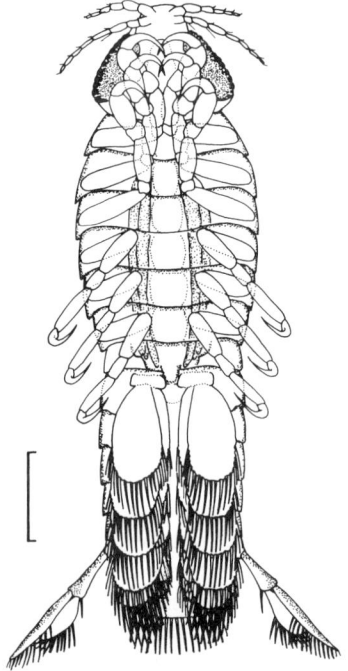

Fig. 15.6. Manca larva of the cymothoid isopod *Anilocra pomacentri* in ventral view. Scale bar: 0.5 mm. From Adlard & Lester, 1995, reproduced with kind permission of CSIRO Publishing (http://www.publish.csiro.au/journals/ajz).

safe attachment while the posterior moult takes place and by the time the anterior moult begins the posterior pereopods have hardened sufficiently to regain their function and ensure security.

Egg laying takes place several days after the completion of the marsupium. Each egg is just over 1 mm in diameter and the female lays 250 – 350 in summer, but less than 100 in the autumn.

Legrand described the aeration of the developing eggs within the marsupium by water currents produced by movements of the mouthparts. He noted that the incubating female does not feed, the internal organs being compressed by the enlarged marsupium. According to Legrand, development of the eggs of *A. physodes* lasts about 20 days.

Adlard & Lester (1995) found that the eggs of *A. pomacentri* take about 26 days to reach the stage that he called Prehatch I. After a further eight days the egg membrane ruptures to release a Prehatch II stage with well-developed eyes and body pigmentation, but lacking the hairs or setae on the pleopods and uropods that are necessary for propulsion of the larva after hatching. The hairless Prehatch II stage moults within the marsupium, producing the setose and rather formidable looking manca larva (Figures 15.1, 15.6). This is released from the marsupium about 10 days after the appearance of the Prehatch II stage. The mancae of *A. pomacentri* are released singly or in small groups over a period of 1 – 3 hours. Three days after this release the female *A. pomacentri* moults and shortly afterwards the presence of blood in the intestinal caeca of the parent indicates that feeding has resumed. A new brood of eggs appears in the marsupium of *A. pomacentri* about 18 days after the release of the mancae.

15.3.5 Infecting new hosts

The free-swimming mancae are highly active and strongly photopositive. Legrand (1952) observed that larvae of *A. physodes* respond to slight differences in lighting by moving in a compact group to the more brightly illuminated zone. Adlard & Lester (1995) observed that the manca of *A. pomacentri* could suspend itself from the surface film by inserting the tips of its pereopods through the meniscus. At this stage the manca has only six pairs of pereopods, the seventh pair being acquired at the next moult.

While free-swimming, the larvae subsist on their stores of yolk and Legrand (1952) kept the mancae of *A. physodes* alive for 18 days in circulating seawater. According to Adlard & Lester (1995), the manca of *A. pomacentri* is infective as soon as it is released from the marsupium, but its ability to attach to a host decreases with time. Eight days after release only 50% of larvae were infective.

Legrand (1952) observed that the mancae of *A. physodes* do not respond to young fishes (wrasse) if the latter remain immobile. However, currents produced by active fishes induce movements of the hook-bearing pereopods that serve to attach the manca larva to any part of the fish that comes into range. The paired fins and the caudal fin are common targets.

The potential host is not insensitive to the attentions of these larvae. Legrand (1952) reported that fishes sometimes execute an abrupt *volte-face* in response to mancae of *A. physodes* targeting the caudal fin, and that these fishes may ingest the parasite. Adlard & Lester (1994, 1995) noted that some newly invaded fishes (*Chromis nitida*) perform erratic movements, including apparent attempts to dislodge *A. pomacentri* by scraping against the substrate or other fishes.

Knowledge of how the manca larva finds its host is limited. Moser & Sakanari (1985) studied the larvae of *Lironeca vulgaris*, a cymothoid found in California in the gill cavity or buccal cavity of local fishes. They identified some behavioural responses of the free-swimming larvae to environmental cues, but the relevance of some of these responses to host location are not obvious. They found that the larvae reduce their swimming speed significantly in the presence of fish mucus and settle more readily in the presence of mucus or a white substrate. The larvae were photopositive when illuminated horizontally, but were geopositive when illuminated from above and when in darkness.

15.3.6 Nature of the food
Feeding takes place soon after attachment of the manca larva to the host. According to Legrand (1952) blood in the digestive tract indicates that *A. physodes* takes a blood meal about 70 hours after attachment. Romestand & Trilles (1976a,b) demonstrated that the latero-oesophageal glands in the cephalon of *A. physodes* secrete an anticoagulant. Similarly Adlard & Lester (1995) observed that the newly attached *A. pomacentri* creates a lesion by rasping away with the mouthparts and penetrates into the muscles beneath the dermis. They found that lesions produced by mancae within the first 48 hours of parasitism ranged from 0.9 to 2.5 mm in diameter. Some lesions exposed the neural spines of the host and haemorrhages from small blood vessels were observed in 87% of newly infected fishes. Adlard & Lester found host striated muscle fibres in the gut caeca of young parasites 48 hours after infection. Host red blood corpuscles were first observed in the caeca after 96 hours.

Adlard & Lester (1995) found that fishes infected with feeding parasites had significantly lower red blood cell counts than uninfected fishes or fishes with non-feeding parasites. Adlard & Lester (1994) observed that the parasite affected growth of the host and that egg production of female fishes was reduced. Their field observations showed that parasitism contributed significantly to the mortality of juvenile fishes.

15.4 GNATHIIDS

Gnathiid larvae spend three periods of their lives gorging themselves on fish blood (Figure 15.2), using mouthparts modified for piercing and sucking, as are those of a mosquito. After each feed, the larva leaves the host, digests the blood meal and undergoes a moult, the last moult producing the non-feeding mature males and females. Unlike the cymothoids, there is no sex change from male to female and sexual dimorphism is striking, the male having a relatively large square head with massive forward-projecting mandibles, while the female has a reduced skeleton and no functional mouthparts (Figure 15.2). *Paragnathia formica* lives communally in burrows, with a single male presiding over a 'harem' of up to 25 females.

Lifetime nutritional intake is limited to the three larval feeds of blood, which must provide resources to support maturation of the males and females, a process which may take many months, and, since gnathiids are viviparous, full development of the embryos inside the body of the female.

The parasitic larvae and the free-living males and females of gnathiids are so different that they were originally assigned to different genera, namely *Praniza*, *Anceus* and *Gnathia* respectively (see Monod, 1926, Mouchet, 1928a). '*Gnathia*' takes priority over '*Praniza*' and '*Anceus*'. However, Stoll (1962) introduced terms derived from the old generic names *Zuphea* and *Praniza*, namely 'zuphée' for the pre-feeding larval stage in which body segmentation is evident and 'pranize' for the bloated post-feeding larval stage in which segmentation is obscured. She used the following abbreviations to represent the larval stages that succeed one another during the gnathiid life cycle: Z1 (pullus), P1, Z2, P2, Z3, P3. Subsequent authors have adopted the same notation and the terms 'zuphea' and 'praniza' (see Upton, 1987a; Tinsley & Reilly, 2002).

There are morphological features that set the gnathiids apart from their fellow isopods. Two pereon segments are fused to the cephalon, not one as is typical of isopods, including the cymothoids. Thus although the anterior tagma of gnathiids is frequently referred to as a 'cephalon' it is technically a cephalothorax.

The limbs of the second pereon segment fused to the cephalon are miniaturised to form an extra pair of mouthparts called gnathopods or pylopods that lie ventral (posterior) to the other mouthparts. A major difference between the genera *Paragnathia* and *Gnathia* concerns these pylopods: those of the male have five podomeres in the former compared with two or three in the latter.

Compared to the isopod norm of seven pairs, the life cycle stages of gnathiids have only five pairs of ambulatory limbs or pereopods, carried by pereon segments 3 – 7. The last pereon segment bears no limbs. The large quantities of blood ingested by parasitic larvae lead to inflation of pereon segments 3 – 5. These segments are also inflated and fused in the non-feeding female to provide space for the accommodation of developing embryos. According to Monod (1926) the embryos are retained inside the ovary (often referred to as a 'uterus'), not inside an external marsupium as in cymothoids. However, in *P. formica* Monod found recognisable vestiges of oostegites in the female on the ventral side of pereon segments 3, 4 and 5.

The five pairs of pleopods on the pleon or abdomen and the single pair of uropods on the telson have no distinctive features, each limb consisting of two flat plates.

15.4.1 Host invasion
According to Monod (1926), the zuphea larva of *Paragnathia formica* sometimes swims slowly, but is capable of extremely rapid bursts of speed as it is propelled like an arrow by its pleopods. New-born, non-feeding larvae survive for an average of 43.1 days (Tinsley & Reilly, 2002). Some of these new-born larvae rapidly infest the host, as indicated by linked peaks in density of new-born larvae and prevalence of parasites on the host. However, the extended survival period of the free-living larvae provides opportunities for infection at least for days and perhaps for longer (Tinsley & Reilly, 2002). As Tinsley & Reilly point out, the fact that females allocate resources to support this longevity suggests that this strategy pays off in terms of extra invasion success. When host density is low, a 'sit-and-wait' strategy adequately provisioned by the females is likely to be better for obtaining a host than resorting to active location.

Fishes exposed for the first time to invading zupheas were observed by Monod (1926) to react vigorously, although they soon became accustomed to the attentions of the parasites. For *P. formica*, Monod recorded a lack of host specificity, which extended experimentally even to the frog *Rana temporaria*, a potential host that would never be encountered naturally.

15.4.2 Attachment to the fish
The zuphea may attach itself anywhere on the surface of the fish where there is ready access to a blood vessel. Fins provide a particularly suitable substrate (Figure 15.7A) and dermal scales do not necessarily constitute an impassable barrier if there is

sufficient space to penetrate between the scales (Figure 15.7B). The buccal and gill cavities also provide a soft substrate.

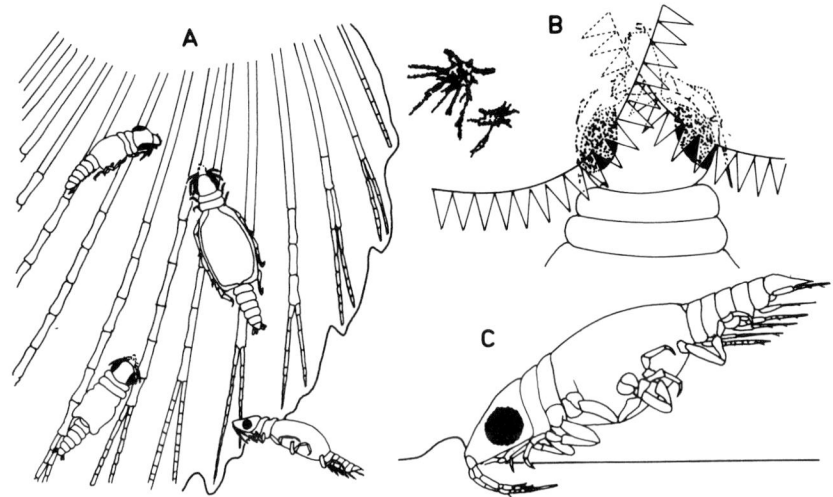

Fig. 15.7. Praniza larvae of *Paragnathia formica* attached to a goby. (A) Pranizas attached to a pectoral fin; (B) the head of a praniza lodged beneath scales; (C) lateral view of a praniza attached to its host. From Monod, 1926.

Larvae on the skin almost always orientate themselves with the head anterior with respect to the fish and, when viewed from the side, it is clear that, unlike the cymothoids, gnathiid larvae attach themselves to the fish by the mouthparts, not by their pereopods (Figure 15.7C). The body of the attached larva is orientated obliquely with respect to the host's surface. The first and second pereopods are folded and directed anteriorly, but even if their tips make contact with the host they do not serve for attachment. Pereopods 3, 4 and 5 are also folded and held against the sides of the body. The occasional beating of the pleopods presumably serves for respiration.

15.4.3 Mouthparts
The mouthparts of the zuphea/praniza must attach the parasite securely to its host and assist in obtaining a blood meal. The form and disposition of the mouthparts reflect this dual function and these extreme adaptations mean that the mouthparts bear little resemblance to those of a typical omnivorous, carnivorous or herbivorous isopod, or indeed to those of adult gnathiids. The presence of an anterior tapering 'rostrum' gives the head a roughly triangular shape (Figure 15.8) and this rostrum encloses a bunch of stylets, comprising the needle-like terminations of the mandibles, the first maxillae, the paragnaths (an extra pair of projections) and the maxillipeds (Figures 15.8, 15.9). The second maxillae are also needle-like (Figure 15.9D), but are too short to take part in this grouping and are thought to make little contribution to attachment and feeding (Figure15.8B). The walls of the cone-shaped 'rostrum' are formed dorsally and laterally by the upper lip or labrum and ventrally by the maxillipeds and pylopods (Figure 15.8). The cone is directed obliquely forwards and ventrally.

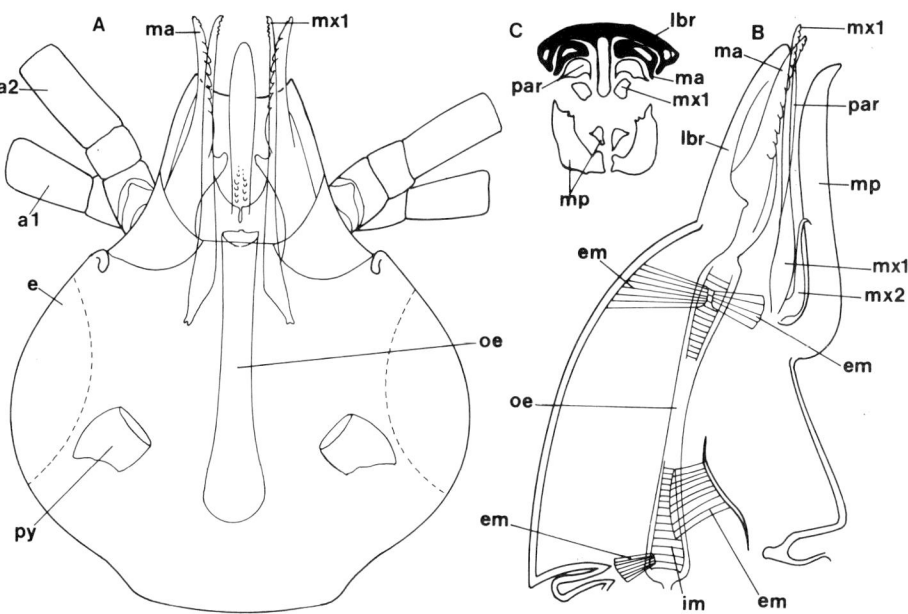

Fig. 15.8. The mouthparts and associated structures of the praniza larva of *Paragnathia formica*. Semi-diagrammatic views of (A) the head in ventral view and (B) a longitudinal (sagittal) section of the head. (C) A transverse section through the distal region of the mouthparts. a1,a2, Bases of first and second antennae; e, eye; em, extrinsic oesophageal muscles; im, intrinsic oesophageal muscles; lbr, labrum; ma, mandible; mp, maxilliped; mx1, mx2, first and second maxillae; oe, oesophagus; par, paragnath; py, basal podomere of pylopod. From Monod, 1926, (A) and (B) modified.

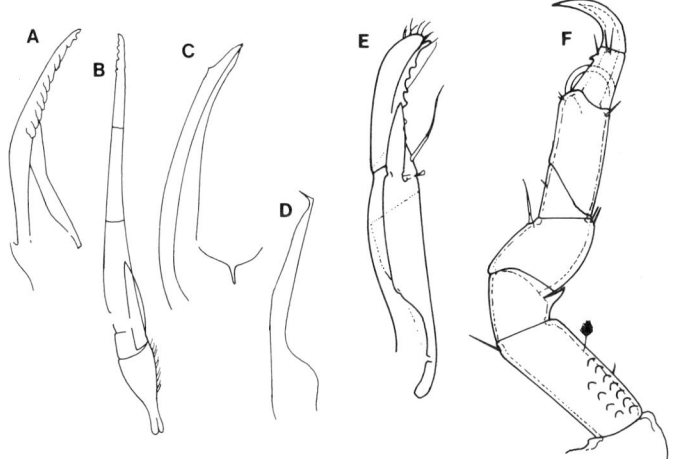

Fig. 15.9. Mouthparts and associated appendages of the praniza larva of *Paragnathia formica*. (A) Mandible; (B) first maxilla; (C) paragnath; (D) second maxilla; (E) maxilliped; (F) pylopod. From Monod, 1926.

The mandibles of *P. formica* have a basal articulation with the labrum and taper to a toothed point, with one edge of the mandible bearing a row of backwardly directed barbs (Figures 15.8, 15.9A). Monod (1926) claimed that the mandibles are more or less immobile and that their principal role is for attachment. He likened them to harpoons, their lateral barbs presumably preventing withdrawal after the mandibles are driven into the skin.

According to Monod (1926) the main organ of perforation in *P. formica* is the first maxilla (Figures 15.8, 15.9B). He described its action as rapid, short stabbing movements, about five in number, which rip open the tissue. He also proposed that the action of the first maxillae serves to keep the feeding wound open and to maintain the free flow of blood from the wound. Perforation is also likely to be the function of the paragnaths (Figure 15.9C), each of which has a single sharp point.

The maxillipeds are structurally more complex (Figure 15.9E). Monod (1926) pointed out that bristles on the terminal palps of these appendages in *P. formica* are well deployed to make contact with the host's surface and may have a sensory role. The other terminal toothed projections of the maxilliped are sharp enough to contribute to perforation and/or to enlarge or maintain the open wound.

In the zuphea/praniza of *P. formica,* the pylopods are uniramous limbs and, according to Monod (1926) have a crucial role in attachment and in perforation (Figure 15.9F). Each has a hooked tip and in this respect they are reminiscent of the hooked pereopods of the cymothoids (see above). These hooks attach the parasite to the host at the level of the apices of the mouthparts protruding from the rostral cone, and the firm anchorage that they provide is important for the performance of the thrusting movements of the first maxillae. On occasions, Monod observed these hooks groping around until they established anchorage in a more anterior location. Then a vigorous traction was exerted, which dragged the rostral cone forwards and deeper into the skin.

The mouthparts of the praniza larva of *Gnathia maxillaris* are similar to those of the larva of *P. formica* (see Davies, 1981).

15.4.4 *Feeding and digestion*

Stoll (1962) observed that the attached feeding period in *P. formica* may be as short as 2 hours and does not exceed two to three days. Davies (1981) observed that the larva of *G. maxillaris* spends 2 – 24 hours attached and feeding.

The oesophagus of *P. formica* is responsible for sucking up host blood. The suction generated by the oesophagus is produced by contraction of intrinsic muscles located in its wall and extrinsic muscles running from the external surface of the oesophagus to locations on the exoskeleton (Figure 15.8B). It seems likely that an anticoagulant is secreted during feeding. In *P. formica* there are glands tentatively described as 'salivary glands' by Monod, situated just behind the cephalon in pranizas and males, but not in females. However, anticoagulant properties of their secretion have not been demonstrated and, in fact, the sites where these glands open are unknown.

The ingested blood shows through the distended cuticle of the body and the pranizas are therefore colourful. They vary from red to yellow, depending on the density of red blood cells ingested with the plasma (Upton, 1987a). According to Mouchet (1928b) the blood of some fishes is rapidly digested, releasing the pigment biliverdin, which produces a vivid green colour. Davies (1982) found that pranizas that had fed recently on shanny (*Lipophrys pholis*) and five-bearded rockling (*Ciliata mustela*) in rock pools at Aberystwyth were easy to distinguish from those that had fed recently on sea scorpion (*Taurulus bubalis*) and corkwing wrasse (*Crenilabrus melops*), because pranizas from the first two fishes had bright red pereons and those from the last two had emerald green pereons.

The oesophagus leads to a tiny 'stomach' (Figure 15.10A), which, according to Juilfs & Wägele (1987), acts as a pump and a filter. The blood meal is then accepted and accommodated in a reservoir called by Monod (1926) the intestinal reservoir and by Juilfs & Wägele (1987) the anterior hindgut. This reservoir expands to fill out much of the body cavity in the region of pereon segments 3 to 5 (Figure 15.10AC). However, it is thought that utilisation of the blood meal does not take place within the reservoir. According to Monod, the contents of the reservoir become progressively more solid in

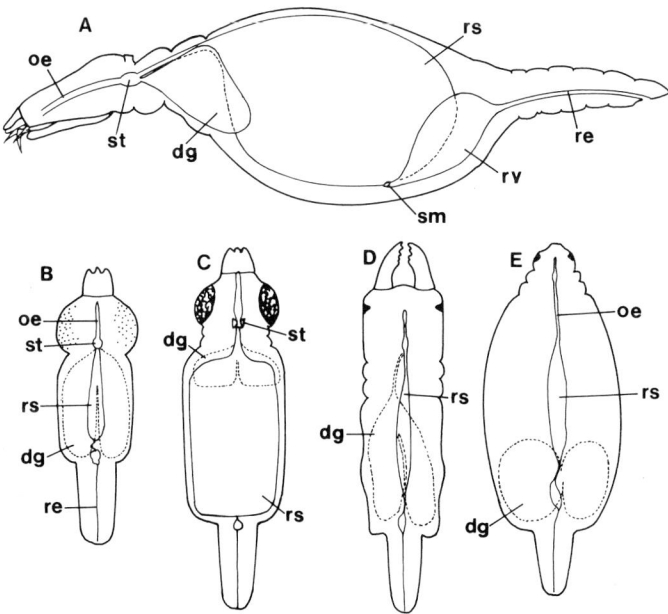

Fig. 15.10. Schematic drawings of the digestive tracts of (A) the praniza of *Gnathia calva*, (B-E) different stages of *Paragnathia formica*. (B) Embryo; (C) praniza; (D) adult male; (E) adult female. dg, Digestive gland; oe, oesophagus; re, rectum; rs, reservoir; rv, rectal vesicle; sm, sphincter muscle; st, stomach. (A) From Juilfs & Wägele, 1987. © Springer-Verlag. Reproduced with kind permission of Springer-Verlag; (B-E) from Monod, 1926.

consistency by an unknown process, but later they regain their fluidity and are transferred via a common duct from the stomach to two blind caeca, called 'enteric organs' by Monod and 'digestive glands' by Juilfs & Wägele (Figure 15.10AC). The contents of the digestive glands are food reserves that can be tapped by the adults, which are incapable of feeding. The reservoirs in the adults are empty (Figure 15.10DE), apart from a residue that persists until death, and, according to Monod, the rectum, although present and leading to the anus on the telson (Figure 15.10), is never functional either in the praniza or in the adult.

In the praniza of *P. formica* the digestive glands are relatively small (Figure, 15.10C) although they have already served a nutritive function in the embryo by absorbing yolk material (Figure 15.10B). In the male the digestive glands are much longer than in the praniza and fill the posterior part of the thorax (Figure 15.10D). In the female the more or less spherical glands are displaced to the rear of the thoracic cavity by the huge 'uterus' (Figure 15.10E) and Monod suspected that the digestive glands sometimes lose their connection with the stomach in these females.

There is another anatomical feature of the gut of gnathiids that may be much more important than its size would indicate. This is the rectal vesicle, a slightly enlarged portion of the hindgut immediately posterior to the blood reservoir and separated from it by a sphincter muscle in *Gnathia calva* (see Juilfs & Wägele, 1987) (Figure 15.10A). At its posterior end the vesicle gives rise to the narrow rectum (Figure 15.10A). Apart from commenting on the thick liquid contents of the vesicle, Monod (1926) was unable to offer an explanation as to its function, but Juilfs & Wägele discovered that the vesicle in *G. calva* is filled with bacteria and lined by an epithelium with a distinctive ultrastructure. Juilfs & Wägele suggested that these bacteria are symbiotic and speculated that they perform a physiological task connected with blood feeding that cannot be performed by the isopod. They remind us that similar relationships have arisen independently between micro-organisms and other blood-feeding fish parasites such as leeches (see Chapter 7). However, if there are any symbiotic interactions between bacteria and the gut contents of gnathiids, the site of these remains unknown. It seems unlikely that they take place in the vesicle because Juilfs & Wägele found no evidence that material similar in appearance to that found in the blood reservoir passes in a posterior direction through the sphincter. We also do not know how these micro-organisms are passed from one generation of isopods to another.

15.4.5 Transmission of micro-organisms

Leeches are undoubtedly important agents in the transmission of blood-dwelling protozoan parasites between fish hosts (Chapter 7). The observations of Davies and co-workers (Davies & Johnston, 1976; Davies, 1982; Davies *et al.*, 1994) on Welsh and Portuguese populations of blennies (the shanny, *Lipophrys pholis*, in Wales and *L. pholis* and Montagu's blenny, *Coryphoblennius galerita*, in Portugal) infected with haemogregarines, show that gnathiids are also transmitters of micro-organisms, and indicate that they play a more significant role in the spread of infection than previously suspected.

At Aberystwyth, the study site in Wales, all shannies over 5 cm in length were infected with *Haemogregarina bigemina*, a protozoan parasite living inside the red blood corpuscles of the fish. However, the infection was first detected in young fish too small to have been bitten by *Oceanobdella blennii*, the only leech known to exist on blennies in Wales (see Chapter 7). Furthermore, this leech was not known at the study site and *H. bigemina* was acquired by these young shannies at a time when the leech was known not to feed. In contrast, praniza larvae of *Gnathia maxillaris* were common at the study site, fed on fish all the year and their intestinal reservoirs (anterior hindgut) contained micro-organisms that were presumed to be developmental stages of *H. bigemina*. Further studies by Davies & Smit (2001) on *H. bigemina* and *Gnathia africana* in South Africa lent even stronger support to the claim that gnathiid larvae, not leeches, transmit the haemogregarines from fish to fish. However, efforts to locate potential transmission stages of the haemogregarines in the salivary glands of pranizas have so far not been successful (Davies & Johnston, 2000; Davies & Smit, 2001), and it seems likely that fishes typically acquire the haemogregarine by eating infected pranizas. As Davies (1982) has pointed out, shannies of all ages readily eat praniza larvae, providing a mechanism for infection of even the smallest fishes.

15.4.6 Predation on gnathiid larvae

Given the opportunity, gnathiid larvae are readily eaten by the natural fish hosts on which gnathiids feed (see above). Gnathiid larvae together with caligid copepods (Chapter 8) also feature prominently in the diet of cleaner fishes. On the Great Barrier Reef in Australia, the bluestreak cleaner wrasse, *Labroides dimidiatus*, cleans 2300 fish from more than 132 species per day and consumes as many as 1200 parasites per day, most of which are juvenile gnathiid isopods (references in Grutter & Hendrikz, 1999). It is not surprising that gnathiids are important target organisms for cleaner fishes, because they are conspicuous and nutritious, containing when fully fed a relatively large quantity of fish blood.

Grutter (1999, 2002) has vividly demonstrated the impact of cleaners on gnathiid populations. She compared burdens of gnathiid pranizas on caged fishes on reefs with cleaner fish and on reefs from which all cleaners had been removed. After 12 days, she found a 3.8 fold increase in gnathiid larvae on reefs without cleaners. Of special interest was the finding that there was no difference in parasite abundance on the two groups of caged fishes at dawn after 12 hours, but when sampled at sunset after 24 hours, there was a 4.5 fold increase in gnathiid abundance on fish without cleaners. This probably reflects the fact that cleaners are only active during the day.

The consumption of gnathiid larvae is not restricted to fish from tropical seas. Potts (1973) observed the following fish species removing gnathiid larvae from parasitised fishes in the Public Aquarium of the Marine Biological Association at Plymouth, U.K.: corkwing wrasse (*Crenilabrus melops*), goldsinny (*Ctenolabrus rupestris*) and the pipefishes *Entelurus aequoreus*, *Syngnathus typhle* and *S. acus*. He found *Gnathia maxillaris* to be prevalent in the tanks. In the wild at Salcombe in Devon, Potts observed only one fish indulging in cleaning activities, namely rock cook (*Centrolabrus exoletus*), but his discovery of praniza larvae among the stomach contents of locally caught wild corkwing wrasse suggests that cleaning symbiosis involving this fish may also occur under natural conditions. Galeote & Otero (1998) observed cleaning behaviour by rock cook in the wild off the south coast of Spain. Free-living copepods were significant in their diet, but the second most important items were gnathiid larvae derived from their cleaning activity. It has been pointed out by Otero & Galeote (1994, in Galeote & Otero, 1998) that at the front of each jaw the rock cook has projecting canine-like teeth, which are well suited to the removal of ectoparasites.

The use of wrasse to control sea lice infestation of farmed Atlantic salmon (*Salmo salar*) has been referred to above (Chapter 8), although Sayer *et al.* (1996) failed to find ectoparasites in the guts of wild caught rock cook from the west coast of Scotland.

Grutter (2002) has considered features of ectoparasites like gnathiids that might counteract the threat from cleaner organisms. Most gnathiid praniza larvae emerge in search of hosts at night and some infect their hosts only at night, thereby avoiding the attentions of diurnal cleaners and diurnal plankton feeders. Those that parasitise diurnally active hosts may deter cleaners in another way. Some praniza larvae have swirling or banded colour patterns, colourful eye-like markings on the dorsal surface and bright yellow pigments (references in Grutter, 2002), which may advertise the presence of noxious or unpalatable chemicals and 'warn off' potential predators. Selection of large gnathiid larvae by cleaners may have led to reduction in body size in gnathiid pranizas (see Grutter, 2002).

Grutter (2002) pointed out that gnathiid larvae minimise the time spent exposed to cleaners on the surface of the host, requiring no more than an hour to feed and then leaving the host and retiring to their hiding places. Mobility may attract predators and feeding larvae minimise this risk by remaining immobile while feeding and rarely move to other sites on the host. If they are disturbed while feeding on teleosts, the larvae readily abandon the host and, according to Grutter (2002), swim rapidly at speeds of 10 to 20 cm/second. Curiously, gnathiid larvae on sharks and rays are more reluctant to abandon their hosts. This raises the interesting question as to whether gnathiids on elasmobranchs suffer less attention from cleaners than those on teleosts.

15.4.7 The free-living reproductive phase in Paragnathia formica
Erosion of salt marsh mud typically produces 'micro-cliffs' up to two metres in height (Upton, 1987b). Fully gorged praniza larvae (P1 – P3) make shallow temporary burrows

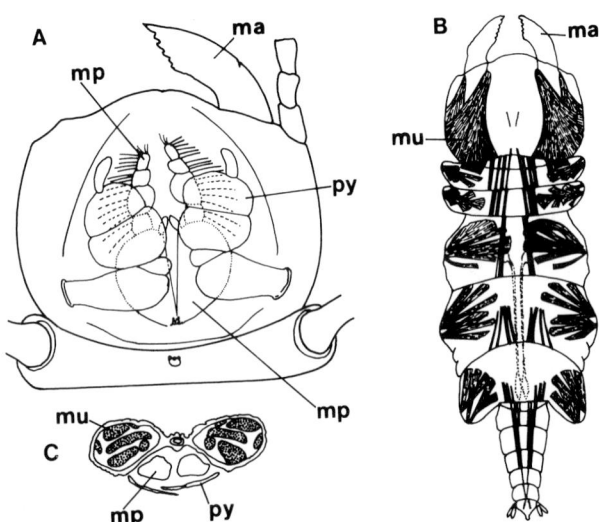

Fig. 15.11. Adult male *Paragnathia formica*. (A) Ventral view of 'cephalon' showing mouthparts. Mandible and base of antenna shown only on the left. (B) Body muscles in dorsal aspect, illustrating massive development of the mandibular muscles. (C) Transverse section through the 'cephalon'. ma, Mandible; mp, maxilliped; mu, jaw muscles; py, pylopod. From Monod, 1926.

in these micro-cliffs, where they digest their blood meals and moult to the next host-seeking zuphea stage, or, in the case of P3, into the adult. At the P3 stage, pranizas can be sexed, and, according to Upton (1987a) P3 males rarely penetrate more than 1 cm into the micro-cliff. The air-filled burrows of adult males are considerably larger, being up to 5 cm long and 1 cm wide at their inner ends. Adult males are always found singly in their burrows and are usually located within narrow entrance necks, 1 – 1.5 mm in diameter.

The extensive development of the head and jaws of the free-living male compared with the female is striking (cf. Figure 15.10D and E). According to Monod (1926), the excessive proportions of the head and jaws of the gnathiid male in relation to the rest of the body are exceptional for the Crustacea and highly reminiscent of some beetles and the soldier caste of termites. The mandibles of the gnathiid male project in an anterior direction like a pair of forceps (Figure 15.11AB). That they are capable of exerting a considerable biting force is evident from the huge size of the adductor muscles, which are housed in the greatly enlarged, more or less quadrangular 'cephalon' (Figure 15.11B). Related to this enlargement is a substantial thickening of the cephalic tegument and an enlargement of one or more of the post-cephalic segments, providing for the cephalon a broad platform, separated from the rest of the body by a 'waist' (Figure 15.11AB). Monod made an even more surprising discovery when he cut transverse sections through the cephalon. He found that in its central region the upper and lower teguments meet and are fused together, creating a massive supporting plate between the two laterally situated tubes housing the mandibular muscles (Figure 15.11C).

Males produce water-borne pheromones that attract young females to the burrow entrances (Upton, 1987a). When within range the female is grasped by the male's jaws and dragged forcibly into the burrow. However, in view of the crushing capability of the jaws with their exceptional musculature, some restraint would seem to be necessary during this manoeuvre, and it seems reasonable to suppose that these remarkable jaws have some additional function.

Monod (1926) suggested that the armoured head and strong jaws might have a defensive/offensive role in competitive interactions with other males. He proposed that the armoured head and strong jaws would provided an effective plug for the small entrance to the burrow (or to the various shelters occupied by males of other gnathiids) and a formidable obstacle for another male or any other organism attempting to enter the gnathiid's retreat. In fact, Upton (1987b) observed that established males use all ten legs to firmly wedge themselves in the narrow entrance neck and easily repel intruders. When viewed laterally, the head is seen to be held at an oblique angle with respect to the long axis of the body, making an effective plug (Figure 15.2), and Monod noted that the head of the male *P. formica* was frequently to be seen obstructing the narrow lumen of the burrow. Monod failed to find any indication that the mandibles play an auxiliary role in copulation and since the male does not feed, they are not used for prey capture or for consumption of food of any kind.

In north Norfolk, males accumulate 'harems' of up to 25 young females in the spring, ready for reproduction during the summer, but at this point there is an additional twist to the life story. Although the females are only nine months old, having been born during the previous autumn, the males dominating these harems are more than a year older than the females, i.e. they belong to the previous generation. This is because of asynchrony in the settlement timing of male and female P3 larvae and a difference in the life spans of the two sexes.

Most females derived from autumn-released Z1 larvae settle as P3 females the following spring and have become adults by the end of July. Their own Z1 young are released by the end of October and the parent females die soon after giving birth. Monod's illustration of the crumpled spent female with the empty prolapsed uterus is reproduced in Figure 15.12. The harem males eject spent females, sometimes just minutes after larval release has ceased (Upton, 1987a). Thus females have a life span of about one year, with no overlap between successive generations of adult females. Autumn-released Z1 larvae destined to become males spend much longer as larvae, overwintering and settling as P3 males the following autumn, when females of their own generation have already reproduced. These males overwinter as adults in enlarged burrows and are ready to breed in the following spring/summer. Death follows in the autumn, the males surviving for just over two years.

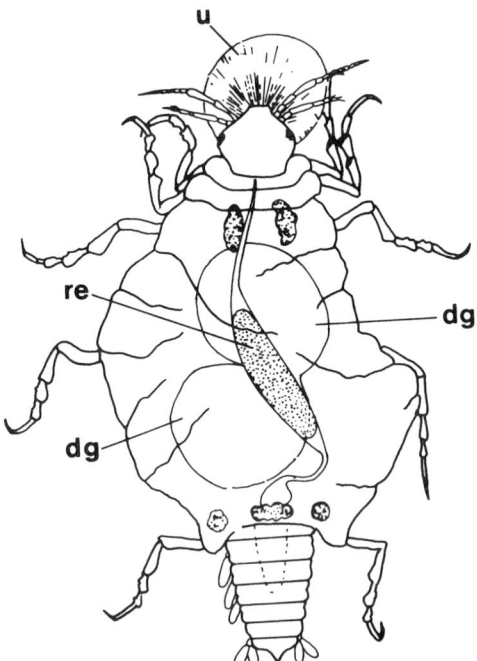

Fig. 15.12. Spent female *Paragnathia formica*. dg, Digestive gland; re, reservoir; u, prolapsed uterus. From Monod, 1926.

Regular tidal inundation combined with rainstorms and evaporation when the mud is exposed will create an environment of fluctuating salinity and Charmantier (1980) found that the males are euryhaline, i.e. can tolerate salinity ranging from 0.1 seawater to 1.6 seawater. The males maintain a hyperosmotic internal environment (with a higher osmotic solute content than the external environment) and their survival seems to depend on physical contact with water, since even in an atmosphere with 100% relative humidity, they die within a few hours if denied direct contact with water. The waterlogged walls of their air-filled burrows presumably provide this contact.

This unusual pattern in which larval females are attracted to and captured by males from the previous generation, may be beneficial for females, since they join males that have already demonstrated their ability to survive and to build and defend a burrow (Upton, 1987a).

The females release Zuphea 1 larvae during autumn high tides. Tinsley & Reilly (2002) used a plankton sampling technique to investigate patterns of larval release, and compared this with prevalence of parasites on common gobies, *Pomatoschistus microps*, in the area. They found evidence that release of larvae is influenced by tide height. During a sampling period of a week, high densities of released larvae occurred on only one high tide, corresponding with a peak in parasite prevalence of 10%. Synchronised massive release of larvae may have evolved in response to the special challenge posed by fish such as the common goby, which are not only hosts for *P. formica* but also predators on the larvae. Simultaneous massive release of the harem progeny on the rising tide, at a time when the host fish is feeding and in greatest abundance, would maximise opportunities for parasitism, while at the same time minimising losses to predation by confusing and saturating the predators. Rapid release of larvae as the tide rises may enhance exploitation of shoals of potential hosts passing the burrow entrances at this time.

Upton (1987a) found that mature females begin to release their larvae within a minute of immersion in seawater. Birth is traumatic – the pylopods open, the vestigial oostegites separate and the uterus ruptures, releasing the larvae from between the anterior pereopods. However, tidal inundation of the micro-cliffs does not always trigger release of larvae. Upton proposed that the burrows normally remain filled with air, even at high tide, the narrow burrow entrance effectively excluding water. Males may play a part in the flooding of these burrows and hence in the release of zuphea larvae, since Upton observed males enlarging burrow entrances until bubbles of air emerged, indicating the influx of water. This male activity was soon followed by the emergence of zuphea larvae from the burrow entrance in groups of five to ten. Tinsley & Reilly (2002) also found evidence that burrows are not always opened on their first submergence.

16

UNIONACEAN MOLLUSCS (NAIADS)

16.1 INTRODUCTION

While walking along a riverbank or skirting the edge of a large pond or lake there is a good chance that we will encounter the shell valves of the swan mussel, *Anodonta cygnea* (Figure 16.1), most probably left high and dry by changing water levels. On finding these impressive shells, fish parasites could not be further from our minds, but, surprising as it may seem, these large mussels spend a brief period of their early lives as parasites of fishes. Moreover, this parasitic phase is obligatory and these mussels depend entirely on freshwater fishes for their survival.

Fig. 16.1. The left shell valve of a large specimen of the swan mussel, *Anodonta cygnea*, from the Czech Republic. Photograph by Sheila Davies, courtesy of Professor John Reynolds.

The swan mussel belongs to a group of molluscs called the Pelecypoda, also known as Bivalvia (bivalves) or Lamellibranchia. Each of these names refers to important anatomical features. 'Bivalvia', of course, refers to the two shell valves, which enclose the soft body and are capable of opening to a limited extent. 'Pelecypoda' recalls the axe-shaped foot, which is capable of protruding between the gaping shell valves (Figure 16.2A). Bivalves are lethargic, but can move in a slow and limited way by extending and contracting the foot. 'Lamellibranchia' refers to the large gills or ctenidia, one of which lies on each side of the foot within the shell valves (Figure 16.2).

16.2 THE ADULT UNIONACEAN

The two shell valves are hinged dorsally and are closed by powerful adductor muscles running between them (Figure 16.2A). When these muscles relax the elasticity of a stretched ligament lying dorsal to the hinge (Figure 16.2AB) causes the valves to gape. A flap of tissue called the mantle lines each shell valve and it is the edge of this flap that

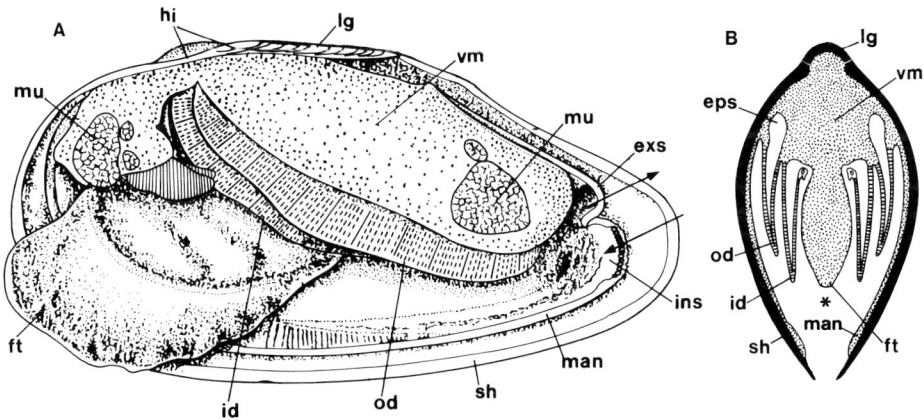

Fig. 16.2. The anatomy of *Anodonta cygnea*. (A) Whole animal in lateral view with the left shell valve and left mantle flap removed. (B) Diagrammatic transverse section of whole animal. eps, Epibranchial space; exs, exhalant siphon; ft, foot; hi, position of hinge; id, inner demibranch of ctenidium; ins, inhalant siphon; lg, ligament; man, mantle flap; mu, adductor muscle; od, outer demibranch of ctenidium; sh, shell; vm, visceral mass, containing internal organs. Star indicates mantle cavity; arrows indicate directions of water flow. From Ellis, 1978; (B) modified.

secretes additional shell material as the mantle grows. The two shell valves enclose a large space called the mantle cavity in which the gills are suspended (Figure 16.2). Each gill or ctenidium consists of two half gills or demibranchs and each of these comprises two sheets or lamellae, each of which has many perforations or ostia creating a net (see Morton, 1958). In transverse section a ctenidium takes the shape of the letter W (Figure 16.2B). The space within each demibranch may be divided into 'water tubes' by vertical partitions running between the inner and outer lamellae (Figure 16.3).

With the valves gaping, water is drawn into the mantle cavity via an elongated slit called the inhalant siphon (Figure 16.2A). This water is propelled through the ostia in the gill lamellae into the spaces inside the demibranchs. As expected the gills extract oxygen from the current, but they also have an essential role to play in feeding. The mollusc is a filter feeder and the net-like gills act like a sieve, collecting edible particles such as algae and suspended organic matter on the outer surfaces of the lamellae. These particles are bound with mucus and transported to the mouth. A remarkable feature is that the flow of water through the animal and the transport of food-laden mucus to the mouth are maintained solely by the activities of thousands of tiny beating cilia covering the gills and other surfaces.

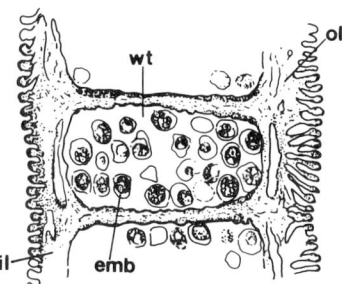

Fig. 16.3. Horizontal section through part of a demibranch of *Anodonta*. emb, Embryo; il, inner lamella of demibranch; ol, outer lamella of demibranch; wt, water tube. Reproduced from Lefevre & Curtis, 1910a, with permission from The Biological Bulletin.

Material that is not suitable for ingestion is diverted before it reaches the mouth, follows ciliated tracts to the edge of the mantle and is ejected as mucous strings called pseudofaeces. Particles entering the gut are subject to digestion and any remaining indigestible or unsuitable material is ejected as true faeces. After passing through the gills, the filtered water flows from the water tubes into an epibranchial space above the gills (Figure 16.2B) and leaves the animal via an exhalant siphon (Figure 16.2A).

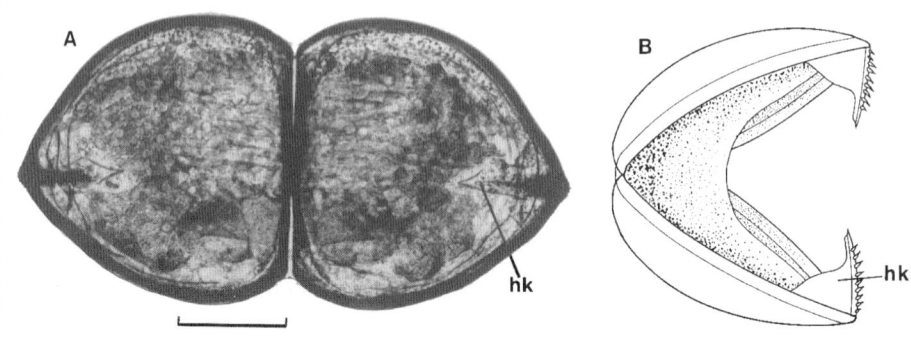

Fig. 16.4. (A) Freshly expelled, living glochidium of *Anodonta* with shell valves fully open (ventral view – hinge dorsal). (B) Sketch of partly open glochidium viewed obliquely. (B) Modified from Paling, 1968. hk, Hook. Scale bar: approximately 100 μm.

These multifunctional gills have one more role – in reproduction. The water tubes act as brood chambers and accommodate the mussel's developing eggs (Figure 16.3), which have been fertilised by sperm carried in by the inhalant gill current. Embryos developing in the water-filled spaces (brood chambers or marsupia) inside the demibranchs become specialised miniature bivalves called glochidia (Figures 16.4,

16.6), which are typically released into the water via the exhalant siphon. However, the glochidia larvae are incapable of continuing their development as free-living organisms and their further progress demands the intervention of a freshwater fish.

One feature of the biology of adult unionaceans has had unfortunate consequences for the survival of these interesting animals. The shell valves of the adult mussels are lined with mother-of-pearl and any foreign body that finds its way between the mantle and the shell is likely to be encapsulated by concentric layers of this material, producing a pearl. Pearls made by some species are of high quality and this has led to serious depletion of mussel stocks, both in North America and in Britain. In North America, extensive exploitation of mussels for a pearl button industry endangered many species at the beginning of the twentieth century (Conover, 1998). Fortunately, the advent of plastic buttons came to the rescue, and the button industry dwindled as a consequence, only to be replaced in the middle of the twentieth century by a new demand for American mussel shell. It was found to be the best material to provide an artificial nucleus for the culture of pearls by oysters. The problems currently facing the British freshwater pearl mussel *Margaritifera margaritifera* will be considered below.

16.3 THE LIFE CYCLE

After ejection from the adult bivalve, survival of the free glochidium depends on making contact with a suitable fish host. When the opportunity arises the glochidium uses the hooks on its shell valves (Figure 16.4B) to clip itself to the fish's body skin, to the fins (Figure 16.5) or sometimes to the gills and, sooner or later, it is completely enclosed by overgrowth of the host's epidermis. Thus, the glochidium is essentially an endoparasite, but since its habitat is superficial it remains visible externally throughout

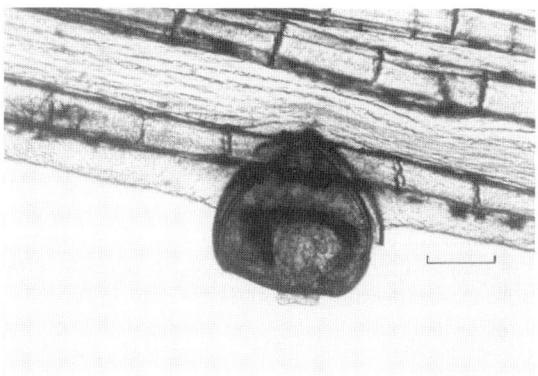

Fig. 16.5. Living glochidium of *Anodonta* freshly attached to the edge of a fin.

its parasitic life and for this reason is included in this book. However, *Mutela bourguignati*, a freshwater mussel belonging to the African Mutelidae, is unique in being a mesoparasite (see Wächtler *et al.*, 2001). The larva is morphologically distinct from the glochidium and is termed a lasidium. The lasidium attaches itself initially to the host's skin by means of hooks, but then develops into a stalk-like outgrowth, projecting from the skin surface distally but embedded proximally in the skin by means of a bifid root. The mollusc develops from the free distal end of the stalk, i.e. outside the host.

Nutrients absorbed from the host tissue support the development of the fully embedded glochidium. Eventually metamorphosis takes place, a process in which the body of the glochidium is reconstituted with the equipment required for life as a filter-feeding bivalve. With metamorphosis complete the young mussel emerges from the host's skin, falls to the bottom and commences life as a free-living organism.

Retention of eggs in a brood chamber is a specialised feature of these bivalves related to their freshwater habitat. Marine bivalves typically release their eggs into the water where they develop into free-living pelagic larvae or veligers, which are carried far and wide by ocean currents. Dispersive larvae of this kind would be a serious handicap in bodies of freshwater where there is a gravitational flow, since they would be carried inexorably downstream, ultimately to the sea, and would have no means of re-colonising upstream waters. The reduced buoyancy afforded by freshwater would also demand greater expenditure of energy from the veliger to keep afloat. It is not surprising therefore that freshwater bivalves have suppressed the planktonic phase, with the exception of the non-parasitic *Dreissena polymorpha*, which is able to survive only in still or slowly flowing water (Ellis, 1978). Nevertheless, the advantages gained by unionaceans by sequestering their offspring during embryogenesis may be offset by substantial losses when glochidia are released, especially in unionaceans living in fast-flowing water (see below). However, those glochidia that do survive long enough to establish themselves as fish parasites are able to exploit the mobility of their fish hosts and achieve dispersal even in an upstream direction.

16.4 BRITISH UNIONACEANS

The Suborder Unionacea of the Order Unionoida (= Naiadida) is represented by two families in Britain, namely the Unionidae and the Margaritiferidae (= Margaritanidae) (see Ellis, 1978). The Unionidae includes *Anodonta cygnea*, *A. anatina* (= *A. piscinalis*), *A.* (= *Pseudanodonta*) *complanata*, *Unio pictorum* and *Unio tumidus*. The Margaritiferidae is represented by a single living species, *Margaritifera margaritifera*. Shells of *M. auricularia* have been dredged from the Thames and have been found in association with Neolithic implements in Interglacial deposits from Trafalgar Square, London, but living animals are now confined to a few localities in Spain, France and Morocco (Araujo & Ramos, 2000). They are likely to decline further because of the absence of their specific host, the sturgeon *Acipenser sturio* (Araujo & Ramos, 2001). The main habitats of the British unionaceans and estimates of their current conservation status are given by Willing (1997). They all share the same life cycle pattern, but there are differences across the board in important traits such as glochidial size and structure, fecundity, host specificity, site of attachment to the host and duration of the parasitic phase.

Bauer (1994) recognised a continuum of reproductive strategies in these unionacean molluscs and, although our British fauna is impoverished compared with those of continental Europe and North America, we do have representatives of the two ends of this continuum, namely *Anodonta* and *Margaritifera*. According to Bauer, *Anodonta* produces relatively small numbers of large 'hooked' glochidia. The apex of each shell valve carries a prominent hook (Figure 16.4B), which is needed for attachment to the relatively tough scales and fins of their hosts. These large glochidia are well developed when released by the parent mussel and therefore require a relatively short parasitic lifetime to complete their development, placing relatively modest demands on the host for resources. Associated with this is the ability to utilise a broad range of hosts and a variety of freshwater habitats. Bauer (1994) also suggested that the parasitic phase in *Anodonta* provides insufficient time for the host to initiate and deploy its immune defences against the parasite, a situation that would also favour exploitation of a range of hosts. At the other end of the spectrum *Margaritifera* displays high fecundity and has relatively small 'hookless' glochidia. According to Bauer (1994), the absence of glochidial hooks is a reflection of the relatively soft nature of their preferred microhabitat, namely the gills. A long development time in the parasitic phase is necessary because of the small initial size of the glochidium and this is associated with a narrow host range and a restricted freshwater habitat.

With emphasis on their parasitic stages, details of the biologies of *Anodonta* and *Margaritifera* are given below and illustrate well the breadth of unionacean life cycles.

16.5 THE SWAN MUSSEL – *ANODONTA CYGNEA*

Normally the sexes are separate in *Anodonta cygnea*, but, according to Ellis (1978), individuals changing sex commonly occur and self-fertilisation is claimed to take place.

16.5.1 *The glochidium*
The two outer demibranchs serve as brood chambers (marsupia) in the Unionidae (Wächtler *et al.*, 2001). Wood (1974ab) has described the anatomy and development of

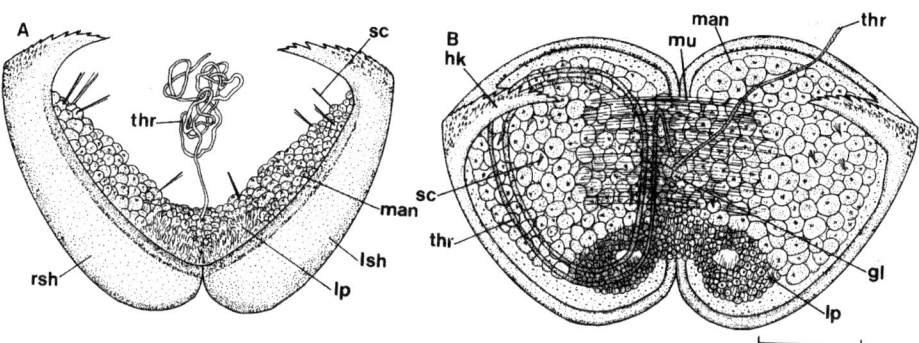

Fig. 16.6. The glochidium of *Anodonta cygnea*, (A) in lateral view and (B) in ventral view. gl, Gland cell producing thread; hk, hook; lp, ciliated lateral pit cells; lsh, left shell valve; man, mantle cells; mu, adductor muscle, closes shell valves; rsh, right shell valve; sc, sensilla; thr, thread. Scale bar: 100 μm. From (A) Wood, 1974b, (B) Wood, 1974a, with permission of Cambridge University Press.

the glochidia of *A. cygnea*. At maturity all the glochidial soft parts are enclosed in a bivalve shell (Figure 16.6). Each shell valve bears a large hook, which is armed with spines of various sizes. Mantle cells line the shell valves and a thread, which may be up to 2 mm long, emerges from a central position between the valves. In addition to the external thread, part of the thread remains inside the body of the glochidium, coiled around beneath the mantle cells lining the right shell valve (Figure 16.6B). So-called lateral pit cells, which are ciliated, straddle the hinge line in the 'posterior' region of the glochidium and sensory hairs (sensilla) project from the mantle cells lining each valve.

According to Wood (1974ab), embryonic development takes place during early autumn and the glochidia are fully formed by December at the latest. Although December glochidia are able to infect fishes (Wood, 1974a), they are not discharged from the gills of the adult bivalve until May or June. Wood (1974a) pointed out that the glochidia are likely to require some nutrients for maintenance through this long period in the brood chambers and, after feeding the adult mussels on C^{14}-labelled algal cells, detected C^{14} in the mantle cells and lateral pit cells of the glochidia. Thus, it is likely that the glochidia absorb nutrients from the adult mussel and Wood speculated that this nourishment might be provided in the form of mucus secreted by the adult's gills.

16.5.2 Host finding and establishment on the host
In order to become successfully established on a fish host the glochidium must be released by the parent mussel, make contact with a fish, make a positive identification, grip the host by closing the shell valves and maintain this attached position until it is enclosed by epidermis.

According to Anders & Wiese (1993), shadows cast by a potential fish host stimulate mature *A. anatina* to release large numbers of glochidia. However, this expulsion may simply be the consequence of the shell closure reflex of adult mussels, initiated by shadows.

The observations of Wood (1974b) on *A. cygnea* show that the glochidial thread is important initially, both as a means of establishing and maintaining the larvae in suspension and for attachment to the host. However, the latter function is not essential since glochidia may grip the host by closing their shell valves without prior attachment by the thread. A powerful adductor muscle running between the two shell valves (Figure 16.6B) is responsible for this action.

Wood (1974b) observed that when released by the adult mussel the glochidia of *A. cygnea* sink to the bottom in still water. Their shell valves are usually wide apart (Figure 16.4A) and the threads of the glochidia may be tangled. Wood noted that they are unable to swim by 'clapping together' the shell valves, in spite of earlier statements that they are able to do this. However, currents produced by fish readily carry the glochidia into suspension. Wood observed that the threads stream out behind the glochidia and that they significantly retard sinking, in a similar way to the long egg stalks of the monogeneans *Leptocotyle minor* and *Hexabothrium appendiculatum* (see Chapters 4 and 6 respectively). Wood concluded that the threads act as 'draglines', keeping the larvae in suspension and increasing the chances of contact between the threads and a fish. Wood found that the threads are sticky and readily become attached to a piece of fin. Moreover, when the fin is moved away, the thread does not break and the glochidium is towed through the water. In fact, the thread is elastic and if stretched achieves an approximately nine-fold increase in length before breaking. The elasticity of the thread and the central position of its root between the widely open shell valves will facilitate contact between the exposed mantle surface and the fish, aided no doubt by the twists and turns of the latter. In this attitude, closure of the shell valves by contraction of the adductor muscle will clamp the larva to the edge of the fin as shown in Figure 16.5.

Wood (1974b) examined glochidia freshly attached to a fin and observed that the hooks on the apices of the valves pierce the host tissue until they meet each other. The force of adduction bends the hooks inwards until they lie parallel, at right angles to their former positions. However, it appears that these hooks are not locked together by their spines, since Wood found that glochidia induced to grasp paper tissue did not keep the valves closed for more than a few minutes.

Wood (1974b) conducted an elegant series of tests to identify the stimuli that are most likely to induce the glochidium of *A. cygnea* to close its shell valves and grip the host. Tension applied to the glochidium by pulling on the thread failed to induce shell valve closure. Wood then used an air bubble attached to the bottom of a dish to prop up a fully open glochidium in such a position that it could be viewed laterally (as in Figure 16.6A). In this attitude the sensory hairs projecting from the mantle surface were readily visible and could be easily stimulated.

Tactile stimuli applied to the sensory hairs or to the mantle cells by touching or stroking with a fine needle were almost totally ineffective in promoting shell valve closure, unless mechanical deformation was severe. On the other hand, touching the hairs with a piece of fish fin almost invariably resulted in a fast and complete contraction of the adductor muscle and closure of the valves. Touching the mantle cells with the fin failed to elicit contraction. Thus, the hairs appear to be the sensory structures responsible for identification of a suitable fish host and the effective stimulus to which hairs respond seems to be chemical rather than tactile.

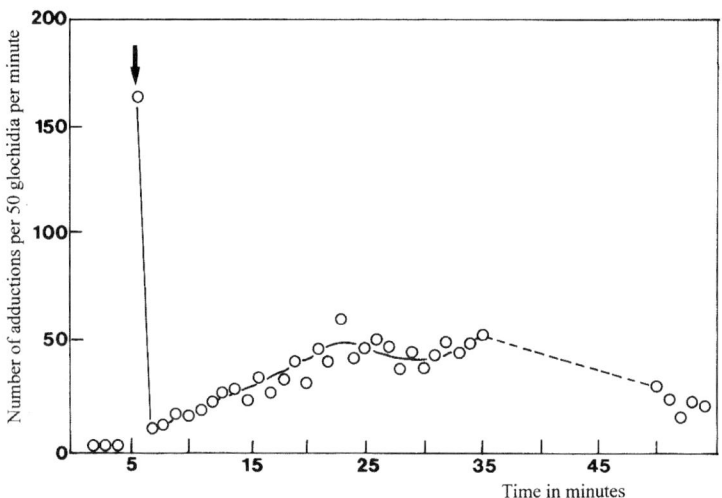

Fig. 16.7. The response of glochidia to mucous fluid prepared from goldfish, *Carassius auratus*. Arrow indicates addition of mucous fluid. Modified from Wood, 1974b.

These conclusions received support from further experiments conducted by Wood (1974b). She placed glochidia of *A. cygnea* on a fin and found that most responded by contracting and relaxing the adductor muscle several times. This indicated that the stimulating substances were associated with the surface of the fish and she made up a test solution comprising mucus, epidermal cells and some scales scraped from a goldfish, *Carassius auratus*. Within the 15 – 20 second period after addition of this mucous test fluid all the glochidia responded by rapidly and repeatedly contracting and relaxing the adductor muscles (Figure 16.7). This initial burst of activity then ceased with the adductor muscle in the relaxed position. After about one minute contractions resumed, but in this delayed phase rarely reached the high level of activity displayed in the initial phase (Figure 16.7). Further tests indicated that different chemical substances promoted the initial and delayed responses and that these chemical substances are small thermostable molecules.

Wood (1974b) concluded that, having made contact with a potential host by chance, the chemosensory hairs or sensilla projecting from the mantle respond to a substance present in fish mucus by snapping shut the shell valves and gripping host tissue between them. If no further appropriate chemical stimulus is received the adductor muscle will relax and the glochidium will become detached. If continued chemical stimulation is experienced, most probably by a different chemical substance, a tonic contraction of the adductor muscle will occur locking the closed valves together and ensuring that the parasite remains attached to the host.

16.5.3 *Host specificity and infection sites*

Dartnall & Walkey (1979) listed a range of freshwater fishes in the United Kingdom that are capable of infection with glochidia (presumably *A. cygnea*) and came to the conclusion that glochidia are non-specific. Giusti *et al.* (1975), working on *A. cygnea* in an Italian lake, reached more or less the same conclusion, reporting glochidia from all the common fishes, except rudd, *Scardinius erythrophthalmus*. However, they found differences in the levels of infestation in the different fish hosts, which they attributed to differences in opportunities for contact between host and parasite dictated by differences in host behaviour patterns. For example, they suggested that the high levels of infestation in pike, *Esox lucius*, relate to the fish's habit of lurking in ambush near the bottom, where swan mussels may be emitting glochidia. It is of interest to note that the cyprinid *Leucaspius delineatus* (the sunbleak or belica), which has been introduced recently into Britain (see Chapter 18), is an unsuitable host (Bauer & Wächtler, 2001).

Work by Dudgeon & Morton (1984) indicated that glochidia might not be entirely non-selective and opportunistic. They studied attachment of glochidia of *A. woodiana*, a mussel from eastern Asia, which has spread across the world, probably on introduced exotic fishes. It has not yet reached Britain, but is not far away, having been discovered in France in 1982 (Watters, 1997). Dudgeon & Morton (1984) selected four potential host fishes commonly found in southern China, namely the mosquitofish, *Gambusia affinis*, and three cyprinids, including *Rhodeus sinensis*. They exposed these fishes experimentally to glochidia of *A. woodiana*. The mosquitofish was found to be particularly susceptible, with a prevalence of 100% and an average of 10 (maximum 39) glochidia per fish. On the other hand, *R. sinensis* was an unsuitable host, only 10% being infected, each with no more than a single parasite. The other two cyprinids had levels of infection intermediate between these two extremes.

The unattractiveness of *R. sinensis* is of special interest, because this fish is a close relative of the bitterling, *R. sericeus*, an exotic freshwater fish that was first noticed in Britain in the 1920s and may have been introduced by liberation of pet fish (Wheeler, 1969). The cloaca of the female bitterling develops into a tubular ovipositor, which is used to insert eggs into the mantle cavity of *Anodonta* and other unionaceans. Here the eggs develop and hatch and the male bitterling defends a territory around the mussel. Consequently, the habits of fish and mollusc force them into close proximity and there is a real danger of parasite overload, which could be fatal, especially for the fry. However, Blažek et al. (1999) and Blažek (2000) reported that bitterling in the Czech Republic were rarely infected and in laboratory studies on the susceptibility of bitterling to the glochidia of *A. cygnea* and *A. anatina*, Dr D. Aldridge (personal communication) showed that the glochidia were sloughed within three days of infestation, while successful metamorphosis to juvenile mussels was achieved on three-spined stickleback (*Gasterosteus aculeatus*), perch (*Perca fluviatilis*) and rudd (*Scardinius erythrophthalmus*). The statement by Wheeler (1969) that the bitterling often acts as a host for glochidia is likely to be erroneous.

R. sinensis has an ovipositor similar to that of *R. sericeus*, implying similar breeding habits and *A. woodiana* is sufficiently common in the Far East to be a likely incubator for the eggs of *R. sinensis*. Like its European relative, *R. sinensis* may have become unsuitable as a host for the glochidia of *A. woodiana* in response to the threat of overloading by the parasites.

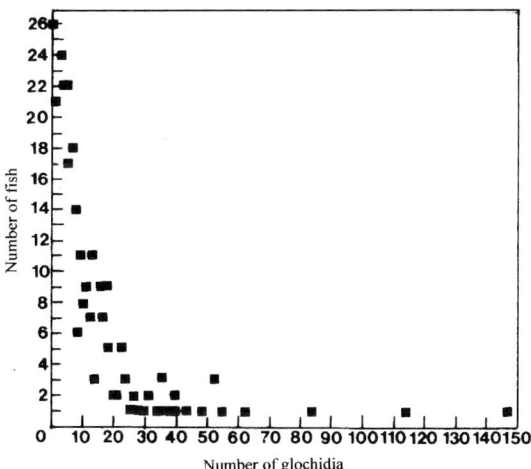

Fig. 16.8. Frequency distribution showing the numbers of glochidia on three-spined sticklebacks, *Gasterosteus aculeatus*. From Dartnall & Walkey, 1979, with permission of Cambridge University Press.

In their study of three-spined sticklebacks, Dartnall & Walkey (1979) identified a general increase in infection intensity with increasing host size, which they attributed, in part, to the increased surface area available for glochidial attachment. They also detected over-dispersion, a common phenomenon in parasite infections in which one or two hosts carry large burdens of parasites while most hosts carry relatively few (Figure 16.8). An obvious contributory factor to this pattern is the fact that glochidia are released *en masse* by the adult mussel. Any susceptible fish in the vicinity at the time of such a release is likely to receive a heavy dose of glochidia, while fishes remote from the event may receive relatively few.

Several authors have commented on the preference of the 'hooked' glochidia of *Anodonta* species for the fins of the host. Dudgeon & Morton (1984) found most glochidia of *A. woodiana* on the pectoral and caudal fins. In experimental infections of common carp, *Cyprinus carpio*, with the hooked glochidia of an American bivalve, *A. cataracta*, Lefevre & Curtis (1910b) observed that the fin margins often became loaded with glochidia packed side by side, with few on the flat surfaces of the fins. The thickened pelvic fins or gonopodia of male mosquitofish (*Gambusia affinis*) are less often parasitised by *A. woodiana* than the unthickened pelvic fins of the female (Dudgeon & Morton, 1984). Lefevre & Curtis (1910b) pointed out that the valves of the glochidium of *A. cataracta* could readily grasp the free edge of a fin, but have difficulty in gaining a hold on a flat surface. Those that do manage to attach themselves to a fin surface did so by gripping one of the raised ridges formed by a fin ray. The few glochidia that do succeed in establishing themselves on the fin surface become enclosed by overgrowth of host epidermis in just the same way as those attached to the fin margins.

Gills may also be infected by *Anodonta* spp.. Dartnall & Walkey (1979) studied a population of three-spined sticklebacks from a pond in Epping Forest. These fishes were naturally infected with *A. cygnea*, with 47.6% of the glochidia on the fins (especially the caudal and pectoral fins), 24.7% on the external surfaces of the head (including the mouth, throat, opercula and eyes) and 18% on the gills and in the buccal cavity. Dartnall & Walkey reported an almost identical distribution of glochidia on sticklebacks infected experimentally in the laboratory.

The proportion of glochidia of *A. cygnea* on the gills varies from one host to another. The gills of tench (*Tinca tinca*), pumpkinseed (*Lepomis gibbosus*) and the big-scale sand smelt (*Atherina boyeri*) were found by Giusti *et al.* (1975) to harbour a higher percentage of glochidia of *A. cygnea* than any of the fins or pairs of fins, while the gills of perch (*Perca fluviatilis*) and pike (*Esox lucius*) accommodated fewer parasites than the fins. The reasons for these differences are obscure, but may reflect behavioural differences of the hosts. The special adaptations of the American unionacean *Lampsilis perovalis* are thought to induce fishes to prey on glochidia (see below) and this seems likely to lead to infection of the gills. Three-spined sticklebacks in Britain are known to ingest glochidia of *A. cygnea* both in the laboratory and in the wild, but in spite of this Dartnall & Walkey (1979) reported relatively few glochidia on the gills (see above).

Dudgeon & Morton (1984) believed that predation on glochidia was likely to promote parasitism of the gills and buccal cavity. They attributed the same opinion to Dartnall & Walkey (1979), in spite of the fact that these latter authors had demonstrated that sticklebacks had few gill parasites and were known to eat glochidia. Dudgeon & Morton found no parasites on the gills of the fishes that they exposed to glochidia of *A. woodiana* and related this to the small size of the hosts and to their lack of predation on glochidia.

Lefevre & Curtis (1910b) observed that the glochidia of the American bivalve *Anodonta cataracta* become attached in large numbers to the gills as well as to the fins of the common carp. However, surprisingly, after about one week all the glochidia had disappeared from the gills, which were then "as clean as though never infected".

Paling (1968) used glochidia of *A. cygnea* as a means of estimating the relative volumes of water flowing over the different gills of brown trout, *Salmo trutta*. In doing so he made the assumptions that glochidia are taken in passively with the host's gill ventilating current and that glochidia passing through the different regions of the gills are equally successful at achieving attachment. This will be considered further below.

16.5.4 *The parasitic stage*

Lefevre & Curtis (1910b) described the enclosure of the glochidium of the American bivalve *Anodonta cataracta* by epidermal proliferation at the fin margin of the common carp (*Cyprinus carpio*) (Figure 16.9). Three and a half hours after attachment of the glochidium, proliferation of the host epidermis was detectable as a mass of tissue containing nuclei. In 24 hours most of the glochidia were more than half covered by the extending host epithelium and, at the end of a 36-hour period, most glochidia were well embedded. The only noticeable further change was a slight increase in opacity that rendered the internal structure of the glochidium less distinct.

Encystment of glochidia appears to be the natural repair process of the host epithelium induced by the mechanical trauma inflicted by the shell valves. Arey (1921) found no evidence for the involvement of any chemical or vital influence from the glochidium, since tiny clips made of metal implanted on a gill filament were overgrown by the host epithelium in exactly the same way as glochidia.

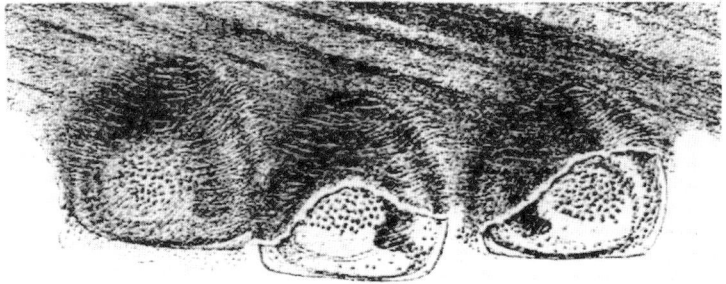

Fig. 16.9. Glochidia of *Anodonta cataracta* about 12 hours after infection. About half of two of the glochidia is embedded, while the third is completely enclosed. From Lefevre & Curtis, 1910b. Copyright © (1910, John Wiley). Reproduced by permission of Wiley-Liss, Inc., a subsidiary of John Wiley & Sons, Inc.

As Arey (1932) has pointed out, vascularisation of the host tissue to facilitate the passage of nutrients from host to parasite does not take place. However, resources are essential to support the extensive developmental and metamorphic changes undergone by the glochidium, as indicated by observations made by Moles (1983) on the effect of glochidia on growth of young coho salmon, *Oncorhynchus kisutch*. He found that as few as 1 – 20 glochidia attached to fry of coho salmon retarded growth and reduced their fat content, while burdens of more than 50 glochidia were lethal.

Arey (1932) identified two sources of nutrients for the developing glochidia of North American unionaceans. The first of these is the plug of host tissue grasped between the two shell valves. According to Arey, the cells of the larval mantle have phagocytic properties and are able to ingest and digest fragments of this plug. In addition, he proposed that digestive secretion released by the mantle cells contributes to the dissolution of the enclosed host tissue. The recycled organs of the glochidium provide the second source of nutrients. The larval adductor muscle undergoes attrition. Arey (1932) claimed that wandering amoeboid cells hasten its disintegration by engulfing muscle fragments and transporting them to the larval mantle, where their digestion is completed. The adult mantle subsequently replaces the larval mantle and the adductor muscles, foot and gut of the adult appear *de novo*. Arey (1932) reminds us that the gut appears at a relatively early stage in the parasitic phase and may play a part in the processing of host-derived and/or recycled material.

The parasitic phase of *Anodonta* typically lasts for 12 – 24 days, but may be longer if the water is cold (Ellis, 1978). Lefevre & Curtis (1910b) described the 'excystment' of North American unionaceans. Prior to emergence the host tissue enclosing the parasite assumed a looser texture and from time to time the encapsulated mussel opened its shell valves and extended the foot, rupturing the enclosing host tissue. This permitted the young mussel to escape and fall to the bottom. Here the newly emerged bivalve was very active, using its foot to creep about. At the end of the first week of free life, marginal growth of the shell was detectable (Figure 16.10). Lefevre & Curtis were unable to keep the young mussel alive for more than six weeks in the laboratory and Ellis (1978) regarded the natural fate of young British mussels as mysterious, since they are rarely found, possibly because they bury themselves in the substrate. North American unionaceans newly freed from the fish may attach themselves temporarily to a stone or to other objects by a special thread or byssus.

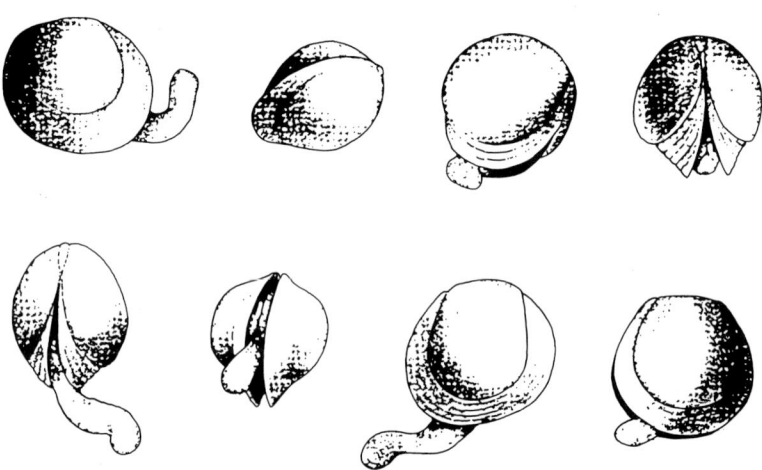

Fig. 16.10. Young of the American unionacean *Lampsilis ligamentinus* (= *L. ligamentina*) one week after liberation from the fish, showing various positions during crawling and new growth of the shell. From Lefevre & Curtis, 1910b. Copyright © (1910, John Wiley). Reproduced by permission of Wiley-Liss Inc., a subsidiary of John Wiley & Sons Inc.

16.6 THE PEARL MUSSEL – *MARGARITIFERA MARGARITIFERA*

The freshwater pearl mussel, *Margaritifera margaritifera*, has some special features. According to Young & Williams (1984a), it is Britain's most massive bivalve, in spite of the fact that it is restricted to waters low in calcium. In addition, it probably has the longest life span of any invertebrate (see Hutchinson, 1979, p. 42). It reaches maturity in Scotland when it is about 12 years old and 6 – 7.5 cm in length (Young & Williams, 1984b) and lives at least 70 – 80 years, with the age of some individuals in Scottish rivers exceeding 100 years (Hastie et al., 2000). This contrasts sharply with similarly sized species of *Anodonta* and *Unio*, which live for 10 – 15 and 13 – 15 years respectively (see references in Young & Williams, 1984b). Sadly, the pearl mussel is in serious decline throughout its extensive Holarctic range (see Cosgrove et al., 2000).

Young & Williams (1984ab) made a detailed study of the reproductive biology of *Margaritifera margaritifera* both in the laboratory and in a small stream, Stac Burn, in Wester Ross, Scotland. The stream is about 3 m wide and 20 – 50 cm deep and is a typical habitat for the pearl mussel, being cool, fast running, oligotrophic and low in calcium.

Young & Williams (1984b) stated that *M. margaritifera* is almost invariably dioecious (separate sexes) with no evidence that sex changes with age. However, Bauer (1987a) detected hermaphrodite individuals in German populations, and he also found evidence of sex changes, mainly, if not exclusively, from females to hermaphrodites. Because of the preponderance of female tissue in these hermaphrodites, Bauer thought it more likely that the relatively small quantities of spermatozoa were retained by the producer and used for self-insemination rather than being released and available for cross-insemination. Self-insemination is strongly suggested by Bauer's finding that sparse populations of mussels have high fertility. He believed that previous workers had failed to find hermaphrodites because of the rarity of these individuals in large populations. Hermaphroditism and sex change have been recorded in *Anodonta* (see above).

Some adults and almost all juveniles bury themselves almost completely in the substrate, but most adults have about one third of their shells exposed. Pearl mussels typically orientate themselves so that their inhalent apertures face into the current (Bauer, 2001). According to Cosgrove et al. (2000), mussels may remain in the same place for the whole of their long lives if the substrate is stable, but can rebury themselves if dislodged and can move slowly over a sandy bottom. An adult pearl mussel is capable of filtering up to 50 litres of water a day (see Ziuganov et al., in Cosgrove et al., 2000). When mussels are abundant their combined filtering efforts are likely to make a significant contribution to clarification of the river water, benefiting other animals including young salmon, *Salmo salar*, and brown or sea trout (*S. trutta*), which act as hosts for the glochidia.

Young & Williams (1984b) found great variability in the density of mussels in Stac Burn. Some areas lacked mussels, but these areas were rare and did not extend for more than 10 – 15 m. Close-packing of mussels, which has been observed in populations elsewhere, was not evident, possibly because of the coarse and variable substrates, which are not conducive to high densities. Removal of mussels by pearl fishermen may also have influenced their density. However, in spite of this, Young & Williams considered that glochidia are effectively carried throughout the stream and that brown trout, which are the main fish hosts in Stac Burn probably encounter glochidia wherever the fish occur in the stream.

16.6.1 *The glochidium*

All four demibranchs act as marsupia in the Margaritiferidae (Wächtler *et al.*, 2001). Compared with glochidia of *Anodonta* (see above) those of *M. margaritifera* are small, 60 by 80 µm. According to Pekkarinen & Valovirta (1996), the glochidial shell of *M. margaritifera* is thin (0.3 – 0.5 µm) and the larval adductor muscle can be seen through it (Figure 16.11). The shell has no hooks, but the membranous edges of the shell valves turn inwards. The exposed surface of the larval mantle is covered with microvilli (minute finger-like projections from the surface cells) and the mantle cells are laden with granules. Granules are absent from the regions of the foot rudiment and the lateral pits. Two tufts of sensilla project from the mantle on each side. Long cilia are present on the foot rudiment and shorter cilia in the regions of the lateral pits. No larval thread has been reported.

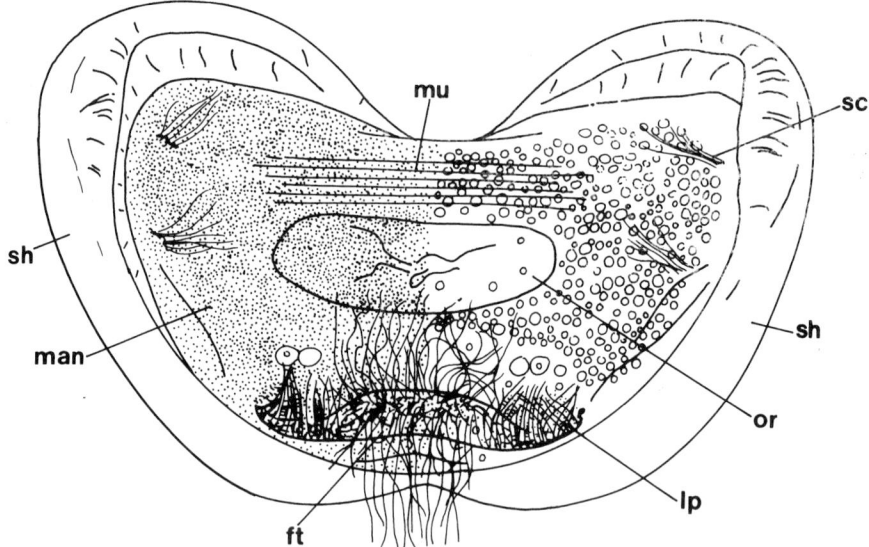

Fig. 16.11. The glochidium of *Margaritifera margaritifera*. ft, Foot rudiment with long cilia; lp, lateral pit with cilia; man, mantle; mu, adductor muscle; or, oral plate with a few cilia; sc, sensilla tuft; sh, shell valve. Granules in mantle cells shown on one side only. From Pekkarinen & Valovirta, 1996, with permission from Schweizerbart-Publishers (http://www.schweizerbart.de).

Young & Williams (1984b) estimated that a single mussel is capable of producing between one million and 17 million glochidia and there seems to be an approximate, direct relationship between size of mussels and the number of glochidia produced. In a study of six pearl mussel populations in Scottish rivers conducted from 1993 to 2002, Hastie & Young (2003) found that annual spawning (deposition of eggs within the gills) occurred from June to July and spat (release of glochidia) from June to September. Differences between the rivers in the timing of these events seem to be related to water temperature, mussels in the warmer rivers tending to spawn and spat earlier. The timing of reproduction from year to year also appears to be influenced by temperature, with delays of several weeks occurring during cold years. The timing of spawning is determined gradually, probably by a thermal summation effect, while spats are typified by a sudden massive release of glochidia over a period of 1 – 2 days, soon after commencement of glochidial release. Moreover, there is a high degree of synchrony between these massive release events within rivers, attributed by Hastie & Young (2003) to an environmental cue. Spats often follow sudden changes in water temperature and/or river water level and Hastie & Young suggested that impaired respiration of the brooding females might trigger release. They also recommended an investigation of the possibility that there is a link between the timing of reproduction in *M. margaritifera* and the availability of host fishes.

More glochidia are shed between 10.00h and 17.00h than overnight. Young & Williams (1984b) suspect that this is also related to the higher temperatures during this period of the day rather than to light levels, since the period of daylight in summer at this northerly latitude is long. They speculate that higher ventilation rates of mussels at the higher temperatures might free more glochidia.

16.6.2 Infection of the host
Young & Williams (1984a) observed that free infective glochidia snap their valves at a rate of 15 – 20 snaps per 100 glochidia per minute. Presumably snapping involves a twitch of the adductor muscle, leading to rapid partial closure and immediate reopening of the valves. The glochidia do not fully close their valves during the first three hours after release unless stressed or stimulated in some way. Touching the glochidial surface between the shell valves elicited immediate closure of the valves, but reopening was never observed.

The snapping rate was found to decline as the glochidia grew older; 24 hours after their release glochidia appeared lifeless. Agitation of the water had no effect on fresh infective glochidia, as might be expected since the parasite lives in turbulent streams. However, mucus, gill tissue or fin tissue from a brown trout led to a significant increase in the snapping rate to between 200 and 300 snaps per 100 glochidia per minute, indicating responsiveness to water-borne chemicals. The presence of gill or fin tissue also induced full closure of the valves, even when not in direct contact with the glochidia. An increase in snapping rate also followed treatment with trout blood, but the response was less vigorous (between 60 and 80 snaps per 100 glochidia per minute). Young & Williams (1984a) concluded that the most likely infection sequence begins when the glochidia detect the trout at close range, perhaps only when drawn into the gill chamber by the gill ventilating current. This leads to an increase in snapping rate and a greater opportunity for gill tissue to contact the glochidial surface between the valves and to stimulate permanent closure and clamping to the gill.

Young & Williams (1984a) observed that all attached glochidia of *M. margaritifera* were on the gill lamellae of the trout, not on the skin or gill rakers. However, the glochidia were not distributed uniformly on the gills. They were more commonly found in the central region of each hemibranch, rather than in the proximal region near the gill arch or in the distal free region of the gill. More parasites than expected occurred on gill pairs 1 and 4. This distribution differed significantly from a theoretical uniform distribution of parasites and from the expected distribution if numbers of parasites settled in proportion to gill area. Young & Williams (1984a) came to the conclusion that glochidia are carried passively to the gills in proportion to the volume of water passing over each gill (they have no means of propulsion or of steering their bodies). However, they believe that glochidia may have difficulty in attaching themselves in regions swept by faster currents, leading to variable attachment success in different regions of the gill and an eventual distribution that does not relate directly either to relative gill area or to relative volumes of water passing over each gill section. This is at odds with the assumption of Paling (1968; see above p. 307) that the glochidia of *Anodonta cygnea* attach wherever they are carried passively by the gill ventilating current and that their distribution on the gills of a trout provide a reliable estimate of the relative volumes of water passing over the different gills.

Young & Williams (1984b) found that 0+ trout (trout in their first year) carried the highest infections in Stac Burn. Brown trout migration patterns are likely to contribute to this. 0+ trout remain in the streams and during their 1+ year typically move to the lochs where they grow and mature. Adult trout spawn in the streams in winter, but usually return quickly to the lochs where most of their adult lives are spent. However, a few fish spend all of their lives in the streams and Young & Williams found some 2+ and 3+ fish with glochidia (see also below).

16.6.3 *Survival in the host*

There are undoubtedly enormous numbers of glochidia widely dispersed throughout Stac Burn, but, in spite of this, Young & Williams (1984b) reported that some 0+ trout were uninfected and those that were infected carried at the most about 1600 glochidia and usually fewer than this number. This raises the question of whether the host's immune system plays a part in the fate of glochidia.

Young & Williams (1984ab) recorded substantial losses of encysted glochidia from wild brown trout between December 1979 and the following May and between September 1980 and May 1981, such that only 5% of the glochidia survived (Young & Williams, 1984b). A similar loss was observed in the laboratory for brown trout and salmon (Young & Williams, 1984a), the fishes losing most of their glochidia within 40 days of attachment. Those remaining then persisted (or perhaps were lost at a very slow rate) until their development was complete after about 290 days. Deaths of heavily infected fishes have occurred in the laboratory and similar mortality in the wild may contribute to the relatively low levels of infestation in the trout of Stac Burn and to the fall in parasite levels overwinter. However, shedding of glochidia is mainly responsible for the latter phenomenon.

Evidence in support of acquired immunity to glochidial infection with some American naiads was published early last century (see below). Based on experimental infections with *M. margaritifera* using German stocks of brown trout, Bauer & Vogel (1987) suggested that initial losses soon after establishment of a primary infection were due to a general host tissue response resulting in sloughing of tissue, while a host humoral response might be responsible for subsequent losses. They also showed that after recovery from a primary infection, the host response to a secondary infection is enhanced. Observations made by Bauer (1987b) on populations in the field indicated that annual exposure to glochidia leads to complete resistance in older fishes. He noted that trout in his rivers matured in their third year, by which time they were usually completely refractory to infection, while all fish in their first year were infected.

Glochidial mortality indicative of an immune response does not always take place. Young *et al.* (1987) experimentally infected fingerling trout in September 1982. They found great variation in the numbers of glochidia on each fish but no evidence of a subsequent decline in numbers. The glochidia on these fishes were permitted to complete their development and after metamorphosis left their hosts. The following September (1983) the same fishes, at that time free of parasites, were experimentally exposed to more glochidia, which were found to establish themselves successfully. A marked decline in numbers of attached glochidia occurred on this occasion, but those that survived continued to grow well. Young *et al.* concluded that older fishes can be successfully re-infected with glochidia and that, if an immune response is induced by an earlier infection, it is weak or transitory.

For more detailed discussion of the immunity question see Jansen *et al.* (2001).

16.6.4 Development of glochidia

In Scotland, glochidia of *M. margaritifera* grow a little after attachment to the host until October/November. Growth then ceases overwinter and resumes in March, with full size being reached in June. For successful development after leaving the host fish, young mussels require a clean substrate of sand or gravel (Cosgrove *et al.*, 2000). Bauer (1979, in Young & Williams, 1984a) observed that in Bavaria glochidia grow quickly enough to complete development in October, when the young mussels are able to leave the fish. He also showed by transfer experiments to more northern streams that this feature is genetically determined and not controlled by local environmental factors such as temperature.

16.6.5 Host specificity

Glochidia of *M. margaritifera* rarely occur on other species of fish. In Stac Burn, Young & Williams (1984b) found that eels, *Anguilla anguilla*, collected in autumn harbour a few glochidia, but these parasites do not persist until spring and it is unlikely that eels travelling overland would carry mussels to new water systems. Minnows, *Phoxinus phoxinus*, are common fishes in many British mussel rivers, but Young & Williams (1984a) found that glochidia of *M. margaritifera* fail to become encapsulated on their gills and are lost within a few hours. Rainbow trout, *Oncorhynchus mykiss*, are also unsuitable hosts. Glochidia become encapsulated but are shed within 24 hours, leaving visible hyperplastic scars on the gills. Salmon parr (*Salmo salar*) proved to be suitable hosts in laboratory tests (see Young & Williams, 1984a). American workers cited by Young & Williams (1984a) have reported susceptibility of salmonid hosts not found in Europe and Young & Williams have raised the possibility that different races of *M. margaritifera* may utilise different host fishes. Thus, *M. margaritifera* shows rather narrow host specificity, focussed on salmonid hosts.

16.6.6 Overview of the life cycle

Young & Williams (1984b) showed that glochidia and early developmental stages of the mussel experience massive losses. The proportion of glochidia lost between emission from the mussel and attachment to the fish was estimated to be 99.9996%. A further 95% of those that successfully attach themselves to a fish fail to survive and, of those that complete their development and leave the host, only 5% become established in the substrate. These huge losses are an important consequence of retaining a freely dispersed glochidial stage in a fast flowing freshwater environment and enormous fecundity is required in compensation. However, this seemingly disastrous and wasteful strategy persists because of the great longevity of the adult and the many opportunities for breeding. Unfortunately, this delicate balance is readily disturbed by human activities.

16.6.7 *Threats to the pearl mussel*

Cosgrove *et al.* (2000), Young *et al.* (2001) and Skinner *et al.* (2003) have reviewed the historic importance of the freshwater pearl mussel and its current status and distribution, with details of habitat requirements and threats to its continued survival.

The occurrence of the freshwater pearl mussel in Britain is said to have been an important stimulus for the Roman invasion of 55 BC. By the 12th century, Scottish pearls had a market in Europe and during the 16th century the pearl industry expanded across Britain and Ireland, with river 'bailiffs' to ensure that valuable pearls were kept for the king. Because of the destructive nature of this exploitation however, the industry was not sustainable. There was a resurgence of interest in the 19th century, but the fishery declined sharply after 15 years, possibly because the mussels became rare. Since then a small-scale pearl industry has survived, but the story from all parts of Britain and Ireland and from every part of its Holarctic range is of steady decline, justifying the listing of the mussel as a seriously endangered species. Pearl fishing is now illegal in Britain and it is an offence to knowingly harm a mussel or its habitat. The precarious position of the species is recognised by its full protection under Schedule 5 of the Wildlife and Countryside Act (1981) and its inclusion in Annexes II and V of the European Union's Habitats and Species Directive and in Appendix III of the Bern Convention. It is included in the IUCN (The World Conservation Union) Invertebrate Red List, where its status is described as vulnerable. The mussel was afforded complete protection in Britain in 1998. It is also on the UK Government list of priority species for which a national Species Action Plan has been prepared to encourage measures for its survival.

Recent surveys reveal that the situation of the pearl mussel is critical, with former populations in England and Wales being virtually extinct with little active recruitment (Young *et al.*, 2001). The situation in Northern Ireland is not much better. The mussel is now extinct in most of the lowlands of Scotland and scarce everywhere, except in a few Highland rivers. With up to half of the world's known remaining functional populations of the mussel now occurring in Scotland, action taken within the UK will have a direct consequence for the global survival of the species. On the positive side, Hastie & Cosgrove (2002) demonstrated that sampling techniques undertaken in fast-flowing rivers tend to under-record numbers of juvenile mussels.

One of the major early causes of decline was illegal pearl-fishing, involving the senseless destruction of mussels in an attempt to find one or two containing saleable pearls. It is possible to open a mussel with a suitable tool in such a way as to permit detection and removal of pearls without destroying the mussel, which can then be returned alive to the river. In the past, industrial pollution has also had a detrimental effect, especially in lowland and southern areas. Eutrophication resulting from agricultural runoff, aquaculture and sewage disposal is a more recent problem, the increase in nutrients encouraging growth of algae and rooted plants. Sedimentation increases, clogging the substrate spaces and leading to death of newly released mussels and established mussels alike.

Pearl mussel populations are also at risk from severe floods. According to Cosgrove *et al.* (2000), significant changes in the hydrological behaviour of Scottish rivers occurred in the late 1980s as a result of climatic effects. These changes included new maximum flood records, increases in frequency of high-flow events and greater annual runoffs, all of which have rendered important mussel populations in Scotland more vulnerable to the effects of large floods than previously.

Another significant threat is the dramatic decline of potential hosts, namely Atlantic salmon and sea trout. A marked decline in salmon catches in north-west Scotland, where most mussel populations occur, took place during the 1990s (Skinner *et al.*, (2003). This decline has been matched by a widespread recruitment failure in the freshwater pearl mussel. Cosgrove *et al.* (2000) point out that a factor in the decline of salmonids is the loss or degradation of suitable spawning and nursery areas and that freshwater mussels have in the past contributed to the maintenance of these sites by way of their filter-feeding activity. Thus, the association between the mussel and the salmonid fishes is more like a mutually advantageous partnership, with the fish providing transport and nourishment for the young mussels without suffering obvious detrimental effects, while the adult mussel improves the fish's environment.

As Cosgrove *et al.* (2000) have pointed out, it may be possible to reduce the impact of threats such as illegal pearl fishing, but the emergence of new threats such as the decline of fish stocks, brings a degree of uncertainty to the success of all the efforts to save this fascinating animal.

16.7 SOME SPECIAL UNIONACEANS

The unionacean fauna of North America is much richer than that of Britain, with nearly 300 species in the United States (mostly in south-eastern rivers) (Conover, 1998), and

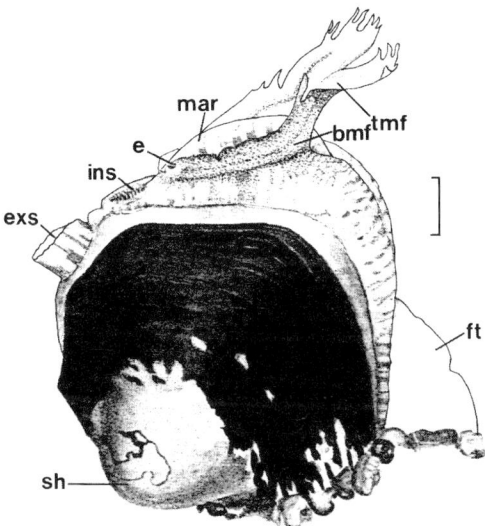

Fig. 16.12. The American unionacean *Lampsilis ventricosa* in the orientation adopted when the mantle flap is undulating. bmf, Base of mantle flap; e, 'eye-spot'; exs, exhalent siphon; ft, foot; ins, inhalent siphon; mar, marsupium; sh, shell; tmf, tail of mantle flap. Scale bar: 1 cm. Reproduced with permission from Kraemer, 1970.

some of these freshwater mussels are remarkably specialised. Some species of *Lampsilis* have elaborate mantle flaps that are capable of undulating at the rate of three undulations per second (Kraemer, 1970; Kraemer & Swanson, 1985). Each flap of *L.ventricosa* has a pigmented spot, which, reinforced by the appearance and movements of the tapering flap, creates the strong impression of a small fish to the human observer (Figure 16.12). This has led to the notion that the flap attracts predatory fish, which are then infected by glochidia released by the bivalve. However, although mantle flaps only develop in mature females, Kraemer (1970) doubted that they serve to attract fish hosts. She thought it more likely that the flap movements keep newly released glochidia in suspension, thereby increasing their chances of making contact with a suitable fish. The glochidia of these bivalves are released first into the mantle cavity via special pores in the water tubes and are discharged through the incurrent siphon, which is adjacent to the mantle flaps (Kat, 1984).

The related bivalve *L. perovalis* has an adaptation that seems much more likely to promote contact between glochidia and the fish host. According to Haag *et al*. (1995) this mollusc extrudes what is called a superconglutinate, a body containing the entire glochidial contents of a single gill. This emerges through the excurrent siphon, to which it is tethered for a time by a mucous cord up to 250 cm long. The resemblance of the superconglutinate to a small fish is reinforced by its colouring, by the presence of an 'eyespot' and by its movements in water currents (Figure 16.13). Haag *et al*. believe that this superconglutinate does act as a lure and that piscivorous fishes such as the freshwater basses *Micropterus coosae* and *M. punctulatus* become infected by ingesting it. This requires confirmation.

16.8 ACQUIRED IMMUNITY TO GLOCHIDIA

As long ago as 1919, Reuling was the first to demonstrate experimentally the development of acquired immunity to a metazoan parasite, using glochidia of American bivalves belonging to the genus *Lampsilis* and large-mouth bass (*Micropterus salmoides*) as hosts. He exposed large-mouth bass to two consecutive infections with glochidia of *Lampsilis luteola*, permitted these infections to complete their development on the gills and leave the host, then exposed the same fishes to a third wave of infection. The fishes initially accepted these glochidia, but sloughed off all of them over the next two days. Reuling went on to show that the immunity generated against *L. luteola* was equally effective against further infections of the same hosts with first *L. ventricosa* and then *L. ligamentina*. Wilson (1914, 1917) reported antagonism between glochidia and copepods, such that fishes with heavy infections of copepods accepted few or no glochidia and *vice versa*.

Reuling (1919) found no evidence that this acquired immunity was attributable to the development of some kind of fibrous barrier on the gills of the host or to an increase in the density of mucous cells on the gills. However, he did find that within a few hours glochidia died and disintegrated in blood serum from immune fishes, while glochidia were still alive after 48 hours in normal (non-immune) serum. Since the gills are richly supplied with blood and these unionaceans are gill parasites, blood-mediated host defence seemed to Reuling to be highly likely.

Rogers & Dimock (2003) studied the American unionacean *Utterbackia imbecillis*, the hooked glochidia of which establish themselves on the gills and the skin of bluegill, *Lepomis macrochirus*. They discovered that the fraction of total attached glochidia that completed development and successfully left the host alive was greater than 45% for the first two infections of an originally naïve fish, but fell to less than 25%

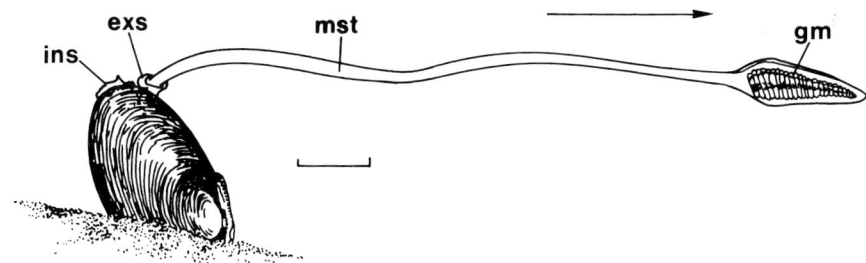

Fig. 16.13. The American unionacean *Lampsilis perovalis* releasing superconglutinate. Arrow, direction of water current; exs, exhalent siphon; gm, glochidial mass; ins, inhalent siphon; mst, mucous strand. Scale bar: 2 cm. From Haag *et al.*, 1995, with permission from Blackwell Publishers.

for the third and fourth infections. The third and fourth infections were also characterised by a marked increase in the percentage of dead and live glochidia shed during the first five days of each infection. However, in spite of the passage of more than 80 years since Reuling (1919) identified and studied the phenomenon of acquired immunity, there is little to add to his interpretation of the mechanism.

17

LAMPREYS

"Salar could not shake off Petromyzon. The lamprey's mouth was stuck firmly to his left side ……..Petromyzon sucked the scales closer to his teeth, and began to rasp away and swallow skin and curd and flesh. He drew blood, and fed contentedly"

Henry Williamson (1935) "Salar the Salmon".

17.1 INTRODUCTION

In 1923, Mr Henry Lamond, who was then Secretary of the Loch Lomond Angling Improvement Association referred to a report received in September 1922 by the Fishery Board for Scotland concerning a mysterious fish 'disease' in the lake. This 'disease' affected a relative of the salmon, the powan or 'freshwater herring', *Coregonus clupeoides* (now *C. lavaretus*), a whitefish that spends its whole life in fresh water, being restricted in Scotland to Loch Lomond and Loch Eck. The 'disease' manifested itself as circular sores about half an inch in diameter scooped out of the back of the fish. 'Diseased' fishes were reported throughout the year and the affliction appeared to be on the increase.

The Board took steps to investigate the problem. It emerged that these sores could not be attributed to infection by any micro-organism or to the feeding activities of any parasitic invertebrate, but were the feeding sites of a river lamprey or lampern (*Lampetra fluviatilis*), a remarkable parasitic fish (Figure 17.1). According to Mr Lamond, river lampreys were present in great numbers in Loch Lomond in 1923. The members of the Board were clearly unaware of an earlier report by Robertson (1875) of lampreys feeding on powan in Loch Lomond. It turns out that this Loch Lomond race of river lampreys has some special features, one of which is their small size, and we will consider these special features later.

The river lamprey is one of three British lampreys. It is intermediate in size between the larger sea lamprey, *Petromyzon marinus*, and the smaller brook lamprey, *Lampetra planeri*. Most river lampreys fall between 30 and 35 cm in length (maximum 50 cm), but individuals of the Loch Lomond race measure 18 – 24 cm (Maitland, 2004). *P. marinus* measures 50 – 70 cm (maximum 86 cm) and *L. planeri* 10 – 15 cm (maximum 18 cm). For information on the current conservation status of lampreys in Britain, their distribution as far as it is known and their ecological requirements the reader is referred to Johns (2002) and Maitland (2003).

Although lampreys resemble eels (Figure 17.1), with which they are sometimes confused (see Robertson, 1875), their similarities extend no further than general appearance and sliminess, and the two are not in any way related. In fact, strictly speaking, agnathans are not true fishes, but are usually referred to as such and for convenience this practice will continue in this book. Lampreys are special because, together with their relatives the hagfishes, they are amongst the most primitive vertebrates alive today, being representatives of a group known as the Agnatha, which literally means 'jawless' (Hubbs & Potter, 1971; Pough *et al.*, 1990; Maitland & Campbell, 1992). Other British fishes (Gnathostomata; see Appendix 1)), and indeed other vertebrates have upper jaws fixed to the skull and hinged opposable lower jaws, permitting a biting action. Agnathans have no jaws; in lampreys the jawless mouth lies in the centre of a tooth-bearing sucker-disk. Other striking features of agnathans are their lack of paired fins, lack of scales, presence of only one nostril and possession of a cartilaginous as opposed to a bony skeleton.

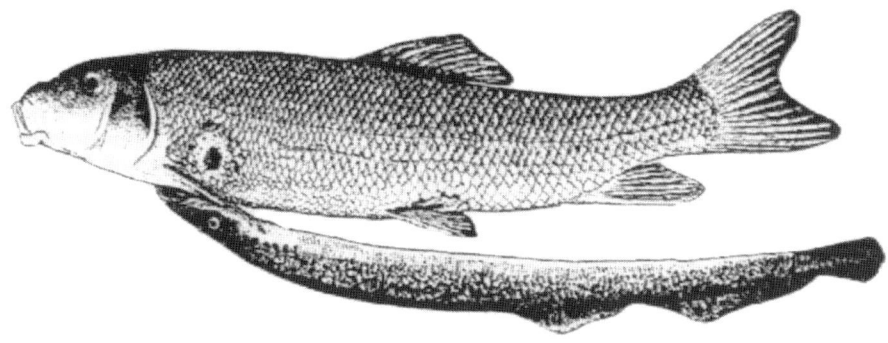

Fig. 17.1 A lamprey attached to a fish. Note feeding scars above the pectoral and pelvic fins made by other lampreys. From Gage & Gage-Day, 1927, reproduced from a photograph in Surface (1898).

Like eels, river and sea lampreys are migratory, spending part of their lives in the sea and part in fresh water, but the patterns of their life cycles are totally different. The eel, *Anguilla anguilla*, is catadromous, spawning in the sea, while the river and sea lampreys are anadromous, spawning in fresh water. Most populations of river and sea lampreys do not feed when in fresh water, but the river lampreys of Loch Lomond (see above) are an exception, as are some populations of sea lamprey (see below).

The hagfishes are entirely marine. Strahan (1963) reviewed the early literature on hagfish feeding and came to the conclusion that they are not parasitic, in spite of their reputation for attacking other fishes. Their interest in other fishes appears to be restricted to dead or dying individuals and all the evidence points strongly to a scavenging life style. Hardy (1959) summarised this life style as follows: "that they feed on all manner of benthic animals and that their habit of eating into dead and dying fish is by no means their only or even their typical mode of life – they just seize on dead fish when the opportunity of such a good meal presents itself".

There is some debate as to whether *L. fluviatilis* and *P. marinus* are predators or parasites (*L. planeri* does not feed as an adult and merits special treatment at the end of the chapter). There are those who view them as predators, since the intestines of *L. fluviatilis* have been found to contain, as well as fish blood and scales, muscle, spines, eggs and tissue from the pyloric caeca and swimbladder of fishes (references in Hardisty & Potter, 1971a). There are reports too of *L. fluviatilis* with intestines full of tiny fishes and they are also reputed to take bottom-dwelling invertebrates such as worms and crustaceans. On the other hand, Gage (1929) dissected many recently fed lampreys taken from hosts maintained in the laboratory and from wild hosts and consistently found host blood in the intestines, and Lennon (1954) regarded the feeding apparatus of the sea lamprey as adapted for obtaining liquid food, principally blood. The anticoagulant properties of the buccal gland secretion of lampreys (see below) are consistent with this. According to Lennon, some flesh is also taken from the host, after its liquefaction at the site of lamprey attachment by another component of the buccal gland secretion. This diet contrasts sharply with that of the predatory/scavenging hagfish and is compatible with a parasitic life style. As Baxter (1956) has pointed out, the absence of buccal glands in hagfishes and the lack of anticoagulant activity in extracts of the buccal epithelium of these fishes strongly suggest that the parasitic habit is confined to the lampreys.

Further support for a parasitic rather than a predatory relationship between lampreys and their fish hosts comes from observations of Maitland (1980) on the interaction between *L. fluviatilis* and powan in Loch Lomond. Maitland found that 26 to 50% of powan carried healed lamprey scars: 55% of scarred fishes carried more than one scar; two scars were common and multiple scarring with up to eight scars was not unusual. Thus, recovery of parasitised powan seems to be good, provided that the lamprey does not penetrate into the body cavity (see Maitland *et al.*, 1994). In fact Slack (1955, in Maitland, 1980) found no evidence of loss of condition in powan scarred by lampreys. However, there are records of serious damage inflicted on host populations by lampreys and these will be considered further below.

There is evidence that scars on the bodies of whales inhabiting the North Pacific Ocean are the feeding sites of the lamprey *Lampetra tridentata* (= *Entosphenus tridentatus*) (see Slijper, 1979).

17.2 THE RIVER LAMPREY – *LAMPETRA FLUVIATILIS*

17.2.1 *The ammocoete*

Adult river lampreys seek out stretches of running water with a bed of stones or gravel where they spawn in shallow nests (see below). The larval lamprey emerging from the egg is already eel-like in shape, but curiously the larva or ammocoete (Figure 17.2) is fundamentally different from the adult lamprey. The main differences concern the mouth and gill regions. An oral hood overhanging the mouth, which is guarded by a ring of finger-like processes or oral cirrhi, creates a blunt 'snout'. There is no oral sucker surrounding the mouth and no teeth. The ammocoete has seven pairs of gill slits like the adult, more reminiscent of the gill slits of a shark or ray than the single pair of operculum-covered gill openings of a teleost fish. However, unlike the adult lamprey, the elongated gill chamber or pharynx is floored by a so-called endostyle, which is glandular and produces a mucoid secretion.

These striking differences between larval and adult lampreys are so great that a relationship between the two stages was originally not suspected and the larval stage was placed in a separate genus, *Ammocoetes*. It was not until 1856 that Müller in Germany made the connection. He was aware that the river lamprey (at that time called *Petromyzon fluviatilis*) was present in his streams only during the spawning season and that none but fully-grown individuals were seen. He went on to collect and incubate eggs from these lampreys and noticed the close similarity between his newly hatched individuals and specimens of *Ammocoetes* from the same streams. He came to the inevitable conclusion that "the supposed genus *Ammocoetes* was in reality founded upon the young of *Petromyzon*". As so often happens in such circumstances, the old generic name was preserved and the larval lamprey became known as an ammocoete.

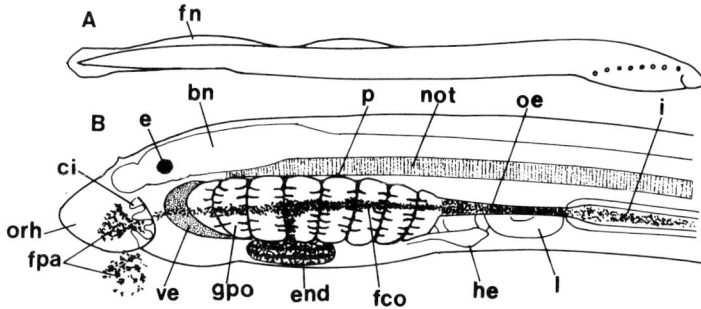

Fig. 17.2. The ammocoete (larval lamprey). (A) Whole animal. (B) Animal preserved while feeding, then rendered translucent after staining. bn, Brain; ci, oral cirrhi; e, eye; end, endostyle; fco, food cord in pharynx; fn, fin fold; fpa, food particles; gpo, gill pouch; he, heart; i, intestine; l, liver; not, stiffening rod (notochord); oe, oesophagus; orh, oral hood; p, pharynx; ve, velum. Modified from Young, 1962.

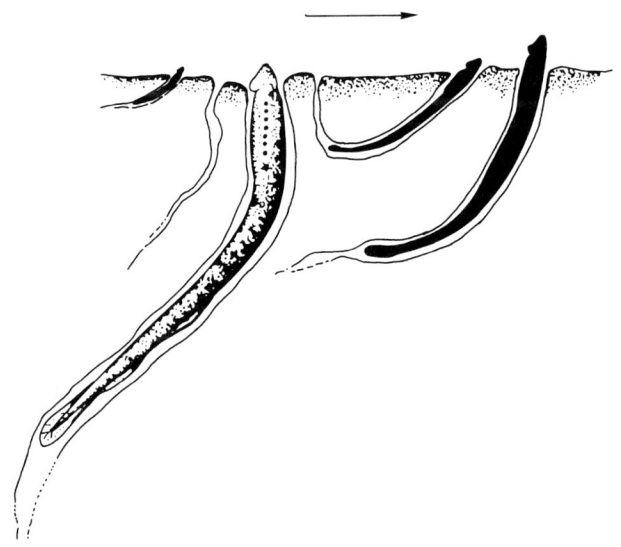

Fig. 17.3. Ammocoetes in their burrows. The arrow shows the direction of the current. From Sterba, 1962, reproduced with permission from Schweizerbart-Publishers (http://www.schweizerbart.de).

The reason for the differences between the ammocoete and the adult lamprey, as might be suspected by the reader, is that the ammocoete is not a parasite. It obtains its nutriment in a unique and remarkable way. The ammocoete lives in a burrow in a sandy substrate (Figure 17.3) and subsists by extracting minute digestible particles (microscopic animals and plants) from a current of water drawn into the mouth from just above the sediment. This current is created, not by cilia as in the adult swan mussel (see Chapter 16), but by the sucking and pumping action of the muscular pharynx, assisted by movements of a pair of muscular flaps (the velum) located at the anterior end of the pharynx (Figures 17.2, 17.4C; Newth, 1930; Hardisty & Potter, 1971b).

Gage (1929) used a magnifier to observe a feeding ammocoete held in a test tube. He administered a milky suspension of starch or flour, which was drawn into the gill chamber via the mouth with each intake of water. Some of this suspension emerged from the seven pairs of gill openings, but the rest was diverted to the oesophagus. Newth (1930) showed that the sticky mucous strands secreted by the endostyle play an important part in collecting the digestible particles. These strands form a conical net through which the food-laden current must pass as it travels from the mouth to the gill slits. Food particles adhere to the mucous strands, which unite to form a single cord. This cord passes down the middle of the pharynx to the oesophagus, which is ciliated, permitting the oesophagus to haul in the cord rapidly and pass it on to the intestine (see Figure 17.2B). The finger-like oral cirrhi provide an effective strainer, removing over-sized and potentially unusable particles from the inflowing current before they reach the pharynx. Gage (1929) observed that when these cirrhi become clogged with unwanted material the gill openings close and muscular contraction of the pharyngeal region forces water out of the pharynx through the mouth, thereby dislodging the detritus and clearing the clogged oral filter.

The endostyle appears to produce excess mucus and some of it issues from the gill openings. Newth (1930) and others have suggested that this mucus serves to bind and stabilise the walls of the burrow, which consequently do not easily collapse.

These tiny ammocoetes seem defenceless and highly vulnerable to predation, but Pfeiffer & Fletcher (1964) reported that North American lampreys and their larvae were rarely found in the stomachs of salmonid fishes. Feeding experiments confirmed the apparently distasteful nature of lampreys to many fishes and Pfeiffer & Fletcher proposed that skin secretions are responsible for this rejection. In the River Teify in Wales, Thomas (1962) found that the brown trout, *Salmo trutta*, rarely contained the remains of ammocoetes, but they were more likely to appear in the diet of eels. Presumably the danger from predators is reduced when the ammocoetes are buried and Gage (1929) reported that exposed ammocoetes appear restless and bury themselves quickly.

An ideal substrate for an ammocoete is an open-structured sediment composed of sand and silt with a clay fraction, preferably located in an eddy or backwater where current velocity is slow and steady and where organic material tends to accumulate (see Hardisty & Potter, 1971b). Sawyer (1959) observed that his American ammocoetes burrow rapidly, first using whip-like tail movements to drive the head vertically into the sediment. This is followed by the smooth disappearance of the larva into the substrate, without further undulation or noticeable muscular activity. Sawyer satisfied his curiosity about this apparently effortless disappearance by permitting his ammocoetes to burrow into a wad of cotton in which their behaviour could be observed. He found that contractions of the segmental muscles of the trunk provide propulsion, in conjunction with the action of the oral hood. When folded the hood provides a pointed probe, which pushes its way into the substrate. When opened out the hood acts as an anchor, like the expanded tip of the foot of a burrowing bivalve mollusc, permitting the body to be pulled into the substrate. Finally, the body is arched to bring the oral hood just above the sediment, so that the animal can draw in a food-bearing current (Figure 17.3). When conditions are right, a funnel-shaped depression, in the centre of which the oral hood may be visible, indicates the presence of a buried ammocoete. In the localities favoured by ammocoetes (see above) conditions often encourage the growth of diatoms, which form an important part of their diet.

Thus, like the gnathiid isopods (Chapter 15) and the unionacean mussels (Chapter 16), *L. fluviatilis* has a 'Jekyll and Hyde' life style, combining a parasitic phase and a free-living phase. However, it is the larval stages of gnathiids and unionacean molluscs that are parasitic, their adults being free-living, while in *L. fluviatilis* the larvae are free-living and the adults parasitic. Hardisty & Potter (1971c) estimated that ammocoetes of *L. fluviatilis* spend 4¼ - 4½ years in the larval free-living phase (see Figure 17.14).

17.2.2 Metamorphosis (transformation)
The metamorphosis or transformation undergone at the end of the ammocoete phase is drastic. Within a period of 4 – 5 weeks the oral disk and teeth develop, the eyes enlarge and differentiate and the pre-orbital region of the head extends (Figure 17.4AB). Changes occur in the gill openings, the fins and the pigmentation. However, the completion of these external changes does not mean that metamorphosis has run its course, because changes inside the body continue to take place. The most significant of

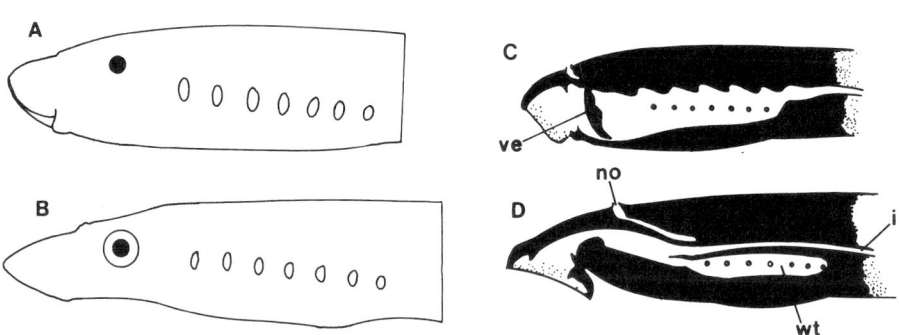

Fig. 17.4 Reorganisation of the head region of a lamprey during metamorphosis (transformation) from ammocoete (AC) to young lamprey (BD). (AB) Changes in external features. (CD) Changes in the foregut. i, Intestine; no, nostril; ve, velum; wt, water tube. (AB) Drawn from a photograph by Hardisty and Potter (1971a); (CD) from Sterba, 1962, reproduced with permission from Schweizerbart-Publishers (http://www.schweizerbart.de).

these is the development of the foregut (Figure 17.4CD), which is essential to permit parasitic feeding and to enable the fish to drink seawater to compensate for the osmotic loss of water to the sea via the gills and skin. Hardisty & Potter (1971a) referred to this period of continuing internal change as the 'macrophthalmia' stage, a term reflecting the relatively large size of the eyes (Figure 17.4B). They regarded the onset of feeding as the end of the 'macrophthalmia' phase and the beginning of life as a juvenile.

After the completion of all changes needed to enable the lampreys to feed parasitically and to cope at least with the osmotic challenge of estuarine conditions, the young lampreys emerge from the sediment at night and are swept downstream, especially during periods of flood (Hardisty & Potter, 1971a), eventually reaching the sea. The lapse of time between the onset of metamorphosis and this downstream movement may be considerable. Most *L. fluviatilis* in the southern part of Britain enter metamorphosis in July or August, but they do not move downstream until the following March, April or even May (Maitland *et al.*, 1984). This variation in the timing of the migration may be related to differences between individuals in the completion of the internal changes essential for survival in the marine environment (see Maitland *et al.*, 1984).

Many river lampreys spend time in the estuaries of major rivers, where a variety of estuarine fishes act as hosts, such as herring (*Clupea harengus*), sprat (*Sprattus sprattus*) and flounder (*Platichthys flesus*) (see Maitland, 2003). Bahr (1952, in Hardisty & Potter, 1971a) regarded the river lamprey as a brackish water species, restricted in the North Sea to coastal waters with a critical salinity of 22‰. However, the ability of some macrophthalmia to survive transfer to full strength seawater shows that they are potentially able to penetrate the marine environment (references in Hardisty & Potter, 1971a).

According to Hardisty & Potter (1971a), *L. fluviatilis* typically does not feed in fresh water, either during the downstream migration after metamorphosis or during the upstream spawning migration that follows the parasitic feeding period spent in estuaries or in the sea. An exception to this will be apparent immediately to the reader. This is the intensive feeding of *L. fluviatilis* on powan (*Coregonus lavaretus*) in Loch Lomond described at the beginning of this chapter. Loch Lomond is apparently unique among British lakes in possessing a large population of adult *L. fluviatilis* feeding in fresh water and this will be considered further below.

17.2.3 Attachment, feeding and breathing in the adult
The oral disk of the lamprey (Figure 17.5) is a unique feature among fishes, recalling the oral sucker of a leech (Chapter 7) and reflecting the parasitic life style of the adult lamprey. The attachment ability of the oral disk is also important during nest building (Figure 17.11), in mating (Figure 17.12) and for ascending waterfalls during the spawning run (see below). As in a leech, suction plays a major role in attachment to the host. However, the exposed surface of the lamprey oral disk is also armed with symmetrically arranged teeth (Figure 17.5), the main function of which may be to prevent the disk sliding over the slimy skin of the host during feeding. According to Hardisty & Potter (1971a), the disk of a newly attached lamprey may slide over the host's surface until a favourable spot for feeding is selected, when the teeth on the disk are presumably engaged.

Since hosts invariably struggle vigorously when a lamprey establishes itself (see Gage, 1929, Lennon, 1954), it is critical that attachment by the oral disk should be tenacious. Norman (1947, in Lennon, 1954) suggested that the best way to appreciate the great efficiency of suctorial attachment by the lamprey was to allow the parasite to attach itself to a hand or arm and then attempt to dislodge the animal under water. He claimed that this would be almost impossible to achieve. It was necessary to remove the limb from the water to promote detachment.

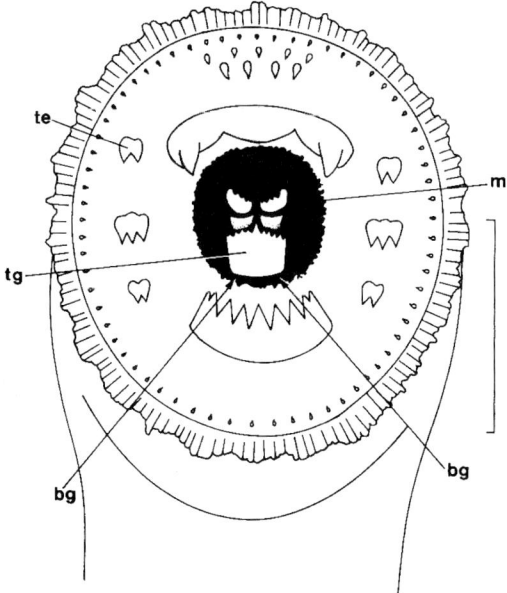

Fig. 17.5. Oral sucking disk of an adult *Lampetra fluviatilis*. bg, Positions of buccal gland openings (from Baxter, 1956); m, mouth; te, teeth of oral disk; tg, tongue armed with teeth. Scale bar: 1 cm. Reproduced from Maitland, 1972, with the permission of The Freshwater Biological Association.

The oral disk enables the lamprey to remain attached to its fish host for long periods, but the act of feeding while attached presents considerable physical challenges. The tooth-bearing tongue, which is located in the oral cavity in the centre of the sucker (Figure 17.5), must be free to abrade the flesh of the host and the animal must be able to ingest blood and tissue fragments from the feeding site. Both of these functions need to be conducted without loss of suction pressure. Two other problems arise. (1) An expanded oral disk would greatly interfere with the hydrodynamics of a swimming lamprey, especially when engaged in the energy demanding business of upstream migration or requiring a turn of speed to make contact with a suitable swimming host. (2) While the oral disk is attached, the gills cannot be ventilated in the conventional fashion by a current of water entering the mouth.

Lampreys have a unique set of behavioural and anatomical adaptations that provide solutions to these problems and ensure smooth functioning of the animal. First, while swimming, the lateral margins of the oral disk are brought together so that the opening of the funnel is reduced to a slit (Lennon, 1954; Lanzing, 1958). The disk so-folded forms a vertical wedge-shaped cutwater, which is much more streamlined than the open funnel (Lennon, 1954). Secondly, in adult lampreys breathing is tidal. Water is driven out of each of the seven pairs of gill pouches (Figure 17.6) by muscular contraction, and each pouch is refilled via the external branchiopore (gill opening) by elastic recoil when the muscles relax (Randall, 1972). Each branchial pouch opens via an internal branchiopore into a water tube or branchial cavity, which ends blindly in a posterior direction but communicates with the pharyngeal cavity, oral passage and buccal cavity in an anterior direction (Figures 17.4D, 17.6). However, water from the gill pouches is normally unable to enter the pharyngeal cavity because there is a valve

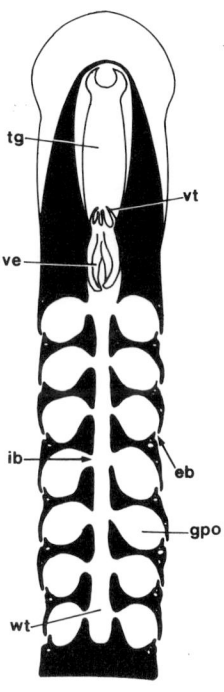

Fig. 17.6. Schematic diagram of a longitudinal facial section through the head region of an adult lamprey. eb, External branchiopore (opening of gill pouch); gpo, gill pouch; ib, internal branchiopore; tg, tongue; ve, velum; vt, velar tentacles; wt, water tube. From Randall, 1972.

(the velum) that effectively closes this route. Thus, in the adult feeding lamprey, breathing is essentially independent of events in the pharyngeal and buccal cavities, and this functional separation appears to be maintained even when the lamprey is not attached either to a host or to the substrate (Dawson, 1905 in Randall, 1972).

The situation is different in the ammocoete, Here the oral intake of water is unobstructed and essential for filter feeding (see Figures 17.2, 17.4C). Gill ventilation in the ammocoete is unidirectional as in most other fishes, not tidal, water entering via the mouth, passing over the gills and leaving via the seven pairs of gill clefts.

Lennon (1954) regarded vision as being of prime importance in selecting and making contact with a suitable host, while Kleerekoper & Mogensen (1963) found evidence that *Petromyzon marinus* uses its olfactory sense to orientate towards fishes, responding in particular to amines secreted by its targets. Lennon (1954) described the attack mounted by *P. marinus* as direct, swift and accurate, the lamprey lifting its head so as to raise the oral disk into a leading position and then opening the disk out from its folded condition ready for application to the host. Most of the feeding sites of *L.*

fluviatilis on the freshwater powan were above the lateral line, between the head and the small dorsally situated adipose fin (Maitland, 1980) (Figure 17.7A). Similar patterns were found on male and female powan and the distributions of scars on the left and right sides of the body were also similar. The patterns of scars on brown trout (*Salmo trutta*), which are much less frequently attacked in Loch Lomond, were not greatly different from those on powan, although possibly more dispersed on the sides of the body (Figure 17.7B).

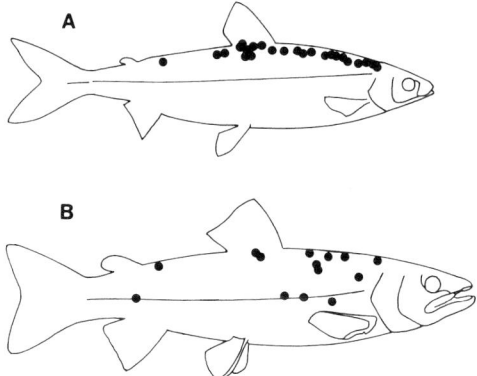

Fig. 17.7. The distribution of lamprey scars on (A) powan, *Coregonus lavaretus*, and (B) brown trout, *Salmo trutta*, collected together in Loch Lomond in 1966. From Maitland, 1980, reproduced by courtesy of Fisheries and Oceans Canada, with the permission of Her Majesty the Queen in Right of Canada, 2004.

The feeding mechanisms of *L. fluviatilis* and *P. marinus* appear to be similar and the following account is based on studies by Lanzing (1958) on *L. fluviatilis* and by Reynolds (1931) and Lennon (1954) on *P. marinus*. After application of the oral disk to

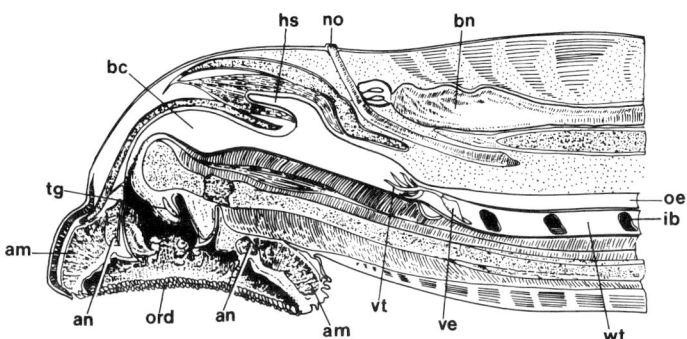

Fig. 17.8. Semi-diagrammatic longitudinal (sagittal) section through the head region of an adult lamprey. am, Annular muscle of oral disk; an, annular cartilage of oral disk; bc, buccal cavity; bn, brain; hs, hydrosinus; ib, internal branchiopore; no, nostril; oe, oesophagus; ord, oral disk; tg, tongue; ve, velum; vt, velar tentacles; wt, water tube. Reproduced with permission from Reynolds, 1931.

the host, the disk margin spreads out over the substrate and the volume of the buccal cavity is reduced by contraction of the annular muscle (Figures 17.8, 17.9A). This forces water past the tongue through the oral passage and into the pharyngeal cavity. Some of this water fills the hydrosinus, an extensible pouch opening from the pharyngeal cavity and lying dorsal to it (Figure 17.8). The tongue then retracts cutting off communication between the buccal and pharyngeal cavities (Figure 17.9B). The sealed buccal cavity then expands and the pressure falls within it. Thus far, the lamprey is attached by suction to the host, but is unable to use the tongue to rasp the host's skin because to do so would flood the buccal cavity and destroy the suction pressure. It is necessary to reduce pharyngeal pressure in order to free the tongue to engage in feeding.

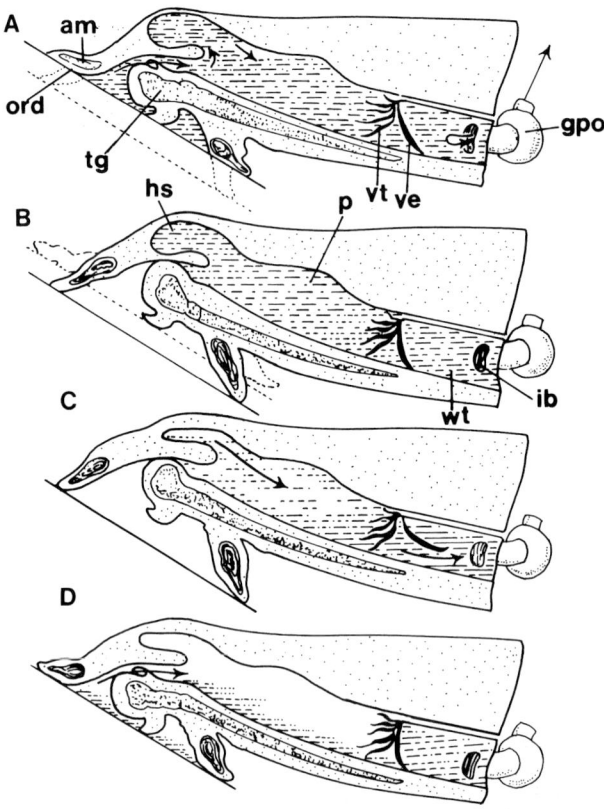

Fig. 17.9. (A – D) The sequence of events during attachment and feeding in the adult lamprey. See text for explanation. Arrows show direction of water currents and decreased density of shading indicates decreasing pressure. am, Annular muscle; gpo, gill pouch; hs, hydrosinus; ib, internal branchiopore; ord, oral disk; p, pharyngeal region; tg, tongue; ve, velum; vt, velar tentacles; wt, water tube. Reproduced with permission from Reynolds, 1931.

This is achieved by contraction of the hydrosinus and perhaps also the pharyngeal cavity, expelling water via the velar valve into the branchial cavity (Figure 17.9C). Expansion of the pharyngeal cavity then leads to a fall in pressure (Figure 17.9D). This permits the tongue to move forward, restoring communication between the buccal and pharyngeal cavities. The pressure in these two chambers is equalised at a level much lower than that outside the animal. In other words suction pressure is maintained, but the tongue is free to abrade host tissue in the manner shown in Figure 17.10.

Views differ on the mechanics of tongue operation and on how the blood and abraded host tissue are transported to the oesophagus, but it is agreed that buccal gland secretion has an important complementary role in feeding (see Hardisty & Potter, 1971a). There is a single gland on each side and the duct from each gland passes forward and opens on a small papilla in the buccal funnel below the tongue (Figure 17.5). Lennon (1954) showed that in *P. marinus* the buccal gland secretion, known as lamphredin, has anticoagulant and histolytic properties and Baxter (1956) demonstrated that the secretion from *L. fluviatilis* delays blood clotting.

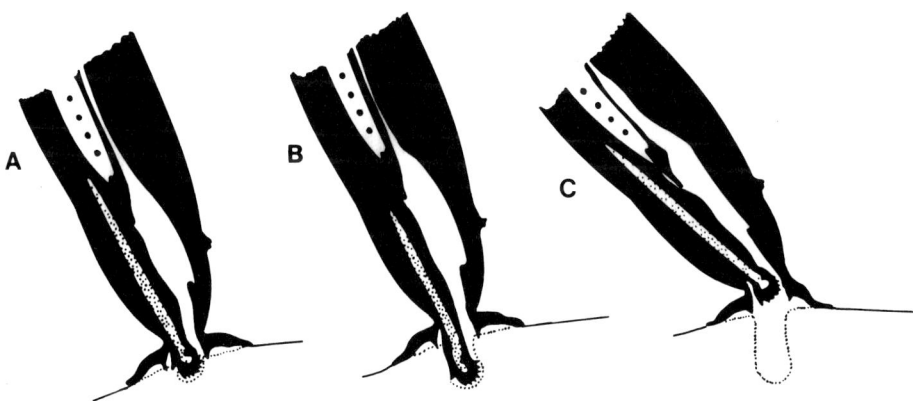

Fig. 17.10. Diagram illustrating successive stages (ABC) in use of the tongue during feeding of the adult lamprey. From Sterba, 1962, with permission from Schweizerbart-Publishers (http://www.schweizerbart.de).

17.2.4 Spawning

It is during the upstream spawning run of lampreys that commercial fisheries are in operation. Substantial fisheries used to exist on some of Britain's larger rivers, such as the Severn, and the deaths of both King Henry I and King John are attributed to consumption of a surfeit of lampreys (see Maitland & Campbell, 1992). Maitland & Campbell suggest that the source of these fatal overdoses may have been the lamprey pies that the citizens of Gloucester presented annually to the sovereign. Apart from a commercial fishery for river lampreys on the River Ouse in Yorkshire, which takes about two tonnes per year for use as bait by anglers (Maitland, 2004), lampreys are no longer of economic significance in the British Isles, but they are still important in parts of Europe. In Finland there are grill houses that sell nothing else but hot grilled river lamprey (see Maitland & Campbell, 1992) and sea lampreys are netted commercially in Portugal, where they fetch high market prices *en route* to restaurants (Maitland, 2004 and personal communication). Apparently the Romans thought highly of lamprey, but because they are scaleless, Jews were forbidden to eat them. Fontaine (1938) gave a detailed account of lamprey fisheries on the French Rivers Loire, Gironde and Rhône, including descriptions of the variety of apparatus used to capture them.

In the River Severn, the first river lampreys on the spawning run reach the weir at Tewkesbury during August, but the peak of the run usually occurs in October and November. Migratory activity appears to be limited to the hours of darkness; the lampreys avoid the light during the hours of daylight by seeking out resting places under rocks or riverbanks. In British rivers, *L. fluviatilis* spawns when the water temperature reaches 10 – 11 ° C, generally from March to April. The 'nest' or redd is essentially an oval depression in the gravel, which may be constructed by up to a dozen or more adults (Maitland, 2003). The oral sucker is used to lift and shift stones and loose material is driven downstream by vigorous body movements with the head attached to a stone. A similar nest constructed by brook lampreys (again a dozen or more adults may be involved; see Maitland, 2003) is illustrated in Figure 17.11.

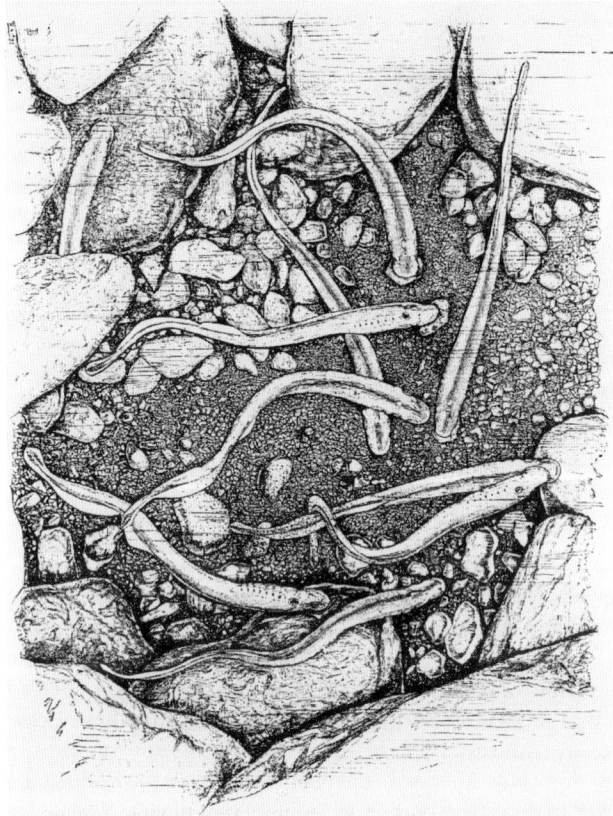

Fig. 17.11. Nest building and spawning of a brook lamprey. From Gage, 1929, from a drawing made at Lincoln Park, USA. by Bashford Dean, April 16th, 1897.

Prior to pairing in *L. fluviatilis* the female is attached by her oral sucker to a stone near the leading edge of the nest and the male normally approaches her from downstream (references in Hardisty & Potter, 1971a). The male slides along the dorso-lateral surface of the female and after attaching his oral sucker to the side of her head curls his tail in a tight spiral around her (Figure 17.12). The coiled tail then slides backwards, possibly helping to squeeze out eggs from the female. The tiny 'penis' of the male (Figure 17.13A) most probably serves to direct a jet of sperm towards the eggs as they are expelled. A swelling at the anterior edge of the female's second dorsal fin and a swelling anterior to the cloaca (Figure 17.13B) prevent the coiled tail of the male from sliding too far backwards and obstructing the female's cloaca.

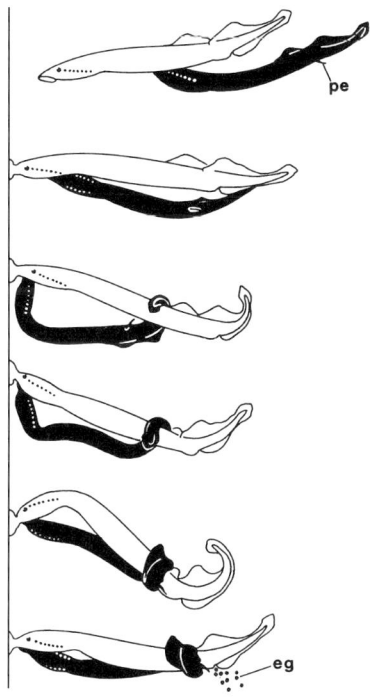

Fig. 17.12. Mating in *Lampetra fluviatilis*. Female: white; male: black. eg, Eggs; pe, 'penis'. From Sterba, 1962, reproduced with permission from Schweizerbart-Publishers (http://www.schweizerbart.de).

Sand particles readily adhere to the sticky outer surface of the eggs and the activities of the spawning lampreys ensure that the eggs with their attached ballast become buried in the substrate of the nest. Only small numbers of eggs are discharged at each spawning, so mating must occur repeatedly over a period of a few days.

As mentioned above, no feeding takes place during upstream migration and spawning (see Figure 17.14). Body reserves accumulated during the parasitic phase are diverted to facilitate development of the gonads and to meet other energy demands. The consequences of this diversion include shrinkage in length, loss of weight, degeneration of the eyes, closure of the foregut, atrophy of the intestine, and changes in the liver (Hardisty & Potter, 1971a). Death following spawning is inevitable. Müller (1856) referred to the dead bodies of lampreys floating in the water and that the ovaries of the females were quite empty. Curiously, the only part of the lamprey to persist after decay is the stiff supporting notochord, the rod-like structure running dorsally down the length of the body (Figure 17.2). When first found, the nature of these worm-like structures was unknown, but their true identity became apparent when partially decayed lampreys were found with the notochord partly in place (see Gage, 1929).

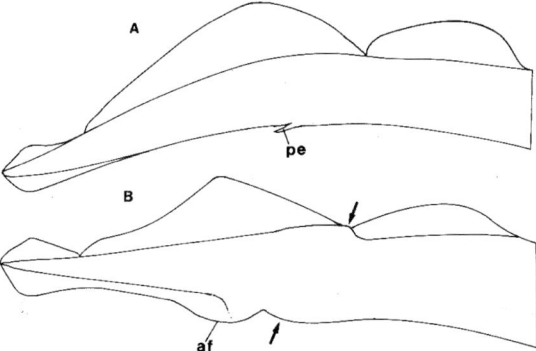

Fig. 17.13. The tail region of (A) a mature male and (B) a mature female of *Lampetra fluviatilis*, showing the secondary sexual characters. Arrows in the female indicate swellings on the leading edge of the second dorsal fin and in front of the cloaca. af, Anal fin fold behind the cloaca; pe, 'penis'. From photographs by Hardisty & Potter, 1971a.

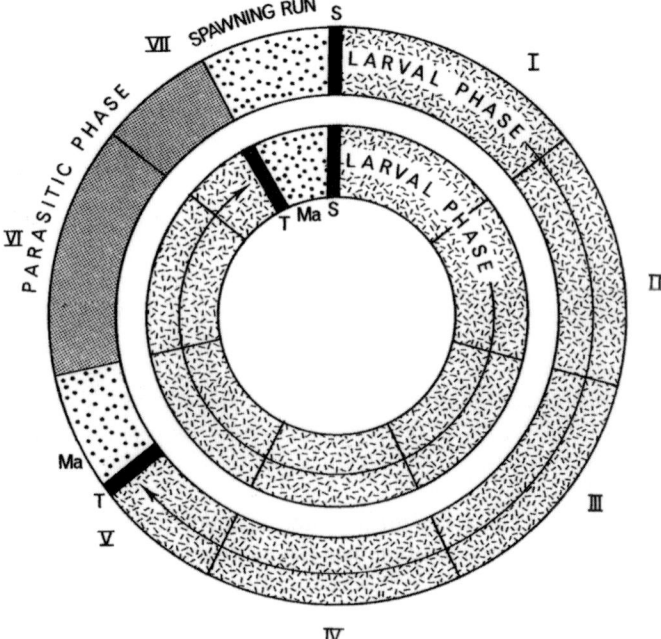

Fig. 17.14. The seven-year life cycles of the anadromous parasitic river lamprey *Lampetra fluviatilis* (outer circle) and the non-parasitic brook lamprey *L. planeri* (inner circle). During periods indicated by black areas (T, transformation and S, spawning) and large dots (Ma, macrophthalmia and spawning run) feeding does not normally take place. From Hardisty & Potter, 1971a, with permission from Professor I. Potter.

Hardisty & Potter (1971c) estimated that the average length of post larval/adult life in *L. fluviatilis* is 2½ - 2¾ years, which if added to the larval period of 4¼ - 4½ years gives an average life span of seven years (see Figure 17.14).

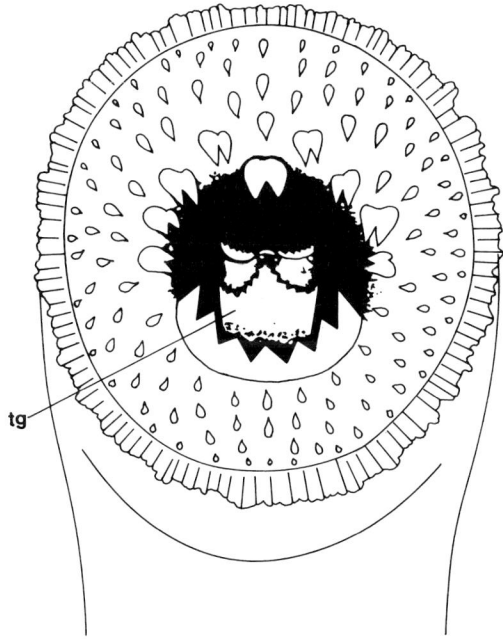

Fig. 17.15. Oral disk of *Petromyzon marinus*. tg, Tongue. Reproduced from Maitland, 1972, with the permission of The Freshwater Biological Association.

17.3 THE SEA LAMPREY – *PETROMYZON MARINUS*

The following is a quotation from Lamond (1923) recounting an experience on Loch Lomond in 1919:

"I happened to be acting as boatman for my friend, Mr D. Fletcher Buchanan, of Glasgow, a well-known salmon angler. While fishing with fly along the northeastern shore of Inch Fad, Mr Buchanan hooked, and I netted for him, an 8½ lb. salmon. Within a few minutes he hooked a 19 lb. fish, a strong male. I had some difficulty in netting him, as the landing net was inconveniently small, and the fish had to be played 'dead' before netting dare be attempted. While the fish was in the last stages of exhaustion we saw quite plainly that a large sea lamprey was fixed to the shoulder of the salmon. It was fully 18 inches in length and was at least 5 inches in girth. It was greyish in colour, with irregular blotches of a brownish tint throughout its length. I successfully netted the salmon, and barely missed getting the lamprey too, but it either abandoned its hold or the rim of the net detached it. We examined the salmon for any markings, but there was none, and I infer that the lamprey attacked the salmon when it was nearly exhausted, and had no time to penetrate the skin, although it had time to affix itself by suction."

It seems that Lamond's sea lamprey was not feeding, but there is abundant evidence that some sea lampreys, including those in Loch Lomond, feed in fresh water like their smaller cousins. Professor Peter Maitland (personal communication) has evidence, in the form of specimens and accounts from anglers, of sea lampreys attached to Atlantic salmon and to sea trout in Loch Lomond. The feeding of lampreys in fresh water will be considered further below.

The sea lamprey is readily distinguished from the river lamprey by its larger size (see above). There are also differences in coloration (see Maitland, 2004) and substantial differences in the number and pattern of teeth on the oral disk (compare Figures 17.5 and 17.15). The general biology of *P. marinus* is similar to that of *L. fluviatilis* outlined above. *P. marinus* has a wide geographical distribution, occurring on both sides of the Atlantic and in the Mediterranean, while *L. fluviatilis* is restricted to western Europe (Hubbs & Potter, 1971). In the British Isles, *P. marinus* is absent from rivers north of the Great Glen of Scotland, but is widespread but not common elsewhere. It is no longer able to inhabit some southern rivers because of pollution or the construction of barriers (Maitland, 1972; Maitland & Campbell, 1992).

There are indications that North American sea lampreys choose feeding sites below the lateral line (Figure 17.1; Maitland, 1980).

Although our knowledge of the marine distribution of anadromous lampreys is limited, the indications are that the larger species like *P. marinus* travel further away from the coastline than smaller ones like *L. fluviatilis*. In fact Matthews & Parker (1950) found *P. marinus* to be common on the basking shark, *Cetorhinus maximus*, although the superficial nature of the marks left by the lampreys indicated that they were not feeding, merely using the basking shark as a resting-place or for transport (phoresy).

17.4 FLEXIBILITY OF LAMPREY LIFE CYCLES

17.4.1 *Loch Lomond lampreys*

Maitland (1980) and Maitland *et al.* (1994) have described the special features of the lamprey populations of the Loch Lomond catchment in Scotland and the following account is condensed from these studies.

Loch Lomond has southern access to the sea via the River Leven and the Clyde estuary. River lampreys feed in the lake during the summer months, predominantly on powan but also on trout. At the end of the summer, feeding ceases and the lampreys migrate into the River Endrick, which drains into the southern end of the lake. Spawning has not been recorded in the other rivers and streams entering the lake. Almost all of the river lampreys in the Loch Lomond system are dwarf forms (see above for measurements). There are other differences: the dwarf forms have larger oral suckers, pre-branchial areas and eyes and they are darker, almost black in colour. The feeding period in the lake is restricted to a few months, compared with 15 – 18 months in typical estuarine-feeding river lampreys, and the dwarf form spawns earlier at a lower temperature.

There is every reason to believe that the dwarf river lampreys of Loch Lomond are land-locked, but a few lampreys of typical size have been found in the River Endrick. Maitland *et al.* (1994) regarded these as typical anadromous individuals, gaining access to the lake via the River Leven from the Clyde.

The sea lamprey is not common in Loch Lomond, but there is evidence that some of them, like the dwarf river lamprey, feed in the lake (see above).

The population of dwarf river lampreys in Loch Lomond appears to be unique in the British Isles, although similar populations are known from a few other large lakes in western Europe. Similarly, although sea lampreys regularly feed in fresh water in North America (see below), there is only one other European record of sea lampreys feeding in fresh water – a report by F.H. Jones (in Maitland *et al.*, 1994) of two feeding sea lampreys in Llandegfedd Reservoir in Wales. This led Maitland (1980) to pose the major question: "why is Loch Lomond apparently unique among British lakes in possessing a large population of lampreys feeding in freshwater?"

17.4.2 Sea lampreys in North America

It seems that *P. marinus* may feed in fresh water if it is forced to pass through large lakes or to undergo long migrations on its way to the sea (Hardisty & Potter, 1971a). Salmon and other fishes were reported by Davis (1967) to be parasitised by newly metamorphosed sea lampreys in a freshwater lake in Maine, USA. It is then but a short step to abandonment of the journey to the sea, and land-locked populations of *P. marinus* that have taken this step occur in Cayuga, Seneca and other lakes in New York State (see Surface, 1898) and in the Great Lakes (see below).

It is this readiness to feed in fresh water together with flexibility in choice of host that has permitted *P. marinus* to take advantage of a unique man-made opportunity in the American Great Lakes and to create a fisheries disaster of enormous proportions (see Smith, 1971). Lake Ontario was originally colonised by *P. marinus* by way of the St. Lawrence River and, according to Pough *et al.* (1990), the lake and the rivers and streams that flow into it provided acceptable conditions for land-locked populations to develop. Hubbs & Pope (1937) reported the presence in abundance of a "considerably dwarfed form" of *P. marinus* in Lake Ontario and commented that the lampreys even tormented swimmers. However, there is no evidence that they injured humans.

Although the other Great Lakes were connected to each other and to Lake Ontario, the sea lamprey was restricted to Lake Ontario until the 1920's. This is because the Niagara Falls between Lakes Ontario and Erie provided an insurmountable barrier. *P. marinus* can negotiate a waterfall at least 2 m high, if the current is not too strong, by inching its way upwards using its oral sucker, but no lamprey could climb 50 m up the face of Niagara. When the Welland Ship Canal was built, bypassing Niagara Falls, the way was open for colonisation of Lake Erie and beyond, although, curiously, it took 100 years (the canal was opened in 1829) for the lamprey to establish itself in Lake Erie. By the late 1940s the sea lamprey had spread throughout the Great Lakes system. It flourished on the freshwater fishes of the lakes, with disastrous results for the local freshwater fishery. In the earlier stages of the invasion, the lake trout, *Salvelinus namaycush*, was heavily parasitised and virtually eliminated from Lakes Huron and Michigan, but lampreys were sufficiently adaptable to turn to other hosts, populations of which also declined sharply (see Smith, 1971 for details).

The economic costs of the sea lamprey invasion of the upper Great Lakes have been and continue to be considerable. An effective lamprey larvicide was discovered in the 1960s (see Smith, 1971) and as long ago as 1980 the annual cost of lamprey control to Canada and the United States was estimated at about $5 million dollars (Talhelm & Bishop, 1980).

17.4.3 *The brook lamprey* – Lampetra planeri

It could be argued that our third British lamprey, the brook lamprey (*Lampetra planeri*) has no place in a book on parasites because it is not parasitic. However, there is no doubt that it had a parasitic ancestor and to omit a brief account of the biology of this lamprey would leave an unacceptable gap in our picture of lamprey evolution. Like land-locked *P. marinus*, the brook lamprey has abandoned the marine phase of its life cycle, but after metamorphosing, *L. planeri* does not feed (Figure 17.14). In the typical parasitic lamprey *L. fluviatilis*, metamorphosis and sexual maturation are separated in time, but these two processes are virtually simultaneous in *L. planeri* (see Hardisty & Potter, 1971c). Upstream migration, spawning and death are usually completed by *L. planeri* within a period of 6 – 9 months from the onset of metamorphosis (Figure 17.14). All of these energy-demanding events must be fuelled by reserves stock-piled during life as an ammocoete. Consequently the duration of this larval, filter-feeding stage must be extended to permit accumulation of sufficient reserves to keep the non-parasitic adult going during its short but energetic life. According to Hardisty & Potter (1971c), the life span of the ammocoete of *L. planeri* is 6¼ - 6½ years, i.e. about two years longer than that of the ammocoete of the parasitic *L. fluviatilis*. Since the adult *L. planeri* lives for only ½ to ¾ of a year, the total life span of *L. planeri* is the same as that of *L. fluviatilis* (see above and Figure 17.14).

The extended larval period of *L. planeri* is reflected just prior to metamorphosis in the large size of this lamprey compared with *L. fluviatilis*. However, in the brook lamprey, the need to mobilise body reserves to achieve early sexual maturity results in a synchronous reduction in weight and length at metamorphosis, so that after metamorphosis the sexually mature brook lamprey is significantly smaller than it was before the start of metamorphosis. This mobilisation of reserves in the maturing brook lamprey even extends to many of the oocytes in the ammocoete ovary, so that fecundity of the brook lamprey is greatly reduced compared with that of *L. fluviatilis*.

There is no doubt that *L. fluviatilis* and *L. planeri* are closely related. It is relatively easy to distinguish adults by their size, but more difficult to distinguish their larval, metamorphosing and macrophthalmia stages (Hardisty & Potter, 1971c). In fact, some earlier workers who were unaware of their divergent life cycles regarded them as belonging to a single species, and they have been considered more recently as ecological races of the same species (references in Hardisty & Potter, 1971c). Nowadays, separate specific status for these two lampreys is generally accepted, but there is no doubt that they are close relatives and that they share a parasitic/anadromous ancestor. Thus, *L. planeri* has evolved from an anadromous, parasitic lamprey, which also gave rise to *L. fluviatilis*, by eliminating the adult feeding and migratory stage with its attendant high mortality and replacing it with an extended larval phase, which is relatively protected (Hardisty & Potter, 1971c). Evidence that the brook lamprey has evolved from a parasitic ancestor is provided by the persistence of buccal glands in the former, although these do not reach full development (Baxter, 1956). Surprisingly, according to Baxter (1956), the buccal gland secretion of the brook lamprey retains anticoagulant activity, in spite of the fact that it is never used, as far as is known.

This evolutionary dichotomy is remarkable enough, but even more remarkable is the fact that it is not a unique development. Many other modern parasitic lampreys have corresponding non-parasitic forms. According to Hardisty & Potter (1971c), common ancestry can be inferred for eight 'paired species' worldwide, each pair comprising a parasitic and a non-parasitic species, the best known couple being *Lampetra fluviatilis/L. planeri*. Two other parasitic species, *Ichthyomyzon bdellium* and *Lampetra japonica* (now called *Lethenteron camtschaticum*), share common ancestry with more than one non-parasitic species, the former having two corresponding brook lampreys and the latter three. There are two additional brook lampreys that cannot be assigned to any extant parasitic species.

It is intriguing that brook lampreys with basically similar life cycles, morphology and physiology should have arisen repeatedly and independently in different river systems over a wide geographical area. However, this evolutionary divergence is not universal and there are three genera of parasitic lampreys that do not have related, non-feeding, non-parasitic counterparts. One of these is *Petromyzon*. Thus, our British lamprey fauna comprises one paired species, *L. fluviatilis/L. planeri*, and one unpaired parasitic lamprey, *P. marinus*.

Perhaps the last word should go to the North Italian (Po) brook lamprey, *Lampetra* (= *Lethenteron*) *zanandreai*. Zanandrea (1957) found that 12 out of 200 ammocoetes of this species collected from a stream near Verona in Italy possessed ovaries in an advanced state of maturity. In other respects these animals were larvae, with an intact endostyle and no post-metamorphic features. This extreme truncation of the life cycle in which sexual maturity is achieved in a larva is known as neoteny.

17.5 OTHER FISHES WITH PARASITIC TENDENCIES

A few 'jawed' fishes (Gnathostomata) have adopted a life style that bears some resemblance to that of lampreys. Clarke & Merrett (1972) and Carrassón *et al*. (1992) identified skin and blubber from whales and dolphins (cetaceans) in the gut of the shark *Centroscymnus coelolepis*. Carrassón *et al*. believed that this tissue was acquired by scavenging, but Clarke & Merrett argued that living whales could possibly be the source of this material. However, *C. coelolepis* feeds predominantly on cephalopods (Carrassón *et al*., 1992) and therefore barely qualifies as a parasite. Similar uncertainty exists about the life style of the snubnosed eel, *Simenchelys parasitica*. This fish is known to ingest flesh by burrowing into the bodies of larger fish like halibut, but a preference for dead and dying fish and hence a scavenging habit seems more likely (Robins & Robins, 1989).

The cookiecutter sharks, *Isistius brasiliensis* and *I. plutodus* are much better qualified as facultative ectoparasites of fishes. These small sharks (maximum length 50 cm and 42 cm respectively; see Compagno, 1984) are well adapted for attaching themselves to larger fishes such as tunny and swordfish or to cetaceans and excising a plug of flesh. Crater wounds, crescent-shaped wounds and scars of similar shapes are evidence of their activities. These sharks have a short snout enabling them to apply the fleshy lips surrounding the mouth to the surface of their host. Jones (1971) showed that, like a lamprey, attachment of the mouth is by suction, probably generated by retraction of a cartilage-supported 'tongue', loss of suction being prevented by the lips. However, unlike lampreys, feeding is not prolonged – while attached, a row of razor-sharp, saw-like teeth in the lower jaw scoops out a conical plug of flesh, aided by twisting of the body, and the shark then pulls free, with the plug cradled in its scoop-like lower jaw and held in place by hook-like upper teeth. Plugs of skin and flesh from large host fishes have been found in the stomachs of freshly caught *I. brasiliensis*.

The last word should go to bathypelagic angler fishes such as *Ceratias holboelli* described on p. 231, in which the males are tiny and truly parasitic on females of their own species.

18

CONCLUSIONS

18.1 CRUSTACEAN DOMINANCE

One of the most obvious features of this book is the preponderance of chapters on crustaceans. Seven of the sixteen chapters on parasites are concerned with crustaceans and five of these seven chapters describe copepods. This abundance and diversity undoubtedly reflects the wealth of opportunity for contact in aquatic space between crustaceans, copepods in particular, and fishes. As stated in Chapter 8, Sir Alistair Hardy (1970) estimated that in terms of absolute numbers of individuals, copepods are the most numerous multicellular animals on the planet. Moreover, the diversity of life cycles and general biology of copepod fish parasites (see below) indicate that parasitism of fishes has arisen several times among the copepods. Other features that may have a bearing on the success of crustaceans as fish parasites are considered below.

18.2 LIFE CYCLES

The life cycles of most fish skin and gill parasites involve only one host. Ectoparasites like monogeneans live more or less permanently on their hosts and mesoparasites are partly embedded in host tissue and hence unable to change hosts or to change their locations on their hosts. Some parasites such as leeches and gnathiid isopods make only short visits to their hosts for feeding and spend the rest of their lives detached.
    Some organisms are parasitic only as adults. The mature female poecilostomatoid copepod *Ergasilus* is the only parasitic stage, the juveniles and males enjoying a free-living existence, feeding most probably on planktonic organisms. Lampreys also spend their early lives as free-living, mud-dwelling, filter feeding ammocoetes, which undergo substantial anatomical change to produce the parasitic adults. In contrast, unionacean molluscs spend only a brief early period of their lives as parasites, later becoming typical, free-living, filter feeding, adult bivalves. Gnathiid adults are also free-living but they do not feed.
    The life cycle of the pennellid copepod *Lernaeocera* is unique, involving two different and obligatory host species. Juvenile stages typically parasitise a flatfish, while the blood feeding adult female lives on the gills of a gadid host. It seems unlikely that both hosts were acquired simultaneously and it is difficult to know which host came first. Perhaps the larval and juvenile stages were originally free-living, completing their cycle via an egg laying adult on a gadid, or perhaps the whole of the life cycle took place originally on a flatfish, the gadid host being acquired later. A shift to a separate, round-bodied fish host for the female parasite, as demanded by the latter scenario, would have advantages in terms of dissemination of parasite eggs and reducing the parasite load on the flatfish host. The fact that *L. lusci* is able to complete its life cycle on a flatfish in aquaria (Slinn, 1970), lends some support to this argument. In the related

pennellid *Lernaeenicus sprattae* the first and second hosts belong to the same species (*Sprattus sprattus*), but, between the two parasitic phases, the males and females are free-living, mating in the open sea. It remains to be determined whether the freshwater cyclopoid copepod *Lernaea cyprinacea* has an extra facultative or obligatory host, possibly belonging to the same or to a different host species.

If the ability of fishes to mobilise their immune system and mount an attack on multicellular parasites infesting their skin or gills proves to be a widespread and effective phenomenon it is tempting to evoke this to explain some fundamental differences in the life cycle patterns of fish parasites. On the one hand we have what might be called 'permanent' parasites, like the monogeneans that spend virtually all of their lives on a single host individual. Some of these parasites, such as *Entobdella soleae* and the caligid copepods, may occasionally move from one host individual to another, and this may provide some respite from attack by the host's immune system. At the other end of the spectrum we have most leeches, crustaceans such as the branchiurans (*Argulus*) and the gnathiid isopods. These 'temporary' parasites spend much of their time separated from their hosts, visiting their hosts briefly for feeding. They are therefore exposed to anti-parasite defences for relatively short periods of time, perhaps too short for the host's immune system to respond effectively. There may be an additional advantage for 'temporary' crustacean parasites since this may permit them to avoid the dangers of moulting on the host (see below). Gnathiids do not moult on their fish host and female ergasilids complete moulting before becoming parasitic.

It is interesting that the leech *Hemibdella soleae*, which appears to have adopted a permanently parasitic life style on the common sole (*Solea solea*), attaches itself to one of the spines of the host's ctenoid scales, a site which may be less provocative to the host's defences than the more typical leech attachment site. There may be some advantage too in changing location on the same host as many 'permanent' parasites do, since there is evidence to indicate that host defences may develop locally at the attachment or feeding site of a parasite (see Kearn, 1999). However, most polyopisthocotylean monogeneans on the gills are sedentary, but, with few exceptions, lack any obvious signs of generating a local host reaction. It is hard to understand how these parasites avoid provoking the host's immune system and it is equally difficult to explain the survival and success of the mesoparasites, which are permanently anchored and unable to leave the host, unable to change location and, to all intents and purposes, as provocative to the host's defences as it is possible to be.

18.3 REPRODUCTIVE BIOLOGY

Given the many different kinds of organisms that have adopted the parasitic way of life on fish skin and gills, it is not surprising that their reproductive biology is so diverse. The protozoans considered in this book multiply asexually. Monogeneans, leeches and the parasitic barnacle *Anelasma* are hermaphrodite and mutual cross-insemination is common in monogeneans and leeches. Unilateral cross insemination and self-insemination are also known in monogeneans. In the viviparous gyrodactylids, generations are telescoped and the first one or two embryos may be produced asexually. Copepods and isopods are dioecious (having separate male and female individuals).

Sperm transfer is by way of spermatophores in many monogeneans, leeches and copepods. This prompts us to ask whether ectoparasitism of fishes somehow favours this mode of insemination or whether it is simply a reflection of the widespread employment of spermatophores in invertebrates. The spermatophores of *Salmincola californiensis* are elaborate, having a cement component that seals the vaginal opening preventing further matings.

In the parasitic copepods, mating between separate males and females may occur on the surface of the host, as in the caligid *Lepeophtheirus salmonis*, in the pennellid *Lernaeocera branchialis* and in the lernaeopodid *Salmincola californiensis*, or in open water, as in the marine pennellid *Lernaeenicus sprattae* and the freshwater ergasilids. Evidence from work on the lernaeopodid *Alella pagelli* has shown that invasive copepodids destined to become males attach themselves only to hosts already carrying females.

In chondracanthid copepods, the male is tiny, structurally simplified compared with the female and permanently attached to the genital region of the female. The permanent presence of the males may discourage further matings and there is a possibility that the males may extract nutriment from the specialised papillae to which they are attached, i.e. the males themselves may be parasitic on the females.

It is among the isopods that we find the most surprising modifications of crustacean reproductive biology. In cymothoids like *Anilocra*, individual parasites are known to change their sex, young parasites developing first into males, which retain the ability to swim freely and change hosts. The same individuals later become sedentary females. However, this transition is not a straightforward one, since the presence of a mature female on the host retards progression of an accompanying male towards the female phase. Loss of the female removes the suppression and the male will become female. The presence of a young male also accelerates the transition of an accompanying parasite in the intermediate stage into a fully mature female.

Males and females of gnathiid isopods are free-living and strikingly different morphologically. The males have a pair of massive jaws, but these are not used for feeding, both sexes taking no food during their adult lives. Each male of *Paragnathia formica* occupies a burrow in the muddy banks of salt marsh channels, where its jaws are most probably used to protect a 'harem' of as many as 25 young females.

These developments in the isopods are intriguing, but it is in branchiurans such as *Argulus* that we find a reproductive system with some features that are so bizarre that they set branchiurans apart from all other crustacean fish parasites. In *Argulus* the two sexes are similar in size. Mating appears to take place on the host, the male using a 'press-stud' arrangement to clip its limbs to those of the female. This is not so remarkable but it seems that there is no penis, each male ejaculatory duct ending blindly. Release of pressurised sperm into the female apparently depends on perforation of each blind ending duct by a spine carried on a female seminal papilla. The structure of the ovary too is exceptional and these features, together with sperm structure and molecular evidence, suggest an alliance between branchiurans and the pentastomids, the latter being parasites of the respiratory tracts mainly of carnivorous reptiles.

18.4 ATTACHMENT AND FEEDING

18.4.1 *Hooks, harpoons and pincers*

Hooks and suction, or combinations of these, have been adopted for attachment by most skin and gill parasites of fishes. Hooks are notably absent in leeches; their posterior suckers are derived from several fused posterior segments. In contrast, it is unlikely that there are any crustaceans parasitic on fishes that do not use hooks for attachment at some stage in their lives. Their use of hooks may be limited to initial attachment to the host, but in ergasilid and chondracanthid copepods and in cymothoid isopods they provide the only means of attachment for the adults. Variations on the theme of hooks in crustaceans are the employment of barbed, harpoon-like mandibles in gnathiid isopods and the use of pincer-like (chelate or sub-chelate) structures in, for example, the copepodid of the pennellid *Lernaeenicus sprattae*. Crustacean hooks and their derivatives develop from one or more pairs of the segmentally repeated, jointed, cephalic or thoracic limbs. The possession of these limbs, requiring little if any evolutionary change to create an effective hook, chela or harpoon, may have been an important 'pre-adaptive' factor contributing to the wide range of crustaceans that have adopted a successful parasitic life style on fishes.

There is however a potential problem for any crustacean adopting a parasitic life style on a fish and that is the need to moult. This not only demands the sloughing off of the old exoskeleton, including that enclosing the limbs, but also requires a period of rest to permit expansion and then hardening of the new exoskeleton. This not only makes the parasite vulnerable to attack by predators, but also threatens to leave the parasite temporarily without rigid attachment organs. Remarkably, crustacean fish parasites have found ways of circumventing the problem, which include the exploitation during development of a secreted tethering frontal filament in caligid, pennellid and lernaeopodid copepods, moulting in two stages (the posterior half of the body first) in the cymothoid isopod *Anilocra*, and undergoing moulting only while separated from the host as in the ergasilid copepods and the gnathiid isopods. In lernaeopodid copepods the importance of the tethering secretion has been greatly enhanced. The adult parasite anchors itself not directly to the host but by barbed hooks at the tip of the second maxillae to a button-like bulla derived from the tethering filament secretion and cemented to a host fin ray or basement membrane.

Monogeneans are almost exclusively skin and gill parasites of fishes and also owe their success and diversity to hooks. The hooks of monogeneans are proteinaceous, secreted structures (sclerites) invariably embedded in a posterior attachment organ or haptor. They provide security of attachment at two levels: small hooklets for attachment to host epidermal cells and larger hamuli capable of penetrating through the host's epidermis into the collagenous dermis (see review by Kearn, 1999). Hooklets, which are usually 10, 14 or 16 in number, are essentially larval structures, but stresses on the haptor imposed by post-larval growth demand that the hooklets are supplemented by the larger hamuli. The hooked regions of the large anterior hamuli of *Entobdella soleae* contain a sulphur-containing protein resembling keratin found in human hair (Lyons, 1966). Keratin is known to be chemically resistant and the presence of similar material in those regions of the hamuli that are embedded in host tissue may be necessary to resist host attack during their lifetime of exposure. Similar properties may be inherent in the rigid, chitin-based, cuticular covering of the limb-derived hooks of crustacean parasites.

Monogeneans such as the dactylogyroideans use hooks for impaling host secondary gill lamellae, but the polyopisthocotyleans have evolved a means of attachment to gills that is unique. Jaw-like clamps with a supporting skeleton of sclerites, closed either by a muscle/tendon/pulley arrangement or by suction, are used to grip rather than pierce one or two secondary gill lamellae. Perhaps clamping is less provocative to host defences than perforation by hooks, although a few hooks often persist in adult polyopisthocotyleans at the posterior border of the haptor.

18.4.2 *Suction*

Most gyrodactylid monogeneans appear to rely entirely on hooks for attachment to skin and/or gills, but other skin-parasitic monogeneans rarely do so. The haptors of many skin parasites are disk-shaped and are capable of generating suction independently or with the aid of hooks. Suction is also a popular option with crustacean parasites and a large part of the anterior region of the body has given rise to a disk-shaped sucker in the caligid copepods and, probably independently, in the bomolochid and taeniacanthid copepods. It is possible, and indeed likely, that these monogeneans and copepods use hooks for attachment when physical stress on them is light. The more energy-demanding suction option may come into play when there is danger of dislodgement, such as when powerful currents are generated by sudden increases in host swimming activity or when a predator attacks.

It is in the branchiurans (*Argulus* spp.) that we encounter one of the most remarkable morphological transformations, not just in fish parasites, but also in the Animal Kingdom. This is the transformation of the proximal podomere of the first maxilla into a powerful sucker, thereby functionally replacing the various hooked appendages, including two barbed hooks at the distal end of the first maxilla. The branchiuran sucker seems to me to come close to qualifying as an 'organ of extreme perfection and complication' as exemplified by the human eye in Darwin's "Origin of Species" (1859), but the steps in the evolutionary change from a limb podomere to a sucker defy the imagination.

Suction is thought to be important in the attachment of the epizoic trichodinid protozoans (ciliates) to the surfaces of their 'hosts', but attachment of the pathogenic flagellate protozoan *Ichthyobodo* is unique, the surface membrane of its attachment plaque fusing with the surface membrane of a host epithelial cell. This forms the basis for the penetration of the host cell by the cytostomal tube, which extracts nutrients from the cell.

Attachment and feeding are intimately linked in the lampreys. The head is modified to such an extent that it is able to attach suctorially to its fish host and, without losing suction, to abrade host tissue with the tongue. This does not interfere with breathing, since the gills are ventilated not via the mouth but tidally via the external openings of the gill pouches.

18.4.3 *Glands*

Glands are associated with the attachment organs of many skin and gill parasites of fishes, including monogeneans, leeches and branchiuran crustaceans. Haptor glands in microbothriid monogeneans secrete cement, which is used to attach the parasites to the hard surfaces of shark denticles, a habit that has led to the loss of hooks. There is evidence that a mixture of two different secretions released by pads on the head of the monogenean *Entobdella soleae* creates an adhesive that adheres strongly to wet, slimy fish skin. The parasite is capable of severing this tenacious bond with great speed. These are properties that render the pads highly effective for rapid leech-like locomotion on the host. Cement is probably important too in attachment of leeches such as *Hemibdella soleae* and possibly in some epizoic peritrich ciliate protozoans (scyphidiids).

Other functions for secretions associated with attachment organs have been suggested. There are claims that they may digest and remove host slime, seal the margins of suckers, act as an anti-inflammatory or even promote host tissue hyperplasia, but these suggestions are purely speculative. However, the importance of these glands should not be underestimated.

18.4.4 *Mesoparasites*

The attachment structures considered above are either non-invasive, like the suckers of leeches and the clamps of monogeneans, or minimise invasion, like the hooks of monogeneans and crustaceans. In contrast, some fish parasites are highly invasive, penetrating deeply into the tissues of skin or gills and developing branched roots that are impossible to remove intact without the most careful dissection. We might expect crustaceans, with their relatively rigid exoskeleton, to be the least likely of the parasites to develop such a flexible deep-rooting system, but in fact this mesoparasitic way of life has evolved more than once in the crustaceans. Convergent evolution has produced remarkably similar, deeply rooted, but distantly related copepods: on the one hand pennellids like *Lernaeocera branchialis* and on the other the cyclopoid *Lernaea cyprinacea*. The embedded peduncular half of the body of the mesoparasitic barnacle *Anelasma squalicola* also bears extensive roots. The poecilostomatoid copepod (chondracanthid) *Lernentoma asellina* does not have a branched root system, but its greatly elongated cephalosome ensures that the antennal region is deeply embedded.

The invasive strategy of these mesoparasites will provide secure attachment, without the continuous expenditure of energy required by superficially attached ectoparasites to maintain suction or to prevent hook disengagement. However, there will be an initial energy demand as the mesoparasite invades host tissue and there may be a prolonged energy requirement to resist host attack.

The mesoparasitic relationships between these crustaceans and their hosts probably arise by active penetration of host tissue by the parasites. In monogeneans, similar mesoparasitic associations appear to have arisen in two other ways. The first of these concerns the dactylogyrine monogenean gill parasite *Dactylogyrus extensus*. In this parasite hyperplasia of the host gill epithelium leads to complete enclosure of the haptor, preventing further locomotion. A possible explanation of this situation is that the parasite/host relationship is relatively new, the host's defences responding strongly to the presence of the parasites, possibly leading ultimately to their death – dead parasites have been found with trapped haptors. However, it is equally possible that hyperplasia is encouraged/stimulated by the parasite as part of an adaptive 'strategy', eliminating energy expenditure for attachment and favouring nutrient uptake by the parasite via the embedded haptor tegument. Dead parasites may simply be old. *D. anchoratus* infecting the same host, does not induce hyperplasia, possibly because it has a much older, more balanced parasite/host relationship or because it moves frequently to new sites and thereby avoids local host reactions.

The second way in which monogeneans have developed a mesoparasitic association concerns *Amphibdella flavolineata*. Its haptor is also embedded in the gill of the host (the electric ray, *Torpedo nobiliana*), but hyperplasia is not evident. In this case the parasite comes from within, since juvenile parasites are endoparasites, living and mating inside the host's heart. The body of the mature or almost mature parasite emerges from the gill, permitting the release of eggs into the gill ventilating current.

Hyperplastic host epidermis also engulfs the glochidium of unionacean molluscs, after initial superficial attachment by hooks on the shell valves. This is an essential development for the survival, growth and metamorphosis of the young mussel.

Attachment and nutrition are intimately linked in mesoparasites. Penetration is so deep in *Lernaeocera branchialis* that the mouth is able to withdraw blood from the host's arterial system or heart. However, in *Lernaea cyprinacea* there is no evidence of blood feeding and in *Lernentoma asellina* and *Anelasma squalicola* the mouth is not embedded in host tissue. This raises the interesting possibility that these parasites are able to absorb nutrients from the host through the embedded surfaces. The elongation of the embedded regions and the extensive branching greatly increases the surface area of the parasite in contact with host tissue, developments which may enhance nutrient uptake. There is an intriguing possibility that the barnacle *Anelasma* may be a semi-parasite, absorbing nutrients from the host via the peduncle and its branches and supplementing this with material ingested from the surrounding water, like its free-living relatives.

Whether the attachment organ of a parasite is embedded in host tissue or attached superficially to the host's surface, it is in intimate contact with the host more or less permanently and certainly for very much longer than any other part of the body. The remoteness from the gut of some regions of the haptor of *Entobdella soleae* together with other clues, points to the possibility that the haptor may absorb nutrients directly from the host (see Kearn, 2002). This seems worthy of investigation in a range of fish ectoparasites.

18.4.5 *Diet*

Some parasites inhabiting the skin and gills of fishes have a restricted diet, while others are less discriminating. Monogenean skin parasites like *Entobdella soleae* appear to ingest only epidermis; glochidia and protozoans such as *Ichthyobodo necator* and *Ichthyophthirius multifiliis* are also restricted to the epithelial layers. In contrast crustaceans such as caligid copepods and lampreys may erode the dermis or even the underlying muscle layers. Curiously, although *Argulus* spp. are very common parasites of fishes in freshwater, we have no clear picture of how they feed and only conflicting statements about the nature of their food. In particular, a protrusible stylet, supplied with glands but not linked to the gut, is probably linked with feeding, but its role is not fully understood.

Many fish skin and gill parasites take in blood. Blood may be only accidentally ingested along with other tissue, as for example in the monogenean gill parasite *Tetraonchus monenteron* or in lampreys, but many parasites ingest only blood, such as polyopisthocotylean monogeneans, leeches, the pennellid copepod *Lernaeocera branchialis* (see above) and cymothoid and gnathiid isopods. These specialised blood feeders are likely to need an anticoagulant secretion.

Skin-dwelling parasites are required to penetrate through the epidermis into the dermis to tap blood vessels, but in the gills blood is close to the surface in the secondary gill lamellae. The most highly specialised blood-feeding gill parasites are the polyopisthocotylean monogeneans, which have evolved clamps on the haptor to attach themselves to secondary gill lamellae. Their adoption of a diet of blood and its ready accessibility in secondary gill lamellae have permitted most of them to abandon locomotion. The consequence of this is that many of them are irrigated unilaterally by the gill ventilating current and develop asymmetrical body shapes, which minimise interference with water flow and reduce stress on the attached clamps.

Surprisingly, blood turns out to be difficult to digest and leeches rely on symbiotic micro-organisms to complete this task. These micro-organisms are housed in special sacs called mycetomes and are passed on from one generation to another. Microrganisms that may have a similar role have been identified in gnathiid isopods. Only one polyopisthocotylean monogenean, *Diclidophora merlangi*, has been systematically examined for gut-dwelling micro-organisms that might have a symbiotic role, but none have been found. If this lack of symbiotes were confirmed in other polyopisthocotyleans, it would indicate their superiority in independent digestive ability among blood-feeding fish parasites.

Because leeches and gnathiids are blood feeders and change hosts, they are potential transmitters of blood-born fish parasites, such as protozoans. Leeches are known to transmit trypanosomes and trypanoplasms. Gnathiid isopods have been implicated in the transmission of haemogregarines.

Although there is doubt about whether *Argulus* spp. ingest blood, they are known to transmit nematode parasites to fishes. Nematode larvae are presumably ingested during feeding and then enter the body cavity. The nematode larvae moult and then accumulate in the suckers of *Argulus* and when the crustacean feeds the infective nematode larvae enter the fish by an unknown route, eventually reaching maturity typically near the swimbladder.

18.5 PATHOGENESIS AND HOST DEFENCES

Epidermal feeding has the advantage that the feeding wound is rapidly sealed by host epidermal cells migrating inwards from the edge of the wound, thereby minimising risk of infection by micro-organisms. Moreover, the eroded cells are soon replaced by mitosis, so that the osmotic integrity of the skin is maintained. Epidermis feeding may also be less provocative to the host's blood-mediated defences, since blood capillaries do not extend into the epidermis. This 'strategy' may be especially suitable for parasites like monogeneans that spend most of their lives on a single host individual. However, this does not mean that these parasites are non-pathogenic. In the wild, infrapopulations are sufficiently low for the host's regenerative powers to keep pace with epidermal erosion by the parasites, but, in the confines of an aquarium or fish farm, infrapopulations steadily increase as a result of repeated reinfection, until epidermis is removed faster than the host can replace it. The osmotic balance of the host can no longer be maintained and infection by pathogenic micro-organisms may occur, leading ultimately to death of the host. Thus one of the most important factors contributing to exponential increase in parasite population levels and hence epidemic disease in farm conditions is the greatly enhanced survival of infective stages in the confined and densely crowded conditions, combined perhaps with reduced effectiveness of the fish's immune system as a consequence of stress. Parasites with broad host specificity such as the monogenean *Neobenedenia melleni* (see below) are of course especially threatening.

Work on fish parasites such as gyrodactylid monogeneans and glochidia of unionacean molluscs, demonstrates that fishes are capable of mounting a defence against metazoan (multicellular) ectoparasites. The fact that some strains of salmon (*Salmo salar*) are able to control the sizes of infrapopulations of the monogenean *Gyrodactylus salaris*, while others fail to do so and are overwhelmed by the parasite, indicates the importance of genetic factors. However, the way in which hosts defend themselves against these parasites is largely unknown. Mucous cells seem likely to have a central role, but there are many other factors that may be involved (see review by Kearn, 1999), and a role for antibodies in partial immunity against metazoan ectoparasites such as monogeneans cannot be excluded (see Rubio-Godoy *et al.*, 2003). The evidence that the ciliate protozoan *Ichthyophthirius multifiliis* is not killed by the host reaction to it, but forced to abandon the host, is of special interest.

Our lack of understanding of how immune mechanisms protect fishes against skin and gill parasites is an important gap in our knowledge of parasite/host interactions. The impetus for more research in this area stems from the damage inflicted on the salmon industry by *G. salaris* in Norway and advances are eagerly awaited.

18.6 HOST SPECIFICITY AND SPECIATION

Another obvious feature of the parasites covered in this book is their high degree of host-specificity, i.e. the restriction of many of them to a relatively narrow range of hosts. Monogeneans, for example, tend to be strictly host specific, while the protozoan *Ichthyobodo necator* is capable of infecting almost any freshwater fish. Even amongst monogeneans varying degrees of host specificity are to be found. Some monogeneans such as *Diclidophora merlangi* and *Tetraonchus monenteron* are found on a single host species, while others are restricted at the family level, such as *Entobdella soleae* on various soles (Soleidae) and *Plectanocotyle gurnardi* on several species of gurnards (Triglidae). In contrast with the narrow specificity of *E. soleae*, its capsalid relative *Neobenedenia melleni* has been recorded in the wild from 27 hosts belonging to 18 genera, 14 families and 3 orders (Bullard *et al.*, 2000) and from many other fishes in captivity (see Whittington & Horton, 1996).

Narrow specificity is perhaps to be expected where the fish host species has morphological, physiological or behavioural differences from its close relatives. Close adaptation of a parasite to a single host species or to a few closely related hosts may be essential to ensure successful host location and infection, and to ensure that the parasite is able to attach itself securely and obtain the food it needs to develop and reproduce in the face of potential threats from the host's anti-parasite defences. The extent and intimacy that may be involved in the adaptation of a skin parasite to its specific fish host has been illustrated by devoting Chapter 3 to the biology of the monogenean *Entobdella soleae* from the common sole (*Solea solea*) and making comparison with a few of the features of the conspecific *Entobdella hippoglossi* parasitising the halibut, *Hippoglossus hippoglossus*. *E. soleae* and *E. hippoglossi* are closely related and their hosts are bony flatfishes. However, it is evident that the physical, chemical and behavioural features of the common sole are sufficiently different from those of the halibut that specialisation for survival on one of these hosts precludes survival of the parasite on the other. This makes it all the more difficult to understand how *Neobenedenia melleni*, a similar monogenean belonging to the same family (Capsalidae), is capable of surviving on such an enormous range of frequently unrelated hosts.

Host specificity is a consequence of host/parasite co-evolution. If a barrier to gene flow isolates two sub-populations of a single host species infected with a single parasite species, and if this barrier also prevents gene flow between the two newly created sub-populations of parasites, then the parasite may co-speciate with its host (Poulin, 1998). This process may occur repeatedly, so that each parasite species will be strictly host specific and host and parasite phylogenies (evolutionary trees) will correspond. This is known as Fahrenholz's Rule. However, host and parasite phylogenies are rarely entirely congruent. Poulin pointed out that the occurrence of one parasite species on more than one host species could result from the continuation of gene flow between parasite populations on different hosts, or as a result of host switching (ecological transfer), in which a parasite colonises a new host species, which may or may not be related. Poulin also raised the possibility that intra-host parasite speciation may take place, in which parts of the parasite population become genetically isolated without an interruption in gene flow in the host population. In the context of monogenean gill parasites, in which several species of the same genus may co-exist on the same host, Euzet & Combes (1980) proposed a mechanism whereby virtually instantaneous intra-host parasite speciation (sympatric speciation) could take place as a result of mutation affecting the copulatory sclerites. However, there is, as yet, no evidence that speciation does take place in this way.

It is worth noting that some genera of fish ectoparasites have been phenomenally successful in generating species. In the Monogenea, *Dactylogyrus* has more than 900 nominal species (see Gibson *et al.*, 1996) and *Gyrodactylus* includes an estimated 402 species (see Bakke *et al.*, 2002). In 1979, Kabata claimed that the copepod genus *Caligus* contained 200 species worldwide. The reasons for the enormous success of the particular combinations of features that define these taxa are not obvious, but their expansion has taken place against the background of the explosive speciation of the bony fishes (teleosts) and in the case of *Dactylogyrus* the radiation of the cyprinid fishes.

18.7 MORPHOLOGY AND MOLECULES

There is a degree of uncertainty in attempting to assess the breadth of host specificity in parasite/host relationships. To take an example from the Monogenea, several distinct species of rays (*Raja*) are infested with skin-parasitic acanthocotylid monogeneans (see Chapter 4). *Acanthocotyle lobianchi* is a parasite of the thornback ray, *Raja clavata*, and is readily distinguishable morphologically from *A. elegans* and *A. greeni*, also parasitising *R. clavata*. Several other common ray species also harbour acanthocotylids, but, apart from possible size differences, their morphology is indistinguishable from that of *A. lobianchi* on *R. clavata*. This implies that the host specificity of *A. lobianchi* is wide, while that of *A. elegans* and *A. greeni* is narrow, each of them being restricted to a single host species. However, work on other parasite/host systems introduces seeds of doubt into this conclusion.

Leeches found on the shanny (*Lipophrys pholis*) and on the butterfish (*Pholis gunnellus*) are morphologically similar and are currently regarded as a single species, namely *Oceanobdella blennii*, but in this case there is experimental evidence that undermines this assertion. Hussein & Knight-Jones (1995) showed that leeches removed from the shanny readily reattached to shanny when given the opportunity. On the other hand leeches removed from butterfish refused to attach themselves to shanny. Thus there are behavioural differences between the leech populations on shanny and those on butterfish, suggesting that the two populations may be genetically distinct, in spite of their morphological similarity. Whether this genetic distinction represents incipient speciation or whether the two populations have already diverged so much that they can no longer interbreed, i.e. are distinct but cryptic species, remains to be determined. But this does prompt us to ask whether '*Acanthocotyle lobianchi*' is a distinct entity or whether it represents a complex of cryptic and/or incipient species.

Modern molecular techniques have the potential to throw light on these and on many similar problems mentioned in this book. These techniques promise to reveal the extent of genetic divergence between populations of parasites on different hosts and thereby offer opportunities to improve our understanding of the phenomenon of host specificity and of the mechanisms underlying parasite speciation.

18.8 HABITAT SELECTION AND NICHE RESTRICTION

The range of habitats occupied by fish ectoparasites is well illustrated by the monogeneans. Some gyrodactylids roam widely over the skin and into the gill chamber, but most monogeneans restrict themselves to the skin or to the gills. However, parasites are rarely distributed uniformly, even in these sites. *Entobdella soleae* is probably able to feed, grow and assemble eggs in any location on the exposed upper or lower body surfaces of its flatfish host, the common sole, but, in reality, it is on the lower surface that adult parasites spend most, possible all, of their lives. However, the habitat of *E. soleae* is relatively broad compared with that of its capsalid relative *Metabenedeniella parva*, which is restricted to the dorsal fin of its Australian host *Diagramma labiosum* (see Horton & Whittington, 1994). Even this parasite is upstaged by another capsalid from the Australian Pacific coast, namely *Benedenia lutjani* on *Lutjanus carponotatus*. According to Whittington & Ernst (2002), immature parasites aggregate on the host's pelvic fins, but do not remain there – migration to the branchiostegal membranes (membranous folds at the posterior border of the operculum) coincides with the advent of sexual maturity and the commencement of egg laying.

The distribution of some parasites on skin or gills may indicate that settlement is not selective. For example, the relative numbers of the lernaeopodid copepod *Salmincola salmoneus* on the gills are related to the surface areas of the gills available for colonisation (see p. 195). However, many monogenean gill parasites are distributed non-randomly. Adult specimens of *Diclidophora merlangi*, for example, prefer gill I of the whiting (*Merlangius merlangus*), although parasites also occur on gills II, III and IV. *Ergenstrema labrosi* exhibits bihabitat infections, some adults being found on the gills and others on the gill rakers of the thick-lipped grey mullet (*Chelon labrosus*). Extreme microhabitat restriction is displayed by *Kuhnia sprostonae*, which, in mackerel (*Scomber scombrus*) from the North Sea, is found only on the tiny pseudobranch (Rohde & Watson, 1985).

Sometimes the site of invasion of the host by an ectoparasite and the definitive site preferred by the adult parasite coincide, but frequently these sites differ, requiring migrations between the two. *Benedenia lutjani* has just been mentioned. Rohde (1980) highlighted the contrasting behaviour patterns of two copepod species belonging to the genus *Caligus* infecting the same host, the tub gurnard (*Trigla lucerna*) in the North

Sea. Larvae and adults of *C. brevicaudatus* have similar distributions on the body surface, while larvae of *C. diaphanus* are located exclusively on the gills and adults typically on the region of the head just covered by the operculum. There is much to be discovered about the ways in which ectoparasites recognise suitable infection sites and the cues that direct their post-invasion migrations.

The phenomenon of niche restriction in parasites has been discussed by (Rohde, 1993), based mainly on observations on monogeneans and other ectoparasites of marine fishes, and by Poulin (1998). The interested reader is referred to these authors for detailed discussion. The selection pressures that may have led to this phenomenon include the following: interspecific competition; enhanced opportunities for mating; maintenance or reinforcement of reproductive barriers between related parasite species; improved reproductive benefits and hence fitness for parasites in parts of their potential range because of variations in abiotic factors (such as differences in the rate and volume of water flow through different regions of the gill cavity). In spite of much debate, these explanations remain largely hypothetical and we can be certain that niche restriction will continue to attract attention from parasitologists.

Parasite behaviour and particularly its role in host finding and niche selection have been explored in few fish skin and gill parasites. This is in spite of its obvious relevance to our understanding of the phenomenon of host specificity, to our knowledge of the defences that fishes generate against external parasites and to our ability to control effectively pathogenic parasites in fish farms. Where behaviour has been studied the discoveries have been exciting: the reader will recall how responses to light and dark and to host-derived cues of a mechanical and chemical nature influence hatching in monogeneans and host location in the leech *Piscicola geometra* and the crustacean *Argulus foliaceus*. These studies provide a taste of what remains to be discovered.

18.9 THE FUTURE OF THE BRITISH PARASITE FAUNA

The parasite fauna is not static. For example, during the few decades of my career as a parasitologist, changes have occurred in the monogenean fauna of marine fishes off the south-west coast of Britain (Chapter 6). *Grubea cochlear* appears to have increased in frequency on it host the mackerel (*Scomber scombrus*) and a parasite that has not yet been identified with certainty has appeared on poor cod (*Trisopterus minutus*). We can only speculate about the possible reasons for these changes, but one of these is the currently newsworthy phenomenon of global warming. If such a change in climate is sustained its consequences are hard to predict, but are likely to influence our coastal and freshwater fish faunas and their parasites.

There is debate about whether global climate change is a consequence or not of human activity, but there can be no doubt that humans are directly responsible for local environmental changes, many of which are detrimental to wildlife, especially but not exclusively in freshwater habitats. To take just one example, the decline of the freshwater pearl mussel in Britain has been influenced by many of these changes, including over-fishing for pearls, increased pollution and agricultural run-off and habitat changes brought about by river-engineering works. This example also emphasises the critical role of the host, since the recent decline of salmon and trout, which support the early parasitic glochidial stage of the mussel, has probably contributed to its demise.

Human activity is also responsible for the introduction into Britain of exotic freshwater fishes, either by accident or by design. Kennedy (1993) analysed in detail the introduction and spread of helminth and crustacean parasites of fishes in the British Isles and the success and failure of these parasites as colonists of new localities. He found that surprisingly few introduced fish helminths and crustaceans have successfully established themselves and that many still show restricted distributions. At the time when Kennedy's review was written, new introductions of fishes had declined but were not eliminated. In fact, two small cyprinids, the topmouth gudgeon, *Pseudorasbora parva*, and the sunbleak, *Leucaspius delineatus*, appeared in Britain in the 1990s (Pinder & Gozlan, 2003). Re-introductions of non-native fishes, in particular ornamental species such as koi and goldfish, also continued, creating opportunities for introduction of new parasites. It is particularly relevant to this book that most of the parasites that have established themselves in Britain following introductions are those with single-host life cycles, i.e. with no intermediate hosts (Kennedy, 1993). Since almost all skin and gill parasites of fishes have such life cycles (see p. 339), these parasites provide a significant proportion of successful colonists and will continue to do so.

The effects that fish introductions might have in the long term on our indigenous fish fauna are hard to predict, as are the consequences of interactions between parasites introduced on alien hosts and our native fishes. A more predictable threat is posed by the importation of foreign stocks of our native fish species. As well as the potential undesirability of alien genotypes, there is the danger of introducing pathogens like the monogenean *Gyrodactylus salaris*, not present currently on our salmon stocks. The consequences of introducing such a pathogen are likely to be devastating for wild and farmed hosts (Chapter 4).

APPENDIX 1

Classified list of fishes mentioned in the text, with scientific and common names.

Scientific names and classification are from Nelson (1994). Families are listed in the same order as in Nelson (1994), with species names following in alphabetical order. The common names of British fishes mostly follow Wheeler (1978). Common names of foreign fishes are taken from Froese & Pauly (2002). Species in square brackets are referred to in the text but are not found in British waters. Fishes restricted to fresh water are shown in bold type. Fishes ranging from fresh water through brackish water to the sea are underlined; this category includes diadromous fishes that regularly migrate between marine and freshwater environments, spawning either in the sea (catadromous fishes) or in fresh water (anadromous fishes). Not indicated are marine or freshwater fishes that occasionally venture into brackish water.

Superclass Agnatha (jawless fishes)
    Class Myxini (hagfishes)[1]
        Order Myxiniformes
            Family Myxinidae   *Myxine glutinosa*, hagfish
    Class Cephalaspidomorphi (lampreys)[1]
        Order Petromyzontiformes
            Family Petromyzontidae
                [***Ichthyomyzon bdellium*, Ohio lamprey**]
                <u>*Lampetra fluviatilis*, lampern, river lamprey</u>
                ***Lampetra planeri*, brook lamprey**
                [<u>*Lampetra tridentata*, Pacific lamprey</u>]
                <u>*Lethenteron camtschaticum*, Arctic lamprey</u>]
                [***Lethenteron zanandreai*, Po brook lamprey**]
                <u>*Petromyzon marinus*, lamprey</u>
Superclass Gnathostomata (fishes with jaws)
    Grade Chondrichthiomorphi
        Class Chondrichthyes (cartilaginous fishes)
            Subclass Holocephali (holocephalans; rat-fishes)
                Order Chimaeriformes
                    Family Chimaeridae   *Chimaera monstrosa*, rat-fish
            Subclass Elasmobranchii (elasmobranchs; sharks and rays)

---

[1] Some authorities regard lampreys and hagfish as related 'cyclostomes'; others do not share this view.

Order Carcharhiniformes
    Family Scyliorhinidae   *Scyliorhinus canicula*, dogfish
                                Scyliorhinus stellaris, nursehound
    Family Triakidae   *Galeorhinus galeus*, tope
                                  Mustelus mustelus, smooth hound

Order Lamniformes
    Family Odontaspididae   [*Carcharias taurus*, sand tiger shark]
    Family Cetorhinidae   *Cetorhinus maximus*, basking shark

Order Squaliformes
    Family Dalatiidae   [*Centroscyllium nigrum*, combtooth dogfish]
                                Centroscymnus coelolepis, Portuguese shark
                                Etmopterus spinax, velvet-belly
                                [*Isistius brasiliensis*, cookiecutter shark]
                                [*Isistius plutodus* largetooth cookiecutter shark]
    Family Squalidae   *Squalus acanthias*, spurdog

Order Squatiniformes
    Family Squatinidae   *Squatina squatina*, monkfish

Order Rajiformes
    Family Torpedinidae   *Torpedo marmorata*, marbled electric ray
                                  Torpedo nobiliana, electric ray
    Family Rajidae   *Raja batis*, skate
                                  Raja brachyura, blonde ray
                                  Raja clavata, thornback ray, roker
                                  Raja microocellata, small-eyed ray
                                  Raja montagui, spotted ray
                                  Raja naevus, cuckoo ray
                                  Raja radiata, starry ray
    Family Dasyatidae   *Dasyatis pastinaca*, stingray
                                  [*Pteroplatytrygon violacea*, pelagic stingray]

Grade Teleostomi
  Class Sarcopterygii (lobe-finned fishes and tetrapods)
    Subclass Coelacanthimorpha
      Order Coelacanthiformes
          Family Coelacanthidae   [*Latimeria chalumnae*, coelacanth]
    Subclass (not named) Porolepimorpha plus Dipnoi
      Order Ceratodontiformes
          Family Ceratodontidae   [***Neoceratodus forsteri*, Australian lungfish**]
  Class Actinopterygii (ray-finned fishes)
    Subclass Chondrostei
      Order Acipenseriformes

APPENDIX 1

     Family Acipenseridae *Acipenser sturio*, sturgeon
Subclass Neopterygii
 Order Semionotiformes
     Family Lepisosteidae **[*Lepisosteus* spp., gar]**
Division Teleostei (teleosts)
 Order Anguilliformes
     Family Anguillidae *Anguilla anguilla*, eel
     Family Synaphobranchidae
       [*Simenchelys parasitica*, snubnosed eel]
 Order Clupeiformes
     Family Engraulidae *Engraulis encrasicolus*, anchovy
     Family Clupeidae *Clupea harengus*, herring
       Sardina pilchardus, pilchard
       Sprattus sprattus, sprat
 Order Cypriniformes
     Family Cyprinidae ***Abramis brama*, bream**
       ***Alburnus alburnus*, bleak**
       ***Barbus barbus*, barbel**
       [*Barbus meridionalis*, southern barbel]
       ***Blicca bjoerkna*, silver bream**
       ***Carassius auratus*, goldfish**
       ***Carassius carassius*, crucian carp**
       ***Cyprinus carpio*, common carp, koi, mirror carp**
       ***Gobio gobio*, gudgeon**
       ***Leucaspius delineatus*, sunbleak, belica**
       ***Leuciscus cephalus*, chub**
       ***Leuciscus idus*, ide**
       ***Leuciscus leuciscus*, dace**
       ***Phoxinus phoxinus*, minnow**
       [*Pimephales promelas*, fathead minnow]
       ***Pseudorasbora parva*, topmouth gudgeon, stone moroko**
       ***Rhodeus sericeus*, bitterling**
       [*Rhodeus sinensis*]
       ***Rutilus rutilus*, roach**
       ***Scardinius erythrophthalmus*, rudd**
       ***Tinca tinca*, tench**
     Family Cobitidae ***Cobitis taenia*, spined loach**
       [*Misgurnus anguillicaudatus*, oriental weatherfish]
       ***Noemacheilus barbatulus*, stone loach**

Order Characiformes
    Family Curimatidae   **[*Semaprochilodus taeniurus*, silver prochilodus]**
    Family Characidae   **[*Paracheirodon innesi*, neon tetra]**
Order Siluriformes
    Family Ictaluridae   **[*Ictalurus punctatus*, channel catfish]**
    Family Clariidae   **[*Clarias gariepinus*, Nile catfish, North African catfish]**
    Family Ariidae   [*Arius graeffei*, blue catfish, lesser salmon catfish]
Order Esociformes
    Family Esocidae   ***Esox lucius*, pike**
Order Osmeriformes
    Family Alepocephalidae
        Alepocephalus bairdii, Baird's smooth-head
    Family Osmeridae   *Osmerus eperlanus*, smelt
Order Salmoniformes
    Family Salmonidae   ***Coregonus lavaretus*, powan**
        [*Oncorhynchus kisutch*, coho salmon]
        Oncorhynchus mykiss (= *Salmo gairdneri*), rainbow trout
        [*Oncorhynchus nerka*, sockeye salmon]
        Salmo salar, salmon, Atlantic salmon
        Salmo trutta, trout (brown trout, sea trout)
        Salvelinus alpinus, Arctic charr
        Salvelinus fontinalis, brook charr
        [*Salvelinus namaycush*, lake trout]
        [*Thymallus arcticus*, Arctic grayling]
        ***Thymallus thymallus*, grayling**
Order Gadiformes
    Family Macrouridae   *Caelorhynchus occa*, swordsnout grenadier
    Family Moridae   *Lepidion eques*, North Atlantic codling
    Family Phycidae   *Ciliata mustela*, five-bearded rockling
        Enchelyopus cimbrius, four-bearded rockling
        Gaidropsarus vulgaris, three-bearded rockling
        Phycis blennoides, forkbeard
    Family Merlucciidae   *Merluccius merluccius*, hake
    Family Gadidae   *Gadus morhua*, cod
        ***Lota lota*, burbot**

Melanogrammus aeglefinus, haddock
Merlangius merlangus, whiting
Micromesistius poutassou, blue whiting
Molva molva, ling
Pollachius pollachius, pollack
Pollachius virens, saithe
Trisopterus esmarkii, Norway pout
Trisopterus luscus, bib, pouting
Trisopterus minutus, poor cod

Order Lophiiformes
    Family Lophiidae — *Lophius piscatorius*, angler
    Family Antennariidae — [*Antennarius striatus*, striated frogfish]
    Family Ceratiidae — *Ceratias holboelli*, deep-sea angler

Order Mugiliformes
    Family Mugilidae — <u>*Chelon labrosus*, thick-lipped grey mullet</u>
    <u>*Liza ramada*, thin-lipped grey mullet</u>

Order Atheriniformes
    Family Atherinidae — <u>*Atherina boyeri*, big-scale sand smelt</u>

Order Beloniformes
    Family Belonidae — *Belone belone*, garfish
    Family Exocoetidae — [*Parexocoetus brachypterus*, sailfin flyingfish]

Order Cyprinodontiformes
    Family Poeciliidae — **[*Gambusia affinis*, mosquitofish]**
    [*Poecilia reticulata*, guppy]
    Family Cyprinodontidae — **[*Orestias agassizii*]**

Order Zeiformes
    Family Zeidae — *Zeus faber*, dory

Order Gasterosteiformes
    Family Gasterosteidae — <u>*Gasterosteus aculeatus*, stickleback, three-spined stickleback</u>
    [*Gasterosteus wheatlandi*, black-spotted stickleback]
    <u>*Pungitius pungitius*, nine-spined stickleback</u>
    Spinachia spinachia, fifteen-spined stickleback
    Family Syngnathidae — *Entelurus aequoreus*, snake pipefish
    Syngnathus acus, greater pipefish
    Syngnathus typhle, deep-snouted pipefish

Order Scorpaeniformes

Family Triglidae	*Chelidonichthys* (*Aspitrigla*) *cuculus*, red gurnard
	Eutrigla gurnardus, grey gurnard
	Trigloporus lastoviza, streaked gurnard
	Trigla lucerna, tub gurnard
Family Cottidae	**Cottus gobio, bullhead**
	Myoxocephalus scorpius, bull-rout
	Taurulus bubalis, sea scorpion
Family Cyclopteridae	*Cyclopterus lumpus*, lumpsucker

Order Perciformes

Family Moronidae	<u>*Dicentrarchus labrax*, bass</u>
Family Acropomatidae	*Polyprion americanus*, wreckfish[2]
Family Serranidae	*Serranus cabrilla*, comber
Family Centrarchidae	***Lepomis gibbosus*, pumpkinseed**
	[*Lepomis macrochirus*, bluegill]
	[*Micropterus coosae*, redeye bass]
	[*Micropterus punctulatus*, spotted bass]
	***Micropterus salmoides*, large-mouth bass**
Family Percidae	***Gymnocephalus cernuus*, ruffe**
	***Perca fluviatilis*, perch**
	***Stizostedion* (= *Sander*) *lucioperca*, zander, pikeperch**
Family Carangidae	*Trachurus trachurus*, scad
Family Lutjanidae	[*Lutjanus carponotatus*, Spanish flag snapper]
Family Haemulidae	[*Diagramma labiosum*]
Family Sparidae	[*Archosargus probatocephalus*, sheepshead seabream]
	[*Chrysophrys* (*Pagrus*) *auratus*, snapper, squirefish]
	[*Diplodus sargus*, white sea-bream]
	Pagellus bogaraveo, red sea-bream
	Pagellus erythrinus, pandora
	Spondyliosoma cantharus, black sea-bream
Family Sciaenidae	*Argyrosomus regius* (= *Sciaena aquila*), meagre
Family Mullidae	[*Mullus barbatus*, red mullet]
	Mullus surmuletus, striped red mullet

---

[2] Provisional. Nelson (1994) indicates that *Polyprion* is better placed in Polyprionidae.

        Family Pomacentridae [*Chromis nitida*, Barrier Reef chromis]
        Family Labridae *Centrolabrus exoletus*, rock cook
        Coris julis, rainbow wrasse
        Crenilabrus melops, corkwing wrasse
        Ctenolabrus rupestris, goldsinny
        [*Labroides dimidiatus*, bluestreak cleaner wrasse]
        Labrus bergylta, ballan wrasse
        Labrus mixtus, cuckoo wrasse
        [*Tautogolabrus adspersus*, cunner]
        Family Pholidae *Pholis gunnellus*, butterfish
        Family Anarhichadidae *Anarhichas lupus*, wolf-fish
        Family Trachinidae *Echiichthys vipera*, lesser weever
        Family Blenniidae *Coryphoblennius galerita*, Montagu's blenny
        Lipophrys pholis, shanny
        Parablennius gattorugine, tompot blenny
        Family Gobiidae *Pomatoschistus lozanoi*
        Pomatoschistus microps, common goby
        Pomatoschistus minutus, sand goby
        Pomatoschistus pictus, painted goby
        Thorogobius ephippiatus, leopard-spotted goby
        Family Scombridae *Scomber scombrus*, mackerel
        Thunnus thynnus, blue-fin tunny
        Family Xiphiidae *Xiphias gladius*, swordfish
Order Pleuronectiformes
        Family Scophthalmidae *Scophthalmus maximus*, turbot
        Zeugopterus punctatus, topknot
        Family Pleuronectidae *Hippoglossoides platessoides*, long rough dab
        Hippoglossus hippoglossus, halibut
        Limanda limanda, dab
        Microstomus kitt, lemon sole
        <u>*Platichthys flesus*, flounder</u>
        <u>[*Platichthys stellatus*, starry flounder]</u>
        Pleuronectes platessa, plaice
        Family Soleidae *Buglossidium luteum*, solenette
        Microchirus variegates, thickback sole
        Pegusa lascaris, sand sole
        [*Solea senegalensis*, Senegalese sole]
        Solea solea, sole, common sole, Dover sole
Order Tetraodontiformes

Family Balistidae	*Balistes carolinensis*, grey triggerfish
Family Tetraodontidae	[*Takifugu rubripes*, tiger puffer]
Family Molidae	*Mola mola*, sun-fish

APPENDIX 2

Classified list of genera of epizoic and parasitic invertebrates mentioned in the text (lampreys are dealt with in Appendix 1).

The classification schemes are intended as a guide to possible relationships and are not definitive - the taxonomy of these groups is in a state of flux, especially since the application of molecular techniques to the study of phylogeny.

Genera found on freshwater hosts are shown in bold type, but note the following qualifiers:
[1] Genus includes some species on marine fishes.
[2] Reproduces in freshwater but may survive in the sea.
[3] Occurs in brackish water.
[4] Not yet recorded in Britain, but hosts present in the British fauna and parasite likely to be found eventually in British waters.

Genera in square brackets are referred to in the text but are not represented in the British fauna.

Chapter 2 - Protozoans (based on Lom & Dykova, 1992)

Phylum Mastigophora

    Class Kinetoplastidea
        Suborder Trypanosomatina
            Family Trypanosomatidae
                ***Trypanosoma*[1]**
        Suborder Bodonina
            Family Bodonidae   ***Ichthyobodo*[1], *Trypanoplasma*[1]**

Phylum Ciliophora

    Class Oligohymenophorea
        Subclass Hymenostomata
            Family Ichthyophthiriidae
                Ichthyophthirius
        Subclass Peritrichia
            Order Peritrichida
                Family Scyphidiidae   ***Ambiphrya*[3], *Riboscyphidia* = *Scyphidia*[1]**
            Order Mobilina
                Family Trichodinidae   ***Paratrichodina*[1], *Trichodina*[1], *Trichodinella*[1], *Tripartiella***

Chapters 3 to 6 - Flatworms (based on Kearn, 1998)

Phylum Platyhelminthes

    Class Tricladida   *Micropharynx*
    Class Monogenea
        Subclass Monopisthocotylea
            Family Capsalidae   [*Benedenia*], *Capsala, Entobdella,*
                                          [*Metabenedeniella*], [*Neobenedenia*],
                                          Trochopus
            Family Acanthocotylidae
                                          Acanthocotyle
            Family Microbothriidae *Leptocotyle, Pseudocotyle*
            Family Anoplodiscidae [*Anoplodiscus*]
            Family Udonellidae   *Udonella*
            Family Gyrodactylidae **Gyrodactylus**[1]**, Gyrodactyloides,**
                                        Isancistrum
            Family Dactylogyridae **Ancyrocephalus**, *Calceostoma,*
                                        Dactylogyrus, *Ergenstrema*[3]**,** *Ligophorus,*
'Dactylogyroid-   **Neodactylogyrus**
eans'   Family Amphibdellidae *Amphibdella, Amphibdelloides*
            Family Tetraonchidae   **Tetraonchus**
            Family Diplectanidae   *Diplectanum, Pseudodiplectanum*

        Subclass Polyopisthocotylea
            Family Hexabothriidae *Hexabothrium, Rajonchocotyle*
            Family Diclidophoridae *Cyclocotyla, Diclidophora,*
                                        Paracyclocotyla
            Family Gastrocotylidae *Gastrocotyle, Pseudaxine*
            Family Plectanocotylidae
                                        Plectanocotyle
'Mazocraeideans' Family Mazocraeidae   *Grubea, Kuhnia*
            Family Microcotylidae   *Atrispinum, Axine, Microcotyle*
            Family Anthocotylidae   *Anthocotyle*
            Family Diplozooidae   **Diplozoon, Eudiplozoon, Paradiplozoon**
            Family Discocotylidae   **Discocotyle**

            Family Polystomatidae [*Oculotrema*], *Polystoma*

Chapter 7 - Annelids (based on Sawyer, 1986b, incorporating implications from Siddall *et al.*, 2001)

Phylum Annelida

    Class Oligochaeta
        Order Acanthobdellidea   [***Acanthobdella***]

Order Hirudinea
    Suborder Rhynchobdellida
        Family Piscicolidae   *Branchellion, Brumptiana, Calliobdella, Hemibdella, Heptacyclus, Oceanobdella,* **Piscicola**[3], *Platybdella, Pontobdella, Sanguinothus*
        Family Glossiphoniidae   [**Haementeria**], ***Hemiclepsis***, [**Placobdella**]
    Suborder Arhynchobdellida
        Family Hirudinidae   ***Hirudo***
Order Branchiobdellida

Chapters 8 to 15 - Crustaceans (based on Roberts & Janovy, 2000)

Phylum Arthropoda

  Subphylum Crustacea
    Class Maxillopoda
      Subclass Copepoda
        Order Siphonostomatoida
          Family Caligidae   *Caligus, Lepeophtheirus*
          Family Pennellidae   *Lernaeenicus, Lernaeocera*
          Family Lernaeopodidae **Achtheres**, *Albionella, Alella, Charopinus, Clavella, Lernaeopoda, Lernaeopodina, Neobrachiella, Ommatokoita, Pseudocharopinus,* **Salmincola**[2], ***Tracheliastes***, *Vanbenedenia*[4]
        Order Cyclopoida
          Family Lernaeidae   ***Lernaea***
        Order Poecilostomatoida
          Family Ergasilidae   ***Ergasilus***[1], ***Neoergasilus***, *Thersitina*[3]
          Family Bomolochidae   *Bomolochus, Holobomolochus*
          Family Taeniacanthidae *Taeniacanthus*
          Family Chondracanthidae
              Acanthochondria, Acanthochondrites, Chondracanthus, Lernentoma,
          Family Philichthyidae   *Leposphilus*[4], *Philichthys*
      Subclass Branchiura
        Family Argulidae   ***Argulus***[1]
      Subclass Cirripedia (barnacles)
        Family Lepadidae   *Anelasma*
    Class Malacostraca
      Order Isopoda
        Family Cymothoidae   *Anilocra, Ceratothoa,* [*Lironeca*], *Nerocila*
        Family Gnathiidae   *Gnathia, Paragnathia*

Chapter 16 - Unionacean molluscs (based on Walker *et al.*, 2001)

Phylum Mollusca

    Class Pelecypoda
        Order Unionoida (Unionacea)
           Superfamily Unionoidea
               Family Unionidae   ***Anodonta***, [***Lampsilis***], ***Pseudanodonta***, ***Unio***
               Family Margaritiferidae ***Margaritifera***
           Superfamily Muteloidea (Etherioidea)
               Family Mutelidae (Iridinidae)
                          [***Mutela***]

GLOSSARY OF FREQUENTLY USED TERMS
(adj.: adjective; pl.: plural form)

Abdomen – see **Pleon**.
Aboral – surface of a 'protozoan' at the opposite end to the 'mouth' (cytostome).
Accessory sclerites – the anteriormost pair of median attachment sclerites in a capsalid monogenean.
Acquired resistance – resistance generated by exposure to infectious organisms or parasites.
Adductor muscle – the closing muscle running between the shell valves of a bivalve mollusc.
Agonistic behaviour – behaviour patterns including all aspects of aggression, including threat and attack, plus the consequent aspects of appeasement and flight.
Ammocoete – larval stage of a lamprey.
Anadromous – migrating from the sea into rivers to spawn.
Antennae – second antennae; the second pair of cephalic (head) appendages of crustaceans.
Antennules – first antennae; the first pair of cephalic (head) appendages of crustaceans.
Antibody – a protein called an immunoglobulin that reacts with a specific antigen and serves as part of an animal's defences against disease.
Antigen – an alien substance (usually a protein or carbohydrate) that when introduced into the body of an animal induces an immune response, including production of specific antibodies.
Appendages – limbs or legs of a crustacean, one pair per segment.
Article – see **Podomere**.
Autoinfection – establishment of a parasite on the same host individual as its parent.

Basement membrane – thin layer separating an epithelium from underlying tissues.
Benthic – bottom living.
Biramous – pertaining to the two-branched crustacean appendage, consisting of endopod and exopod.
Blood plasma – the watery matrix in which the blood cells are suspended.
Blood serum – blood plasma from which the clotting factor has been removed.
Brood chamber – see **Marsupium**.
Buccal cavity – mouth cavity.
Bulla – chitinous anchor of lernaeopodid copepods, embedded in the surface of the host fish and attached to the maxillary arms of the copepod.

Caecum (pl. **Caeca**) – a blind-ending tube or sac in the digestive system, e.g. the tubular gut caeca of a monogenean.
Capitulum – the exposed region of the mesoparasitic barnacle *Anelasma*.
Carapace (also **Dorsal shield**) – the part of the exoskeleton of a crustacean covering the cephalothorax.
Carbohydrase – an enzyme that digests carbohydrates.

Catadromous – migrating from fresh water to the sea to spawn.
Catecholamines – chemical messengers released as part of the response to stress.
Caudal – pertaining to the tail.
Cell division – see **Mitosis**.
Cephalon – crustacean head, comprising five fused segments.
Cephalosome – anterior tagma of crustaceans comprising the cephalon fused to the first thoracic (maxilliped-bearing) segment.
Cephalothorax – anterior tagma of crustaceans comprising the cephalon fused to more than one thoracic segment.
Chalimus – a developmental stage of copepod crustaceans, following the copepodid stage.
Chela (pl. **Chelae**; adj. **Chelate**) – pincer-like adaptation of a crustacean appendage in which the terminal podomere closes against a distal extension of the penultimate podomere; see also **Subchela**.
Chemoperception – the ability of an animal to detect and respond to chemical stimuli.
Chemoreceptor – a sense organ responding to chemical stimuli.
Chemosensory – pertaining to the detection of chemical stimuli.
Chitin (adj. **Chitinous**) – a polysaccharide found in the cuticle (exoskeleton) of crustaceans.
Chloride cell – cell concerned with salt balance found on fish gills.
Cholinesterase – an enzyme that destroys excess acetylcholine, a nerve transmitter substance.
Chromatophore – a cell containing pigment.
Cilium (pl. **Cilia**) – a hair-like outgrowth from a cell surface, capable of whip-like beating movement.
Clamp – the sclerite-supported haptoral organ used by polyopisthocotylean monogeneans for attachment to secondary gill lamellae.
Claspers – male copulatory organs of elasmobranch fishes.
Cleaner organisms – fishes or crustaceans that remove ectoparasites from fishes.
Clitellum – glandular pad on the body surface of a leech involved in cocoon assembly.
Cloaca – combined opening of the gut and the urinogenital system.
Cocoon – a protective covering enclosing the egg or eggs of a leech.
Co-evolution – evolution (speciation) of host and parasite taking place in parallel; see also **Fahrenholz's Rule, Phylogenetic speciation**.
Collagen – a fibrous protein.
Commensalism – an association between two organisms (commensals) in which one benefits and the other is neither harmed nor helped.
Complement – a group of blood proteins involved in removal of intruding foreign material.
Contractile vacuole – an organelle in a protozoan cell that periodically expands and then expels its contents to the exterior. Appears to be concerned with osmoregulation, expelling excess water.
Convergent evolution – similarity between unrelated organisms as a result of exposure to similar selection pressures.
Copepodid – the developmental stage of copepod crustaceans following the nauplius.
Cortisol – a hormone released as part of the response to stress.
Crop – a sac for storing ingested blood in the gut of a leech.

Cross insemination – donation of sperm from one individual to another or mutual exchange of sperm in hermaphrodites.
Cryptic species – a species that is genetically distinct but not morphologically distinct from its siblings.
Ctenidium – the gill of a bivalve mollusc.
Ctenoid scale - fish scale with a spiny margin.
Cuticle – surface layer secreted by the epidermis.
Cypris larva – the larva of a barnacle.
Cytoplasm – the contents of a cell surrounding the nucleus.
Cytoskeleton – supporting microtubules and microfilaments of a cell.
Cytostome (adj. **Cytostomal**) – the 'mouth' of a single-celled 'protozoan'.

Denticle – (1) placoid scale of an elasmobranch fish; (2) part of the cytoskeleton of a trichodinid ciliate 'protozoan'
Dermis – the inner layer of the skin beneath the epidermis of a vertebrate.
Desmosome – specialised connection between two neighbouring cells.
Diadromous – migratory.
Diapause – a period of arrested growth and development.
Diporpa – juvenile parasitic stage of a diplozoid monogenean before fusion with another individual.
Dorsal shield – see **Carapace**.

Ecdysis – see **Moulting**
Ectoparasite – a parasite that lives on the surface (skin and gills) of its host.
Egg sac – see **Ovisac**
Egg string – see **Ovisac**
Embryogenesis – the development of the embryo.
Endoparasite – a parasite that lives entirely inside its host.
Endopeptidase – a type of protein-splitting enzyme that attacks peptide bonds between amino acids inside a peptide chain.
Endopod – the inner branch of a biramous crustacean appendage.
Epidermal cell – epithelial cell of the epidermis, the outer layer of skin of a vertebrate.
Epithelium (pl. **Epithelia**) – sheet of cells tightly bound together, covering an external or lining an internal surface, e.g. epidermis.
Epizoic – pertaining to an animal that lives attached to another but having no parasitic relationship.
Erythrocyte – red blood corpuscle.
Eukaryotes (adj. **Eukaryotic**) – plants and animals as distinct from bacteria and blue-green algae.
Euryhaline – tolerating a wide range of salinities, from fresh water to full seawater.
Eutrophication (adj. **Eutrophic**) – enrichment of a body of water by pollution with excess organic and mineral nutrients.
Exhalant siphon – tube through which water leaves the mantle cavity of a bivalve mollusc.
Exopeptidase – a type of protein-splitting enzyme that attacks the terminal bonds in a peptide chain.
Exopod – the outer branch of a biramous crustacean appendage.

Exoskeleton – the hard outer covering of a crustacean and other arthropods.

Fahrenholz's Rule – the principle that phylogenies of parasites and their hosts evolve in parallel (co-evolve), so that related hosts have related parasites.
Fibroblast – a cell secreting components of the extracellular matrix such as collagen.
Flagellum (pl. **Flagella**) – a hair-like outgrowth from a cell similar in structure to a cilium. Undulating movements of flagella used to propel organisms.
Frontal filament – a thread secreted by the frontal gland of copepods used for tethering the parasite to the host.

Genital atrium – cavity in the body wall of an animal into which the genital ducts open.
Genotype – the genetic constitution of an individual.
Geotaxis (adj. **Geotactic**) – response of an animal to gravity: downward movement, positive geotaxis; upward movement, negative geotaxis.
Germarium – part of the ovary producing egg cells in platyhelminths.
Gill arch – the arch in fishes supporting two rows of primary gill lamellae (inner and outer hemibranchs) of a single gill.
Gill filaments – see **Primary gill lamellae**.
Gill mucosa – see **Mucosa**.
Gill rakers – row of projections from the gill arch preventing debris from reaching the respiratory surfaces.
Glochidium (pl. **Glochidia**) – the larva of a unionacean mollusc.
Glycoproteins – proteins containing sugars as part of the molecule, as in mucins.
Gnathopod (also **Pylopod**) – limbs of the second pereon segment in gnathiid isopods.
Goblet cell – a mucus-secreting epithelial cell.
Gonochorism – having separate male and female individuals.

Haematin – a black/brown iron-containing pigment derived from breakdown of haemoglobin.
Haemocoel – the blood space of an invertebrate, including arthropods and molluscs.
Haemoglobin – the oxygen-carrying pigment in erythrocytes (red blood corpuscles).
Haemolysis – the disintegration of erythrocytes (red blood corpuscles).
Hamulus (pl. **Hamuli**) – large hooks in the haptor of a monogenean.
Haptor – hook-bearing posterior attachment organ of a monogenean.
Helminth – general term used for parasitic worms, including platyhelminths and nematodes.
Hemibranch – half gill.
Hermaphrodite – an individual with both male and female reproductive organs.
Hooklets – tiny hooks up to 16 in number, typically used for pinning the haptor of a monogenean to fish epidermal cells.
Hormone – a chemical messenger usually transported in the blood.
Host specificity – host range of a parasite.
Host switching – establishment of a parasite on a new host unrelated to, but sharing the same environment as, its original host.
Humoral – of or relating to a body fluid.

Hyperplasia (adj. **Hyperplastic**) – excessive cell multiplication induced by some parasites.

Immunoglobulin – see **Antibody**.
Inflammation – a local response to injury or damage involving dilatation of blood vessels and invasion by leucocytes.
Infrapopulation – all individuals of a single parasite species in/on one host individual.
Innate resistance – inherited resistance to an infectious organism or parasite.
Integument – a covering structure or layer (see also **Tegument**).
Intensity – number of individuals of a parasite species in a single infected host.
Interbranchial septum – partition between two hemibranchs of a gill.
Interstitial fauna – animals living between sand grains on the shore.

Kelt – spent or spawned adult salmon.
Keratin – a tough fibrous protein.
Keratocyte – see **Epidermal cell**; **Malpighian cell**.

Labium (adj. **Labial**) – the lower lip of a crustacean derived from fusion of paragnaths.
Labrum – the upper lip of a crustacean.
Lasidium – mesoparasitic larva of the African freshwater mussel *Mutela bourguignati*.
Lateral line – site of sensory cells of the acoustico-lateralis system of fishes, concerned with perception of pressure waves in water.
Leucocyte – white blood cell.
Lipase – a fat-digesting enzyme.
Loculus (pl. **Loculi**) – suctorial subdivision of the monogenean haptor.
Lymph – the interstitial fluid in the lymphatic system and around the tissues of vertebrates.
Lymphocyte – small uninucleate white blood cell involved in antigen-specific immune reactions.
Lysozyme – an enzyme with antibacterial properties found in animal secretions.

Macrophage – a large uninucleate white blood cell capable of ingesting by phagocytosis and destroying foreign material or components of damaged cells etc.
Macrophthalmia – period of continuing internal change in metamorphosis of an ammocoete into a juvenile lamprey.
Maiden – salmon returning from first sojourn at sea and prior to spawning run.
Malpighian cell – epidermal cell.
Manca (pl. **Mancae**) – larva of a cymothoid isopod.
Mandibles – the third pair of cephalic (head) appendages of crustaceans.
Mantle – flap-like lateral extensions of the body enclosing a space in molluscs (the mantle cavity) and the barnacle *Anelasma*.
Marsupium (pl. **Marsupia**) – brood pouch formed by oostegites in isopods and enclosed within the gill of unionacean molluscs.
Maxillae – second maxillae; the fifth and last pair of cephalic (head) appendages of crustaceans.
Maxillipeds – the first pair of thoracic appendages of crustaceans located on the last segment of the cephalosome.

Maxillules – first maxillae; the fourth pair of cephalic (head) appendages of crustaceans.
Mechanoreceptor – a sense organ responding to mechanical stimuli.
Melanin – black or brown pigment.
Melanocyte – a cell producing melanosomes.
Melanosome – an intracellular organelle (granule) in which melanin is synthesised.
Mended kelt – salmon returning to sea after surviving the first spawning run.
Mesoparasite – a parasite that lives partly embedded in its host and partly exposed on the skin or gills.
Metamere – see **Segment**.
Metamorphosis (also **Transformation**) – change in form and structure during development, involving a radical remodelling of the organism
Microtubules – long slender tubules forming part of the cytoskeleton of a cell.
Mitosis – cell division in a eukaryotic organism.
Monocyte – large white blood cell, a precursor of a macrophage.
Monophyletic – pertaining to a group of organisms containing all the descendants of a particular common ancestor.
Moulting (also **Ecdysis**) – the shedding of the exoskeleton in crustaceans.
Mucin – glycoprotein basis of mucus.
Mucocysts – dischargeable bodies (organelles) near the surface of *Ichthyophthirius multifiliis*.
Mucosa – any epithelium that secretes mucus, e.g. gill mucosa.
Mucous cell – see **Goblet cell**.
Mycetome – organ associated with the gut (e.g. in leeches), housing symbiotic micro-organisms assumed to be involved in digestion of ingested blood.

Naïve host – a host that has not previously been exposed to a parasite.
Nauplius (pl. **Nauplii**) – the first larval stage of a crustacean, typically with three functional pairs of appendages (antennules, antennae, mandibles).
Neutrophil – white blood cell, the granules of which stain only with neutral stains.

Olfactory – pertaining to the sense of smell.
Oligotrophic – pertaining to a water body with a poor nutrient supply and little organic production.
Oncomiracidium (pl. **Oncomiracidia**) – larva of a monogenean.
Oostegites – thoracic limb plates forming a brood pouch or marsupium in isopods.
Ootype – or egg mould, where the egg of a monogenean is assembled.
Operculum – (1) the gill cover of a fish; (2) the detachable lid of a flatworm egg or leech cocoon.
Osmoregulation – the control of osmotic potential or water/salt balance in an organism.
Osmosis (adj. **Osmotic**) – movement of water from a solution with low solute concentration to one of high solute concentration via a semi-permeable membrane; see also **Osmoregulation**.
Ovary – germarium plus vitellarium in platyhelminths.
Ovigerous – ready to lay eggs.
Oviparous –reproduction involving the laying of eggs.
Oviposition – egg laying.

Ovisac (**Egg string**; **Egg sac**) – sac attached externally to a copepod and containing fertilised eggs.

Paragnaths – a pair of small lobes on the ventral surface of a crustacean cephalosome between the bases of the mandibles and the maxillules.
Paraphyletic – pertaining to a group of organisms that excludes one or more descendants of a particular single common ancestor.
Parasite – As used in this book: unicellular or multicellular eukaryotic animal that derives benefit from a symbiotic relationship at the expense of its partner (the host).
Parr – young salmon, from the end of its first summer to migration as a smolt.
Pavement cell – exposed cell on the surface of the epidermis.
Peduncle – (1) the stalk joining the body to the haptor in monogeneans; (2) the stalk joining a clamp to the haptor in monogeneans; (3) the embedded region of the body of the barnacle *Anelasma*.
Pelagic – living in the mass of seawater.
Pereon (also Peraeon) – see **Thorax**. Term commonly in use for the thorax of an isopod.
Pereopods (also Pereiopods) – see **Thoracopods**. Term commonly in use for the thoracic limbs of isopods.
Phagocyte (adj. **Phagocytic**) – a cell such as a macrophage that is capable of ingesting foreign material or damaged cells.
Pharynx – the region of the gut between the mouth and the oesophagus; may be glandular and/or muscular.
Pheromone – a chemical messenger released by an organism.
Phoresy – form of symbiosis in which one animal transports another, with no physiological dependency between them.
Photonegative – moving away from the light.
Photopositive – moving towards the light.
Phototaxis (adj. **Phototactic**) – the movement of an organism towards (positive phototaxis) or away (negative phototaxis) from a directional light stimulus.
Phylogenetic speciation – see **Co-evolution**.
Phylogeny (adj. **Phylogenetic**) – evolutionary history.
Pinocytosis – the active engulfing of very small particles or liquids by cells.
Plankton – drifting animals inhabiting the surface waters of the sea or a lake.
Pleon – abdomen of a crustacean.
Pleopods – paired abdominal appendages of a crustacean.
Podomere (also **Article**) – a division of an arthropod appendage separated from the next division by an articulation.
Polypeptide – a chain of amino acids.
Praniza – post-feeding larval stage of a gnathiid isopod.
Prevalence – percentage of a fish population infected with a parasite species.
Primary gill lamellae (also **Gill filaments**) – plates projecting from the gill arch and bearing secondary gill lamellae.
Protandry – maturation of the male gonads before the female gonads in a hermaphrodite individual.
Protease (adj. **Proteolytic**) – a protein-digesting enzyme.

Protogyny – maturation of the female gonads before the male gonads in a hermaphrodite individual.
Pseudobranch – gill remnant on the inner surface of the operculum in fishes.
Pseudohaptor – attachment disk anterior to the true haptor of acanthocotylid monogeneans and derived from the body.
Pylopod – see **Gnathopod**.

Red blood corpuscle – see **Erythrocyte**.

Salmon parr – young salmon from the end of the first year to migration as a smolt.
Sclerites – hard structures in the body of an invertebrate; may be hooks, tubes or plates.
Scleroprotein (also **Sclerotin**) – a chemically resistant and physically strong, tanned protein.
Scopula – adhesive aboral disk of a scyphidiid ciliate 'protozoan'.
Secondary gill lamellae – thin plates projecting from a primary gill lamella; the site of gaseous exchange.
Segment – a division (somite or metamere) of the body of an annelid or arthropod.
Self insemination – introduction of sperm from a hermaphrodite individual into its own body.
Seminal receptacle (also **Spermatheca**) – chamber for the receipt and storage of sperm in the female reproductive system.
Seminal vesicle (also **Vesicula seminalis**) – a vesicle for storage of sperm in a male animal.
Semi-parasite – an organism obtaining part of its food as a parasite.
Sensillum (pl. **Sensilla**) – hair-like sense organ; may be a modified cilium as in platyhelminths.
Serum – fluid component of blood.
Seta (pl. **Setae**) – slender, often feather-like processes on a crustacean, especially on the appendages.
Setules – the slender side branches on a plumose seta.
Smolt – fully silvered juvenile salmon migrating to sea.
Somite – see **Segment**.
Spermatheca (adj. **Spermathecal**) – see **Seminal receptacle**.
Spermatophore – a capsule or mass of jelly containing sperm transferred during mating.
Sphincter – a ring of muscle surrounding an opening.
Spiracle – exterior opening in elasmobranch fishes connected with breathing.
Squamodisc – supplementary friction pads associated with the haptor of diplectanid monogeneans.
Squamous cell – see **Pavement cell**.
Stenohaline – unable to survive in salt water.
Subchela (adj. **Subchelate**) – a crustacean prehensile claw in which the terminal podomere folds back on the penultimate podomere.
Suprapopulation – all of the parasites of a single species living in an ecosystem, including all developmental stages.
Swimming legs – see **Thoracopods**.
Symbiont – see **Symbiote**.

Symbiosis (adj. **Symbiotic**) – living in association with another organism.
Symbiote (also **Symbiont**) – partner in a symbiotic association.
Synonymy – term used in systematics denoting a different name for the same species or variety.
Systematics (also **Taxonomy**) – the branch of biology concerned with classification of organisms.

Tagma – a major region of the crustacean body defined by a common function.
Taxon (pl. **Taxa**) – any grouping used to classify organisms, e.g. species, order.
Taxonomy – see **Systematics**.
Tegument – surface covering of a multicellular organism. (see also **Integument**).
Tergum (pl. **Terga**) – the plate covering the dorsal surface of a segment.
Terminal web – a net of tonofilaments beneath the surface of pavement cells.
Theront – infective stage of the ciliate 'protozoan' *Ichthyophthirius mulifiliis*.
Thoracopods (see also **Pereopods**) – paired thoracic appendages of a crustacean.
Thorax (see also **Pereon**) – the part of the body in a crustacean between the head and the abdomen.
Tomont – encysted stage of the ciliate 'protozoan' *Ichthyophthirius mulifiliis*.
Tonofilaments - slender unbranched filaments in an epidermal cell.
Transformation – see **Metamorphosis**.
Trophont – feeding stage of the ciliate 'protozoan' *Ichthyophthirius mulifiliis*.

Uniramous – pertaining to crustacean appendages with only one branch (endopod or exopod).
Urea – a major end product of protein metabolism in many vertebrates.
Uropods – posteriormost pair of pleopods of a crustacean. May form a tail fin with the terminal telson.

Vas deferens – duct that transports sperm in the male reproductive system.
Velum – pair of muscular flaps (valves) at the anterior end of the lamprey pharynx.
Vesicula seminalis – see **Seminal vesicle**.
Vitellarium – part of the ovary producing vitelline cells in platyhelminths.
Vitelline cells - provide egg shell and food for developing platyhelminth embryos, but make no genetic contribution.
Viviparous – giving birth to living young.

White blood cells – see **Leucocytes**.

Zuphea – pre-feeding larval stage of a gnathiid isopod.

REFERENCES

Abdelhalim, A.I., Lewis, J.W. & Boxshall, G.A. (1991). The life-cycle of *Ergasilus sieboldi* Nordmann (Copepoda: Poecilostomatoida), parasitic on British freshwater fish. *Journal of Natural History* 25: 559 – 582.

Abdelhalim, A.I., Lewis, J.W. & Boxshall, G.A. (1993). The external morphology of adult female ergasilid copepods (Copepoda: Poecilostomatoida): a comparison between *Ergasilus* and *Neoergasilus*. *Systematic Parasitology* 24: 45 – 52.

Adlard, R.D. & Lester, R.J.G. (1994). Dynamics of the interaction between the parasitic isopod, *Anilocra pomacentri*, and the coral reef fish, *Chromis nitida*. *Parasitology* 109: 311 – 324.

Adlard, R.D. & Lester, R.J.G. (1995). The life cycle and biology of *Anilocra pomacentri* (Isopoda: Cymothoidae), an ectoparasitic isopod of the coral reef fish, *Chromis nitida* (Perciformes: Pomacentridae). *Australian Journal of Zoology* 43: 271 – 281.

Ahne, W. (1985). *Argulus foliaceus* L. and *Piscicola geometra* L. as mechanical vectors of spring viraemia of carp virus (SVCV). *Journal of Fish Diseases* 8: 241 – 242.

Alexander, R. M. (1967). *Functional Design in Fishes*. Hutchinson, London.

Alexander, R.M. (1975). *The Chordates*. Cambridge University Press, London.

Alston, S., Boxshall, G.A. & Lewis, J.W. (1993). A redescription of adult females of *Ergasilus briani* Markewitsch, 1933 (Copepoda: Poecilostomatoida). *Systematic Parasitology* 24: 217 – 227.

Alston, S., Boxshall, G.A. & Lewis, J.W. (1996). The life-cycle of *Ergasilus briani* Markewitsch, 1993 (Copepoda: Poecilostomatoida). *Systematic Parasitology* 35: 79 – 110.

Anders, K. & Wiese, V. (1993). Glochidia of the freshwater mussel, *Anodonta anatina*, affecting the anadromous European smelt (*Osmerus eperlanus*) from the Eider estuary, Germany. *Journal of Fish Biology* 42: 411 – 419.

Anderson, M. (1981a). *Ergenstrema labrosi* sp. nov. (Monogenea) on the gills of the thick-lipped grey mullet *Chelon labrosus* at Plymouth. *Journal of the Marine Biological Association of the United Kingdom* 61: 827 – 832.

Anderson, M. (1981b). The change with host age of the composition of the ancyrocephaline (monogenean) populations of parasites on thick-lipped grey mullets at Plymouth. *Journal of the Marine Biological Association of the United Kingdom* 61: 833 – 842.

Anderson, R.M. (1974). An analysis of the influence of host morphometric features on the population dynamics of *Diplozoon paradoxum* (Nordmann, 1832). *Journal of Animal Ecology* 43: 873 – 887.

Anonymous. (1990). Report on the Norwegian meeting on impacts of aquaculture on wild stocks. *North Atlantic Salmon Conservation Organization, Council Paper* 28: 1 – 9.

Anstensrud, M. (1989). Experimental studies of the reproductive behaviour of the parasitic copepod *Lernaeocera branchialis* (Pennellidae). *Journal of the Marine Biological Association of the United Kingdom* 69: 465 – 476.

Anstensrud, M. (1990). Moulting and mating in *Lepeophtheirus pectoralis* (Copepoda: Caligidae). *Journal of the Marine Biological Association of the United Kingdom* 70: 269 – 281.

Anstensrud, M. & Schram, T.A. (1988). Host and site selection by larval stages and adults of the parasitic copepod *Lernaeenicus sprattae* (Sowerby) (Copepoda, Pennellidae) in the Oslofjord. *Hydrobiologia* 167/168: 587 – 595.

Apakupakul, K., Siddall, M.E. & Burreson, E.M. (1999). Higher level relationships of leeches (Annelida: Clitellata: Euhirudinea) based on morphology and gene sequences. *Molecular Phylogenetics and Evolution* 12: 350 – 359.

Araujo, R. & Ramos, M.A. (2000). Status and conservation of the giant European freshwater pearl mussel (*Margaritifera auricularia*) (Spengler, 1793) (Bivalvia: Unionoidea). *Biological Conservation* 96: 233 – 239.

Araujo, R. & Ramos, M.A. (2001). Life-history data on the virtually unknown *Margaritifera auricularia*. In: *Ecology and Evolution of the Freshwater Mussels Unionoida*. Ecological Studies, vol. 145, Bauer, G. & Wächtler, K. (eds.), pp. 143 – 152, Springer-Verlag, Berlin, Heidelberg.

Arey, L.B. (1921). An experimental study on glochidia and the factors underlying encystment. *Journal of Experimental Zoology* 33: 463 – 499.

Arey, L.B. (1932). The nutrition of glochidia during metamorphosis. A microscopical study of the sources and manner of utilization of nutritive substances. *Journal of Morphology* 53: 201 – 221.

Ausden, M., Banks, B., Donnison, E., Howe, M., Nixon, A., Phillips, D., Wicks, D. & Wynne, C. (2002). The status, conservation and use of the medicinal leech. *British Wildlife* 13: 229 – 238.

Avenant-Oldewage, A. & Swanepoel, J.H. (1993). the male reproductive system and mechanism of sperm transfer in *Argulus japonicus* (Crustacea: Branchiura). *Journal of Morphology* 215: 51 – 63.

Baird, W. (1850). *The Natural History of the British Entomostraca*. Ray Society, no. 8, London.

Bakke, T.A., Harris, P.D., Jansen, P.A. & Hansen, L.P. (1992). Host specificity and dispersal strategy in gyrodactylid monogeneans, with particular reference to *Gyrodactylus salaris* (Platyhelminthes, Monogenea). *Diseases of Aquatic Organisms* 13: 63 – 74.

Bakke, T.A., Harris, P.D. & Cable, J. (2002). Host specificity dynamics: observations on gyrodactylid monogeneans. *International Journal for Parasitology* 32: 281 – 308.

Bakke, T.A. & MacKenzie, K. (1993). Comparative susceptibility of native Scottish and Norwegian stocks of Atlantic salmon, *Salmo salar* L., to *Gyrodactylus salaris* Malmberg: laboratory experiments. *Fisheries Research* 17: 69 – 85.

Ball, I.R. & Khan, R.A. (1976). On *Micropharynx parasitica* Jägerskiöld, a marine planarian ectoparasitic on thorny skate, *Raja radiata* Donovan, from the North Atlantic Ocean. *Journal of Fish Biology* 8: 419 – 426.

Bauer, G. (1987a). Reproductive strategy of the freshwater pearl mussel *Margaritifera margaritifera*. *Journal of Animal Ecology* 56: 691 – 704.

Bauer, G. (1987b). The parasitic stage of the freshwater pearl mussel (*Margaritifera margaritifera* L.). II. Susceptibility of brown trout. *Archiv für Hydrobiologie* 76: 403 – 412.

Bauer, G. (1994). The adaptive value of offspring size among freshwater mussels (Bivalvia; Unionoidea). *Journal of Animal Ecology* 63: 933 – 944.

Bauer, G. (2001). Framework and driving forces for the evolution of naiad life histories. In: *Ecology and Evolution of the Freshwater Mussels Unionoida*. Ecological Studies, vol. 145, Bauer, G. & Wächtler, K. (eds.), pp. 233 – 255, Springer-Verlag, Berlin, Heidelberg.

Bauer, G. & Vogel, C. (1987). The parasitic stage of the freshwater pearl mussel (*Margaritifera margaritifera* L.). I. Host response to glochidiosis. *Archiv für Hydrobiologie* 76: 393 – 402.

Bauer, G. & Wächtler, K. (2001). Ecology and Evolution of the naiads. In: *Ecology and Evolution of the Freshwater Mussels Unionoida*. Ecological Studies, vol. 145, Bauer, G. & Wächtler, K. (eds.), pp. 383 – 388, Springer-Verlag, Berlin, Heidelberg.

Bauer, O.N. (1959). The ecology of parasites of freshwater fish. (Relationship between parasite and environment). In: *Parasites of Freshwater Fish and the Biological Basis of their Control*. Bulletin of the State Scientific Research Institute of Lake and River Fisheries, vol. 49, pp. 3 – 215, Leningrad. English translation by Israel Program for Scientific Translations, Jerusalem, 1962.

Baxter, E.W. (1956). Observations on the buccal glands of lampreys (Petromyzonidae). *Proceedings of the Zoological Society of London*, 127: 95 – 118.

Baylis, H.A. & Jones, E.I. (1933). Some records of parasitic worms from marine fishes at Plymouth. *Journal of the Marine Biological Association of the United Kingdom* 18: 627 – 634.

Baynes, S.M., Howell, B.R., Beard, T.W. & Hallam, J.D. (1994). A description of spawning behaviour of captive Dover sole, *Solea solea* (L.). *Netherlands Journal of Sea Research* 32: 271 – 275.

Berland, B. (1993). Salmon lice on wild salmon (*Salmo salar* L.) in western Norway. In: *Pathogens of Wild and Farmed Fish: Sea Lice*. Boxshall, G.A. & Defaye, D. (eds.), pp. 179 – 187, Ellis Horwood, London.

Berland, B. & Margolis, L. (1983). The early history of "lakselus" and some nomenclatural questions relating to copepod parasites of salmon. *Sarsia* 68: 281 – 288.

Bertin, L. (1958). Organes de la respiration aquatique. In: *Traité de Zoologie*. Grassé, P. (ed.), vol. 13, fasc. 2, pp. 1303 – 1341, Masson, Paris.

Bielecki, A. (1999). The role of eye-like spots of parasitic leeches (Hirudinea, Piscicolidae) in searching for fish hosts. *Wiadomosci Parazytologiczne* 45: 339 – 349. (In Polish). English summary in *Helminthological Abstracts* 69: 458 – 459.

Bird, N.T. (1968). Effects of mating on subsequent development of a parasitic copepod. *Journal of Parasitology* 54: 1194 – 1196.

Black, G.A. (1982). Gills as an attachment site for *Salmincola edwardsii* (Copepoda: Lernaeopodidae). *Journal of Parasitology* 68: 1172 – 1173.

Blažek, R.D. (2000). Distribution and seasonal occurrence of glochidia on the host fishes. *Acta Parasitologica* 45: 267 – 268.

Blažek, R., Jurajda, P., Koubková, B. & Gelnar, M. (1999). Distribution of glochidial larval stages of uniocean (*sic*) mussels (Mollusca: Unionidae) within the populations of the host fishes. *Helminthologia* 36: 128.

Boeger, W.A., Kritsky, D.C. & Pie, M.R. (2003). Context of diversification of the viviparous Gyrodactylidae (Platyhelminthes, Monogenoidea). *Zoologica Scripta* 32: 437 – 448.

Bovet, J. (1967). Contribution à la morphologie et à la biologie de *Diplozoon paradoxum* v. Nordmann, 1832. *Bulletin de la Société Neuchâteloise des Sciences naturelles* 90: 63 – 159.

Bower-Shore, C. (1940). An investigation of the common fish louse, *Argulus foliaceus* (Linn.). *Parasitology* 32: 361 – 371.

Boxshall, G.A. (1974). The developmental stages of *Lepeophtheirus pectoralis* (Müller, 1776) (Copepoda: Caligidae). *Journal of Natural History* 8: 681 – 700.

REFERENCES

Boxshall, G.A. (1976). The host specificity of *Lepeophtheirus pectoralis* (Müller, 1776) (Copepoda: Caligidae). *Journal of Fish Biology* 8: 255 – 264.

Boxshall, G.A. & Halsey, S.H. (2004). *An Introduction to Copepod Diversity*. Part 1. The Ray Society, no. 166, London.

Brandal, P.O., Egidius, E. & Romslo, I. (1976). Host blood: a major food component for the parasitic copepod *Lepeophtheirus salmonis*, Kroyeri, 1838 (Crustacea: Caligidae). *Norwegian Journal of Zoology* 24: 341 –343.

Bron, J.E., Sommerville, C., Jones, M. & Rae, G.H. (1991). The settlement and attachment of early stages of the salmon louse, *Lepeophtheirus salmonis* (Copepoda: Caligidae) on the salmon host, *Salmo salar*. *Journal of Zoology* 224: 201 – 212.

Bron, J.E., Sommerville, C. & Rae, G. (1993a). Aspects of the behaviour of the copepodid larvae of the salmon louse *Lepeophtheirus salmonis* (Krøyer, 1837). In: *Pathogens of Wild and Farmed Fish: Sea Lice*. Boxshall, G.A. & Defaye, D. (eds.), pp. 125 – 141, Ellis Horwood, London.

Bron, J.E., Sommerville, C. & Rae, G.H. (1993b). The functional morphology of the alimentary canal of larval stages of the parasitic copepod *Lepeophtheirus salmonis*. *Journal of Zoology* 230: 207 – 220.

Bruce, N.L. & Bowman, T.E. (1989). Species of the parasitic isopod genera *Ceratothoa* and *Glossobius* (Crustacea: Cymothoidae) from the mouths of flying fishes and halfbeaks (Beloniformes). *Smithsonian Contributions to Zoology* 489: 1 – 28.

Brumpt, E. (1900a). Reproduction des hirudinées. *Mémoires de la Société zoologique de France* 13: 286 – 430.

Brumpt, E. (1900b). Reproduction des hirudinées. Formation du cocon chez *Piscicola* et *Herpobdella*. *Bulletin de la Société zoologique de France* 25: 47 – 51.

Bruno, D.W., Collins, C.M., Cunningham, C.O. & MacKenzie, K. (2001). *Gyrodactyloides bychowskii* (Monogenea: Gyrodactylidae) from sea-caged Atlantic salmon *Salmo salar* in Scotland: occurrence and ribosomal RNA sequence analysis. *Diseases of Aquatic Organisms* 45: 191 – 196.

Buchmann, K. & Bresciani, J. (1998). Microenvironment of *Gyrodactylus derjavini* on rainbow trout *Oncorhynchus mykiss*: association between mucous cell density in skin and site selection. *Parasitology Research* 84: 17 – 24.

Buchmann, K., Lindenstrøm, T. & Sigh, J. (1999). Partial cross protection against *Ichthyophthirius multifiliis* in *Gyrodactylus derjavini* immunized rainbow trout. *Journal of Helminthology* 73: 189 – 195.

Buchmann, K. & Nielsen, M.E. (1999). Chemoattraction of *Ichthyophthirius multifiliis* (Ciliophora) theronts to host molecules. *International Journal for Parasitology* 29: 1415 – 1423.

Buchmann, K. & Uldal, A. (1997). *Gyrodactylus derjavini* infections in four salmonids: comparative host susceptibility and site selection of parasites. *Diseases of Aquatic Organisms* 28: 201 – 209.

Bullard, S.A., Benz, G.W., Overstreet, R.M., Williams, Jr., E.H. & Hemdal, J. (2000). Six new host records and an updated list of wild hosts for *Neobenedenia melleni* (MacCallum) (Monogenea: Capsalidae). *Comparative Parasitology* 67: 190 – 196.

Bullock, A.M., Marks, R. & Roberts, R.J. (1978a). The cell kinetics of teleost fish epidermis: epidermal mitotic activity in relation to wound healing at varying temperatures in plaice (*Pleuronectes* platessa). *Journal of Zoology* 185: 197 – 204.

Bullock, A.M., Marks, R. & Roberts, R.J. (1978b). The cell kinetics of teleost fish epidermis: mitotic activity of the normal epidermis at varying temperatures in plaice (*Pleuronectes platessa*). *Journal of Zoology* 184: 423 – 428.

Bullock, A.M., Phillips, S.E., Gordon, J.D.M. & Roberts, R.J. (1986). *Sarcotaces* sp., a parasitic copepod infection in two deep-sea fishes, *Lepidion eques* and *Coelorhynchus occa*. *Journal of the Marine Biological Association of the United Kingdom* 66: 835 – 843.

Burreson, E.M. (1995). Phylum Annelida: Hirudinea as vectors and disease agents. In: *Fish Diseases and Disorders*. Vol. 1, *Protozoan and Metazoan Infections*. Woo, P.T.K. (ed.), pp. 599 – 629, CAB International, Wallingford.

Cable, J., Tinsley, R.C. & Harris, P.D. (2002a). Survival, feeding and embryo development of *Gyrodactylus gasterostei* (Monogenea: Gyrodactylidae). *Parasitology* 124: 53 – 68.

Cable, J., Scott, E.C.G., Tinsley, R.C. & Harris, P.D. (2002b). Behavior favoring transmission in the viviparous monogenean *Gyrodactylus turnbulli*. *Journal of Parasitology* 88: 183 –184.

Caillet, C. & Raibaut, A. (1979). Observations expérimentales sur la sexualité du copépode caligoïde *Clavellodes macrotrachelus* (Brian, 1906), parasite branchial du sar *Diplodus sargus* (Linné, 1758). *Comptes rendus hebdomadaires des Séances de l'Académie des Sciences, Série D* 288: 223 – 226.

Campbell, A.D. (1971). The occurrence of *Argulus* (Crustacea: Branchiura) in Scotland. *Journal of Fish Biology* 3: 145 – 146.

Canning, E.U., Cox, F.E.G., Croll, N.A. & Lyons, K.M. (1973). The natural history of Slapton Ley nature reserve: VI. Studies on the parasites. *Field Studies* 3: 681 – 717.

Carrassón, M., Stefanescu, C. & Cartes, J.E. (1992). Diets and bathymetric distributions of two bathyal sharks of the Catalan deep sea (western Mediterranean). *Marine Ecology Progress Series* 82: 21 – 30.

Carvalho-Varela, M. & Cunha-Ferreira, V. (1987). Helminth parasites of the common sole, *Solea solea*, and the Senegalese sole, *Solea senegalensis*, on the Portuguese continental coast. *Aquaculture* 67: 135 – 138.

Charmantier, G. (1980). Etude écophysiologique des crustacés isopods Gnathiidae: osmoregulation et résistance à la desiccation des mâles de *Paragnathia formica* (Hesse, 1864). *Journal of Experimental Marine Biology and Ecology* 43: 161 – 171.

Chigasaki, M., Nakane, M., Ogawa, K. & Wakabayashi, H. (2000). Standardized method for experimental infection of tiger puffer *Takifugu rubripes* with oncomiracidia of *Heterobothrium okamotoi* (Monogenea: Diclidophoridae) with some data on oncomiracidial biology. *Fish Pathology* 35: 215 – 221.

Chubb, J.C. (1964). A preliminary comparison of the parasite fauna of the fish of Llyn Padarn, Caernarvonshire, an oligotrophic lake, and Llyn Tegid (Bala Lake), a late oligotrophic or early mesotrophic lake. *Wiadomosci Parazytoligiczne* 10: 499 – 510.

Chubb, J.C. (1965). Report on the parasites of freshwater fishes of Lancashire and Cheshire. In: Thirty-fifth Report of the Lancashire and Cheshire Fauna Committee, pp. 1 – 5.

Chubb, J.C. (1970a). The parasites of the three spined stickleback *Gasterosteus aculeatus* (L.) in an oligotrophic lake, Llyn Padarn, North Wales. *Journal of Parasitology* 56, Section II, 56.

Chubb, J.C. (1970b). The parasite fauna of British freshwater fish. In: *Aspects of Fish Parasitology*. Taylor, A.E.R. & Muller, R. (eds.), Symposia of the British Society for Parasitology, vol. 8, pp. 119 – 144, Blackwell, Oxford.

Clark, N.C. (1902). *Argulus foliaceus*. A contribution to the life history. *Proceedings of the South London Entomological and Natural History Society* 1902, 12 – 21.

Clark, T.G. & Dickerson, H.W. (1997). Antibody-mediated effects on parasite behavior: evidence of a novel mechanism of immunity against a parasitic protist. *Parasitology Today* 13: 477 – 480.

Clark, T.G., Lin, T. & Dickerson, H.W. (1995). Surface immobilization antigens of *Ichthyophthirius multifiliis*: their role in protective immunity. *Annual Review of Fish Diseases* 5: 113 – 131.

Clarke, M.R. & Merrett, N. (1972). The significance of squid, whale and other remains from the stomachs of bottom-living fish. *Journal of the Marine Biological Association of the United Kingdom* 52: 599 – 603.

Compagno, L.J.V. (1984). *FAO Species Catalogue*. Vol. 4, *Sharks of the World*. Part 1, *Hexanchiformes to Lamniformes*. FAO Fisheries Synopsis no. 125, FAO, Rome.

Conover, A. (1998). To reproduce mussels go fishing. *Smithsonian* 28: 65 – 71.

Cosgrove, P., Hastie, L. & Young, M. (2000). Freshwater pearl mussels in peril. *British Wildlife* 11: 340 – 347.

Cressey, R.F. (1983). Crustaceans as parasites of other organisms. In: *The Biology of Crustacea.* Vol. 6. *Pathobiology.* Provenzano, A.J., Jr. (ed.), pp. 251 – 273, Academic Press, New York.

Cross, M.L. & Matthews, R.A. (1992). Ichthyophthiriasis in carp, *Cyprinus carpio* L.: fate of parasites in immunized fish. *Journal of Fish Diseases* 15: 497 – 505.

Cuénot, L. (1912). Contribution à la faune du Bassin d'Arcachon. VI. – Argulides. *Bulletin de la Station Biologique d'Arcachon* 14c: 117 – 127.

Dartnall, H.J.G. (1973). Parasites of the nine-spined stickleback *Pungitius pungitius* (L.). *Journal of Fish Biology* 5: 505 – 509.

Dartnall, H.J.G. & Walkey, M. (1979). The distribution of glochidia of the Swan mussel, *Anodonta cygnea* (Mollusca) on the three-spined stickleback *Gasterosteus aculeatus* (Pisces). *Journal of Zoology* 189: 31 – 37.

Darwin, C. (1851). *A Monograph of the Sub-class Cirripedia. The Lepadidae; or Pedunculated Cirripedes.* Ray Society, London.

Darwin, C. (1859). *The Origin of Species by means of Natural Selection.* John Murray, London. (Facsimile of the First Edition, Harvard University Press, Cambridge, USA).

Davies, A.J. (1981). A scanning electron microscope study of the praniza larva of *Gnathia maxillaris* Montagu (Crustacea, Isopoda, Gnathiidae), with special reference to the mouthparts. *Journal of Natural History* 15: 545 – 554.

Davies, A.J. (1982). Further studies on *Haemogregarina bigemina* Laveran & Mesnil, the marine fish *Blennius pholis* L., and the isopod *Gnathia maxillaris* Montagu. *Journal of Protozoology* 29: 576 – 583.

Davies, A.J., Eiras, J.C. & Austin, R.T.E. (1994). Investigations into the transmission of *Haemogregarina bigemina* Laveran & Mesnil, 1901 (Apicomplexa: Adeleorina) between intertidal fishes in Portugal. *Journal of Fish Diseases* 17: 283 – 289.

Davies, A.J. & Johnston, M.R.L. (1976). The biology of *Haemogregarina bigemina* Laveran & Mesnil, a parasite of the marine fish *Blennius pholis* Linnaeus. *Journal of Protozoology* 23: 315 – 320.

Davies, A.J. & Johnston, M.R.L. (2000). The biology of some intraerythrocytic parasites of fishes, amphibia and reptiles. *Advances in Parasitology* 45: 1 – 107.

REFERENCES

Davies, A.J. & Smit, N.J. (2001). The life cycle of *Haemogregarina bigemina* (Adeleina: Haemogregarinidae) in South African hosts. *Folia Parasitologica* 48: 169 – 177.

Davis, R.M. (1967). Parasitism by newly transformed and anadromous sea lampreys on landlocked salmon and other fishes in a coastal Maine lake. *Transactions of the American Fisheries Society* 96: 11 – 16.

Dawes, B. (1946). *The Trematoda*. Cambridge University Press.

Dawes, B. (1947). *The Trematoda of British Fishes*. Ray Society, London.

Dedie, O. (1940). Etude de *Salmincola mattheyi* n. sp. copépode parasite de l'omble-chevalier (*Salmo salvelinus* L.). *Revue Suisse de Zoologie* 47: 1 – 63.

Denham, K.L. & Long, J. (1999). Occurrence of *Gyrodactylus thymalli* Zitnan, 1960 on grayling, *Thymallus thymallus* (L.), in England. *Journal of Fish Diseases* 22: 247 – 252.

Denton, E.J. (1961). Some recently discovered buoyancy mechanisms in marine animals. *Proceedings of the Royal Society*, Series A 265: 366 – 370.

Denton, E. (1971). Reflectors in fishes. *Scientific American* 224: 65 – 72.

De Veen, J.F. (1967). On the phenomenon of soles (*Solea solea* L.) swimming at the surface. *Journal du Conseil Permanent International pour l'Exploration de la Mer* 31: 207 – 236.

Diamant, A. (1987). Ultrastructure and pathogenesis of *Ichthyobodo* sp. from wild common dab, *Limanda limanda* L., in the North Sea. *Journal of Fish Diseases* 10: 241 – 247.

Donnelly, R.E. & Reynolds, J.D. (1994). Occurrence and distribution of the parasitic copepod *Leposphilus labrei* on corkwing wrasse (*Crenilabrus melops*) from Mulroy Bay, Ireland. *Journal of Parasitology* 80: 331 – 332.

Dudgeon, D. & Morton, B. (1984). Site selection and attachment duration of *Anodonta woodiana* (Bivalvia: Unionacea) glochidia on fish hosts. *Journal of Zoology* 204: 355 – 362.

Dugatkin, L.A., FitzGerald, G.J. & Lavoie, J. (1994). Juvenile three-spined sticklebacks avoid parasitized conspecifics. *Environmental Biology of Fishes* 39: 215 – 218.

Edney, E.B. (1954). *British Woodlice with Keys to the Species*. Synopses of the British Fauna, no. 9, Linnean Society of London.

Einszporn, T. (1965). Nutrition of *Ergasilus sieboldi* Nordmann. II. The uptake of food and the food material. *Acta Parasitologica* 13: 373 – 380.

El Gharbi, S., Rousset, V. & Raibaut, A. (1985). Biologie du copépode *Lernaeenicus sprattae* (Sowerby, 1806) et ses actions pathogènes sur les populations de sardines des côtes du Languedoc-Roussillon. *Revue des Travaux. Institut des Pêches Maritimes* 47: 191 – 201.

Elliott, D.G. (2000a). Integumentary system. In: *The Laboratory Fish*. Ostrander, G.K. (ed.), pp. 95 – 108, Academic Press, San Diego.

Elliott, D.G. (2000b). Integumentary system. In: *The Laboratory Fish*. Ostrander, G.K. (ed.), pp. 271 –306, Academic Press, San Diego.

Elliott, J.M. & Mann, K.H. (1979). *A Key to the British Freshwater Leeches with Notes on Their Life Cycles and Ecology*. Freshwater Biological Association Scientific Publication No. 40.

Ellis, A.E. (1978). *British Freshwater Bivalve Mollusca. Keys and Notes for the Identification of the Species.* Synopses of the British Fauna, no. 11, Kormack, D.M. (ed.), Academic Press, London.

Ellis, A.E. (1981). Stress and the modulation of defence mechanisms in fish. In: *Stress and Fish*. Pickering, A.D. (ed.), pp. 147 – 169, Academic Press, London, New York.

Ellis, A.E. (1982). Differences between the immune mechanisms of fish and higher vertebrates. In: *Microbial Diseases of Fish*. Roberts, R.J. (ed.), pp. 1 – 29, Academic Press, London.

Ellis, E. (1965). *The Broads*. Collins, London.

El-Naggar, M.M., El-Naggar, A.A. & Kearn, G.C. (2004). Swimming in *Gyrodactylus rysavyi* (Monogenea: Gyrodactylidae) from the Nile catfish, *Clarias gariepinus*. *Acta Parasitologica* 49: 102 – 107.

El-Naggar, M.M. & Kearn, G.C. (1980). Ultrastructural observations on the anterior adhesive apparatus in the monogeneans *Dactylogyrus amphibothrium* Wagener, 1857 and *D. hemiamphibothrium* Ergens, 1956. *Zeitschrift für Parasitenkunde* 61: 223 – 241.

El-Rashidy, E. & Boxshall, G.A. (1999). Ergasilid copepods (Poecilostomatoida) from the gills of primitive Mugilidae (grey mullets). *Systematic Parasitology* 42: 161 – 186.

El Saby, M.K. (1930). On the complemental male of *Chondracanthus depressus* (T. Scott). *Proceedings and Transactions of the Liverpool Biological Society* 45: 110 – 115.

Ergens, R. (1976). Variability of hard parts of opisthaptor of two species of *Gyrodactylus* Nordmann, 1832 (Monogenoidea) from *Phoxinus phoxinus* (L.). *Folia Parasitologica* 23: 111 – 126. .

Ergens, R. & Gelnar, M. (1985). Experimental verification of the effect of temperature on the size of hard parts of opisthaptor of *Gyrodactylus katherineri* Malmberg, 1964 (Monogenea). *Folia Parasitologica* 32: 377 – 380. .

REFERENCES

Eschmeyer, W.N. (1998). *Catalog of Fishes,* volume 1, *Introductory Materials, Species of Fishes (A – L).* Special Publication No. 1 of the Center for Biodiversity Research and Information, California Academy of Sciences, San Fransisco.

Euzet, L. (1989). Ecologie et parasitologie. *Bulletin d'Ecologie* 20: 277 – 280.

Euzet, L. & Combes, C. (1980). Les problèmes de l'espèce chez les animaux parasites. In: *Les Problèmes de l'Espèce dans le Règne Animal.* Bocquet, C., Genermont, J. & Lamotte, M. (eds.), *Mémoires de la Société Zoologique de France* 40: 239 – 285.

Euzet, L. & Maillard, C. (1976). The mechanism of attachment to the host of some Hexabothriidae (Monogenea). *Proceedings of the Institute of Biology and Pedology, Far-East Science Centre, Vladivostok* 34: 115 – 122. (In Russian).

Evelyn, T.P.T. (1996). Infection and disease. In: *The Fish Immune System: Organism, Pathogen, and Environment.* Iwama, G. & Nakanishi, T. (eds.), pp. 339 – 366, Academic Press, London.

Ewing, M.S., Ewing, S.A. & Kocan, K.M. (1988). *Ichthyophthirius* (Ciliophora): population studies suggest reproduction in host epithelium. *Journal of Protozoology* 35: 549 – 552.

Ewing, M.S. & Kocan, K.M. (1986). *Ichthyophthirius multifiliis* (Ciliophora) development in gill epithelium. *Journal of Protozoology* 33: 369 – 374.

Ewing, M.S. & Kocan, K.M. (1987). *Ichthyophthirius multifiliis* (Ciliophora) exit from gill epithelium. *Journal of Protozoology* 34: 309 – 312.

Ewing, M.S. & Kocan, K.M. (1992). Invasion and development strategies of *Ichthyophthirius multifiliis,* a parasitic ciliate of fish. *Parasitology Today* 8: 204 – 208.

Ewing, M.S., Kocan, K.M. & Ewing, S.A. (1983). *Ichthyophthirius multifiliis*: morphology of the cyst wall. *Transactions of the American Microscopical Society* 102: 122 – 128.

Ewing, M.S., Kocan, K.M. & Ewing, S.A. (1985). *Ichthyophthirius multifiliis* (Ciliophora) invasion of gill epithelium. *Journal of Protozoology* 32: 305 – 310.

Ewing, M.S., Lynn, M.E. & Ewing, S.A. (1986). Critical periods in development of *Ichthyophthirius multifiliis* (Ciliophora) populations. *Journal of Protozoology* 33: 388 – 391.

Fast, M.D., Burka, J.F., Johnson, S.C. & Ross, N.W. (2003). Enzymes released from *Lepeophtheirus salmonis* in response to mucus from different salmonids. *Journal of Parasitology* 89: 7 – 13.

Fasten, N. (1921). Studies on parasitic copepods of the genus *Salmincola. American Naturalist* 55: 449 – 456.

Ferraz, E., Shinn, A.P. & Sommerville, C. (1994). *Gyrodactylus gemini* n. sp. (Monogenea: Gyrodactylidae), a parasite of *Semaprochilodus taeniurus* (Steindachner) from the Venezuelan Amazon. *Systematic Parasitology* 29: 217 – 222.

Firth, K.J., Johnson, S.C. & Ross, N.W. (2000). Characterization of proteases in the skin mucus of Atlantic salmon (*Salmo salar*) infected with the salmon louse (*Lepeophtheirus salmonis*) and in the whole-body louse homogenate. *Journal of Parasitology* 86: 1199 – 1205.

Fontaine, M. (1938). La lamproie marine. Sa pêche et son importance économique. *Bulletin de la Société d'Océanographie de France* no. 97: 1681 – 1687.

Frankland, H.M.T. (1955). The life history and bionomics of *Diclidophora denticulata* (Trematoda: Monogenea). *Parasitology* 45: 313 – 351.

Friend, G.F. (1941). The life-history and ecology of the salmon gill-maggot *Salmincola salmonea* (L.) (Copepod Crustacean). *Transactions of the Royal Society of Edinburgh* 60: 503 – 541.

Froese, R. & Pauly, D. (Eds.) (2002). *FishBase*. World Wide Web electronic publication. www.fishbase.org, 24 May 2002.

Frost, W.E. (1928). The nauplius larva of *Anelasma squalicola* (Lovén). *Journal of the Marine Biological Association of the United Kingdom* 15: 125 – 128.

Fryer, G. (1968a). The parasitic Crustacea of African freshwater fishes; their biology and distribution. *Journal of Zoology* 156: 45 – 95.

Fryer, G. (1968b). The parasitic copepod *Lernaea cyprinacea* L. in Britain. *Journal of Natural History* 2: 531 – 533.

Fryer, G. (1969). The parasitic copepod *Ergasilus sieboldi* Nordmann, new to Britain. *Naturalist, Hull* no. 909: 49 – 51.

Fryer, G. (1978). A remarkable inland brackish-water crustacean fauna from the Lower Aire Valley, Yorkshire: a conundrum for the ecologist. *Naturalist, Hull* 103: 83 – 94.

Fryer, G. (1981). The copepod *Salmincola edwardsii* as a parasite of *Salvelinus alpinus* in Britain, and a consideration of the so-called relict fauna of Ennerdale Water. *Journal of Zoology* 193: 253 – 268.

Fryer, G. (1982). *The Parasitic Copepoda and Branchiura of British Freshwater Fishes. A Handbook and Key.* Freshwater Biological Association, Scientific Publication no. 46, Ambleside, England.

Fryer, G. & Andrews, C. (1983). The parasitic copepod *Ergasilus briani* Markewitsch in Yorkshire: an addition to the British fauna. *Naturalist, Hull* 108: 7 – 10.

REFERENCES

Gage, S.H. (1929). Lampreys and their ways. *Scientific Monthly* 27: 401 – 416.

Gage, S.H. & Gage-Day, M. (1927). The anti-coagulating action of the secretion of the buccal glands of the lampreys (*Petromyzon*, *Lampetra* and *Entosphenus*). *Science* 66: 282 – 284.

Galarowicz, T. & Cochran, P.A. (1991). Response by the parasitic crustacean *Argulus japonicus* to host chemical clues. *Journal of Freshwater Ecology* 6: 455 – 456.

Galeote, M.D. & Otero, J.G. (1998). Cleaning behaviour of rock cook, *Centrolabrus exoletus* (Labridae), in Tarifa (Gibraltar Strait Area). *Cybium* 22: 57 – 68.

Gannicott, A.M. & Tinsley, R.C. (1997). Egg hatching in the monogenean gill parasite *Discocotyle sagittata* from the rainbow trout (*Oncorhynchus mykiss*). *Parasitology* 114: 569 – 579.

Gannicott, A.M. & Tinsley, R.C. (1998). Larval survival characteristics and behaviour of the gill monogenean *Discocotyle sagittata*. *Parasitology* 117: 491 – 498.

Gaze, W.H. & Wootten, R. (1998). Ectoparasitic species of the genus *Trichodina* (Ciliophora: Peritrichida) parasitising British freshwater fish. *Folia Parasitologica* 45: 177 – 190.

Geisslinger, M. (1987). Observations on the caudal cilium of the tomite of *Ichthyophthirius multifiliis* Fouquet, 1876. *Journal of Protozoology* 34: 180 – 182.

Gibson, D.I., Timofeeva, T.A. & Gerasev, P.I. (1996). A catalogue of the nominal species of the monogenean genus *Dactylogyrus* Diesing, 1850 and their host genera. *Systematic Parasitology* 35: 3 – 48.

Gibson, R.N. & Tong, L.J. (1969). Observations on the biology of the marine leech *Oceanobdella blennii*. *Journal of the Marine Biological Association of the United Kingdom* 49: 433 – 438.

Giusti, F., Castagnolo, L., Moretti Farina, L. & Renzoni, A. (1975). The reproductive cycle and the glochidium of *Anodonta cygnea* L. from Lago Trasimeno (central Italy). *Monitore Zoologico Italiano* 9: 99 – 118.

Grabda, J. (1963). Life cycle and morphogenesis of *Lernaea cyprinacea* L. *Acta Parasitologica Polonica* 11: 169 – 198.

Gradwell, N. (1972). Behaviors of the leech, *Placobdella*, and transducer recordings of suctorial pressures. *Canadian Journal of Zoology* 50: 1325 – 1332.

Grant, A.N. (2002). Medicines for sea lice. *Pest Management Science* 58: 521 – 527.

Greenwood, P.H. (1963). J.R. Norman's *A History of Fishes*. Ernest Benn, London.

Gresty, K.A., Boxshall, G.A. & Nagasawa, K. (1993). The fine structure and function of the cephalic appendages of the branchiuran parasite, *Argulus japonicus* Thiele. *Philosophical Transactions of the Royal Society, London B* 339: 119 – 135.

Grutter, A.S. (1999). Cleaner fish really do clean. *Nature* 398: 672 – 673.

Grutter, A.S. (2002). Cleaning symbioses from the parasites' perspective. *Parasitology* 124: S65 – S81.

Grutter, A.S. & Hendrikz, J. (1999). Diurnal variation in the abundance of juvenile parasitic gnathiid isopods on coral reef fish: implications for parasite-cleaner fish interactions. *Coral Reefs* 18: 187 – 191.

Guberlet, J.E. (1933). Notes on some Onchocotylinae from Naples with a description of a new species. *Pubblicazione della Stazione Zoologica di Napoli* 12: 323 – 336.

Gurney, R. (1913). Some notes on the parasitic copepod *Thersitina gasterostei*, Pagenstecher. *Annals and Magazine of Natural History* 12: 415 – 424.

Gurney, R. (1934). The development of certain parasitic Copepoda of the families Caligidae and Clavellidae. *Proceedings of the Zoological Society of London* 1934: 177 – 217.

Gurney, R. (1947). Some notes on parasitic Copepoda. *Journal of the Marine Biological Association of the United Kingdom* 27: 133 – 137.

Gusev, A.V. (1985). Class Monogenea. In: [*Keys to Parasites of the Freshwater Fish Fauna of the USSR*]. Vol. 2. [*Parasitic Metazoa*]. Bauer, O.N. (ed.), pp. 10 –253, Nauka, Leningrad (in Russian).

Haag, W.R., Butler, R.S. & Hartfield, P.D. (1995). An extraordinary reproductive strategy in freshwater bivalves: prey mimicry to facilitate larval dispersal. *Freshwater Biology* 34: 471 – 476.

Hakalahti, T. & Valtonen, E.T. (2003). Population structure and recruitment of the ectoparasite *Argulus coregoni* Thorell (Crustacea: Branchiura) on a fish farm. *Parasitology* 127: 79 – 85.

Hakalahti, T., Pasternak, A.F. & Valtonen, E.T. (2004). Seasonal dynamics of egg laying and egg-laying strategy of the ectoparasite *Argulus coregoni* (Crustacea: Branchiura). *Parasitology* 128: 655 – 660.

Halton, D.W. (1974). Hemoglobin absorption in the gut of a monogenetic trematode, *Diclidophora merlangi*. *Journal of Parasitology* 60: 59 – 66.

Halton, D.W. (1976). *Diclidophora merlangi*: sloughing and renewal of hematin cells. *Experimental Parasitology* 40: 41 – 47.

Halton, D.W., Maule, A.G., Mair, G.R., & Shaw, C. (1998). Monogenean neuromusculature: some structural and functional correlates. *International Journal for Parasitology* 28: 1609 – 1623.

Halton, D.W., Morris, G.P. & Hardcastle, A. (1974). Gland cells associated with the alimentary tract of a monogenean, *Diclidophora merlangi*. *International Journal for Parasitology* 4: 589 – 599.

Hara, T.J. (2000). Chemoreception. In: *The Laboratory Fish*. Ostrander, G.K. (ed.), pp. 245 – 249, Academic Press, San Diego.

Harant, H. & Grassé, P. (1959). Classe des Annélides Achètes ou Hirudinées ou Sangsues. In: *Traité de Zoologie. Anatomie, Systématique, Biologie*. Vol. 5, Premier Fascicule. Grassé P. (ed.), Masson, Paris.

Harding, W.A. (1910). A revision of the British leeches. *Parasitology* 3: 130 – 201.

Hardisty, M.W. & Potter, I.C. (1971a). The general biology of adult lampreys. In: *The Biology of Lampreys*. Vol. 1, Hardisty, M.W. & Potter, I.C. (eds.), pp. 127 – 206, Academic Press, London.

Hardisty, M.W. & Potter, I.C. (1971b). The behaviour, ecology and growth of larval lampreys. In: *The Biology of Lampreys*. Vol. 1, Hardisty, M.W. & Potter, I.C. (eds.), pp. 85 – 125, Academic Press, London.

Hardisty, M.W. & Potter, I.C. (1971c). Paired species. In: *The Biology of Lampreys*. Vol. 1, Hardisty, M.W. & Potter, I.C. (eds.), pp. 249 –277, Academic Press, London.

Hardy, A.H. (1959). *The Open Sea: Its Natural History*. Part II. *Fish & Fisheries*. Collins, London.

Hardy, A. (1970). *The Open Sea. The World of Plankton*. Collins, London.

Harmer, S.F. & Shipley, A.E. (Eds.) (1909). *The Cambridge Natural History*, Volume 4. Macmillan, London.

Harris, P.D. (1982). Variations in the mechanism of attachment in the Gyrodactyloidea (Monogenea). *Parasitology* 85: lviii.

Harris, P.D. (1985). Species of *Gyrodactylus* von Nordmann, 1832 (Monogenea: Gyrodactylidae) from freshwater fishes in southern England, with a description of *Gyrodactylus rogatensis* sp. nov. from the bullhead *Cottus gobio* L. *Journal of Natural History* 19, 791 – 809.

Harris, P.D. (1993). Interactions between reproduction and population biology in gyrodactylid monogeneans – a review. *Bulletin Français de la Pêche et de la Pisciculture* 328: 47 – 65.

Harris, P.D., Jansen, P.A. & Bakke, T.A. (1994). The population age structure and reproductive biology of *Gyrodactylus salaris* Malmberg (Monogenea). *Parasitology* 108: 167 – 173.

Harris, P.D., Soleng, A. & Bakke, T.A. (1998). Killing of *Gyrodactylus salaris* (Platyhelminthes, Monogenea) mediated by host complement. *Parasitology* 117: 137 – 143.

Hastie, L.C. & Cosgrove, P.J. (2002). Intensive searching for mussels in a fast-flowing river: an estimation of sampling bias. *Journal of Conchology* 37: 309 – 316.

Hastie, L.C. & Young, M.R. (2003). Timing of spawning and glochidial release in Scottish freshwater pearl mussel (*Margaritifera margaritifera*) populations. *Freshwater Biology* 48: 2107 – 2117.

Hastie, L.C., Young, M.R., Boon, P.J., Cosgrove, P.J. & Henninger, B. (2000). Sizes, densities and age structure of Scottish *Margaritifera margaritifera* (L.) populations. *Aquatic Conservation: Marine and Freshwater Ecosystems* 10: 229 – 247.

Hayward, P.J. & Ryland, J.S. (2000). *Handbook of the Marine Fauna of North-West Europe*. Oxford University Press.

Heegaard, P. (1947). Contribution to the phylogeny of the arthropods. Copepoda. *Spolia Zoologica Musei Hauniensis* 8: 1 – 236.

Helfman, G.S., Collette, B.B. & Facey, D.E. (1997). *The Diversity of Fishes*. Blackwell Science, Oxford.

Herter, K. (1929). Studien über Reizphysiologie und Parasitismus bei Fisch- und Entenegeln. *Sitzungsberichte der Gesellschaft naturforschender Freunde zu Berlin* 1929: 142 – 184.

Herter, K. (1968). *Der Medizinische Blutegel und Seine Verwandten*. A. Ziemsen Verlag, Wittenberg Lutherstadt.

Heuch, P.A. (1995). Experimental evidence for aggregation of salmon louse copepodids (*Lepeophtheirus salmonis*) in step salinity gradients. *Journal of the Marine Biological Association of the United Kingdom* 75: 927 – 939.

Heuch, P.A. (2002). Salmon lice: discovering the fascinating biology of an aquaculture pest. *Marine Biological Association News* no. 27: 8.

Heuch, P.A. & Karlsen, H.E. (1997). Detection of infrasonic water oscillations by copepodids of *Lepeophtheirus salmonis* (Copepoda: Caligida). *Journal of Plankton Research* 19: 735 – 747.

Heuch, P.A., Parsons, A. & Boxaspen, K. (1995). Diel vertical migration: a possible host-finding mechanism in salmon louse (*Lepeophtheirus salmonis*) copepodids? *Canadian Journal of Fisheries and Aquatic Sciences* 52: 681 – 689.

Hickling, C.F. (1963). On the small deep-sea shark *Etmopterus spinax* L., and its cirripede parasite *Anelasma squalicola* (Lovén). *Journal of the Linnean Society (Zoology)* 45: 17 – 24.

Hines, R.S. & Spira, D.T. (1974a) Ichthyophthiriasis in the mirror carp *Cyprinus carpio* (L.) V. Acquired immunity. *Journal of Fish Biology* 6: 373 – 378.

Hines, R.S. & Spira, D.T. (1974b). Ichthyophthiriasis in the mirror carp *Cyprinus carpio* (L.) III. Pathology. *Journal of Fish Biology* 6: 189 – 196.

Höglund, J. & Thulin, J. (1988). The external morphology of the parasitic copepod *Holobomolochus confusus* (Stock). *Zoologica Scripta* 17: 371 – 379.

Horton, M.A. & Whittington, I.D. (1994). A new species of *Metabenedeniella* (Monogenea: Capsalidae) from the dorsal fin of *Diagramma pictum* (Perciformes: Haemulidae) from the Great Barrier Reef, Australia with a revision of the genus. *Journal of Parasitology* 80: 998 – 1007.

Horton, T. (2000). *Ceratothoa steindachneri* (Isopoda: Cymothoidae) new to British waters with a key to north-east Atlantic and Mediterranean *Ceratothoa*. *Journal of the Marine Biological Association of the United Kingdom* 80: 1041 – 1052.

Horton, T. (2001). Weever fish parasites. Marine Biological Association News 25: 4.

Hubbs, C.L. & Pope, T.E.B. (1937). The spread of the sea lamprey through the Great Lakes. *Transactions of the American Fisheries Society* 66: 172 – 176.

Hubbs, C.L. & Potter, I.C. (1971). Distribution, phylogeny and taxonomy. In: *The Biology of Lampreys*. Vol. 1, Hardisty, M.W. & Potter, I.C. (eds.), pp. 1 – 65, Academic Press, London.

Hussain, N.A. & Knight-Jones, E.W. (1995). Fish and fish-leeches on rocky shores around Britain. *Journal of the Marine Biological Association of the United Kingdom* 75: 311 – 322.

Hutchinson, G.E. (1979). *An Introduction to Population Ecology*. Yale University Press, London.

Huys, R. & Boxshall, G.A. (1991). *Copepod Evolution*. Ray Society, no. 159, London.

Huyse, T., Audenaert, V. & Volckaert, F.A.M. (2003). Speciation and host-parasite relationships in the parasite genus *Gyrodactylus* (Monogenea, Platyhelminthes) infecting gobies of the genus *Pomatoschistus* (Gobiidae, Teleostei). *International Journal for Parasitology* 33: 1679 – 1689.

Huyse, T. & Volckaert, F.A.M. (2002). Identification of a host-associated species complex using molecular and morphometric analyses, with the description of *Gyrodactylus rugiensoides* n. sp. (Gyrodactylidae, Monogenea). *International Journal for Parasitology* 32: 907 – 919.

Ikuta, K. & Makioka, T. (1997). Structure of the adult ovary and oogenesis in *Argulus japonicus* Thiele (Crustacea: Branchiura). *Journal of Morphology* 231: 29 – 39.

Ivanov, A.V. (1952). Morphology of *Udonella caligorum* Johnston, 1835, and the position of Udonellidae in the systematics of platyhelminths. *Parazitologicheskii Sbornik* 14: 112 – 163. English translation, 1981, edited by Simmons, J.E., Hargis, W.J. & Zwerner, D.E., University of California at Berkeley, Berkeley, USA.

Izawa, K. (1969). Life history of *Caligus spinosus* Yamaguti, 1939 obtained from cultured yellow tail, *Seriola quinqueradiata* T. & S. (Crustacea: Caligoida). *Report of Faculty of Fisheries, Prefectural University of Mie* 6: 127 – 157.

Izawa, K. (1973). On the development of parasitic Copepoda. I. *Sarcotaces pacificus* Komai (Cyclopoida: Philichthyidae). *Publications of the Seto Marine Biological Laboratory* 21: 77 – 86.

Jackson, C.J. & Marcogliese, D.J. (1995). An unique association between *Argulus alosae* (Branchiura) and *Mysis stenolepis* (Mysidacea). *Crustaceana* 68: 910 – 912.

Jansen, W., Bauer, G. & Zahner-Meike, E. (2001). Glochidial mortality in freshwater mussels. In: *Ecology and Evolution of the Freshwater Mussels Unionoida*. Ecological Studies, vol. 145, Bauer, G. & Wächtler, K. (eds.), pp. 186 – 211, Springer-Verlag, Berlin, Heidelberg.

Jaschke, W. (1933). Beiträge zur Kenntnis der symbiontischen Einrichtungen bei Hirudineen und Ixodiden. *Zeitschrift für Parasitenkunde* 5: 515 – 541.

Jennings, J.B. & van der Lande, V.M. (1967). Histochemical and bacteriological studies on digestion in nine species of leeches (Annelida: Hirudinea). *Biological Bulletin* 133: 166 – 183.

Johannessen, A. (1978). Early stages of *Lepeophtheirus salmonis* (Copepoda Caligidae). *Sarsia* 63: 169 – 176.

Johns, M. (2002). Lamprey: relicts from the past. *British Wildlife* 13: 381 – 388.

Johnsen, B.O. (1978). The effect of an attack by the parasite *Gyrodactylus salaris* on the population of salmon parr in the river Lakselva, Misvaer in northern Norway. *Astarte* 11: 7 – 9.

Johnsen, B.O. & Jensen, A.J. (1986). Infestations of Atlantic salmon, *Salmo salar*, by *Gyrodactylus salaris* in Norwegian rivers. *Journal of Fish Biology* 29: 233 – 241.

Johnsen, B.O. & Jensen, A.J. (1988). Introduction and establishment of *Gyrodactylus salaris* Malmberg, 1957, on Atlantic salmon, *Salmo salar* L., fry and parr in the river Vefsna, northern Norway. *Journal of Fish Diseases* 11: 35 – 45.

Johnson, S.C. & Albright, L.J. (1991). Development, growth, and survival of *Lepeophtheirus salmonis* (Copepoda: Caligidae) under laboratory conditions. *Journal of the Marine Biological Association of the United Kingdom* 71: 425 – 436.

Johnstone, J. & Frost, W.E. (1927). The cirripede fish parasite *Anelasma squalicola* (Lovén): its general morphology. *35th Annual Report of the Lancashire Sea Fisheries Laboratory for 1926*, 29 – 91.

REFERENCES

Jones, E.C. (1971). *Isistius brasiliensis*, a squaloid shark, the probable cause of crater wounds on fishes and cetaceans. *Fishery Bulletin* 69: 791 – 798.

Jones, M.W., Sommerville, C. & Bron, J. (1990). The histopathology associated with the juvenile stages of *Lepeophtheirus salmonis* on the Atlantic salmon, *Salmo salar* L. *Journal of Fish Diseases* 13: 303 – 310.

Joyon, L. & Lom, J. (1969). Etude cytologique, systématique et pathologique d'*Ichtyobodo necator* (Henneguy, 1883) Pinto, 1928 (zooflagelle). *Journal of Protozoology* 16: 703 – 719.

Juilfs, H.B. & Wägele, J.W. (1987). Symbiontic bacteria in the gut of the blood-sucking Antarctic fish parasite *Gnathia calva* (Crustacea: Isopoda). *Marine Biology* 95: 493 – 499.

Jurine, M. (1806). Mémoire sur l'Argule foliacé (*Argulus foliaceus*). *Annales du Muséum d'Histoire naturelle* 7: 431 – 458.

Justine, J.-L. (1998). Non-monophyly of the monogeneans? *International Journal for Parasitology* 28: 1653 – 1657.

Kaattari, S.L. & Piganelli, J.D. (1996). The specific immune system: humoral defense. In: *The Fish Immune System: Organism, Pathogen, and Environment*. Iwama, G. & Nakanishi, T. (eds.), pp. 207 –254, Academic Press, London.

Kabata, Z. (1957). *Lernaeocera obtusa* n. sp., a hitherto undescribed parasite of the haddock (*Gadus aeglefinus* L.). *Journal of the Marine Biological Association of the United Kingdom* 36: 569 – 592.

Kabata, Z. (1958). *Lernaeocera obtusa* n.sp. Its biology and its effects on the haddock. *Marine Research* no. 3: 3 –26.

Kabata, Z. (1960). Observations on *Clavella* (Copepoda) parasitic on some British gadoids. *Crustaceana* 1: 342 – 352.

Kabata, Z. (1962). The mouth and the mouth-parts of *Lernaeocera branchialis* (L.), a parasitic copepod. *Crustaceana* 3: 311 – 317.

Kabata, Z. (1969a). *Phrixocephalus cincinnatus* Wilson, 1908 (Copepoda: Lernaeoceridae): morphology, metamorphosis, and host-parasite relationship. *Journal of the Fisheries Research Board of Canada* 26: 921 – 934.

Kabata, Z. (1969b). Four Lernaeopodidae (Copepoda) parasitic on fishes from Newfoundland and West Greenland. *Journal of the Fisheries Research Board of Canada* 26: 311 – 324.

Kabata, Z. (1969c). Revision of the genus *Salmincola* Wilson, 1915 (Copepoda: Lernaeopodidae). *Journal of the Fisheries Research Board of Canada* 26: 2987 – 3041.

Kabata, Z. (1970). *Crustacea as Enemies of Fishes.* Book 1 of: *Diseases of Fishes.* Snieszko, S.F. & Axelrod, H.R. (eds.), T.F.H. Publications, Jersey City, USA.

Kabata, Z. (1972). Developmental stages of *Caligus clemensi* (Copepoda: Caligidae). *Journal of the Fisheries Research Board of Canada* 29: 1571 – 1593.

Kabata, Z. (1973). Distribution of *Udonella caligorum* Johnston, 1835 (Monogenea: Udonellidae) on *Caligus elongatus* Nordmann, 1832 (Copepoda: Caligidae). *Journal of the Fisheries Research Board of Canada* 30: 1793 – 1798.

Kabata, Z. (1974). Mouth and mode of feeding of Caligidae (Copepoda), parasites of fishes, as determined by light and scanning electron microscopy. *Journal of the Fisheries Research Board of Canada* 31: 1583 – 1588.

Kabata, Z. (1979). *Parasitic Copepoda of British Fishes.* Ray Society, no. 152, London.

Kabata, Z. (1981). Copepoda (Crustacea) parasitic on fishes: problems and perspectives. *Advances in Parasitology* 19: 1 – 71.

Kabata, Z. (1982). The evolution of host-parasite systems between fishes and Copepoda. In: *Parasites – Their World and Ours.* Mettrick D.F. & Desser S.S. (eds.), Elsevier Biomedical Press.

Kabata, Z. (1992). *Copepods Parasitic on Fishes.* Synopses of the British Fauna (New Series), no. 47. Kermack D.M., Barnes R.S.K. & Crothers, J.H. (eds.), Universal Book Services/Dr W. Backhuys, Oegstgeest, The Netherlands.

Kabata, Z. & Cousens, B. (1973). Life cycle of *Salmincola californiensis* (Dana 1852) (Copepoda: Lernaeopodidae). *Journal of the Fisheries Research Board of Canada* 30: 881 – 903.

Kabata, Z. & Cousens, B. (1977). Host-parasite relationships between sockeye salmon, *Oncorhynchus nerka,* and *Salmincola californiensis* (Copepoda: Lernaeopodidae). *Journal of the Fisheries Research Board of Canada* 34: 191 – 202.

Kabata, Z. & Hewitt, G.C. (1971). Locomotory mechanisms in Caligidae (Crustacea: Copepoda). *Journal of the Fisheries Research Board of Canada* 28: 1143 – 1151.

Kane, M.B. (1966). Parasites of Irish fishes. *Scientific Proceedings Royal Dublin Society, B.* 1: 205 – 220.

Kat, P.W. (1984). Parasitism and the Unionacea (Bivalvia). *Biological Reviews* 59: 189 – 207.

Kearn, G.C. (1963a). Feeding in some monogenean skin parasites: *Entobdella soleae* on *Solea solea* and *Acanthocotyle* sp. on *Raia clavata. Journal of the Marine Biological Association of the United Kingdom* 43: 749 – 766.

REFERENCES

Kearn, G.C. (1963b). The oncomiracidium of *Capsala martinieri*, a monogenean parasite of the sun fish (*Mola mola*). *Parasitology* 53: 449 – 453.

Kearn, G.C. (1965). The biology of *Leptocotyle minor*, a skin parasite of the dogfish, *Scyliorhinus canicula*. *Parasitology* 55: 473 – 480.

Kearn, G.C. (1966). The adhesive mechanism of the monogenean parasite *Tetraonchus monenteron* from the gills of the pike (*Esox lucius*). *Parasitology* 56: 505 – 510.

Kearn, G.C. (1967a). Experiments on host-finding and host-specificity in the monogenean skin parasite *Entobdella soleae*. *Parasitology* 57: 585 – 605.

Kearn, G.C. (1967b). The life-cycles and larval development of some acanthocotylids (Monogenea) from Plymouth rays. *Parasitology* 57: 157 – 167.

Kearn, G.C. (1968). The development of the adhesive organs of some diplectanid, tetraonchid and dactylogyrid gill parasites (Monogenea). *Parasitology* 58: 149 – 163.

Kearn, G.C. (1970). The production, transfer and assimilation of spermatophores by *Entobdella soleae*, a monogenean skin parasite of the common sole. *Parasitology* 60: 301 – 311.

Kearn, G.C. (1971a). The physiology and behaviour of the monogenean skin parasite *Entobdella soleae* in relation to its host (*Solea solea*). In: *Ecology and Physiology of Parasites, a Symposium*, Fallis A.M. (ed.), pp. 161 – 187, University of Toronto Press.

Kearn, G.C. (1971b). The attachment site, invasion route and larval development of *Trochopus pini*, a monogenean from the gills of *Trigla hirundo*. *Parasitology* 63: 513 – 525.

Kearn, G.C. (1974). Nocturnal hatching in the monogenean skin parasite *Entobdella hippoglossi* from the halibut, *Hippoglossus hippoglossus*. *Parasitology* 68: 161 – 172.

Kearn, G.C. (1975). The mode of hatching of the monogenean *Entobdella soleae*, a skin parasite of the common sole (*Solea solea*). *Parasitology* 71: 419 – 431.

Kearn, G.C. (1976). Body surface of fishes. In: *Ecological Aspects of Parasitology*. Kennedy C.R. (ed.), pp. 185 – 208, North-Holland Publishing Company, Amsterdam.

Kearn, G.C. (1982). Rapid hatching induced by light intensity reduction in the monogenean *Entobdella diadema*. *Journal of Parasitology* 68: 171 – 172.

Kearn, G.C. (1984). The migration of the monogenean *Entobdella soleae* on the surface of its host, *Solea solea*. *International Journal for Parasitology* 14: 63 – 69.

Kearn, G.C. (1986a). The eggs of monogeneans. *Advances in Parasitology* 25: 175 – 273.

Kearn, G.C. (1986b). Role of chemical substances from fish hosts in hatching and host-finding in monogeneans. *Journal of Chemical Ecology* 12: 1651 – 1658. .

Kearn, G.C. (1987). Locomotion in the gill-parasitic monogenean *Tetraonchus monenteron*. *Journal of Parasitology* 73: 224 – 225.

Kearn, G.C. (1988). The monogenean skin parasite *Entobdella soleae*: movement of adults and juveniles from host to host (*Solea solea*). *International Journal for Parasitology* 18: 313 –319.

Kearn, G.C. (1994). Evolutionary expansion of the Monogenea. *International Journal for Parasitology* 24: 1227 – 1271.

Kearn, G.C. (1998). *Parasitism and the Platyhelminths*. Chapman & Hall, London.

Kearn, G.C. (1999). The survival of monogenean (platyhelminth) parasites on fish skin. *Parasitology* 119: S57 – S88.

Kearn, G.C. (2002). *Entobdella soleae* – pointers to the future. *International Journal for Parasitology* 32: 367 – 372.

Kearn, G.C., Al-Sehaibani, M.A., Whittington, I.D., Evans-Gowing, R. & Cribb, B.W. (1996). Swallowing of sea water and its role in egestion in the monogenean *Entobdella soleae*, a skin parasite of the common sole (*Solea solea*), with observations on other monogeneans and on a freshwater temnocephalan. *Journal of Natural History* 30: 637 – 646.

Kearn, G.C. & Gowing, R. (1989). Glands and sensilla associated with the haptor of the gill-parasitic monogenean *Tetraonchus monenteron*. *International Journal for Parasitology* 19: 673 – 679.

Kearn, G.C. & Gowing, R. (1990). Vestigial marginal hooklets in the oncomiracidium of the microbothriid monogenean *Leptocotyle minor*. *Parasitology Research* 76: 406 – 408.

Kearn, G.C., James, R. & Evans-Gowing, R. (1993). Insemination and population density in *Entobdella soleae*, a monogenean skin parasite of the common sole, *Solea solea*. *International Journal for Parasitology* 23: 891 – 899.

Kearn, G.C. & Macdonald, S. (1976). The chemical nature of host hatching factors in the monogenean skin parasites *Entobdella soleae* and *Acanthocotyle lobianchi*. *International Journal for Parasitology* 6: 457 – 466.

Kearn, G., Whittington, I. & Evans-Gowing, R. (2002). Striated (stratified) muscle and the operation of the ventral hamuli of the ancyrocephaline monogenean *Chauhanellus australis*, from the gills of the blue catfish *Arius graeffei*. In: *Taxonomy, Ecology and Evolution of Metazoan Parasites*. Published in honour of Louis Euzet. Combes, C., Jourdane, J. (eds.), vol. 1, pp. 381 – 396, University of Perpignan Press.

Kennedy, C.R. (1974). A checklist of British and Irish freshwater fish parasites with notes on their distribution. *Journal of Fish Biology* 6: 613 – 644.

Kennedy, C.R. (1975a). The natural history of Slapton Ley Nature Reserve. VIII. The parasites of fish, with special reference to their use as a source of information about the aquatic community. *Field Studies* 4: 177 – 189.

Kennedy, C.R. (1975b). The distribution of some crustacean fish parasites in Britain in relation to the introduction and movement of freshwater fish. *Journal of the Institute of Fishery Management* 6: 36 – 41.

Kennedy, C.R. (1993). Introductions, spread and colonization of new localities by fish helminth and crustacean parasites in the British Isles: a perspective and appraisal. *Journal of Fish Biology* 43: 287 – 301.

Kennedy, C.R. & Di Cave D. (1998). *Gyrodactylus anguillae* (Monogenea): the story of an appearance and a disappearance. *Folia Parasitologica* 45: 77 – 78.

Khotenovskii, I.A. (1985). *Fauna of the USSR. Monogenea. Suborder Octomacrinea Khotenovsky*. Izdatel'stvo "Nauka", Leningrad.

Kleerekoper, H. & Mogensen, J. (1963). Role of olfaction in the orientation of *Petromyzon marinus*. I. Response to a single amine in prey's body odor. *Physiological Zoology* 36: 347 – 360.

Knight-Jones, E.W. (1940). The occurrence of a marine leech, *Abranchus blennii* n. sp., resembling *A. sexoculatus* (Malm), in North Wales. *Journal of the Marine Biological Association of the United Kingdom* 24: 533 – 541.

Knight-Jones, E.W. (1962). The systematics of marine leeches. In: *Leeches (Hirudinea): their Structure, Physiology, Ecology and Embryology*. Mann, K.H. (ed.), pp. 169 – 186, Pergamon Press, Oxford.

Kollatsch, D. (1959). Untersuchungen über die Biologie und Okologie der Karpfenlaus (*Argulus foliaceus* L.). *Zoologische Beiträge* 5: 1 – 36.

Kraemer, L.R. (1970). The mantle flap in three species of *Lampsilis* (Pelecypoda: Unionidae). *Malacologia* 10: 225 – 282.

Kraemer, L.R. & Swanson, C.M. (1985). Functional morphology of "eyespots" of mantle flaps of *Lampsilis* (Bivalvia: Unionacea): evidence for their role as effectors, and basis for hypothesis regarding pigment distribution in bivalve tissues. *Malacologia* 26: 241 – 251.

Kruuk, H. (1963). Diurnal periodicity in the activity of the common sole, *Solea vulgaris* Quensel. *Netherlands Journal of Sea Research* 2: 1 – 28.

Lamond, H. (1923). Some notes on two of the fishes of Loch Lomond: the powan and the lamprey. Fishery Board for Scotland, Salmon Investigations, 1922, no. II, pp. 3 –10.

Lanzing, W.J.R. (1958). Structure and function of the suction apparatus of the lamprey. *Proceedings K. Nederlandse Akademie van Wetenschappen.* Section C. Biological and Medical Sciences C61: 300 – 307.

Lapage, G. (1958). *Parasitic Animals.* Heffer & Sons, Cambridge.

Larsen, A.H., Bresciani, J. & Buchmann, K. (2002). Interactions between ecto- and endoparasites in trout *Salmo trutta. Veterinary Parasitology* 103: 167 – 173.

Lefevre, G. & Curtis, W.C. (1910a). The marsupium of the Unionidae. *Biological Bulletin* 19: 31 – 34.

Lefevre, G. & Curtis, W.C. (1910b). Reproduction and parasitism in the Unionidae. *Journal of Experimental Zoology* 9: 79 – 115.

Legrand, J.-J. (1952). Contribution à l'étude expérimentale et statistique de la biologie d'*Anilocra physodes* L. (Crustacé Isopode Cymothoidé). *Archives de Zoologie Expérimentale et Générale* 89: 1 – 56.

Leigh-Sharpe, W.H. (1935). Two copepods (*Lernaeenicus*) parasitic on *Clupea. Journal of the Marine Biological Association of the United Kingdom* 27: 270 – 275.

Lennon, R.E. (1954). Feeding mechanism of the sea lamprey and its effect on host fishes. *Fisheries Bulletin of the United States Fish & Wildlife Service,* 56: 247 – 293.

Lescher-Moutoué F. (1979). Présence en France du copépode Ergasilidae *Neoergasilus japonicus* (Harada). *Crustaceana* 37: 109 – 112.

Lester, R.J.G. (1972). Attachment of *Gyrodactylus* to *Gasterosteus* and host response. *Journal of Parasitology* 58: 717 – 722.

Letch, C.A. (1980). The life-cycle of *Trypanosoma cobitis* Mitrophanow 1883. *Parasitology* 80: 163 – 169.

Lindenstrøm, T. & Buchmann, K. (2000). Acquired resistance in rainbow trout against *Gyrodactylus derjavini. Journal of Helminthology* 74: 155 – 160.

Littlewood, D.T.J., Rohde, K. & Clough, K.A. (1998). The phyogenetic position of *Udonella* (Platyhelminthes). *International Journal for Parasitology* 28: 1241 – 1250.

Llewellyn, J. (1941). A description of the anatomy of the monogenetic trematode *Choricotyle chrysophryi* van Beneden & Hesse. *Parasitology* 33: 397 – 405.

Llewellyn, J. (1954). Observations on the food and the gut pigment of the Polyopisthocotylea (Trematoda: Monogenea). *Parasitology* 44: 428 –437.

Llewellyn, J. (1956a). The adhesive mechanisms of monogenetic trematodes: the attachment of *Plectanocotyle gurnardi* (v. Ben. & Hesse) to the gills of *Trigla*. *Journal of the Marine Biological Association of the United Kingdom* 35: 507 – 514.

Llewellyn, J. (1956b). The host-specificity, micro-ecology, adhesive attitudes, and comparative morphology of some trematode gill parasites. *Journal of the Marine Biological Association of the United Kingdom* 35: 113 – 127.

Llewellyn, J. (1957). The mechanism of the attachment of *Kuhnia scombri* (Kuhn, 1829) (Trematoda: Monogenea) to the gills of its host *Scomber scombrus* L., including a note on the taxonomy of the parasite. *Parasitology* 47: 30 – 39.

Llewellyn, J. (1958). The adhesive mechanisms of monogenetic trematodes: the attachment of species of the Diclidophoridae to the gills of gadoid fishes. *Journal of the Marine Biological Association of the United Kingdom* 37: 67 – 79.

Llewellyn, J. (1959). The larval development of two species of gastrocotylid trematode parasites from the gills of *Trachurus trachurus*. *Journal of the Marine Biological Association of the United Kingdom* 38: 461 – 467.

Llewellyn, J. (1960). Amphibdellid (monogenean) parasites of electric rays (Torpedinidae). *Journal of the Marine Biological Association of the United Kingdom* 39: 561 – 589.

Llewellyn, J. (1962). The life histories and population dynamics of monogenean gill parasites of *Trachurus trachurus* (L.). *Journal of the Marine Biological Association of the United Kingdom* 42: 587 – 600.

Llewellyn, J. (1963). Larvae and larval development of monogeneans. *Advances in Parasitology* 1: 287 – 326.

Llewellyn, J. (1964). The effects of the host and its habits on the morphology and life-cycle of a monogenean parasite. In: *Parasitic Worms and Aquatic Conditions*. Ergens, R. & Rysavy, B. (eds.), pp. 147 – 152, Czechoslovak Academy of Sciences, Prague.

Llewellyn, J. (1965). The evolution of parasitic platyhelminths. In : *Evolution of Parasites*. Taylor, A. (ed.). Third Symposium of the British Society for Parasitology, pp. 47 – 78, Blackwell, Oxford.

Llewellyn, J. (1966). The effects of fish hosts upon the body shapes of their monogenean parasites. *Proceedings of the First International Congress of Parasitology* I: 543 – 545. .

Llewellyn, J. (1970). Taxonomy, genetics and evolution of parasites. Monogenea. *Journal of Parasitology* 56: 493 – 504.

Llewellyn, J. (1972). Phylum Platyhelminthes. In: *Textbook of Zoology. Invertebrates.* Marshall, A.J. & Williams, W.D. (eds.), pp. 212 – 227, Macmillan, London.

Llewellyn, J. (1981). Evolution of viviparity and invasion by adults. In: *Biology of monogeneans. Parasitology* 82: 64 – 68.

Llewellyn, J. (1982). Host-specificity and corresponding evolution in monogenean flatworms and vertebrates. *Mémoires du Muséum national d'Histoire naturelle* 123: 289 – 301.

Llewellyn, J. (1983). Sperm transfer in the monogenean gill parasite *Gastrocotyle trachuri*. *Proceedings of the Royal Society of London, B* 219: 439 – 446.

Llewellyn, J. (1984). The biology of *Isancistrum subulatae* n. sp., a monogenean parasitic on the squid, *Alloteuthis subulata*, at Plymouth. *Journal of the Marine Biological Association of the United Kingdom* 64: 285–302.

Llewellyn, J. & Anderson, M. (1984). The functional morphology of the copulatory apparatus of *Ergenstrema labrosi* and *Ligophorus angustus*, monogenean gill parasites of *Chelon labrosus*. *Parasitology* 88: 1 – 7.

Llewellyn, J. & Euzet, L. (1964). Spermatophores in the monogenean *Entobdella diadema* Monticelli from the skin of sting-rays, with a note on the taxonomy of the parasite. *Parasitology* 54: 337 – 344.

Llewellyn, J., Green, J.E. & Kearn, G.C. (1984). A check-list of monogenean (platyhelminth) parasites of Plymouth hosts. *Journal of the Marine Biological Association of the United Kingdom* 64: 881 – 887.

Llewellyn, J., Macdonald, S. & Green, J.E. (1980). Host-specificity and speciation in diclidophoran (monogenean) gill parasites of trisopteran (gadoid) fishes at Plymouth. *Journal of the Marine Biological Association of the United Kingdom* 60: 73 – 79.

Llewellyn, J. & Owen, I.L. (1960). The attachment of the monogenean *Discocotyle sagittata* Leuckart to the gills of *Salmo trutta* L. *Parasitology* 50: 51 – 59.

Llewellyn, J. & Tully, C.M. (1969). A comparison of speciation in diclidophorinean monogenean gill parasites and in their fish hosts. *Journal of the Fisheries Research Board of Canada* 26: 1063 – 1074.

Llewellyn, L.C. (1965). Some aspects of the biology of the marine leech *Hemibdella soleae*. *Proceedings of the Zoological Society of London* 145: 509 – 528.

Llewellyn, L.C. & Knight-Jones, E.W. (1984). A new genus and species of marine leech from British coastal waters. *Journal of the Marine Biological Association of the United Kingdom* 64: 919 – 934.

REFERENCES

Lobb, C.J. & Clem, W. (1981). The metabolic relationships of the immunoglobulins in fish serum, cutaneous mucus, and bile. *Journal of Immunology* 127:1525 – 1529.

Lom, J. (1958). A contribution to the systematics and morphology of endoparasitic trichodinids from amphibians, with a proposal of uniform specific characteristics. *Journal of Protozoology* 5: 251 – 263.

Lom, J. & Čerasovová, A. (1974). Host-finding in invasive stages of *Ichthyophthirius multifiliis*. *Journal of Protozoology* 21: 457.

Lom, J. & Corliss, J.O. (1968). Observations on the fine structure of two species of the peritrich ciliate genus *Scyphidia* and on their mode of attachment to their host. *Transactions of the American Microscopical Society* 87: 493 – 509.

Lom, J. & Dykova, I. (1992). *Developments in Aquaculture & Fisheries Science. 26. Protozoan Parasites of Fishes*. Elsevier, London, Amsterdam.

Long, D.J. & Waggoner, B.M. (1993). The ectoparasitic barnacle *Anelasma* (Cirripedia, Thoracica, Lepadomorpha) on the shark *Centroscyllium nigrum* (Chondrichthyes, Squalidae) from the Pacific sub-Antarctic. *Systematic Parasitology* 26: 133 – 136.

Longshaw, M., Pursglove, M. & Shinn, A.P. (2003). *Gyrodactylus quadratidigitus* n. sp. (Monogenea: Gyrodactylidae), a parasite of the leopard-spotted goby *Thorogobius ephippiatus* (Lowe) from the south-western coast of the UK. *Systematic Parasitology* 55: 151 – 157.

Lyndon, A.R. & Vidal-Martinez, V.M. (1994). The microhabitat and morphology of *Grubea cochlear* on the gills of mackerel from Lyme Bay, southern England. *Journal of the Marine Biological Association of the United Kingdom* 74: 731 – 734.

Lyons, K.M. (1966). The chemical nature and evolutionary significance of monogenean attachment sclerites. *Parasitology* 56: 63 – 100.

Lyons, K.M. (1969). Compound sensilla in monogenean skin parasites. *Parasitology* 59: 625 – 636.

Macdonald, S. (1974). Host skin mucus as a hatching stimulant in *Acanthocotyle lobianchi*, a monogenean from the skin of *Raja* spp. *Parasitology* 68: 331 – 338.

Macdonald, S. (1975). Hatching rhythms in three species of *Diclidophora* (Monogenea) with observations on host behaviour. *Parasitology* 71: 211 – 228.

Macdonald, S. & Caley, J. (1975). Sexual reproduction in the monogenean *Diclidophora merlangi*: tissue penetration by sperms. *Zeitschrift für Parasitenkunde* 45: 323 – 334.

Macdonald, S. & Jones, A. (1978). Egg-laying and hatching rhythms in the monogenean *Diplozoon homoion gracile* from the southern barbel (*Barbus meridionalis*). *Journal of Helminthology* 52: 23 – 28.

Macdonald, S. & Llewellyn, J. (1980). Reproduction in *Acanthocotyle greeni* n. sp. (Monogenean) from the skin of *Raia* spp. at Plymouth. *Journal of the Marine Biological Association of the United Kingdom* 60: 81 – 88.

MacKenzie, K. (1969). *Scyphidia (Gerda) adunconucleata* n. sp. and *Trichodina borealis* (Dogiel, 1940) Shulman et Shulman – Albova, 1953 (Protozoa Ciliata) from young plaice in Scottish waters. *Journal of Fish Biology* 1: 239 – 247.

MacKenzie, K. (1970). *Gyrodactylus unicopula* Glukhova, 1955, from young plaice *Pleuronectes platessa* L. with notes on the ecology of the parasite. *Journal of Fish Biology* 2: 23 – 34.

MacLennan, R.F. (1937). Growth in the ciliate *Ichthyophthirius*. I. Maturity and encystment. *Journal of Experimental Zoology* 76: 423 – 440.

Madsen, N. (1964). The anatomy of *Argulus foliaceus* Linné. *Lunds Universitets Arsskrift* 59: 1 – 31.

Maitland, P.S. (1972). *A Key to the Freshwater Fishes of the British Isles with Notes on their Distribution and Ecology*. Freshwater Biological Association, Scientific Publication no. 27.

Maitland, P.S. (1980). Scarring of whitefish (*Coregonus lavaretus*) by European river lamprey (*Lampetra fluviatilis*) in Loch Lomond, Scotland. *Canadian Journal of Fisheries and Aquatic Sciences* 37: 1981 – 1988.

Maitland, P.S. (2003). *Ecology of the River, Brook and Sea Lamprey*. Conserving Natura 2000 Rivers Ecology Series no. 5. English Nature, Peterborough.

Maitland, P.S. (2004). *Keys to the Freshwater Fish of Britain and Ireland*. Freshwater Biological Association, Ambleside, U.K.

Maitland, P.S. & Campbell, R.N. (1992). *Freshwater Fishes of the British Isles*, HarperCollins, London.

Maitland, P.S., Morris, K.H. & East K. (1994). The ecology of lampreys (Petromyzonidae) in the Loch Lomond area. *Hydrobiologia* 290: 105 – 120.

Maitland, P.S., Morris, K.H., East, K., Schoonoord, M.P., van der Wal, B. & Potter, I.C. (1984). The estuarine biology of the river lamprey, *Lampetra fluviatilis*, in the Firth of Forth, Scotland, with particular reference to size composition and feeding. *Journal of Zoology*, 203: 211 – 225.

Malecha, J. (1984a). Cycle biologique de l'hirudinée rhynchobdelle *Piscicola geometra* L. *Hydrobiologia* 118: 237 – 243.

Malecha, J. (1984b). Influence des facteurs externes sur l'activité reproductrice de l'hirudinée rhynchobdelle *Piscicola geometra* L. *Hydrobiologia* 118: 245 – 254.

Malecha, J. & Vinckier, D. (1983). Formation du cocon chez l'hirudinée rhynchobdelle *Piscicola geometra* L. *Archives de Biologie* 94: 183 – 205.

Malmberg, G. (1970). The excretory systems and the marginal hooks as a basis for the systematics of *Gyrodactylus* (Trematoda, Monogenea). *Archiv för Zoologi* 23: 1 – 235.

Mann, K.H. (1962). *Leeches (Hirudinea). Their Structure, Physiology, Ecology and Embryology*. Pergamon Press, New York.

Marine Biological Association (1957). *Plymouth Marine Fauna*. Marine Biological Association of the United Kingdom, Plymouth, U.K.

Marshall, N.B. (1971). *Explorations in the life of fishes*. Harvard University Press, Cambridge, Massachusetts.

Martin, M.F. (1932). On the morphology and classification of *Argulus* (Crustacea). *Proceedings of the Zoological Society of London* 1932: 771 – 806.

Matejusová, I., Koubková, B., D'Amelio, S. & Cunningham, C.O. (2001) Genetic characterization of six species of diplozoids (Monogenea; Diplozoidae). *Parasitology* 123: 465 – 474.

Matthews, L.H. & Parker, H.W. (1950). Notes on the anatomy and biology of the basking shark (*Cetorhinus maximus* Gunner). *Proceedings of the Zoological Society of London* 120: 535 – 576.

Matthews, R.A. (1994). *Ichthyophthirius multifiliis* Fouquet, 1876: infection and protective rsepsonse within the fish host. In: *Parasitic Diseases of Fish*. Pike, A.W. & Lewis, J.W. (eds.), pp. 17 – 42, Samara Publishing, Tresaith.

Matthews, R.A. & Matthews, B.F. (1984). *Ichthyophthirius multifiliis*, Fouquet in juvenile carp: the infection process. *Parasitology* 89: xxxiii.

Maule, A.G., Halton, D.W., Allen, J.M. & Fairweather, I. (1989). Studies on motility *in vitro* of an ectoparasitic monogenean, *Diclidophora merlangi*. *Parasitology* 98: 85 – 93.

McCartney, J.B., Fortner, G.W. & Hansen, M.F. (1985). Scanning electron microscopic studies of the life cycle of *Ichthyophthirius multifiliis*. *Journal of Parasitology* 71: 218 – 226.

Metzger, A. (1869). On the male and female of the genus *Lernaea* before the commencement of the so-called retrograde metamorphosis. *Annals and Magazine of Natural History* Series 4, 3: 154 –157 (translated from the German by W.S. Dallas).

Mikheev, V.N., Mikheev, A.V., Pasternak, A.F. & Valtonen, E.T. (2000). Light-mediated host searching strategies in a fish ectoparasite, *Argulus foliaceus* L. (Crustacea: Branchiura). *Parasitology* 120: 409 – 416.

Mikheev, V.N., Pasternak, A.F., Valtonen, E.T. & Lankinen, Y. (2001). Spatial distribution and hatching of overwintered eggs of a fish ectoparasite, *Argulus coregoni* (Crustacea: Branchiura). *Diseases of Aquatic Organisms* 46: 123 – 128.

Mikheev, V.N., Valtonen, E.T. & Rintämaki-Kinnunen, P. (1998). Host searching in *Argulus foliaceus* L. (Crustacea: Branchiura): the role of vision and selectivity. *Parasitology* 116: 425 – 430.

Minchin, E.A. (1909). Observations on the flagellates parasitic in the blood of freshwater fishes. *Proceedings of the Zoological Society of London* 2: 30.

Mishra, T.N. & Chubb, J.C. (1969). The parasite fauna of the fish of the Shropshire Union Canal, Cheshire. *Journal of Zoology* 157: 213 – 224.

Moles, A. (1983). Effect of parasitism by mussel glochidia on growth of coho salmon. *Transactions of the American Fisheries Society* 112: 201 – 204.

Mollaret, I., Jamieson, B.G.M., Adlard, R.D., Hugall, A., Lecointre, G., Chombard, C. & Justine, J.-L. (1997). Phylogenetic analysis of the Monogenea and their relationships with Digenea and Eucestoda inferred from 28S rDNA sequences. *Molecular and Biochemical Parasitology* 90: 433 – 438.

Möller, H. & Anders, K. (1983). *Krankheiten und Parasiten der Meeresfische*. Verlag Heino Möller, Kiel.

Molnár, K. & Székely, C. (1998). Occurrence of skrjabillanid nematodes in fishes of Hungary and in the intermediate host, *Argulus foliaceus* L. *Acta Veterinaria Hungarica* 46: 451 – 463.

Monod, T. (1926). Les Gnathiidae. Essai monographique. (Morphologie, Biologie, Systématique). *Mémoires de la Société des Sciences naturelles du Maroc*. No. 13: 1 – 668.

Monod, T. (1928). Les argulidés du musée du Congo. *Revue de Zoologie et de Botanique Africaines* 16: 242 – 274.

Monticelli, F.S. 1899 Il genere '*Acanthocotyle*'. *Archives de Parasitologie* 2: 75 – 120.

Monticelli, F.S. (1902). A proposito di una specie del genere *Epibdella*. *Boletino della Società di Naturalisti i Napoli* 15: 137 – 145.

Moore, J.D., Ototake, M. & Nakanishi, T. (1998). Particulate antigen uptake during immersion immunisation of fish: the effectiveness of prolonged exposure and the roles of skin and gill. *Fish and Shellfish Immunology* 8: 393 – 407.

Moravec, F. (1994). *Parasitic Nematodes of Freshwater Fishes of Europe*. Academia, Prague.

Morris, G.P. & Halton, D.W. (1975). The occurrence of bacteria and mycoplasm-like organisms in a monogenean parasite, *Diclidophora merlangi*. *International Journal for Parasitology* 5: 495 – 498.

Morton, J.E. (1958). *Molluscs*. Hutchinson & Co., London.

Moser, M. & Sakanari, J. (1985). Aspects of host location in the juvenile isopod *Lironeca vulgaris* (Stimpson, 1857). *Journal of Parasitology* 71: 464 – 468.

Mouchet, S. (1928a). Note sur le cycle évolutif des Gnathiidae. *Bulletin de la Société Zoologique de France* 53: 392 – 400.

Mouchet, S. (1928b). Contribution à l'étude de la digestion chez les Gnathiidae. *Bulletin de la Société Zoologique de France* 53: 442 – 452.

Mugridge, R.E.R., Stallybrass, H.G. & Hollman, A. (1982). *Neoergasilus japonicus* (Crustacea: Ergasilidae). A parasitic copepod new to Britain. *Journal of Zoology* 197: 551 – 557.

Müller, M.A. (1856). On the development of the lampreys. *Annals and Magazine of Natural History*, 18: 298 – 301.

Müller, O.F. (1776). Zoologiae Danicae prodromus, seu animalium Daniae et Norvegiae indigernarum characteres, nomina, et synonma imprimis popularium Havniae.

Mustafa, A., Speare, D.J., Daley, J., Conboy, G. A. & Burka, J.F. (2000). Enhanced susceptibility of seawater cultured rainbow trout, *Oncorhynchus mykiss* (Walbaum), to the microsporidian *Loma salmonae* during a primary infection with the sea louse, *Lepeophtheirus salmonis*. *Journal of Fish Diseases* 23: 337 – 341.

Naylor, E. (1972). *British Marine Isopods. Keys and Notes for the Identification of the Species.* Synopses of the British Fauna, no. 3, The Linnean Society of London, Academic Press, London.

Needham, T. & Wootten, R. (1978). The parasitology of teleosts. In: *Fish Pathology*. Roberts, R.J. (ed.), pp. 144 – 182, Baillière Tindall, London.

Nelson, J.S. (1994). *Fishes of the World*. Third edition, John Wiley & Sons, New York.

Newth, H.G. (1930). The feeding of ammocoetes. *Nature* 126: 94 – 95.

Nichols, K.C. (1975). Observations on little-known flatworms: *Udonella*. *International Journal for Parasitology* 5: 475 – 482.

Noble, E.R. & Noble, G.A. (1971). *Parasitology. The Biology of Animal Parasites*. Third edition, Lea & Febiger, Philadelphia.

Oliver, G. (1968). Recherches sur les Diplectanidae (Monogenea) parasites de téléostéens du Golfe du Lion. I. - Diplectaninae Monticelli, 1903. *Vie et Milieu*, Série A: Biologie Marine 19: 95 – 138.

Oliver, G. (1980). Description de *Pseudodiplectanum kearni* n. sp. (Monogenea, Diplectanidae), des côtes européennes, parasite d'un poisson pleuronectiforme. *Bulletin du Muséum nationale d'Histoire naturelle* 2: 691 – 695.

Olivier, P.A.S., Dippenaar, S.M., Khalil, L.F. & Mokgalong, N.M. (2000). Observations on a lesser-known monogenean, *Udonella myliobati*, from a copepod parasite, *Lepeophtheirus natalensis*, parasitizing the spotted ragged-tooth shark, *Carcharias taurus*, from South African waters. *Onderstepoort Journal of Veterinary Research* 67: 135 – 140.

Olson, K.R. (2000a). Gross functional anatomy. Respiratory system. In: *The Laboratory Fish*. Ostrander, G.K. (ed.), pp. 151 – 159, Academic Press, San Diego.

Olson, K.R. (2000b). Microscopic functional anatomy. Respiratory system. In: *The Laboratory Fish*. Ostrander, G.K. (ed.), pp. 357 – 367, Academic Press, San Diego.

Olson, P.D. & Littlewood, D.T.J. (2002). Phylogenetics of the Monogenea – evidence from a medley of molecules. *International Journal for Parasitology* 32: 233 – 244.

Owen, I.L. (1963). The attachment of the monogenean *Diplozoon paradoxum* to the gills of *Rutilus rutilus* L. II. Structure and mechanism of the adhesive apparatus. *Parasitology* 53: 463 – 468.

Owen, I.L. (1970). The oncomiracidium of the monogenean *Discocotyle sagittata*. *Parasitology* 61: 279 – 292.

Paling, J.E. (1966). The attachment of the monogenean *Diplectanum aequans* (Wagener) Diesing to the gills of *Morone labrax* L. *Parasitology* 56: 493 – 503.

Paling, J.E. (1968). A method of estimating the relative volumes of water flowing over the different gills of freshwater fish. *Journal of Experimental Biology* 48: 533 – 544.

Paling, J.E. (1969). The manner of infection of trout gills by the monogenean parasite *Discocotyle sagittata*. *Journal of Zoology* 159: 293 – 309.

Parker, R.R., Kabata, Z., Margolis, L. & Dean, M.D. (1968). A review and description of *Caligus curtus* Müller, 1785 (Caligidae: Copepoda), type species of its genus. *Journal of the Fisheries Research Board of Canada* 25: 1923 – 1969.

Parker, T.J. & Haswell, W.A. (1961). *A Text-Book of Zoology*. Sixth edition, vol. 1, Macmillan, London.

REFERENCES

Pascoe, P.L. (1987). Monogenean parasites of deep-sea fishes from the Rockall Trough (N.E. Atlantic) including a new species. *Journal of the Marine Biological Association of the United Kingdom* 67: 603 – 622.

Pavlovskii, E.N. (Ed.) (1962). *Key to Parasites of Freshwater Fish of the USSR. Key to Fauna of the USSR.* No. 80, Izdatel'stvo Akademii Nauk SSSR, Moscow, Leningrad. English translation by Israel Program for Scientific Translations, Jerusalem, 1964.

Pearse, V., Pearse, J., Buchsbaum, M. & Buchsbaum, R. (1987). *Living Invertebrates.* Blackwell Scientific Publications, Palo Alto, California.

Pekkarinen, M. & Valovirta, I. (1996). Anatomy of the glochidia of the freshwater pearl mussel, *Margaritifera margaritifera* (L.). *Archiv für Hydrobiologie* 137: 411 – 423.

Peleteiro, M.C. & Richards, R.H. (1988). Immunocytochemical studies on immunoglobulin-containing cells in the epidermis of the rainbow trout *Salmo gairdneri* Richardson: influence of bath vaccination. *Journal of Fish Biology* 32: 845 – 858.

Pfeiffer, W. & Fletcher, T.F. (1964). Club cells and granular cells in the skin of lamprey. *Journal of the Fisheries Research Board of Canada*, 21: 1083 –1088.

Pickering, A.D., Strong, A.J. & Pollard, J. (1985). Differences in the susceptibility of brown trout, *Salmo trutta* L., and American brook trout, *Salvelinus fontinalis* (Mitchill), to infestation by the peritrich ciliate, *Scyphidia* sp. *Journal of Fish Biology* 26: 201 – 208.

Pike, A.W., Mackenzie, K. & Rowand, A. (1993). Ultrastructure of the frontal filament in chalimus larvae of *Caligus elongatus* and *Lepeophtheirus salmonis* from Atlantic salmon, *Salmo salar*. In: *Pathogens of Wild and Farmed Fish: Sea Lice.* Boxshall, G.A. & Defaye, D. (eds.), pp. 99 – 113, Ellis Horwood, London.

Pike, A.W. & Wadsworth, S.L. (1999). Sealice on salmonids: their biology and control. *Advances in Parasitology* 44: 233 – 337.

Pinder, A.C. & Gozlan, R.E. (2003). Sunbleak and topmouth gudgeon – two new additions to Britain's freshwater fishes. *British Wildlife* 15: 77 – 83.

Pottinger, T.G., Pickering, A.D. & Blackstock, N. (1984). Ectoparasite induced changes in epidermal mucification of the brown trout, *Salmo trutta* L. *Journal of Fish Biology*, 25: 123 – 128.

Potts, G.W. (1973). Cleaning symbiosis among British fish with special reference to *Crenilabrus melops* (Labridae). *Journal of the Marine Biological Association of the United Kingdom* 53: 1 – 10.

Pough, F.H., Heiser, J.B. & McFarland, W.N. (1990). *Vertebrate Life.* Third edition, Macmillan Publishing Company, New York.

Poulin, R. (1998). *Evolutionary Ecology of Parasites*. Chapman & Hall, London.

Poulin, R. (1999). Parasitism and shoal size in juvenile sticklebacks: conflicting selection pressures from different ectoparasites? *Ethology* 105: 959 – 968.

Poulin, R., Curtis, M.A. & Rau, M.E. (1990). Responses of the fish ectoparasite *Salmincola edwardsii* (Copepoda) to stimulation, and their implication for host-finding. *Parasitology* 100: 417 – 421.

Poulin, R. & FitzGerald, G.J. (1989a). A possible explanation for the aggregated distribution of *Argulus canadensis* Wilson, 1916 (Crustacea: Branchiura) on juvenile sticklebacks (Gasterosteidae). *Journal of Parasitology* 75: 58 – 60.

Poulin, R. & FitzGerald, G.J. (1989b). Risk of parasitism and microhabitat selection in juvenile sticklebacks. *Canadian Journal of Zoology* 67: 14 – 18.

Poulin, R. & FitzGerald, G.J. (1989c). Shoaling as an anti-ectoparasite mechanism in juvenile sticklebacks (*Gasterosteus* spp.). *Behavioural* Ecology *and Sociobiology* 24: 251 – 255.

Poulin, R. & FitzGerald, G.J. (1989d). Male-biased sex ratio in *Argulus canadensis* Wilson, 1916 (Crustacea: Branchiura) ectoparasitic on sticklebacks. *Canadian Journal of Zoology* 67: 2078 – 2080.

Quignard, J.-P. (1968). Rapport entre la présence d'une "gibbosité frontale" chez les Labridae (Poissons, Téléostéens) et le parasite *Leposphilus labrei* Hesse, 1866 (Copépode Philichthyidae). *Annales de Parasitologie* 43: 51 – 57.

Randall, D.J. (1972). Respiration. In: *The Biology of Lampreys*. Vol. 2, Hardisty, M.W. & Potter, I.C. (eds.), pp. 287 – 306, Academic Press, New York.

Rasheed, A.R.A.M. (1983). *Diplozoon homoion* Bychowsky & Nagibina, 1959 on the roach *Rutilus rutilus* (L.). of Llyn Tegid, Wales. *Parasitology* 87: xxx.

Rasheed, A.R.A.M. (1984). *Diplozoon homoion* Bychowsky and Nagibina, 1959 and *D. paradoxum* Nordmann, 1832 on the gills of cyprinid fishes in the British Isles. *Parasitology* 89: lxxiii.

Rauther, M. (1937). Kiemen der Anamnier. Fische. In: *Handbuch der Vergleichende Anatomie der Wirbeltiere, de Bolk, etc.* Vol. 3, pp. 224 – 251, Berlin and Vienna.

Reuling, F.H. (1919). Acquired immunity to an animal parasite. *Journal of Infectious Diseases* 24: 337 – 346.

Reynolds, T.E. (1931). Hydrostatics of the suctorial mouth of the lamprey. *University of California Publications in Zoology* 37: 15 – 34.

Richards, G.R. & Chubb, J.C. (1996). Host response to initial and challenge infections, following treatment, of *Gyrodactylus bullatarudis* and *G. turnbulli* (Monogenea) on the guppy (*Poecilia reticulata*). *Parasitology Research* 82: 242 – 247.

Riley, J., Banaja, A.A. & James, J.L. (1978). The phylogenetic relationships of the Pentastomida: the case for their inclusion within the Crustacea. *International Journal for Parasitology* 8: 245 – 254.

Ritchie, G., Mordue, A.J., Pike, A.W. & Rae, G.H. (1993). The reproductive output of *Lepeophtheirus salmonis* adult females in relation to seasonal variability of temperature and photoperiod. In: *Pathogens of Wild and Farmed Fish: Sea Lice.* Boxshall, G.A. & Defaye, D. (eds.), pp. 153 – 165, Ellis Horwood, London.

Roberts, L.S. & Janovy, J. Jr. (2000). *Foundations of Parasitology.* Sixth edition, McGraw-Hill, Boston.

Robertson, D. (1875). On *Petromyzon fluviatilis*, and its mode of preying on *Coregonus clupeoides*. *Proceedings of the Natural History Society of Glasgow* 2: 61 – 62.

Robertson, D.A. (1985). A review of *Ichthyobodo necator* (Henneguy, 1883) an important and damaging fish parasite. In: *Recent Advances in Aquaculture.* Muir, J.F. & Roberts, R.J. (eds.), vol. 2, pp. 1 – 30, Croom Helm, London.

Robertson, D.A., Roberts, R.J. & Bullock, A.M. (1981). Pathogenesis and autoradiographic studies of the epidermis of salmonids infested with *Ichtyobodo necator* (Henneguy, 1883). *Journal of Fish Diseases* 4: 113 – 125.

Robertson, M. (1907). Studies on a trypanosome found in the alimentary canal of *Pontobdella muricata*. *Proceedings of the Royal Physical Society of Edinburgh* 17: 83 – 108.

Robertson, M. (1912). Transmission of flagellates living in the blood of certain freshwater fishes. *Philosophical Transactions of the Royal Society* 202: 29 – 50.

Robertson, M. (1927). Notes on certain points in the cytology of *Trypanosoma raiae* and *Bodo caudatus*. *Parasitology* 19: 375 – 393.

Robins, C.H. & Robins, C.R. (1989). Family Synaphobranchidae. In: *Fishes of the Western North Atlantic.* Part 9, volume 1, *Orders Anguilliformes and Saccopharyngiformes.* Böhlke, E.B. (ed.). Memoir of the Sears Foundation for Marine Research, no. 1, Yale University, New Haven.

Rogers, C.L. & Dimock, R.V. Jr. (2003). Acquired resistance of bluegill sunfish *Lepomis macrochirus* to glochidia larvae of the freshwater mussel *Utterbackia imbecillis* (Bivalvia: Unionidae) after multiple infections. *Journal of Parasitology* 89: 51 – 56.

Rohde, K. (1980). Comparative studies on microhabitat utilization by ectoparasites of some marine fishes from the North Sea and Papua New Guinea. *Zoologischer Anzeiger* 204: 27 – 63.

Rohde, K. (1993). *Ecology of Marine Parasites*. Second edition, CAB International, Wallingford, U.K.

Rohde, K. (1994). The minor groups of parasitic Platyhelminthes. *Advances in Parasitology* 33: 145 – 234.

Rohde, K. & Watson, N. (1985). Morphology, microhabitats and geographical variation of *Kuhnia* spp. (Monogenea: Polyopisthocotylea). *International Journal for Parasitology* 15: 569 – 586.

Romestand, B. & Trilles, J.-P. (1976a). Production d'une substance anticoagulante par les glandes exocrines céphalothoraciques des isopodes Cymothoidae *Meinertia oestroides* (Risso, 1826) et *Anilocra physodes* (L., 1758) (Isopoda, Flabellifera, Cymothoidae). *Comptes Rendus Hebdomadaires des Séances de l'Académie des Sciences, Paris* (Sér D) 282: 663 – 665. .

Romestand, B. & Trilles, J.-P. (1976b). Au sujet d'une substance à activité antithrombinique mise en évidence dans les glandes latéro-oesophagiennes de *Meinertia oestroides* (Risso, 1826) (Isopoda, Flabellifera, Cymothoidae; parasite de poissons). *Zeitschrift für Parasitenkunde* 50: 87 – 92.

Roubal, F. R. & Bullock, A.M. (1987). Differences between the host-parasite interface of *Ichthyobodo necator* (Henneguy, 1883) on the skin and gills of salmonids. *Journal of Fish Diseases* 10: 237 – 240.

Roubal, F.R., Bullock, A.M., Robertson, D.A. & Roberts, R.J. (1987). Ultrastructural aspects of infestation by *Ichthyobodo necator* (Henneguy, 1883) on the skin and gills of the salmonids *Salmo salar* L. and *Salmo gairdneri* Richardson. *Journal of Fish Diseases* 10: 181 – 192.

Roubal, F.R. & Quartararo, N. (1992). Observations on the pigmentation of the monogeneans, *Anoplodiscus* spp. (Family Anoplodiscidae) in different microhabitats on their sparid teleost hosts. *International Journal for Parasitology* 22: 459 – 464.

Rousset, V., Raibaut, A., Manier, J.-F. & Coste, F. (1978). Reproduction et sexualité des copépodes parasites de poissons. I. L'appareil reproducteur de *Chondracanthus angustatus* Heller, 1865: anatomie, histologie et spermiogenèse. *Zeitschrift für Parasitenkunde* 55: 73 – 89.

Rubio-Godoy, M., Sigh, J., Buchmann, K. & Tinsley, R.C. (2003). Immunization of rainbow trout *Oncorhynchus mykiss* against *Discocotyle sagittata* (Monogenea). *Diseases of Aquatic Organisms* 55: 23 – 30.

Rubio-Godoy, M. & Tinsley, R.C. (2002). Trickle and single infection with *Discocotyle sagittata* (Monogenea: Polyopisthocotylea): effect of exposure mode on parasite abundance and development. *Folia Parasitologica* 49: 269 – 278.

Rushton-Mellor, S.K. (1992). Discovery of the fish louse, *Argulus japonicus* Thiele (Crustacea: Branchiura), in Britain. *Aquaculture and Fisheries Management* 23: 269 – 271.

Rushton-Mellor, S.K. & Boxshall, G.A. (1994). The developmental sequence of *Argulus foliaceus* (Crustacea: Branchiura). *Journal of Natural History* 28: 763 – 785.

Ruszkowski, J.S. (1931). Sur la découverte d'un ectoparasite *Amphibdella torpedinis* dans le coeur des torpilles. *Pubblicazioni della Stazione Zoologica di Napoli* 11: 161 – 167.

Sawyer, P.J. (1959). Burrowing activities of the larval lampreys. *Copeia* no. 3: 256 – 257.

Sawyer, R.T. (1986a). *Leech Biology and Behaviour*. Vol. I. *Anatomy, Physiology, and Behaviour*. Clarendon Press, Oxford.

Sawyer, R.T. (1986b). *Leech Biology and Behaviour*. Vol. II. *Feeding Biology, Ecology and Systematics*. Clarendon Press, Oxford.

Sayer, M.D.J., Gibson, R.N. & Atkinson, R.J.A. (1996). Seasonal, sexual and geographical variation in the biology of goldsinny, corkwing and rock cook on the west coast of Scotland. In: *Wrasse: Biology and Use in Aquaculture*. Sayer, M.D.J. *et al*. (eds.), pp. 13 – 46, Fishing News Books, Oxford.

Schmidt, G.D. (1992). *Essentials of Parasitology*. Fifth edition, W.C. Brown, Dubuque, USA.

Schmitt, W.L. (1965). *Crustaceans*. Ann Arbor Science Library.

Schram, T.A. (1979). The life history of the eye-maggot of the sprat, *Lernaeenicus spratttae* (Sowerby) (Copepoda, Lernaeoceridae). *Sarsia* 64: 279 – 316.

Schram, T.A. (1993). Supplementary descriptions of the developmental stages of *Lepeophtheirus salmonis* (Krøyer, 1837) (Copepoda: Caligidae). In: *Pathogens of Wild and Farmed Fish: Sea Lice*. Boxshall, G.A. & Defaye, D. (eds.), pp. 30 – 47, Ellis Horwood, London.

Schram, T.A. (2000). The egg string attachment mechanism in salmon lice *Lepeophtheirus salmonis* (Copepoda: Caligidae). *Contributions to Zoology* 69: 21 – 29.

Schram, T.A. & Anstensrud, M. (1985). *Lernaeenicus sprattae* (Sowerby) larvae in the Oslofjord plankton and some laboratory experiments with the nauplius and copepodid (Copepoda, Pennellidae). *Sarsia* 70: 127 – 134.

Scott, A. (1901). Lepeophtheirus *and* Lernaea. Liverpool Marine Biology Committee Memoir no. VI, Williams and Norgate, London.

Scott, T. (1901). Notes on some parasites of fishes. *Report of the Fishery Board for Scotland* 19, part 3: 120 – 153.

Scott, T. & Scott, A. (1913). *The British Parasitic Copepoda.* Vol. I. *Text.* Ray Society, London.

Secombes, C.J. (1996). The non-specific immune system: cellular defenses. In: *The Fish Immune System: Organism, Pathogen, and Environment.* Iwama, G. & Nakanishi, T. (eds.), pp. 63 – 103, Academic Press, London.

Shafir, A. & van As, J.G.. (1986). Laying, development and hatching of eggs of the fish ectoparasite *Argulus japonicus* (Crustacea: Branchiura). *Journal of Zoology* 210: 401 – 414.

Shariff, M. & Roberts, R.J. (1989). The experimental histopathology of *Lernaea polymorpha* Yu, 1938 infection in naïve *Aristichthys nobilis* (Richardson) and a comparison with the lesion in naturally infected clinically resistant fish. *Journal of Fish Diseases* 12: 405 – 414.

Shephard, K.L. (1994). Functions for fish mucus. *Reviews in Fish Biology and Fisheries* 4: 401 – 429.

Shields, R.J. & Goode, R.P. (1978). Host rejection of *Lernaea cyprinacea* L. (Copepoda). *Crustaceana* 35: 301 – 307.

Shimura, S. & Inoue, K. (1984). Toxic effects of extract from the mouth-parts of *Argulus coregoni* Thorell (Crustacea: Branchiura). *Bulletin of the Japanese Society of Scientific Fisheries* 50: 729.

Shinn, A.P., Bron, J.E., Sommerville, C. & Gibson, D.I. (2003). Comments on the mechanism of attachment of the monogenean genus *Gyrodactylus*. *Invertebrate Biology* 122: 1 – 11.

Shinn, A.P., Gibson, D.I. & Sommerville, C. (2001). Morphometric discrimination of *Gyrodactylus salaris* Malmberg (Monogenea) from species of *Gyrodactylus* parasitising British salmonids using novel parameters. *Journal of Fish Diseases* 24: 83 – 97.

Shinn, A.P., Sommerville, C. & Gibson, D.I. (1995). Distribution and characterization of species of *Gyrodactylus* Nordmann, 1832 (Monogenea) parasitizing salmonids in the UK, and their discrimination from *G. salaris* Malmberg, 1957. *Journal of Natural History* 29: 1383 – 1402.

Shinn, A.P., Sommerville, C. & Gibson, D.I. (1998). The application of chaetotaxy in the discrimination of *Gyrodactylus salaris* Malmberg, 1957 (Gyrodactylidae: Monogenea) from species of the genus parasitising British salmonids. *International Journal for Parasitology* 28: 805 – 814.

Siddall, M.E., Apakupakul, K., Burreson, E.M., Coates, K.A., Erséus, C., Gelder, S.R., Källersjö, M. & Trapido-Rosenthal, H. (2001). Validating Livanow: molecular data agree that leeches, branchiobdellidans, and *Acanthobdella peledina* form a monophyletic group of oligochaetes. *Molecular Phylogenetics and Evolution* 21: 346 – 351.

Skinner, A., Young, M. & Hastie, L. (2003). *Ecology of the freshwater pearl mussel.* Conserving Natura 2000 Rivers Ecology Series no. 2, English Nature, Peterborough.

Slijper, E.J. (1979). *Whales.* Second edition, Hutchinson, London.

Slinn, D.J. (1970). An infestation of adult *Lernaeocera* (Copepoda) on wild sole, *Solea solea*, kept under hatchery conditions. *Journal of the Marine Biological Association of the United Kingdom* 50: 787 – 800.

Smith, B.R. (1971). Sea lampreys in the Great Lakes of North America. In: *The Biology of Lampreys.* Vol. 1, Hardisty, M.W. & Potter, I.C. (eds.), pp. 207 – 247, Academic Press, London.

Smith, J.W. (1969). The distribution of one monogenean and two copepod parasites of whiting, *Merlangius merlangus* (L.), caught in British waters. *Nytt Magasin for Zoologi* 17: 57 – 63.

Sproston, N.G. (1942). The developmental stages of *Lernaeocera branchialis* (Linn.). *Journal of the Marine Biological Association of the United Kingdom* 25: 441 – 466.

Sproston, N.G. (1945). The genus *Kuhnia* n.g. (Trematoda: Monogenea). Examination of the value of specific characters including factors of relative growth. *Parasitology* 37: 176 – 190.

Sproston, N.G. (1946). A synopsis of the monogenetic trematodes. *Transactions of the Zoological Society of London* 25: 185 – 600.

Srivastava, L.P. & James, B.L. (1967). The morphology and occurrence of *Gyrodactylus medius* Kathariner, 1894 (Monogenoidea) from *Onos mustelus* (L.). *Journal of Natural History* 1: 481 – 489.

Stammer, J. (1959). Beiträge zur Morphologie, Biologie und Bekämpfung der Karpfenläuse. *Zeitschrift für Parasitenkunde* 19: 135 – 208.

Sterba, G. (1962). Die Neunaugen (Petromyzonidae). In: *Handbuch der Binnenfischerei Mitteleuropas.* Demoll, R. & Maier, H.N. (eds.), pp. 263 – 352, E. Schweizerbart, Stuttgart.

Sterud, E., Harris, P.D. & Bakke, T.A. (1998). The influence of *Gyrodactylus salaris* Malmberg, 1957 (Monogenea) on the epidermis of Atlantic salmon, *Salmo salar* L., and brook trout, *Salvelinus fontinalis* (Mitchill): experimental studies. *Journal of Fish Diseases* 21: 257 – 263.

Stoll, C. (1962). Cycle évolutif de *Paragnathia formica* (Hesse) (Isopode – Gnathiidae). *Cahiers de Biologie Marine* 3: 401 – 416.

Strahan, R. (1963). The behaviour of myxinoids. *Acta Zoologica* 24: 73 – 102.

Surface, H.A. (1898). The lampreys of central New York. *Bulletin of the United States Fisheries Commission* 17: 209 – 215.

Sutherland, D.R. & Wittrock, D.D. (1986). Surface topography of the branchiuran *Argulus appendiculosus* Wilson, 1907 as revealed by scanning electron microscopy. *Parasitology Research* 72: 405 – 415.

Swanepoel, J.H. & Avenant-Oldewage, A. (1992). Comments on the morphology of the pre-oral spine in *Argulus* (Crustacea: Branchiura). *Journal of Morphology* 212: 155 – 162.

Talhelm, D.R. & Bishop, R.C. (1980). Benefits and costs of sea lamprey (*Petromyzon marinus*) control in the Great Lakes: some preliminary results. *Canadian Journal of Fisheries and Aquatic Science* 37: 2169 – 2174.

Thomas, J.D. (1962). The food and growth of brown trout (*Salmo trutta* L.) and its feeding relationships with the salmon parr (*Salmo salar* L.) and the eel (*Anguilla anguilla* (L.)) in the River Teify, west Wales. *Journal of Animal Ecology* 31: 175 – 205.

Timofeeva, T.A. (1977). [The distribution of *Udonella caligorum* Johnston in the waters of East Murman]. In: *Biology of the Northern Waters of the European Part of the USSR*. Apatity, Publishing House of the Kol'skiy Branch of the Academy of Sciences of the USSR, pp. 54 – 59. (In Russian).

Tinsley, M.C. & Reilly, S.D. (2002). Reproductive ecology of the saltmarsh-dwelling marine ectoparasite *Paragnathia formica* (Crustacea: Isopoda). *Journal of the Marine Biological Association of the United Kingdom* 82: 79 – 84.

Tirard, C., Berrebi, P., Raibaut, A. & Frenaud, F. (1992). Parasites as biological markers: evolutionary relationships in the heterospecific combination of helminths (monogeneans) and teleosts (Gadidae). *Biological Journal of the Linnean Society* 47: 173 – 182.

Todal, J.A., Karlsbakk, E., Isaksen, T.E., Plarre, H., Urawa, S., Mouton, A., Hoel, E., Koren, C.W.R. & Nylund, A. (2004). *Ichthyobodo necator* (Kinetoplastida) – a complex of sibling species. *Diseases of Aquatic Organisms* 58: 9 – 16.

Tripathi, Y.R. (1948). A new species of ciliate, *Trichodina branchicola*, from some fishes at Plymouth. *Journal of the Marine Biological Association of the United Kingdom* 27: 440 – 450.

Tsai, J. (1996). Cell renewal in the epidermis of the loach *Misgurnus anguillicaudatus* (Cypriniformes). *Journal of Zoology* 239: 591 – 599.

Tully, O. & Nolan, D.T. (2002). A review of the population biology and host-parasite interactions of the sea louse *Lepeophtheirus salmonis* (Copepoda: Caligidae). *Parasitology* 124: S165 – S182.

Turner, W. & Wilson, H.S. (1862). On the structure of the *Chondracanthus Lophii*, with observations on its larval form. *Transactions of the Royal Society of Edinburgh* 23: 1861 – 1864.

Upton, N.P.D. (1987a). Asynchronous male and female life cycles in the sexually dimorphic, harem-forming isopod *Paragnathia formica* (Crustacea: Isopoda). *Journal of Zoology* 212: 677 – 690.

Upton, N.P.D. (1987b). Gregarious larval settlement within a restricted intertidal zone and sex differences in subsequent mortality in the polygynous saltmarsh isopod *Paragnathia formica* (Crustacea: Isopoda). *Journal of the Marine Biological Association of the United Kingdom* 67: 663 – 678.

Valtonen, E.T., Holmes, J.C. & Koskivaara, M. (1997). Eutrophication, pollution and fragmentation: effects on parasite communities in roach (*Rutilus rutilus*) and perch (*Perca fluviatilis*) in four lakes in central Finland. *Canadian Journal of Fisheries and Aquatic Sciences* 54: 572 – 585.

Van As, J.G. & Basson, L. (1987). Host specificity of trichodinid ectoparasites of freshwater fish. *Parasitology Today* 3: 88 – 90.

Van Beneden, P.-J. (1856). Note sur un trématode nouveau du maigre d'Europe. *Bulletin de l'Académie Royale de Belgique. Classe des Sciences* 23: 502 – 508.

Van Beneden, P.-J. (1858). *Mémoire sur les Vers Intestinaux*. Baillière et fils, Paris.

Van Beneden, P.-J. & Hesse, E. (1864). Recherches sur les bdellodes (hirudinées) et les trématodes marins. *Mémoires de l'Académie Royale des Sciences, des lettres et des beaux-arts de Belgique* 34: 1 – 142.

Van Damme, P.A., Ollevier, F. & Hamerlynck, O. (1994). Pathogenicity of *Lernaeocera lusci* and *L. branchialis* in bib and whiting in the North Sea. *Diseases of Aquatic Organisms* 19: 61 – 65.

Van der Lande, V.M. (1968). Esterase activity in certain glands of leeches (Annelida: Hirudinea). *Comparative Biochemistry & Physiology* 25: 447 – 456.

Van Duijn, C. (1956). *Diseases of Fishes*. Water Life, London.

Wächtler, K., Dreher-Mansur, M.C. & Richter, T. (2001). Larval types and early postlarval biology in naiads. In: *Ecology and Evolution of the Freshwater Mussels Unionoida*. Ecological Studies, vol. 145, Bauer, G. & Wächtler, K. (eds.), pp. 93 – 125, Springer-Verlag, Berlin, Heidelberg.

Walker, K.F., Byrne, M., Hickey, C.W. & Roper, D.S. (2001). Freshwater mussels (Hyriidae) of Australia. In: *Ecology and Evolution of the Freshwater Mussels Unionoida*. Ecological Studies, vol. 145, Bauer, G. & Wächtler, K. (eds.), pp. 5 – 31, Springer-Verlag, Berlin, Heidelberg.

Walkey, M., Lewis, D.B. & Dartnall, H.J.G. (1970). Observations on the host-parasite relations of *Thersitina gasterostei* (Crustacea: Copepoda). *Journal of Zoology* 162: 371 – 381.

Watters, G.T. (1997). A synthesis and review of the expanding range of the Asian freshwater mussel *Anodonta woodiana* (Lea, 1834) (Bivalvia: Unionidae). *The Veliger* 40: 152 – 156.

Wendelaar Bonga, S.E. (1997). The stress response in fish. *Physiological Reviews* 77: 591 – 625.

Wheeler, A. (1969). *The Fishes of the British Isles and North-West Europe*, Macmillan, London.

Wheeler, A. (1977). The origin and distribution of the freshwater fishes of the British Isles. *Journal of Biogeography* 4: 1 – 24.

Wheeler, A. (1978). *Key to the Fishes of Northern Europe*, Frederick Warne, London.

Whitfield, P.J., Pilcher, M.W., Grant, H.J. & Riley, J. (1988). Experimental studies on the development of *Lernaeocera branchialis* (Copepoda: Pennellidae): population processes from egg production to maturation on the flatfish host. *Hydrobiologia* 167/168: 579 – 586.

Whittington, I.D. (1987a). Hatching in two monogenean parasites from the common dogfish (*Scyliorhinus canicula*): the polyopisthocotylean gill parasite, *Hexabothrium appendiculatum* and the microbothriid skin parasite, *Leptocotyle minor*. *Journal of the Marine Biological Association of the United Kingdom* 67: 729 – 756.

Whittington, I.D. (1987b). Studies on the behaviour of the oncomiracidia of the monogenean parasites *Hexabothrium appendiculatum* and *Leptocotyle minor* from the common dogfish, *Scyliorhinus canicula*. *Journal of the Marine Biological Association of the United Kingdom* 67: 773 – 784.

Whittington, I.D. & Ernst, I. (2002). Migration, site-specificity and development of *Benedenia lutjani* (Monogenea: Capsalidae) on the surface of its host, *Lutjanus carponotatus* (Pisces: Lutjanidae). *Parasitology* 124: 423 – 434.

Whittington, I.D. & Horton, M.A. (1996). A revision of *Neobenedenia* Yamaguti 1963 (Monogenea: Capsalidae) including a redescription of *N. melleni* (MacCallum, 1927) Yamaguti, 1963. *Journal of Natural History* 30: 1113 – 1156.

Whittington I.D. & Kearn, G.C. (1988). Rapid hatching of mechanically-disturbed eggs of the monogenean gill parasite *Diclidophora luscae*, with observations on sedimentation of egg bundles. *International Journal for Parasitology* 18: 847 – 852.

Whittington, I.D. & Kearn, G.C. (1989). Rapid hatching induced by light intensity reduction in the polyopisthocotylean monogenean *Plectanocotyle gurnardi* from the gills of gurnards (Triglidae), with observations on the anatomy and behaviour of the oncomiracidium. *Journal of the Marine Biological Association of the United Kingdom* 69: 609 – 624.

Whittington, I.D. & Kearn, G.C. (1990). A comparative study of the anatomy and behaviour of the oncomiracidia of the related monogenean gill parasites *Kuhnia scombri*, *Kuhnia sprostonae* and *Grubea* *cochlear* from the mackerel, *Scomber scombrus*. *Journal of the Marine Biological Association of the United Kingdom* 70: 21 – 32.

Williamson, H. (1935). *Salar the Salmon*. Faber and Faber Limited, London.

Willing, M. (1997). Fresh- and brackish-water molluscs: some current conservation issues. *British Wildlife* 8: 151 – 159.

Wilson, C.B. (1902). North American parasitic copepods of the family Argulidae, with a bibliography of the group and a systematic review of all known species. *Proceedings of the United States National Museum* 25: 635 – 742.

Wilson, C.B. (1914). Copepod parasites of fresh-water fishes and their economic relations to mussel glochidia. *Bulletin of the U.S. Bureau of Fisheries* 34: 333 – 374.

Wilson, C.B. (1917). The economic relations, anatomy, and life history of the genus *Lernaea*. *Bulletin of the U.S. Bureau of Fisheries* 35, document 854: 165 – 198.

Winch, J. (1983). The biology of *Atrispinum labracis* n. comb. (Monogenea) on the gills of the bass, *Dicentrarchus labrax*. *Journal of the Marine Biological Association of the United Kingdom* 63: 915 – 927.

Wingstrand, K.G. (1972). Comparative spermatology of a pentastomid, *Raillietiella hemidactyli*, and a branchiuran crustacean, *Argulus foliaceus*, with a discussion of pentastomid relationships. *Kongelige Danske Videnskabernes Selskab Biologiske Skrifter* 19: 1 – 72.

Wiskin, M. (1970). The oncomiracidium and post-oncomiracidial development of the hexabothriid monogenean *Rajonchocotyle emarginata*. *Parasitology* 60: 457 – 479.

Wood, E.M. (1974a). Development and morphology of the glochidium larva of *Anodonta cygnea* (Mollusca: Bivalvia). *Journal of Zoology* 173: 1 – 13.

Wood, E.M. (1974b). Some mechanisms involved in host recognition and attachment of the glochidium larva of *Anodonta cygnea* (Mollusca: Bivalvia). *Journal of Zoology* 173: 15 – 30.

Wootten, R. & Smith, J.W. (1980). Studies on the parasite fauna of juvenile Atlantic salmon, *Salmo salar* L., cultured in fresh water in eastern Scotland. *Zeitschrift für Parasitenkunde* 63: 221 – 231.

Wootten, R., Smith, J.W. & Needham, E.A. (1982). Aspects of the biology of the parasitic copepods *Lepeophtheirus salmonis* and *Caligus elongatus* on farmed salmonids, and their treatment. *Proceedings of the Royal Society of Edinburgh* 81B: 185 – 197.

Wootton, R. J. (1976). *The Biology of the Sticklebacks*. Academic Press, New York.

Yano, T. (1996). The nonspecific immune system: humoral defense. In: *The Fish Immune System: Organism, Pathogen, and Environment.* Iwama, G. & Nakanishi, T. (eds.), pp. 105 – 157, Academic Press, London.

Yazdani, G.M. & Alexander, R.M. (1967). Respiratory currents of flatfish. *Nature* 213: 96 – 97.

Yeomans, W.E., Chubb, J.C. & Sweeting, R.A. (1997). Use of protozoan communities for pollution monitoring. *Parassitologia* 39: 201 – 212.

Young, J.Z. (1962). *Life of Vertebrates.* Second edition, Oxford University Press.

Young, M.R., Cosgrove, P.J. & Hastie, L.C. (2001). The extent of, and causes for, the decline of a highly threatened naiad: *Margaritifera margaritifera.* In: *Ecology and Evolution of the Freshwater Mussels Unionoida.* Bauer, G. & Wächtler, K. (eds.), Ecological Studies, vol. 145, pp. 337 – 357, Springer-Verlag, Berlin.

Young, M., Purser, J. & Al-Mousawi, B. (1987). Infection and successful reinfection of brown trout [*Salmo trutta* (L)] with glochidia of *Margaritifera margaritifera* (L.). *American Malacological Bulletin* 5: 125 – 128.

Young, M. & Williams, J. (1984a). The reproductive biology of the freshwater pearl mussel *Margaritifera margaritifera* (Linn.) in Scotland. II. Laboratory Studies. *Archiv für Hydrobiologie* 100: 29 – 43.

Young, M. & Williams, J. (1984b). The reproductive biology of the freshwater pearl mussel *Margaritifera margaritifera* (Linn.) in Scotland. I. Field studies. *Archiv für Hydrobiologie* 99: 405 – 422.

Zanandrea, G. (1957). Neoteny in a lamprey. *Nature* 179: 925 – 926.

Ziętara, M.S. & Lumme, J. (2002). Speciation by host switch and adaptive radiation in a fish parasite genus *Gyrodactylus* (Monogenea, Gyrodactylidae). *Evolution* 56: 2445 – 2458.

INDEX OF SCIENTIFIC AND COMMON NAMES

Page references in bold italic typeface refer to location of tables.

Abramis brama, see bream
Acanthobdella peledina 138-139
Acanthochondria cornuta 232-233
   soleae 229
   species 228
Acanthochondrites species 228
Acanthocotyle elegans 62, 64-65, 348
   greeni 62, 64, 65, 348
   lobianchi 62-64, 65-66, 348, 349
Achtheres percarum 207
Acipenser sturio, see sturgeon
Acusicola spinuloderma 223
Aeromonas hydrophila 143
Afrolernaea 212
Albionella globosa ***206***
Alburnus alburnus, see bleak
Alella pagelli 204, ***206***, 341
Alepocephalus bairdii, see Baird's smooth-head
Alloteuthis 70
Ambiphrya 27
Amphibdella flavolineata ***84***, 93-96, 344
Amphibdelloides maccallumi ***84***, 86, 87-89, 90-91, 98-100
Anarhichas lupus, see wolf-fish
anchor worm, see *Lernaea cyprinacea*
anchovy 188
Ancyrocephalus paradoxus ***84***
Anelasma species 265-273, 340
   squalicola 267-273, 344, 345
angler ***133***, 134
Anguilla anguilla, see eel
Anilocra physodes 275, 277, 279-283, 284
   pomacentri 278, 283, 284
   species 277-284, 341, 342
Anodonta anatina 300, 302, 305
   cataracta 306, 307
   complanata 300
   cygnea 296, 300, 301-308, 312
   general 275, 300-301
   woodiana 304, 305
Anoplodiscus cirrusspiralis 102
Antennarius striatus, see frogfish
Anthocotyle merluccii 113-114
Archosargus probatocephalus, see sheepshead seabream
Argulus alosae 262
   appendiculatus 239
   arcassonensis 241
   coregoni 239, 240-241, 251, 258, 263
   foliaceus 237, 239, 240, 241, 242, 244-248, 251, 251-253, 257-258, 258-260, 261, 262, 263, 350
   funduli 246-247, 248-249, 261-262
   japonicus 239, 240, 241, 242-244, 247, 249-251, 253-255, 256-257, 258
   species 155, 341, 343, 345, 346
   viridis 255-256
Argyrosomus regius, see meagre
Aristichthys nobilis, see carp, bighead
Arius graeffei, see catfish, blue
Aspitrigla cuculus, see gurnard, red
Atherina boyeri, see big-scale sand smelt
Atrispinum labracis ***104***, 116, 130
Axine belones ***104***, 114, 116-119, 120

Baird's smooth-head ***104***, 111
Balistes carolinensis, see triggerfish
ballan wrasse 69, 234
barbel 4
   southern 125
Barbus barbus, see barbel
   meridionalis, see barbel, southern
Barrier Reef chromis 278, 283
basking shark 334
bass 3, ***84***, 92, ***104***
   large-mouth, see *Micropterus salmoides*
Belone belone, see garfish
Benedenia lutjani 349, 350
   sciaenae 57
bib ***104***, 124, 178, 184, 186
big-scale sand smelt 306
bitterling 5, 213, 304, 305
bleak 4, 213
blenny, Montagu's 290
   tompot ***24***
Blicca bjoerkna, see bream, silver
bluegill 317
blue whiting ***104***
Bomolochus bellones 226
   soleae 226
Branchellion borealis ***133***
   torpedinis 132, ***133***, 149
bream 3, ***24***, 25, ***85***, ***104***, 113, 116, 120, 153, 216, 217, 218, 241
   silver 4
Brumptiana lineata 132, ***133***, 140
Buglossidium luteum, see solenette
bullhead 6, ***79***, 137
bull-rout ***133***, 135
burbot 4
butterfish 132, ***133***, 348

Calceostoma sp. 83, **84**
Caligus brevicaudatus 350
   centrodonti 69
   clemensi 160
   curtus 172
   diaphanus 350
   elongatus 69, 155, 174-177
   labracis 69
   sp. 161, 162
Calliobdella lophii **133**, 134
   nodulifera **133**
   punctata **133**
Capsala martinieri 61-62
Carassius auratus, see goldfish
   carassius, see carp, Crucian
Carcharias taurus 69
carp, bighead 211, 213
   common 19-20, **24**, 25, 30, 35, 83, 84, **85**,
      92-93, 213, 216, 218, 241, 247, 262, 306,
      307 (see also carp, mirror; koi)
   Crucian, 4, 208, 212, 213, 216
   mirror 241
catfish, blue 88
   channel 32, 33, 34
   Nile 76
Centrolabrus exoletus, see rock cook
Centroscyllium nigrum, see dogfish,
   combtooth
Centroscymnus coelolepis 337
Ceratias holboelli 231, 338
Ceratothoa steindachneri 275, 279
Cetorhinus maximus, see basking shark
Charopinus dalmanni **206**
charr, Arctic 197, 205, 240
   brook 5, 27, 197
Chauhanellus australis 88
Chelidonichthys cuculus, see gurnard, red
Chelon labrosus, see mullet, thick-lipped grey
Chimaera monstrosa, see rat-fish
Chondracanthus angustatus 231-232
   lophii 231
   merluccii 229
   species 228
   zei 229, 230
Chromis nitida, see Barrier Reef chromis
Chrysophrys auratus, see snapper
chub **85**, 239
Ciliata mustela, see rockling, five-bearded
Clarias gariepinus 76
Clavella adunca **206**
   stellata **206**
Clupea harengus, see herring
Cobitis taenia, see loach, spined
cod 28, 68, 178, 184, 185, 186, **206**, 226, 228
   poor **104**, 105, 350
coelacanth 37
Coelorinchus occa, see swordsnout grenadier
Colobomatus bergyltae 235
comber 226

Concinnocotyla australis 114
Coregonus lavaretus, see powan
Coris julis, see wrasse, rainbow
Coryphoblennius galerita 290
Costia necatrix, see *Ichthyobodo necator*
Cottus gobio, see bullhead
Crenilabrus melops, see wrasse, corkwing
Ctenolabrus rupestris, see goldsinny
Cyclocotyla chrysophryi **104**, 111
Cyclops 210
Cyclopterus lumpus, see lumpsucker
Cyprinus carpio, see carp, common

dab 28, 51, 173-174, **173**, **174**
   long rough 232, 233
dace 3, **85**
Dactylogyrus amphibothrium 84, **85**, 97
   anchoratus **85**, 93, 344-345
   auriculatus 84, **85**, 97
   extensus 84, **85**, 92-93, 97, 344
   hemiamphibothrium 84, **85**
   sphyrna 97
   vastator 84
Daniconema anguillae 263
Dasyatis pastinaca, see stingray
Diagramma labiosum 349
Dicentrarchus labrax, see bass
Diclidophora denticulata **104**, 126, 128
   esmarkii **104**, 105
   luscae **104**, 114, 121, 124-125
   merlangi **104**, 110, 115, 116, 121, 127, 128, 130,
      206, 346, 347, 349
   minor **104**
   palmata **104**
   phycidis **104**, 110
   pollachii **104**
   species 108-111, 116, 121, 243
Diplectanum aequans **84**, 92
Diplodus sargus 205
Diplozoon paradoxum **104**, 113, 116, 120, 126, 129
Discocotyle sagittata **104**, 104, 113, 114, 116, 120,
   123, 126, 128
dogfish 66, 68, **104**, 122-123, **206**
   combtooth 267
dory 229, 230
Dreissena polymorpha 300

Echiichthys vipera, see weever, lesser
eel 3, 4, 77, **79**, 80, 217, 263, 313, 318, 319, 322
   snubnosed 337
Enchelyopus cimbrius, see rockling, four-bearded
Engraulis encrasicolus, see anchovy
Entelurus aequoreus 291
Entobdella diadema 39, 58-59, 60, 123
   hippoglossi 38, 58, 60, 347, 348
   soleae 39-58, 59-60, 62, 63, 65, 66, 73, 75,
      139, 146, 149, 150, 340, 342, 343, 345,
      347-348, 349
Ergasilus briani 214, 215, 216-217, 218, 220-

INDEX OF NAMES

225
gibbus 215, 217, 218
lizae 216, 217-218, 218
nanus, see *Ergasilus lizae*
sieboldi 214, 215, 216, 218, 220, 225
Ergenstrema labrosi **84**, 85, 100-102, 349
mugilis **84**, 85, 101,102
Esox lucius, see pike
Etmopterus spinax, see velvet-belly
Eutrigla gurnardus, see gurnard, grey

flounder 3, 77, 160, 161, 166, 173-174, *173*, *174*, 179-180, 181, 186, 324
starry 70
forkbeard *104*
frogfish 236

Gadus morhua, see cod
Gaidropsarus vulgaris, see rockling, three-bearded
Galeorhinus galeus, see tope
garfish 104, 116, 226
Gambusia affinis, see mosquitofish
Gasterosteus aculeatus, see stickleback, three-spined
wheatlandi, see stickleback, black-spotted
Gastrocotyle trachuri **104**, 105, 113, 117, 125, 130
Gnathia africana 290
calva 290
dentata 275
maxillaris 275-276, 288, 290, 291
oxyuraea 275-276
vorax 275-276
Gobio gobio, see gudgeon
goby, common 277, 295
leopard-spotted *79*
goldfish *24*, 153, 213, 304, 351
goldsinny 177, 291
grayling *78*, *84*, 206, 240
Grubea cochlear **104**, 105, 120, 350
gudgeon **85**, 137, 217
topmouth 351
guppy 74, 75, 76-77
gurnard, grey *104*
red *104*, 241
streaked *104*
tub *23*, 83, 349
gurnards, general 233
Gymnocephalus cernuus, see ruffe
Gyrodactyloides bychowskii 79
Gyrodactylus alviga 70
anguillae 79, 80
bullatarudis 75, 77
derjavini 75, *79*
gasterostei 73, 76, 77, 78, *79*, 80
gemini 74
longidactykus 80

micropsi 80
pungitii 71, 79, *79*, 80
rarus 73
rogatensis 78, *79*
rugiensis 79, 80
rugiensoides 80
rysavyi 76
salaris 17, 70, 75, 77, 78, 80-81, 347, 351
turnbulli 74, 75, 76
wageneri 80

haddock 185, 186, **206**
Haementeria ghilianii 132
Haemogregarina bigemina 290-291
hake **104**, 113, **206**, 229
halibut 38, 58, 59, 60, 347, 348
Hemibdella soleae **133**, 134-135, 138, 141, 149-150, 340, 344
Hemiclepsis marginata **133**, 135, 136-137, 139, 141, 143, 153
Heptacyclus myoxocephali **133**
herring 187, 191, 324
Heterobothrium okamotoi 126
Hexabothrium appendiculatum **104**, 104, 122-123, 302
Hippoglossoides platessoides, see dab, long rough
Hippoglossus hippoglossus, see halibut
hippopotamus 38, 60
Hirudo medicinalis 132, **133**, 135, 137, 141, 142-143, 151-152
Holobomolochus confusus 226, 227, 228

Ichthyobodo necator 22, 27-30, 343, 345, 347
Ichthyomyzon bdellium 337
Ichthyophthirius multifiliis 22, 30-36, 345, 347
Ictalurus punctatus, see catfish, channel
ide 213
Isancistrum 70, 71, 74, 80
Isistius brasiliensis 337-338
plutodus 337

killifish 30
koi 241, 351
Kuhnia scombri **104**, 113, 115, 116, 120
sprostonae **104**, 127, 349

Labrus bergylta, see wrasse, ballan
mixtus, see wrasse, cuckoo
lampern, see lamprey, river
Lampetra fluviatilis, see lamprey, river
japonica, see *Lethenteron camtschaticum*
planeri, see lamprey, brook
tridentata 320
zanandreai, see *Lethenteron zanandreai*
lamprey, brook 318, 319, 330, 336-337
river 3, 5, 318, 319, 320, 320-333, 334, 336, 337

sea 3, 318, 319, 326, 327, 329, 333-334, 335, 337
lampreys in North America 335
Lampsilis ligamentina 316
    luteola 316
    perovalis 306, 316
    ventricosa 316
Labroides dimidiatus, see wrasse, bluestreak cleaner
Latimeria chalumnae, see coelacanth
Lepeophtheirus myliobati 69
    pectoralis 155, 159, 160, 161, 162, 163, 164-166, 166, 167, 168, 169, 170, 172, 173-174, 221
    salmonis 155, 159, 163, 165, 166, 167, 168, 169, 170, 171, 174-177, 197, 341
Lepidion eques, see North Atlantic codling
Lepomis gibbosus, see pumpkinseed
    macrochirus, see bluegill
Leposphilus labrei 235, 235-236
Leptocotyle minor 66-68, 122, 123, 302
Lernaea cyprinacea 208-213, 340, 344, 345
    polymorpha 211, 213
Lernaeenicus encrasicoli 188
    sprattae 186, 187, 188-194, 340, 341, 342
Lernaeocera branchialis 180-186, 188, 214, 341, 344, 345
    lusci 184, 185, 186, 339
    minuta 185
    obtusa 185-186
Lernaeopoda galei **206**
Lernaeopodina longimana 196, **206**
    longibrachia 187, 196
Lernentoma asellina 233-234, 344, 345
    species 228
Lethenteron camtschaticum 337
    zanandreai 337
Leucaspius delineatus, see sunbleak
Leuciscus cephalus, see chub
    idus, see ide
    leuciscus, see dace
Ligophorus angustus 84, **84**, 100, 101
Limanda limanda, see dab
ling **104**
Limnocalanus macrurus 205
Lipophrys pholis, see shanny
Lironeca vulgaris 283-284
Liza ramada, see mullet, thin-lipped grey
loach (*Misgurnus anguillicaudatus*) 9
loach, spined 4
    stone **79**, 80, 137
Loma salmonae 176
Lophius piscatorius, see angler
Lota lota, see burbot
Lucionema balatonense 263
Lumbricus terrestris 131
lumpsucker 179, 186
lungfish 114
Lutjanus carponotatus 349

mackerel 6, **104**, 105, 113, 120, 127, 349, 350
Margaritifera auricularia 300
    margaritifera 299, 300, 308-315, 350
meagre 57, 83, **84**
Melanogrammus aeglefinus, see haddock
Merlangius merlangus, see whiting
Merluccius merluccius, see hake
Metabenedeniella parva 349
Microbothrium apiculatum 67
Microchirus variegatus, see sole, thickback
Microcotyle donavini **104**, 116
Micromesistius poutassou, see blue whiting
Micropharynx parasitica 37
Micropterus coosae 316
    punctulatus 316
    salmoides 239, 316
Microstomus kitt, see sole, lemon
minnow 24, **79**, 80, **85**, 313
Misgurnus anguillicaudatus 9
Mola mola, see sun-fish
Molnaria intestinalis 263
Molva molva, see ling
monkfish 66
mosquitofish 304, 306
mullet, red 241
    striped red 275
    thick-lipped grey 3, 84, **84**, 101, 217, 349
    thin-lipped grey 3, **84**, 85, 101
Mullus barbatus, see mullet, red
    surmuletus, see mullet, striped red
Mustelus mustelus, see smooth hound
Mutela bourguignati 299
Myoxocephalus scorpius, see bull-rout
Mysis relicta 205
    stenolepis 262

Neobenedenia melleni 346, 348
Neobrachiella merluccii, **206**
Neoceratodus forsteri, see lungfish
Neodactylogyrus crucifer **85**, 93, 97
Neoergasilus japonicus 216, 218-219, 223-224
Nerocila neapolitana 275
Noemacheilus barbatulus, see loach, stone
North Atlantic codling 235, 236
Norway pout **104**, 105
Notonecta 262
nursehound **104**

Oceanobdella blennii 132, **133**, 140, 290, 348
    microstoma **133**, 134
    sexoculata **133**
Oculotrema hippopotami 38
Ommatokoita elongata 187, **206**
Oncorhynchus kisutch, see salmon, coho
    mykiss, see trout, rainbow
    nerka, see salmon, sockeye
Orestias agassizii, see killifish

Pagellus bogaraveo, see sea-bream, red
  erythrinus, see pandora
pandora **206**
Parablennius gattorugine, see blenny, tompot
Paracyclocotyla cherbonnieri ***104***, 111-112
Paradiplozoon homoion ***104***, 113, 125
Paraergasilus rylovi 225
Paragnathia formica 277, 284, 285, 288, 290, 292-295, 341
Paratrichodina 25
pearl mussel, see *Margaritifera margaritifera*
Pegusa lascaris, see sole, sand
Perca fluviatilis, see perch
perch 3, 15, ***84***, 153, 207, 218, 245-248, 261, 305, 306
Petromyzon marinus, see lamprey, sea
Philichthys xiphiae 214, 234, 235
Pholis gunnellus, see butterfish
Phoxinus phoxinus, see minnow
Phrixocephalus cincinnatus 187
Phycis blennoides, see forkbeard
pike 3, ***79***, ***84***, 89, 96, 304, 306
pilchard 187, 193
Piscicola geometra ***133***, 135-136, 137, 139, 140, 141, 143-149, 150-151, 152, 350
Placobdella parasitica 137-138
plaice 6, 7, 18, ***24***, 27, 60, ***79***, 132, ***133***, 135, 173-174, ***173***, ***174***
Platichthys flesus, see flounder
  stellatus, see flounder, starry
Platybdella anarrichae ***133***
Plectanocotyle gurnardi ***104***, 105, 110, 112-113, 114, 116, 120-121, 123-124, 126-127, 347
Pleuronectes platessa, see plaice
Poecilia reticulata, see guppy
Pollachius pollachius, see pollack
  virens, see saithe
pollack ***104***
Polystoma integerrimum 103
Pomatoschistus lozanoi 80
  microps 79, 80, 277
  minutus 79, 80
  pictus 80
Pontobdella muricata 133, ***133***, 139-140, 153
  vosmaeri 133, ***133***
powan 318, 320, 324, 327
Pseudaxine trachuri ***104***, 105, 117, 125
Pseudocharopinus bicaudatus ***206***
Pseudocotyle squatinae 66
Pseudodactylogyrus 80
Pseudodiplectanum kearni ***84***, 92
Pseudomonas sp. 143
Pseudorasbora parva, see gudgeon, topmouth
Pteroplatytrygon violacea 39
puffer, tiger 126
pumpkinseed 5, 306
Pungitius pungitius, see stickleback, nine-spined

Raja clavata, see ray, thornback
  species 37, 51, 60, 62, ***104***, ***133***, 153, ***206***, 348
Rajonchocotyle emarginata ***104***, 105, 115, 126
Rana temporaria 103, 285
rat-fish 199
ray, electric ***84***, 85, 86, 93-96, 344
  marbled electric 95
  thornback 62, 64, 65, 66, ***104***, 126, ***206***, 348
Rhodeus sericeus, see bitterling
  sinensis 304, 305
Riboscyphidia, see scyphidiids
roach 3, ***85***, ***104***, 113, 125, 213, 216, 217, 218, 241, 245-248, 261
rock cook 134, 177, 236, 291
rockling, five-bearded ***24***, ***79***, 226, 275, 288
  four-bearded 226
  three-bearded ***24***
rudd ***85***, 216, 217, 218, 241, 262, 263, 304, 305
ruffe 4, 84, ***85***
Rutilus rutilus, see roach

saithe ***104***, 126
Sacculina 266, 271
Salmincola californiensis 196, 197, 197-198, 198-203, 341
  edwardsii 197, 198, 205
  gordoni, 206
  mattheyi 197
  salmoneus 196-197, 198, 205, 206, 349
  thymalli 206
Salmo salar, see salmon, Atlantic
Salmo trutta, see trout, brown; trout, sea
salmon, Atlantic 3, 4, 17, 26, 28, 70, 75, 77, 78, ***79***, 81, 155, 166, 168-169, 174-177, 196-197, 198, 205, 291, 309, 313, 315, 333, 334, 347, 350 (see also kelt; maiden; parr; salmon and gill maggot, life cycle; smolt)
  coho 176, 307
  sockeye 197
Salvelinus alpinus, see charr, Arctic
  fontinalis, see charr, brook
  namaycush, see trout, lake
Sanguinothus pinnarum 133, ***133***
Saprolegnia 81
Sarcotaces pacificus 236
  sp. 235, 236
Sardina pilchardus, see pilchard
scad ***104***, 105, 113, 117, 125
Scardinius erythrophthalmus, see rudd
Sciaena aquila, see meagre
Scomber scombrus, see mackerel
Scophthalmus maximus, see turbot
Scyliorhinus canicula, see dogfish
Scyliorhinus stellaris, see nursehound
Scyphidia spp., see scyphidiids
sea-bream, black ***206***
  red ***104***, 111, ***206***

sea scorpion *24*, 132, ***133***, 275, 289
Semaprochilodus taeniurus 74
Serranus cabrilla, see comber
shanny *24*, 132, ***133***, 149, 275, 288, 290, 348
sheepshead seabream 18
Simenchelys parasitica, see eel, snubnosed
Skrjabillanus scardinii 262, 263
smooth hound ***206***
snapper 102
sole, common Frontispiece, 7, 39-58, 59, 60, 62, 66, 75, ***133***, 134-135, 138, 149, 150, 173-174, ***173***, ***174***, 186, 226, 229, 340, 347, 348
    lemon 51, 186
    sand 39, 135
    Senegalese 39
    thickback ***84***, 92, 135
Solea senegalensis, see sole, Senegalese
Solea solea, see sole, common
solenette 51, 60, 135
Spinachia spinachia, see stickleback, fifteen-spined
Spondyliosoma cantharus, see sea-bream, black
sprat 187, 188-194, 324
Sprattus sprattus 187, 340
spurdog 67, ***206***
Squalonchocotyle catenulata 106
Squalus acanthias, see spurdog
Squatina squatina, see monkfish
stickleback, black-spotted 248-249
      fifteen-spined ***24***
      nine-spined ***24***, 78, ***79***, 80, 219
      three-spined 3, ***24***, 25, 73, 77, 78, ***79***, 80, ***133***, 137, 213, 219, 246-247, 248-249, 261, 305, 306
stingray 11, 39, 58-59, 60, 123
Stizostedion lucioperca, see zander
sturgeon 300
sunbleak 304, 351
sun-fish 61, 62
swan mussel, see *Anodonta cygnea*
swordfish 234
swordsnout grenadier 235, 236
Syngnathus acus 291
    typhle 291

Taeniacanthus species 226
Takifugu rubripes, see puffer, tiger
Taurulus bubalis, see sea scorpion
tench 213, 216, 217, 218, 225, 240, 306
Tetraonchus borealis ***84***
    monenteron ***84***, 86, 89-92, 96-97, 345, 347
Thersitina gasterostei 216, 219-220, 249
Thorogobius ephippiatus, see goby, leopard-spotted
Thunnus thynnus, see tunny
Thymallus thymallus, see grayling

tope ***206***
topknot 226
Tinca tinca, see tench
Torpedo marmorata, see ray, marbled electric
    nobiliana, see ray, electric
Tracheliastes polycolpus 207
Trachurus trachurus, see scad
Trichodinella epizootica 25
triggerfish 241
Trigla lucerna, see gurnard, tub
Trigloporus lastoviza, see gurnard, streaked
Tripartiella 25
Trisopterus esmarkii, see Norway pout
    luscus, see bib
    minutus, see cod, poor
Trochopus pini 83
trout, brown 4, 5, 15, ***24***, 25, 27, 75, 78, ***79***, ***104***, 113, 114, 120, 123, 126, 206, 216, 240, 307, 309, 311-313, 322, 327, 350
    lake 335
    rainbow 5, 17, ***24***, 25, ***104***, 105, 116, 123, 126, 127, 176, 206, 251, 313
    sea 155, 174, 177, 206, 309, 315, 334
Trypanoplasma 153
Trypanosoma cotti ***133***
    raiae 153
tunny 6
turbot 8, 173, ***173***, ***174***

Udonella caligorum 69-70
    myliobati 69
Unio pictorum 300
    tumidus 300
Utterbackia imbecillis 316-317

Vanbenedenia kroeyeri 199
velvet-belly 267, 270-271
Vibrio anguillarum 17

weever, lesser 275
whiting ***104***, 110, 127, 128, 178, 179-180, 184, 185, 186, ***206***, 349
wolf-fish ***133***
wrasse, ballan 69, ***104***, 278
    bluestreak cleaner 291
    corkwing 133, 177, 235, 236, 275, 278, 289, 291
    cuckoo 177
    rainbow 236

Xiphias gladius, see swordfish

zander 5, 263
Zeugopterus punctatus, see topknot
Zeus faber, see dory

SUBJECT INDEX

Page references in bold italic typeface refer to location of tables. Entries entirely in bold typeface indicate positions of life cycle diagrams in the text.

acanthobdellidans 131
acanthocotylids 62-66
acid-base balance 12
acoustico-lateralis system, see lateral line
agnathans 1, 2, 37, 70, 318-319
amines 326
ammocoete larvae 320-323, 336, 337, 339
amphibdellids 85
amphibians 213
anadromous fishes 3
anchor of *Lernaea*, see holdfast of *Lernaea*
ancyrocephalines *84*, 84-85, 100-102
Anilocra, life cycle Fig. 15.1, p. 274
antennae, first, in *Argulus* 242, 260
    in caligids 155, 160, 167, 169
    in cymothoids 278
    in free-living barnacles 265
antennae, second, in *Argulus* 242, 258
    in caligids 160, 164, 167, 170 (see also mouthparts)
    in chondracanthids 229, 231, 232, 233 (see also mouthparts)
    in cymothoids 278
    in ergasilids 215, 218, 219, 223, 225 (see also mouthparts)
    in lernaeopodids 198, 199, 201, 203
    in pennellids 181, 182, 183, 188, 189
antibodies 16, 36, 75, 347
anticoagulants 132, 142, 251, 284, 288, 320, 329, 345
antigens 16, 36, 75
Argulus spp., general morphology 238-239 (see also branchiurans)
arhynchobdellidans 132
asymmetry in polyopisthocotyleans 116-120, 346
attachment, general 341-345
    of *Acanthocotyle* 66
    of *Amphibdelloides* 87-88, 90-91
    of *Anelasma* 269
    of *Argulus* 241-244
    of bomolochids/taeniacanthids 227
    of caligids 159-160
    of chondracanthids 229
    of cymothoids 279
    of *Diclidophora* spp. 108-111
    of *Diplectanum* 92
    of *Entobdella soleae* 40-43
    of female ergasilids 223
    of glochidia 303
    of gyrodactylids 72-73
    of hexabothriids 107-108
    of *Ichthyobodo* 29
    of lampreys, see suction, in lampreys
    of microbothriids 66-67
    of *Plectanocotyle* 112-113
    of pranizas 286, 288
    of scyphidiids 27
    of *Tetraonchus* 89-92
    of *Trochopus* 83
    of udonellids 68-69
    to epidermal cells 28, 29, 72-73
autoinfection 65, 74, 167

bacteria 153 (see also symbiotic micro-organisms)
barnacles 265-273, 340 (see also *Anelasma*)
    introduction 265-267
basement membrane 10
behaviour of parasites, general 350
bivalves, general 296-297
blood plasma 14
    serum 17
blood/brain barrier 19
bomolochids 214, 226-228, 343
bony fishes, see teleosts
branchiae 132
branchiobdellidans 131
branchiurans 237-264, 340, 341, 343 (see also *Argulus*)
    introduction 237-238
breathing, in *Entobdella soleae* 47-48, 58
    in lampreys 325-326
British fish fauna, *Argulus* spp. 239-241
    caligid copepods 155
    chondracanthids 228
    dactylogyrines 83-84, *85*
    dactylogyroideans 83-85, *84*
    ergasilids 215-220
    future of 350-351
    gyrodactylids 78 –80, *79*
    isopods 275-277
    leeches 132-135, *133*
    lernaeopodids 196, 205-207
    origins 3-5
    pennellids 185-186, 187-188
    polyopisthocotyleans 103-105, **104**
    Trichodina spp. 23, *24*, 25
    unionaceans 300
British parasite fauna, future of 350-351
buccal force pump 14

bucco-intestinal canal 115
bulla 195, 201, 202, **206**, 342
burrows of gnathiids 284, 292-293, 294, 295
byssus 308

caligids 155-177, 340, 342, 343, 345 (see also *Caligus*; *Lepeophtheirus*)
   introduction 155-159
Caligus, number of species 348
camouflage in skin parasites 62, 161, 280
capsalids 61-62, 82, 83 (see also *Entobdella*)
carapace, of *Argulus* 237, 238, 239, 252, 258
   of caligids 156, 159, 160
   of free-living barnacles 265
catadromous fishes 3
catecholamines 20
catfish, general 218
cementary glands of *Anelasma* 273
cephalon 156
cephalosome 223, 229, 230, 233, 278
cephalothorax 156, 159-160, 186, 192, 195, 199, 202, 222, 223, 227, 285
chalimus, of caligids 158-159
   of lernaeopodids 200-202, 203
   of pennellids 181, 182, 184, 188, 189, 191
chelae 181, 182, 188, 195, 202, 342
chemoperception 31, 51, 76, 140, 161, 169, 184, 205, 247, 248, 303, 304, 311
chloride cells 15, 16, 29
cholinesterase 137
chondracanthids 214, 228-234, 341 (see also *Acanthochondria*; *Chondracanthus*; *Lernentoma*)
chondrichthyans 70 (see also elasmobranchs; holocephalans)
chromatophores 10, 11
circulatory system of fishes 95-96, 184, 186
cirripedes, see barnacles
clamps 106, 108-114, 116-121, 127, 128, 129, 343, 345
claspers **206**
cleaner, fishes/crustaceans 21, 57, 62, 82, 133-134, 177, 291-292
clitellum 131, 148
cloaca **206**
club cells 10
cocoon of leeches, assembly 147-149
   general 131, 145, 149-150, 151-152
coelacanths 2, 37
co-evolution of hosts and parasites 37, 103, 348 (see also phylogenetic speciation)
collagen 10, 11, 14, 213
colours in free-living ergasilid stages 224-225
commensalism 27
complement 17
control of sea lice 176-177
convergent evolution 122, 208, 227, 243
cookiecutter sharks 337

copepodid, of caligids 158, 168-170, 173-174, *173*, *174*
   of chondracanthids 232, 233
   of ergasilids 214, 221-222, 224, 225
   of *Lernaea* 210
   of lernaeopodids 197-198, 198-199, 200, 203, 204-205
   of pennellids 181, 182, 188, 190, 191
   of philichthyids 236
copepodite, see copepodid
copepods 154-236, 340
   and phoresy 68-70
   dominance 339
   introduction 154-155
copulatory organs, and speciation 348
   of *Argulus* 251-257
   of *Amphibdelloides* 98-100
   of dactylogyrines 98
   of *Ergenstrema* 100-101
coregonid fishes 6
cortisol 20
costiasis, see pathology caused by *Ichthyobodo*
cottid fishes 80, *133*
'cough' reflex 110
C-reactive protein 17
crustaceans, dominance 339
   general 344
cryptic species 28, 80, 132, 349
ctenidium 297
cuticle 9, 11
cyclopoid copepods 208-213 (see also *Lernaea*)
cyclopoid stage of *Lernaea* 210
cyclostomes 7, 10
cymothoids 112, 275, 277-284, 345 (see also *Anilocra*)
cyprinid fishes 5, 80, 84, 207, 213, 217, 258, 348, 351
cypriniform fishes 3, 5
cypris larva 265, 267
cytokines 20
cytoskeleton 22-23
cytostome 26, 29, 31, 32

dactylogyrines 82, 83-84, 92-93, 98 (see also *Dactylogyrus*; *Neodactylogyrus*)
dactylogyroideans 82, 83-102, 342
Dactylogyrus, British species 83-84, **85**
   number of species 82, 348
damage to fisheries by sea lampreys 335
Darwin, Charles 265, 267
denticles, in elasmobranch fishes 11, 67, 343
   in trichodinid protozoans 23
dermis of fishes 10-11, 73
desmosomes 9
development, of *Acanthocotyle* 66
   of *Argulus*, 258-261
   of caligids 167-173
   of chondracanthids 232-233

SUBJECT INDEX 427

of *Ergasilus briani* 220-225
of gyrodactylids 71-72
of *Lernaea* 200-211
of *Lernaeenicus* 188-193
of *Lernaeocera* 180-181
of lernaeopodids 197-202
of polyopisthocotyleans 117
of *Salmincola* 198-202
(see also metamorphosis)
diapause 125
diplozoids 103, 113, 114, 115, 119, 120, 125, 128 (see also *Diplozoon paradoxum*, *Paradiplozoon homoion*)
Diplozoon, derivation of name 129
diporpa 128, 129
discocotylids 103 (see also *Discocotyle*)
dwarf males in chondracanthids 230-232

ecdysis, see moulting
ecological transfer, see host switching
ecology of ancyrocephalines from mullets 101-102
ectoparasites 1, 18, 339
egestion in *Entobdella soleae* 46
egg assembly in *Entobdella soleae* 46-47
egg retention by cymothoids 283
egg sacs of copepods 158, 167, 178, 180, 193, 197, 205, 224, 228, 235
eggs, of *Anelasma* 272
of *Argulus* 257-258
of cymothoids 279
of lampreys 331
of leeches 145
of monogeneans 39-40, 67-68, 70, 101, 121-125
(see also egg assembly; hatching; laying; egg retention; egg sacs)
elasmobranchs 2, 5, 7, 8, 11, 16, 37, 60, 85, 107, 292 (see also sharks, rays)
endoparasites 1, 17, 93-96, 214, 234, 299, 344
endopeptidase 141, 143, 144
endostyle 320, 322, 337
Ennerdale Water 205
***Entobdella soleae*, life cycle Fig. 3.2, p. 40**
epidermis of fishes 8-10, 11, 18, 304, 307
epithelial cells of fishes, see epidermis of fishes
epizoic animals 22-27, 266, 343, 344
ergasilids 214-225, 339, 340, 342
***Ergasilus briani*, life cycle Fig. 12.1, p. 215**
euryhaline fishes 3
exopeptidase 143, 144
exotic fishes, see fishes, introduction into Britain
eye-maggots 186-194 (see also *Lernaeenicus*)
eye of fishes 186, 192-194, **206** (see also eye-maggots; *Lernaeenicus*)

Fahrenholz's rule 103, 348
feeding, general 345-346
in *Acanthocotyle* 62
in *Amphibdella* 96
in *Anelasma* 269-271
in *Argulus* 249-251, 345
in caligids 161-163
in cookiecutter sharks 337-338
in cymothoids 283, 284
in *Entobdella soleae* 43-46
in free-living ergasilid stages 224-225
in glochidia 307-308
in gnathiids 284, 288-290, 292
in *Ichthyobodo* 30
in *Ichthyophthirius* 33-34
in lampreys 319-320, 325, 326-329
in leeches 141-145, 150
in *Leptocotyle* 67
in *Lernaea* 211-212
in nauplius of *Ergasilus* 221
in parasitic ergasilid stages 225
in polyopisthocotyleans 114-116, 126
in *Salmincola* 197
in udonellids 69
on blood 96, 114-116, 125, 126, 131-132, 136, 141-145, 163, 193, 184, 186, 196, 197, 212, 225, 276, 283, 284, 320, 345-346
on epidermis 30, 37, 43-46, 62, 67, 69, 73, 78, 81, 163, 170, 171, 176, 197, 225, 346
fibroblasts 19
filament, frontal 157, 160, 170-173, 181, 182, 189, 195, 199-202, 342
filter-feeding, in adult unionaceans 297-298, 309
in ammocoetes 322
fins 5-6, 7, 54, 123
and *Anelasma* 267
and *Argulus* 240, 246
and caligids 161, 163, 166, 167, 168, 169, 172
and ergasilids 218, 223, 224
and isopods 283, 285
and leeches **133**, 133, 141, 149
and *Lernaeenicus* 187, 188, 192
and lernaeopodids 195, 197, 199, 205, 207
and monogeneans **79**, 55, 102, 349
and protozoans 32, 34
and unionaceans 299, 301, 302-303, 306, 307, 311
fish farms 105, 116, 152, 346, 349, 350, 351 (see also salmon farming industry)
fish fauna, British, boreal (Arctic) element 3
Lusitanian element 3
fish stocks, decline of 315, 351
fish louse, see sea louse
fisheries for lampreys 329-330
fishes, anadromous 3, 4-5
bony, see teleosts
British fauna 2-5
cartilaginous, see elasmobranchs

catadromous 3
diadromous 3
euryhaline 3, 4, 80
external features 5-8
gills, see gills of fishes
immune system 16-20, see also immunity
introduction into Britain 5, 351
olfactory organs, see olfactory organs of fishes
skin, see skin of fishes
stenohaline 3, 4, 5
flagella 27-30
flatfish 7-8
flatworms general 37
fusion in diplozoids 129

gadid fishes *133*, *206*
gasterosteid fishes 80
genito-intestinal canal 115-116
gill, arch 12-16, 111, 118, 120, 127, 184, 186, 205, *206*
　filaments, see primary gill lamellae
　lamellae, see primary gill lamellae; secondary gill lamellae
　maggot, see *Salmincola*
　rakers 15, 101-102, 111, *206*, 225, 349
gills of fishes 7-8, 12-16
glaciation 3-4, 80
global warming 350
glochidia, development of 313
　escape from host 308
　general 298-299, 299-300, 345, 346
　of *Anodonta* 300-301, 301-307
　of *Margaritifera* 301, 310-314, 350
　release of 302, 310-311
glycocalyx 11
glycoproteins 11
gnathiids 275, 284-295, 339, 340, 341, 342, 345 (see also *Gnathia*; *Paragnathia*)
gnathopods, see pylopods
gnathostomes 1
goblet cells 10, 11, 15, 16, 17
Great Lakes 335
gurnards *104*, 112, 123, 127, 347
gyrodactylids 61, 70-81, 82, 83, 340, 343, 346, 349 (see also *Gyrodactylus*)
***Gyrodactylus*, life cycle Fig. 4.10, p. 72**
　British species *79*
　number of species 70, 82, 348

haematin 43, 96, 114, 115, 126, 197
haemogregarines 153, 290-291, 346
haemolysin 17
hagfishes 318, 319
hamulus 40, 41-42, 72, 73, 74, 82, 85-95, 97, 108, 342
haptor 40-43, 51, 62, 64, 66, 67, 68, 69, 70, 71, 72, 73, 75, 76, 78, 82, 83, 85-95, 96-98, 105-106, 107-114, 116-121, 128, 129, 344, 345
harems of gnathiids 284, 293, 341
hatching, in *Acanthocotyle* 62-66
　in *Argulus* 258
　in copepods, see egg sacs
　in *Entobdella* spp. 48-51, 58, 58-59
　in leeches 149-150, 151
　in *Leptocotyle minor* 68
　in polyopisthocotyleans 122-125
hementin 132
hemibranch 12
hexabothriids 107-108, 114 (see also *Hexabothrium*; *Rajonchocotyle*)
hirudin 132
holdfast of *Lernaea* 208, 211, 213
holocephalans 2, 37
hooks, general 341-343
hooklets of monogeneans 43, 51, 62, 66, 68, 71, 72-73, 74, 85, 88-89, 92, 93, 95, 97, 103, 105, 106, 108, 342
hormones of the host 125
host defences against parasites, see immunity in fishes; host reaction
host finding, in *Argulus* 244-249
　in caligids 168-170
　in cymothoids 283-284
　in *Entobdella* 49-52
　in ergasilids 225
　in gnathiids 285
　in lampreys 326
　in leeches 140-141
　in *Salmincola* 198
　in unionaceans 302-304, 311-312
host reaction 93, 195, 202, 230, 344-345 (see also immunity in fishes)
　to glochidia 307
host recognition, see host finding
host specificity 5, 27, 39, 59-60, 62, 70, 77, 77-78, 104-105, 135, 173-174, 191, 213, 304-305, 313, 346, 347, 347-349, 350
host switching 59, 70, 80, 84, 85, 348
hyperplasia 30, 93, 307, 344, 345
hypodermis of fishes 11

ich, see white spot disease
immunity, in fishes 16-20, 74-75, 78, 92-93, 101, 340, 346-347, 347, 350 (see also host reaction)
　acquired, 36, 75, 213, 312-313, 316-317
　innate 78, 101
　to caligids 176
　to *Ichthyophthirius* 36
　to unionaceans 312-313, 316-317
immunoglobulin 16, 17, 18 (see also antibodies)
immunologically privileged sites 19
immunosuppression 20
inflammatory response 18, 176, 213

integument 234, 264, 269
interbranchial septum 15, 16, 93, 120
interstitial fauna 37, 48, 154
iodine worms 236
ionocytes, see chloride cells
isopods 274-295, 340, 341 (see also
    cymothoids; gnathiids)
    introduction 274-275

kelt 4
    mended 4, 205
keratin 10, 342
keratocytes, see epidermis of fishes

lacunar system in *Anelasma* 269, 271-272
lamprey, brook, life cycle Fig. 17.14, p. 332
lamprey, river, life cycle Fig. 17.14, p. 332
lampreys 1, 4, 7, 318-338, 339, 343, 345 (see
    also *Lampetra*; *Petromyzon*)
    feeding in fresh water, see Loch Lomond;
        Great Lakes
    introduction 318-320
    land-locked fishes 4-5, 334, 335
lasidium 299
lateral line 8, 234, 236
laying, in ancyrocephalines from mullets 101
    in *Argulus* 257-258
    in copepods, see egg sacs
    in *Entobdella soleae* 47
    in leeches 147-149
lectin 17
leeches 131-153, 340, 341-342, 343, 345, 346
    (see also *Hemibdella*; *Hemiclepsis*;
    Hirudo; *Piscicola*)
    introduction 131-132
***Lepeophtheirus pectoralis*, life cycle Fig. 8.1, p. 156**
***Lernaea cyprinacea*, life cycle Fig. 11.2, p. 209**
Lernaea cyprinacea, general 210
***Lernaeocera branchialis*, life cycle Fig. 9.3, p. 180**
Lernaeocera, general 178-186, 208, 210, 339
lernaeopodids 195-207, 341, 342
leucocytes 10, 17, 18
life cycle, of bomolochids 228
    of caligids 158-159
    of chondracanthids 232-233
    of cymothoids 275
    of *Entobdella soleae* 39-40
    of eye-maggot 188-193
    of gnathiids 275
    of gyrodactylids 71-72
    of leeches 149-152
    of *Lernaea* 210-213
    of *Lernaeocera* 179-181, 184-185, 339
    of *Salmincola* 205
    of *Sarcotaces* 236

of unionaceans 299-300
life cycles, general 339-340
limbs of crustaceans 154
Loch Lomond 318, 320, 324, 334-335
locomotion, in *Argulus* 241, 244
    in caligids 160
    in *Entobdella* 43
    in *Gyrodactylus* 75, 76
    in leeches 139-140
    in *Leptocotyle* 67
    in polyopisthocotyleans 119, 120, 128
    in *Salmincola* 203
    in *Tetraonchus* 96-97
longevity of unionaceans 309
lungfishes 2
lymph 141
lymphocytes 16, 17
lysozyme 17

macrophages 18
macrophthalmia 323-324, 336
maiden 4, 205
Malpighian cells, see epidermis of fishes
manca larva 279, 283, 284 (see also predation)
mandibles, in *Argulus* 249, 258
    in caligids, see mouthparts
    in gnathiid males 284, 293
    in lernaeopodids 200
    in pennellids, see mouthparts
    in pranizas 288
mantle, in *Anelasma* 268-269, 272, 273
    in unionaceans 297, 302, 303, 308, 316
mantle cavity in unionaceans 297, 305
Margaritifera, general 300-301
marsupium, in cymothoids 279, 282-283
    in unionaceans 298, 301, 310
mate guarding 164, 182
mating, general 98
    in *Amphibdella* 95
    in *Argulus* 252-257
    in caligids 164-166
    in chondracanthids 230-232
    in cymothoids 281-282
    in *Entobdella soleae* 46
    in lampreys 329-331
    in leeches 145-147
    in *Lernaeocera* 182-184
    in poecilostomatoids 214
    in polyopisthocotyleans 127-130
    in *Salmincola* 203
maxillae, first, in *Argulus* 237, 241 259-261 (see
    also suction, in *Argulus*)
    in pranizas 288
maxillae, second, in *Argulus* 242, 260
    in caligids 159, 160, 169
    in ergasilids 224, 229
    in lernaeopodids 195, 196, 198, 199, 200, 201, 202, 203

maxillipeds, in caligids 157, 158, 160, 164, 165, 168, 169, 171, 172
   in chondracanthids 229, 233
   in lernaeopodids 199, 201, 202, 203
   in pennellids 183, 184
   in pranizas 288
   sexual dimorphism 214, 222-223
melanin 19, 69
melanocytes 19
melanosome 19
mesoparasites 1, 17, 18, 95, 185, 210, 211, 233-234, 271, 299, 339, 340, 344-345
metamorphosis 7, 185, 300, 323-324, 336
metanauplius of *Argulus* 258-259
microbothriids 66-68, 343
migration on host, in *Acanthocotyle elegans* 64-65
   in caligids 350
   in *Entobdella soleae* 52-58
   in polyopisthocotyleans 126-127, 128
molluscs, cephalopod 2, 80
monocytes 18
monogeneans, general 37-130, 340, 342-343
   introduction 37-38
monopisthocotyleans, general 37, 103, 105, 106, 108, 114, 115, 120, 126
morphology of cymothoids 278-279
mother-of-pearl 299
moulting 157, 158, 164, 166, 171-173, 183, 195, 233, 282, 342
mouthparts, of *Anelasma* 269
   of *Argulus* 238, 249
   of caligids 157, 162, 167
   of chondracanthids 228, 229, 230
   of cymothoids 278-279
   of ergasilids 221-222, 224
   of gnathiids 284, 285, 286-288
   of *Lernentoma* 233
   of pennellids 182-183
   of *Sarcotaces* 236
mucins, see glycoproteins
mucocyst 31, 32, 33
mucous cells of fishes 27, 29, 30, 32, 74-75, 347 (see also goblet cells)
mucus of fish skin and gills 11, 17, 32, 36, 284, 304, 311
mycetomes 143, 346

naiads, see unionaceans; *Anodonta*; *Margaritifera*
nasal cavities of fishes **206**, 226, 228
nauplius, of *Anelasma* 272
   of caligids 70, 158, 167-168
   of chondracanthids 232
   of ergasilids 214, 221, 225
   of free-living barnacles 265
   of *Lernaea* 210
   of lernaeopodids 197

   of pennellids 180, 181, 188, 190
   of philichthyids 236
nematodes 75, 262-263, 346
Neodactylogyrus, British species 83, 85
   haptor 92
   number of species 82
neoteny 337
neuropeptides 20
neutrophils 18
niche restriction 127-8, 350 (see also site selection)

olfactory organs of fishes 8, 102
oncomiracidium, general 96, 105, 122-125, 126-127
   of *Acanthocotyle* 64
   of *Capsala martinieri* 61, 62
   of *Entobdella soleae* 48-52, 62
oostegites 279, 282, 285, 295
opercular suction pump 14
operculum, of fishes 7, 12, 16, 110, 111, 126, 127, 205, 206, **206**, 226, 219, 306, 349, 350
   of leech cocoon 148, 149
osmoregulation 12, 18, 19, 81
osmoregulation failure 36, 45, 81, 176, 346
ovary of *Argulus* 253-254
over-dispersion 305
ovisacs, see egg sacs of copepods

paired species in lampreys 336-337
Paragnathia formica, life cycle Fig. 15.2, p. 276
paragnaths 288
parasite, definition of 1
parasites introduced into Britain 216, 218, 239, 241
parr 4, 28, 80, 81, 313
pathology caused, by *Anelasma* 270
   by *Anilocra* 284
   by *Argulus* 240
   by caligids 163, 175, 176
   by *Entobdella soleae* 45
   by ergasilids 216
   by eye-maggots 193-194
   by glochidia 307
   by *Gyrodactylus salaris* 81
   by *Ichthyobodo* 30
   by *Ichthyophthirius* 30, 36
   by leeches 152
   by *Lernaea* 213
   by polyopisthocotyleans 116
   by *Salmincola* 202
   by trichodinids 26
pavement cells 9, 15-16
pearl mussel, distribution in Britain 314
   fishing 314
   historical 314
   threats to 314-315, 350
   (see also *Margaritifera margaritifera*)
pearls 299

pennellids 178-194, 208, 339 (see also *Lernaeenicus*; *Lernaeocera*)
pentastomids 263-264, 341
percid fishes 80, 84, 258
perciform fishes 218
pereopods 274, 279, 281, 282, 283, 285, 286, 288, 295
peritrich ciliates, see trichodinids; scyphidiids
phagocytes 18
pheromones 58, 184, 205, 293
philichthyids 234-236 (see also *Leposphilus*; *Philichthys*; *Sarcotaces*)
phoresy 70, 112, 262, 334
phylogenetic speciation 60, 80
pigment cells, see chromatophores
pillar cells 14
pinocytosis 73, 115
piroplasms 153
placoid scales, see denticles
pleopods 279, 280, 283, 285, 286
pleuronectid fishes 155
poecilostomatoids, general 155 (footnote), 214
polyopisthocotyleans 103-130, 342, 345, 346
    introduction 37, 82, 103
polystomatids 37, 60, 80, 103, 106, 113
praniza larva, derivation of name 284-285
predation, on *Argulus* 248, 261-262
    on *Entobdella* eggs 48
    on glochidia 306, 316
    on lampreys 322
    on manca larvae 283
    on praniza larvae 291-292, 295
primary gill lamellae 12, 14-16, 32, 83, 96, 111, 112, 113, 118-121, 128, 181, 197, 199, 223
proboscis of leech 131, 132
protandry 46, 95, 145
protogyny 74
protozoans 22-36, 340 (see also *Ichthyobodo*; *Ichthyophthirius*; *Trichodina*)
    ciliate 22-27, 30-36
    epizoic 22-27
    flagellate 27-30
    introduction 22
pseudobranch 127, 350
pseudohaptor 66
pullus 285
pulsatile vesicles 135
pylopods 285, 286, 288, 295

rays 2, 5, 7, (see also *Raja* spp.; ray, thornback)
    general 132, 133, 226
reactive nitrogen species 18
reactive oxygen species 18
relationships of branchiurans 263-264
reproductive biology, general 340-341
resistance, acquired, see immunity, acquired

resistance, innate, see immunity, innate
respiratory burst 18
response of parasites, to chemical stimuli 169, 247, 303, 311, 326
    to current direction 168
    to gravity 49, 284
    to haloclines 169
    to hydrodynamic stimuli 198, 247
    to light 49, 50, 168, 190, 245-247, 283, 284
    to shadows 59, 123, 124, 140, 168, 198, 244, 302
    to touch 55, 311
    to water turbulence 140, 168, 283
    (see also hatching)
rhynchobdellidans 132
rodlet cells 10, 18

sacciform cells 10, 11
salinity preference of monogeneans 101
salinity tolerance of parasites 205, 206, 217-218, 219, 294
salivary glands, of leeches 141-142, 143, 144
    of gnathiids 288, 291
salmon and gill maggot, life cycle Fig. 10.8, p. 204
salmon louse, see sea louse
salmon farming industry 80-81, 155, 175-177, 347
salmonid fishes 4, 6, 20, 75, 77, 78, 81, 174-177, 258, 313, 315
salt balance, see osmoregulation
scales, ctenoid 11, 52, 54-55, 134, 138, 340
    of fishes 10-11, 137, 199, 210, 211, 285, 304
    placoid, see denticles, in elasmobranch fishes
scavengers 319, 337
sclerotin 47, 48, 149
scombrid fishes 6
scopula 27
scyphidiids 22, 26, 27, 344
sea louse 155, 174-177
secondary gill lamellae 12-16, 85, 96-97, 108-110, 112, 113, 118, 119, 120, 121, 343, 345
secretions, adhesive 27, 43, 67, 69, 70, 97, 120, 137-139, 170, 343-344 (see also bulla; filament, frontal)
semi-parasite 271, 345
sense organs of parasites 75-76, 160-161, 167-168, 303
sensilla 55, 76, 92, 258, 304 (see also sensory hairs)
sensory hairs 303
septal channel 1
sex change, in cymothoids 279-281, 341
    in unionaceans 301, 309
sharks 2, 5, *206*
shoaling and *Argulus* 248

sibling species, see cryptic species
site selection, general 349-350
    in caligids 166-167
    in eye-maggots 193
    in glochidia 311-312
    in polyopisthocotyleans 127
    in *Salmincola* 197, 198
skate leeches = skate suckers, 133
skin of fishes 8-11
smelts 4
smolt 4, 28, 169, 175, 205
smoltification 4
soles, general 347
sparid fishes 204
spawning of lampreys 330-331
speciation of parasites 132, 173, 348, 349
spermatophores 46, 57, 59, 146-147, 164-165,
    166, 183, 203, 210, 231, 237, 255, 340-
    341
spiracle 7, 8, **206**
spring viraemia of carp 262
squamodiscs 92
squamous cells, see pavement cells
squid 70
stenohaline fishes 3, 5, 80
sticklebacks, general 4
stress in fishes 20, 346
strigil 162
sturgeons 2, 37, 70
stylet of *Argulus* 237, 249-251
sub-chelae 222, 223, 342
sucker, of *Argulus*, development 259-261 (see
    also suction, in *Argulus*)
suction, general 106, 343
    in *Acanthocotyle* 66
    in *Argulus* 237, 241, 242-244
    in bomolochids/taeniacanthids 227
    in caligids 156, 159-160
    in *Capsala* 62
    in cookiecutter sharks 337
    in *Cyclocotyla* 111-112
    in diclidophorids 108-111
    in *Entobdella* 40-43
    in hexabothriids 107-108
    in lampreys 324-325, 327-329
    in leeches 131, 137-139
    in trichodinids 26
superconglutinate 316
swimbladder 5, 6, 7
symbiotic micro-organisms 115-116, 143-145,
    290, 346
sympatric speciation, see speciation of
    parasites

tadpoles 213
taeniacanthids 214, 226-228, 343
tegument 96 (see also integument)
teleosts 1, 2, 5-16, 37, 60, 70, 348

telotrochs 27
terminal web 9, 72
theront 30-32
thoracopods, in *Anelasma* 269
    in *Argulus* 238, 239, 240, 252, 254-256,
        258, 259, 260, 263
    in bomolochids/taeniacanthids 227
    in caligids 157, 160, 164, 165, 168
    in chondracanthids 232
    in ergasilids 218, 218-219, 221, 222, 223, 224
    in free-living barnacles 265
    in *Lernaea* 211
    in lernaeopodids 198,
    in pennellids 181, 183, 184
    (see also pereopods)
tomont 35-36
tongue worms, see pentastomids
tonofilaments 9
transformation, see metamorphosis;
    macrophthalmia
transmission of disease, by *Argulus* 262, 346
    by gnathiids 290-291, 346
    by leeches 152-153, 346
transmission, in *Acanthocotyle* 62-64
    in *Entobdella soleae* 48-51, 55-56
    in gyrodactylids 75-77
    in *Leptocotyle* 67-68
    in polyopisthocotyleans 121-127
    in udonellids 70
Trichodina species, British **24**
trichodinids 22-27, 343
trophont 32-35
trypanoplasms 153, 346
***Trypanosoma cobitis*, life cycle Fig. 7.12, p. 152**
trypanosomes 152 –153, 346
turbellarians 37, 68

udonellids 68-70
unionaceans 296-317, 339, 345
    introduction 296-299
    of North America 315-317
urea, role as hatching stimulant 65-66, 68, 123

viruses 153, 262
viviparity 71-72, 74, 284

whales 320, 337
white blood cells (see leucocytes;
    lymphocytes; macrophages; monocytes;
    neutrophils; phagocytes)
white spot disease, see pathology caused, by
    Ichthyophthirius
woodlice 274-275, 282
wound healing in fish epidermis 18, 307

zuphea, derivation of name 284-285
    birth 295